Multiphase Environmental Chemistry in the Atmosphere

ACS SYMPOSIUM SERIES **1299**

Multiphase Environmental Chemistry in the Atmosphere

Sherri W. Hunt, Editor
US Environmental Protection Agency
Washington, DC

Alexander Laskin, Editor
Purdue University
West Lafayette, Indiana

Sergey A. Nizkorodov, Editor
University of California, Irvine
Irvine, California

Sponsored by the
ACS Division of Environmental Chemistry, Inc.

American Chemical Society, Washington, DC

Distributed in print by Oxford University Press

Library of Congress Cataloging-in-Publication Data

Names: Hunt, Sherri W., editor. | Laskin, Alexander, editor. | Nizkorodov, Sergey A., editor.
Title: Multiphase environmental chemistry in the atmosphere / Sherri W. Hunt (US Environmental Protection Agency), Alexander Laskin (Purdue University), Sergey A. Nizkorodov (University of California, Irvine), editors.
Description: Washington, DC : American Chemical Society, [2018] | Series: ACS symposium series ; 1299 | Includes bibliographical references and index.
Identifiers: LCCN 2018037222 (print) | LCCN 2018049429 (ebook) | ISBN 9780841233621 (ebook) | ISBN 9780841233638 (print)
Subjects: LCSH: Atmospheric chemistry. | Environmental chemistry.
Classification: LCC QC879.6 (ebook) | LCC QC879.6 .M85 2018 (print) | DDC 551.51/1--dc23
LC record available at https://lccn.loc.gov/2018037222

The paper used in this publication meets the minimum requirements of American National Standard for Information Sciences—Permanence of Paper for Printed Library Materials, ANSI Z39.48n1984.

Distributed in print by Oxford University Press

Foreword

The ACS Symposium Series was first published in 1974 to provide a mechanism for publishing symposia quickly in book form. The purpose of the series is to publish timely, comprehensive books developed from the ACS sponsored symposia based on current scientific research. Occasionally, books are developed from symposia sponsored by other organizations when the topic is of keen interest to the chemistry audience.

Before agreeing to publish a book, the proposed table of contents is reviewed for appropriate and comprehensive coverage and for interest to the audience. Some papers may be excluded to better focus the book; others may be added to provide comprehensiveness. When appropriate, overview or introductory chapters are added. Drafts of chapters are peer-reviewed prior to final acceptance or rejection, and manuscripts are prepared in camera-ready format.

As a rule, only original research papers and original review papers are included in the volumes. Verbatim reproductions of previous published papers are not accepted.

ACS Books Department

Contents

Modeling Multiphase Chemistry

Chemistry and Characterization of Fires

Toxicity and Impacts of Aerosols

Indexes

Preface

Degrading environmental quality and accelerating climate change resulted from rapid developments in the global economy, and both have become major challenges worldwide. Of particular concern are increased primary emissions and secondary formation of atmospheric aerosols, which have profound effects on the Earth's radiation balance, biogeochemical cycles, and human health. Their effects are predicted to be even more severe in the near and distant future as the amount of aerosols increases with both primary emissions and temperature. Understanding and mitigating environmental impacts of aerosols relies on the fundamental knowledge of their sources, chemical composition, and evolution of their physio-chemical properties during their atmospheric life cycle. Comprehensive studies of a diverse mixture of natural and anthropogenic aerosols require multi-dimensional measurements and complementary applications of novel experimental and modeling approaches.

This book highlights new cross-disciplinary advances in aerosol chemistry that involve more than one phase, for example, unique chemical processes occurring on gas-solid and liquid-solid interfaces. Predictive understanding of these processes is critical for more accurate modeling of the atmospheric environment. Because of the staggering rate this field is advancing, many of these new ideas are only present in narrowly focused scientific journals in spite of their relevance to researchers across scientific disciplines.

This eBook is based on a selection of presentations given at the very successful symposium "Multiphase Chemistry of Atmospheric Aerosols" held at the 2017 ACS Fall meeting and attended by a large number of researchers. This symposium provided an excellent opportunity to hear about multiple aspects of atmospheric multiphase chemistry from a diverse spectrum of presenters, including laboratory and field experimentalists and modelers. Similarly, by presenting the material in a single edited book, we hope to encourage cross-disciplinary thinking among these topics so that more scientists can imagine solutions to the challenges of understanding and mitigating the effects of atmospheric aerosols.

The chapter authors begin with introductory material addressing scientists who may work in a broad range of disciplines, and then move to more specific details for the experts in the field. Therefore, this book should be an excellent resource for those just starting in the field of atmospheric chemistry and for those who want to initiate new research directions with a mix of basics and some of the newest advances.

All three editors, and many of the chapter authors, have been impacted by the engaging and encouraging atmospheric chemistry researchers in the United States and elsewhere. We would like to thank these colleagues and many friends for inspiring discussions about atmospheric chemistry over the years. We would like to especially thank Prof. Barbara Finlayson-Pitts from the University of California, Irvine, for giving us a breadth of knowledge and an appreciation of the power of interdisciplinary research and discussions. We would also like to thank our families, who motivate us to work towards improving air quality for everyone.

Sherri W. Hunt
Physical Scientist
Office of Research and Development
US Environmental Protection Agency
Washington, DC 20004
202-564-4486 (telephone)
hunt.sherri@epa.gov (e-mail)

Alexander Laskin
Professor
Department of Chemistry
Purdue University
West Lafayette, IN 47907
765-494-5243 (telephone)
alaskin@purdue.edu (e-mail)

Sergey A. Nizkorodov
Professor
Department of Chemistry
University of California
Irvine, CA 92697
(949) 824-1262 (telephone)
nizkorod@uci.edu (e-mail)

Chapter 1

Editors' Perspective on Multiphase Chemistry in the Atmosphere

Sherri W. Hunt,[1,*] Alexander Laskin,[2] and Sergey A. Nizkorodov[3]

[1]US Environmental Protection Agency, Washington, DC 20004, United States
[2]Department of Chemistry, Purdue University, West Lafayette, Indiana 47907, United States
[3]Department of Chemistry, University of California, Irvine, California 92697, United States
[*]E-mail: hunt.sherri@epa.gov

The chapter outlines authors' view on the importance of multiphase chemistry in the atmosphere. The chapter starts with a short description of the development of atmospheric chemistry from gas-phase spectroscopy and kinetics into a highly interdisciplinary research on complex multiphase processes. Despite the impressive progress over the years in our understanding of multiphase processes, the air pollution and climate problems are far from being solved. Further research on multiphase chemical reactions and the interactions across phases is needed in order to understand the properties and effects of complex mixtures of trace gases and particles that surround us.

Some of us can still remember the days when science was broken up into traditional areas, with well-defined boundaries between physics, chemistry, biology, medicine, etc. As scientific progress has been made, and more challenging problems have been recognized and investigated, researchers have shifted into areas that creatively combine multiple traditional fields. In atmospheric chemistry, the transition to a multidisciplinary research approach naturally happened decades ago. The early atmospheric chemistry research focused on understanding problems that are relatively simple by today's standards, for example, studying photochemistry and kinetics of small gas-phase molecules such as aldehydes and ketones (*1*). As our understanding of atmospheric chemistry grew, more challenging problems involving chemistry occurring on surfaces and in condensed-phases had to be attacked. The current research frontier in atmospheric chemistry involves understanding the behavior of molecules that exist and react in different phases. The term "multiphase chemistry" was coined by atmospheric scientists to denote "chemical reactions, transport processes, and transformations between gaseous, liquid, and solid matter" (*2*). The inherent complexity of multiphase processes has motivated the development of novel platforms for laboratory and field measurements, as well as advanced modeling approaches.

The chemistry of the atmosphere is driven by trace chemical species that can be found in the gas phase, solid and liquid particles, cloud and fog droplets, and also on environmental surfaces that are in direct contact with the atmosphere. Chemical species that appear in the atmosphere as a result of human activities, and have adverse effects on the human health and ecosystems, are classified as air pollutants. Air pollution has been a problem in cities for hundreds of years, and it has become an acute problem with rapid industrialization and growth of megacities (*3*). One of the most famous air pollution episodes is the Great Smog of London in 1952, which was a result of smoke and a stagnant air mass over the city, resulting in thousands of deaths over several days (*4*). The presence of air pollutants in the atmosphere affects not only humans but also the entire biosphere. In fact, the research into the hazardous chemical compounds in Los Angeles photochemical smog by Prof. Arie Haagen-Smit and others was driven in part by a critical need to understand the effect of the air pollution on plants and ecosystems in the area (*5*). Extreme air pollution episodes leading to health issues and damage to vegetation have prompted the adoption of strict air pollution regulations in many countries. For example, the United States established the National Ambient Air Quality Standards (NAAQS), under authority of the Clean Air Act; these standards have resulted in dramatic improvements in the air quality in many regions across the country (*6*). Improved air quality has led to massively positive effects on the economy due to the reduction in the number of air pollution-related hospital visits, reduction in the number of days missed from school and work, and improved crop production. In terms of the impact on the economy, the benefits of the Clean Air Act have exceeded the cost by a wide margin making it one of the most successful pieces of legislation passed in the United States in the 20th century (*7*).

Despite the impressive progress over the years, the air pollution problem is far from being solved. The State of Global Air 2018 gives a comprehensive overview of air quality and health levels as well as trends since 1990 (*8*). According to the report, air pollution is the fourth leading risk factor for mortality globally. Millions of people still die each year from diseases associated with their exposure to air pollution and even more suffer from a range of short- and long-term health effects from the development of asthma to a decrease in cognitive function. Airborne particles, or particulate matter, represent an especially potent agent of the adverse health effects of polluted air. The particles with an aerodynamic diameter below

2.5 μm ($PM_{2.5}$) were the sixth-ranking mortality risk factor in 2015, with exposure to $PM_{2.5}$ accounting for 4.1 million deaths that year (*8*).

Global and regional climate patterns are also directly affected by air pollutants. The warming effect of greenhouse gases on the climate is well known, and is relatively easy to describe in quantitative terms. Particles have a more uncertain and poorly understood effect on the climate because of their ability to absorb and scatter solar radiation and modify properties of clouds in a very complex way (*9*). Part of the challenge in accurately modeling the climate effects of particles is the impressive amount of diversity in their chemical and physical properties. For example, soot particles absorb sunlight very strongly (*10*) but are poor cloud nuclei because of their hydrophobicity. In contrast, particles dominated by ammonium sulfate, a major $PM_{2.5}$ constituent in urban areas, do not absorb sunlight and are highly hygroscopic. Biomass burning smoke occupies a position in between these two extremes as it contains light-absorbing "brown carbon" constituents (*11*, *12*), which have a range of hygroscopic properties.

Another challenge for predicting climate and health effects of air pollutants is the highly dynamic nature of the atmospheric environment. The composition of the atmosphere at any given time and at any given place can be described by the species present, their concentrations in different phases, and their physical and chemical properties. The atmospheric composition is largely determined by what is emitted by people and nature and by the complex motion of the atmospheric masses. However, it also depends on how fast species are removed from the atmosphere and transform to yield new products through a combination of chemical and physical processes, involving multiple phases. These "aging processes" occur on local scales as air masses travel across a city and on a global scale as they travel around the globe. These transformations result in the formation of additional gas-phase species, and new particles, as well as in the removal of both gases and particles by precipitation (wet deposition) or by Earth surfaces (dry deposition). Chemical aging processes for particles include reactive uptake of gas-phase oxidants (*13*), coupled gas-particle partitioning and oxidation (*14*), and surface and condensed-phase photochemical processes driven by sunlight (*15*). The aging processes involve multiple phases, and they are excellent examples of the multiphase chemistry described in this book.

Previous improvements in air quality in the United States and in many developed countries have been made possible by the advancement of our fundamental knowledge of gas-phase chemistry that came from extensive laboratory experiments on photooxidation of hydrocarbons (*3*). Our ability to understand the air pollution system and predict how it responds to various chemical and physical perturbations, such as an increase in the emission of a particular class of compounds or increase in global temperature, has enabled the implementation of effective strategies to reduce the concentration of air pollutants. However, gas-phase chemistry alone cannot encompass the full complexity of atmospheric processes – a famous example of this is the dramatic failure of gas-phase mechanisms to predict the existence of the ozone hole over the Antarctic. This phenomenon could only be described by including the efficient releases of photochemically active chlorine from its more abundant and less active condensed-phase reservoir species (*16*). As improvements in understanding have been achieved, the problem has become more challenging. Through further advancements in scientific knowledge of the composition and multiphase chemistry of the atmosphere, we will be better prepared for the future, and be able to more effectively plan the economic developments to ensure that all people have access to clean air (Figure 1).

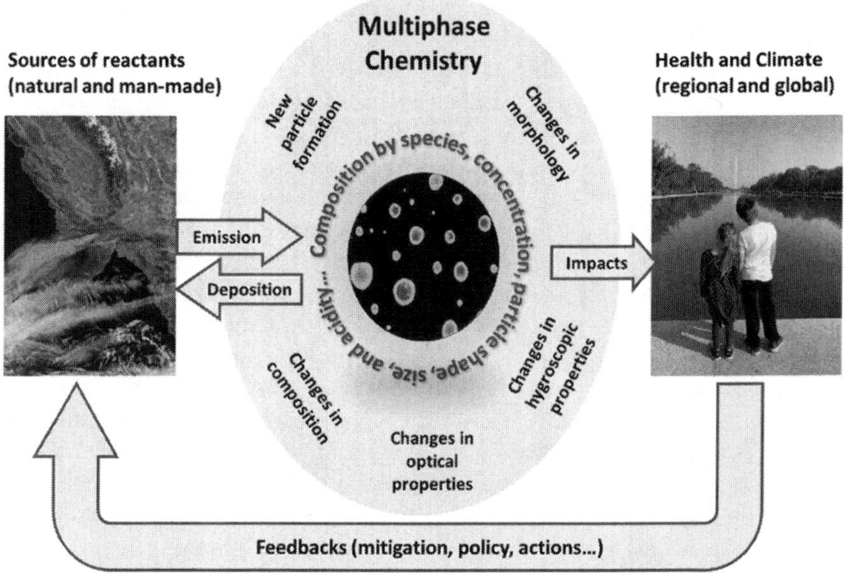

Figure 1. Multiphase chemistry changes the chemical nature and physical properties of the compounds emitted by natural and anthropogenic sources. The air pollutants aged by multiphase processes affect both humans and ecosystems, which in turn affects the emission and deposition of additional air pollutants. Understanding multiphase chemistry is critical for assessing the local and global effects of air pollutants. The image on the left is credited to NASA, and image on the right was taken by one of the authors. The image in the middle is a false colored SEM image illustrating extent of the multi-phase variability in individual atmospheric particles, where green and blue colors depict their organic and inorganic components, respectively.

At any given time, the composition of the atmosphere affects the health of people and the environment, as well as the entire climate system. These impacts can be assessed with a variety of metrics. Mortality and morbidity are the usual metrics for health effects in people, while the health of the environment may be measured by impacts on weather and climate, clouds and precipitation, the amount of light that reaches the Earth's surface, and the diversity and vitality of ecosystems. Understanding the atmospheric composition and how it might respond to changes in emissions, climate, or deposition enables the development of useful models and practical solutions for improving the air quality. While the models and measurement tools are now more complex, the overall goal and process remain to understand what is in the air and how it will respond to environmental changes. With improved fundamental knowledge, scientists can inform decision-makers on a range of issues, from impacts of individual actions to new policies regarding emissions, city development, or energy production.

With a goal of introducing some of the most recent advances in multiphase chemistry and its importance in the atmosphere, the chapters in this book are organized according to several themes. The first section focuses on recently discovered chemical reactions and processes. These include understanding several intermediates and pathways that were not considered in the early air quality studies. Several important physical characteristics of aerosols are described in the second section. These characteristics may impact the chemical reactions that take place within the aerosols or on their surfaces. They also elucidate how solar radiation may interact with aerosol particles. The third section includes two chapters highlighting recent model advances that begin to incorporate multiphase processes. The final two sections shift focus to two topics that are currently receiving a great deal of attention. Three chapters focus on the characterization of emissions from fires and the multiphase chemistry that occurs as these emissions move through the atmosphere. Finally, the last section includes chapters that describe the use of new techniques developed to understand the strength of potential health impacts or toxicity of particles based on a measure of their oxidative potential - this indicates their ability to produce oxidation reactions, which have negative impacts in the body.

None of the topics are covered exhaustively in this collection, but rather each is introduced along with recent findings and references to additional work. By presenting these topics together, additional connections can be identified across them, leading to even more multidisciplinary multiphase chemistry knowledge.

Acknowledgments

The views expressed in this document are solely those of the authors and do not necessarily reflect those of the U.S. Environmental Protection Agency, University of California, or Purdue University.

References

1. Calvert, J. G.; Pitts, J. N. *Photochemistry*; John Wiley & Sons: New York, London, Sydney, 1966; p 899.
2. Pöschl, U.; Shiraiwa, M. Multiphase Chemistry at the Atmosphere–Biosphere Interface Influencing Climate and Public Health in the Anthropocene. *Chem. Rev.* **2015**, *115*, 4440–4475.
3. Finlayson-Pitts, B. J.; Pitts, J. N., *Chemistry of the Upper and Lower Atmosphere*; Academic Press: San Diego, CA, 2000.
4. Wilkins, E. T. Air Pollution Aspects of the London Fog of December 1952 - Discussion. *Q. J. R. Meteorol. Soc.* **1954**, *80*, 267–271.

5. Haagen-Smit, A. J.; Darley, E. F.; Zaitlin, M.; Hull, H.; Noble, W. Investigation on injury to plants from air pollution in the Los Angeles area. *Plant Physiol.* **1952**, *27*, 18–34.
6. U.S. Environmental Protection Agency. Our Nation's Air, Status and Trends Through 2016. https://gispub.epa.gov/air/trendsreport/2017/ (accessed May 23, 2018).
7. U.S. Environmental Protection Agency. The Benefits and Costs of the Clean Air Act from 1990 to 2020. https://www.epa.gov/sites/production/files/2015-07/documents/fullreport_rev_a.pdf (accessed May 23, 2018).
8. State of Global Air 2018. Special Report by the Health Effects Institute. https://www.stateofglobalair.org (accessed May 23, 2018).
9. IPCC. *IPCC, 2013: Climate Change 2013: The Physical Science Basis. Contribution of Working Group I to the Fifth Assessment Report of the Intergovernmental Panel on Climate Change*; Cambridge, United Kingdom, and New York, NY, USA, 2013; p 1535.
10. Bond, T.; Bergstrom, R. Light absorption by carbonaceous particles: an investigative review. *Aerosol Sci. Technol.* **2006**, *40*, 27–67.
11. Laskin, A.; Laskin, J.; Nizkorodov, S. A. Chemistry of Atmospheric Brown Carbon. *Chem. Rev.* **2015**, *115*, 4335–4382.
12. Andreae, M. O.; Gelencser, A. Black carbon or brown carbon? The nature of light-absorbing carbonaceous aerosols. *Atmos. Chem. Phys.* **2006**, *6*, 3131–3148.
13. Rudich, Y.; Donahue, N. M.; Mentel, T. F. Aging of organic aerosol: bridging the gap between laboratory and field studies. *Annu. Rev. Phys. Chem.* **2007**, *58*, 321–352.
14. Donahue, N. M.; Robinson, A. L.; Pandis, S. N. Atmospheric organic particulate matter: From smoke to secondary organic aerosol. *Atmos. Environ.* **2008**, *43*, 94–106.
15. George, C.; Ammann, M.; D'Anna, B.; Donaldson, D. J.; Nizkorodov, S. A. Heterogeneous Photochemistry in the Atmosphere. *Chem. Rev.* **2015**, *115*, 4218–4258.
16. Molina, M. J.; Tso, T. L.; Molina, L. T.; Wang, F. C. Y. Antarctic stratospheric chemistry of chlorine nitrate, hydrogen chloride, and ice: release of active chlorine. *Science* **1987**, *238*, 1253–7.

Measuring Multiphase Chemistry

Chapter 2

Impact of Multiphase Chemistry on Nanoparticle Growth and Composition

Michael J. Apsokardu, Peijun Tu, Yue Wu, and Murray V. Johnston*

Department of Chemistry and Biochemistry, University of Delaware, Newark, Delaware 19716, United States
***E-mail: mvj@udel.edu**

Secondary organic aerosol (SOA) is formed by the oxidation of volatile organic compounds (VOCs) in the atmosphere to give non-volatile products (NVOCs) that condense onto particles and semi-volatile products (SVOCs) that equilibrate between the gas and particle phases. If SVOCs undergo an accretion reaction in the particle phase to form non-volatile oligomers, then SOA formation can be strongly enhanced by the movement of additional SVOC molecules from the gas phase to the particle phase. In the research summarized here, the impact of multiphase chemistry on nanoparticle growth and composition is studied through a combination of experimental measurements and modeling the kinetics of particle growth. Modeling the kinetics under atmospherically relevant conditions in a boreal forest shows that particle-phase reactions can grow nanoparticles at a rate that is faster than what is expected from condensation and partitioning alone. The growth rate enhancement is significant for particles greater than about 20 nm in diameter and its magnitude increases with increasing particle size. In a similar way, the non-volatile oligomers from this reaction make up an increasing fraction of the nanoparticle mass as the particle size increases. The modeled particle compositions are consistent with experimental measurements of chemical composition as a function of particle size for aerosol formed by oxidation of two atmospherically-relevant precursors: β-pinene and decamethylcyclopentasiloxane (D_5). Reaction rate constants on the order of 10^{-3} to 10^{-2} $M^{-1}s^{-1}$ are needed for particle growth by multiphase chemistry to compete favorably with condensational growth by non-volatile molecules formed directly in the gas phase, which implicates hydroperoxides and/or peroxyacids as key reactants in these processes. The atmospheric implications of multiphase chemistry are discussed in the context of both climate and human health.

Introduction

Atmospheric aerosols influence Earth's energy balance either indirectly through cloud formation or by directly scattering incoming solar radiation (*1–3*). A significant fraction of airborne particles originate from gas-to-particle conversion during new particle formation (NPF) (*4*). Volatile compounds emitted into the atmosphere from biogenic and anthropogenic sources are oxidized to produce semi-volatile and non-volatile products. Particle formation begins when these products, in combination with other gas-phase species, come together to form clusters on the order of 1-2 nm in diameter that are able to spontaneously grow to much larger sizes (*4–7*). Depending on chemical composition, once the particles have grown to a size of 50-100 nm in diameter, they are able to serve as cloud condensation nuclei (CCN). The probability that a newly formed nanoparticle will grow to a CCN relevant size depends on its growth rate relative to the loss rate from e.g. coagulation or scavenging by pre-existing aerosols (*8*). Uncertainties in predicting the conditions that favor CCN formation make it challenging to accurately predict future impacts of radiative forcing (*9*).

Atmospheric aerosols also impact human health. Pollution, broadly defined, is estimated to have been responsible for about 16% of all deaths world-wide in 2015, and of this the greatest contributor by far is air pollution (*10–13*). Mortality linked to pollution dwarfs contributions from other causes e.g. AIDS/malaria/tuberculosis, alcohol/drug abuse, and murders/wars. Only tobacco smoking comes close. Human mortality and morbidity both show a strong correlation with $PM_{2.5}$, which is defined as the airborne mass concentration of all particlulate matter 2.5 micrometers in diameter or smaller (*13*). These particles can efficiently penetrate into the alveolar region of the lungs, and upon deposition, cause respiratory distress, allergic reactions, and cardiovascular disease (*14*). While the correlation between $PM_{2.5}$ and human health appears to be independent of chemical composition, some noteworthy chemical species within $PM_{2.5}$ having known health-related responses are carcinogens (e.g. polycyclic aromatics), allergens (e.g. virons and bacteria), and reactive oxygen species (ROS) such as hydrogen peroxide and related compounds that induce oxidative stress on tissues and cells (*15*). Chemical mechanisms of nanoparticle growth are at the nexus of climate and health impacts since they determine how quickly the size and mass of these particles increase.

How Do Ambient Nanoparticles Grow?

Figure 1 summarizes the chemical species and processes that drive ambient nanoparticle formation and growth. The three main chemical species associated with these processes are sulfuric acid (red dots in Figure 1), a neutralizing base, typically ammonia (orange dots in Figure 1), and organic molecules (green and purple dots in Figure 1). The growth rate due to sulfuric acid along with neutralizing base is accurately predicted by experimental measurements of the gas-phase mixing ratio and particle-phase composition using a condensational growth model (*16–19*), though sulfuric acid represents only a minor fraction of the total growth rate of ambient particles (*8*, *20–22*). Nanoparticle composition and growth rate are dominated by organic matter (*16–18*, *23–27*), and though significant molecular insight has been gained (*6*, *28–30*), current growth models for organic matter appear to be incomplete (*31*, *32*).

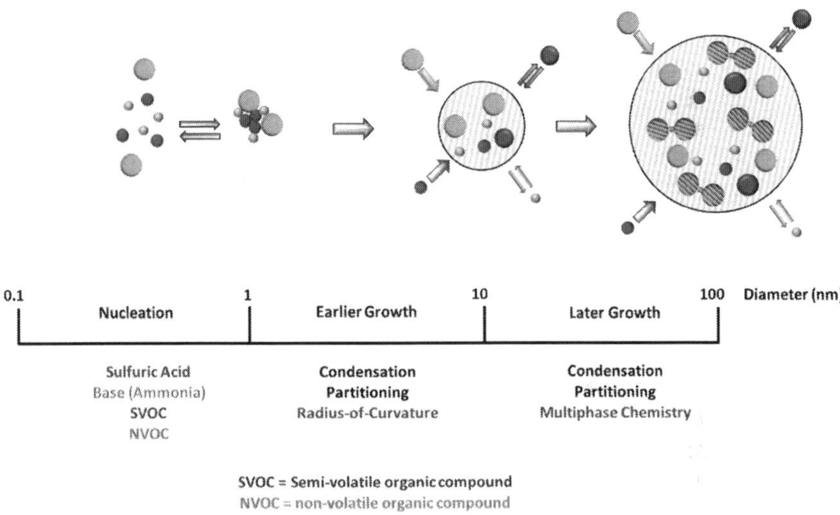

Figure 1. Summary of molecules and processes relevant to particle formation and growth in the atmosphere. (see color insert)

Gas-phase oxidation of volatile organic compounds occurs through a complex set of reaction pathways in both the gas and particle phases to yield particle-phase products that often number in the hundreds or thousands (*33–36*). Absorptive partitioning (*37*, *38*) of gas-phase products to the particle phase forms secondary organic aerosol (SOA), which includes both non-volatile (NVOC, green dots in Figure 1) and semi-volatile (SVOC, purple dots in Figure 1) organic compounds (*27*, *39*). The distribution of organic molecules between the gas and particle phases is described by absorptive partitioning theory (*37*). When precursor molecules are oxidized in the gas phase, the products partition to the particle phase causing particle growth. NVOC molecules have a negligible evaporation rate once they partition to the particle phase. These molecules cause particle growth at a rate given by their condensation rate from the gas phase. SVOC molecules have a substantial evaporation rate and therefore cause particle growth at a rate much slower than their condensation rate.

Figure 1 is divided into the three main size regimes associated with particle formation and growth. Initially, gas-phase molecules come together to form a stable molecular cluster, typically on the order of 1-2 nm in diameter, that is able to spontaneously grow to larger sizes. Although cluster formation is not discussed in this chapter, several reviews are available in the literature (*6, 7, 40–42*). Newly formed particles subsequently grow by a combination of condensation and partitioning. In the 'Earlier Growth' stage of the particle in Figure 1, condensation is dominant owing to the impact of radius-of-curvature on molecular volatility. The equilibrium vapor pressure over a curved surface is higher than that over a flat surface. This is because a molecule on a curved surface has less interaction with the condensed phase and is more easily able to escape into the gas phase. This process by which the equilibrium vapor pressure is modified to give P_d is described by the Kelvin equation:

$$(1)\quad P_d = P_0\, e^{\left[(2\sigma V_{M,p}) / \left(\frac{d}{2} RT\right)\right]}$$

where P_0 is the saturation vapor pressure over a flat surface, σ is the surface tension, $V_{M,p}$ is the average molar volume of the molecules already existing in the particle phase, d is the particle diameter, R is the universal gas constant, and T is the temperature. The subscript d, shown here with P_d , and for other variables hereafter, represents the particle diameter and denotes that the variable is particle size dependent. For 10 nm diameter aqueous droplets, molecular vapor pressures are about 20% greater than those over a flat surface (*43*), and volatility quickly increases with decreasing particle diameter. The strong dependence of molecular volatility on particle diameter is thought to be the reason that particle growth rates increase quickly with increasing particle diameter in the low nanometer size range (*21*). As will be discussed later in this chapter, multiphase chemistry becomes increasingly important with increasing particle diameter. In the 'Later Growth' stage of Figure 1, multiphase chemistry is depicted by the partitioning of two SVOC molecules to the particle phase, where they react to form a non-volatile dimer product (green-purple, striped dots). In actuality, reactions in the particle phase can involve both NVOC and SVOC precursors, though reactions involving SVOC have the greatest potential to enhance particle growth.

Molecular analysis of SOA has shown the presence of compounds that were produced by multiphase chemistry. Some examples include oligomers in biogenic SOA formed by accretion reactions (*44–46*), imine related species formed by the reaction of dicarbonyls with ammonia or amines (*47–50*), and organosulfates (*51–55*). As discussed later in this chapter, reactions such as these increase the aerosol yield by forming additional SOA beyond what would be expected from partitioning alone, if they form non-volatile products from semi-volatile reactants in the particle phase (*56, 57*). Experimental measurements have shown that oligomers can constitute up to about 50% of the mass of SOA produced from biogenic precursors in laboratory reactors, though it is not clear how much of the oligomeric matter is produced from semi-volatile vs. non-volatile precursors (*58*).

Modeling Particle Growth and the Impact of Multiphase Chemistry

Predicting the way in which multiphase chemistry may contribute to nanoparticle growth and composition can be understood in part through kinetic modeling. Such modeling calculations are performed iteratively over a short time period dt by summing the change in total particle volume from each individual species, thereby allowing for growth rate and chemical composition to be tracked as particle diameter increases. The growth calculations discussed below are representative of what might occur in a boreal forest, where numerous ambient new particle formation experiments have been performed (*6*). Table 1 gives the mixing ratios, volatilities, and gas-phase diffusion coefficients of each chemical species included in the calculation. The temperature and relative humidity are assumed to be 282K and 50% respectively. Basic aspects of the growth calculation are summarized below. A more detailed description is given eslewhere (*59*).

Table 1. Mixing ratios and relevant physical properties of chemical species included in particle growth calculations. (T = 282K, RH = 50%)

	Gas-Phase Mixing Ratio, $C_{i,g}$		*Saturation Concen-tration*	*Molar Mass*	*Density*	*Gas-Phase Diffusion Coefficient*
	ng m^{-3}	*pptv*	*log C* (μg m^{-3})*	*g mol^{-1}*	*g cm^{-3}*	*cm^2 s^{-1}*
Sulfuric Acid	0.5	0.12	-	98	1.8	0.1
Ammonia	1	1.25	-	17	0.7	0.2
SVOC	3-11	0.4-1.4	10^0	200	1.2	0.05
NVOC	3	0.4	10^{-4}	200	1.2	0.05
DIMER	-	-	-	400	1.2	-

For condensational growth by a non-volatile species i, the collision rate determines the number of molecules that are taken into the particle during time period dt, thereby incrementing the particle volume by dV_i:

$$(2)\quad \frac{dV_i}{dt} = \frac{c_i}{4}\gamma\pi d^2 C_{i,g}\,\beta_d V_{M,i}$$

where c_i is the mean thermal velocity, γ is the uptake coefficient, d is the particle diameter, $C_{i,g}$ is the gas-phase mixing ratio of species i, β_d is the Loyalka correction factor for mass transport to a spherical particle with diameter d (*43*), and $V_{M,i}$ is the molar volume of species i. Eq. 2 is used to calculate condensational growth by sulfuric acid and NVOC. Inherent to this equation is the assumption that the surface accommodation coefficient is 1. From a calculation standpoint, an accommodation coefficient less than 1 would require proportionally higher gas-phase mixing ratios to reproduce the results shown here. In addition, it is assumed that two ammonia molecules are taken into the particle for every sulfuric acid molecule that condenses. Because the ammonia mixing ratio is so high in this calculation, neutralization is assumed to keep up with condensation.

For growth by a semi-volatile species (SVOC), the volume change of the particle caused by species i is given by:

$$(3)\quad \frac{V_i}{V_{p,n}} = \left(\frac{C_{i,g}}{P_d}\right)\left(\frac{V_{M,i}}{V_{M,p}}\right)\frac{1}{\zeta_i}$$

where ζ_i is the activity coefficient for SVOC in the particle phase (assumed to be 1 in these calculations) and $V_{p,n}$ is the total volume of the particle. For each incremental increase in particle volume due to condensational growth (Eq. 2), SVOC becomes dilute in the particle phase and is no longer in equilibrium with the gas phase. Molecules of SVOC that migrate to the particle phase must be re-calculated after each incremental volume increase to re-establish gas-particle equilibrium. Inherent to Eq. 3 is the assumption that equilibrium is reached during the time increment over which V_i is calculated. For most calculations described below, uptake of SVOC proceeds at a rate much lower than the collision rate.

Molecules that are taken into the particle may undergo accretion reactions. Accretion chemistry is illustrated primarily in the context of two SVOC monomers that react in the particle phase to form a non-volatile dimer (DIMER). The rate at which SVOC monomers react to produce DIMER is given by:

$$(4)\quad \frac{d[DIMER]}{dt} = -\frac{d[SVOC]}{2\,dt} = k_{DIMER}[SVOC]^2$$

where k_{DIMER} is the second order rate constant, and $[SVOC]$ is the particle phase concentration established by partitioning between the gas and particle phases. Gas-particle equilibrium of SVOC is lost when the reaction occurs. Therefore, additional molecules must be taken into the particle to re-establish gas-particle equilibrium. The result is that the rate of change in total particle volume is greater than what would be predicted for partitioning alone.

Figure 2 compares particle growth rates that are calculated for two different scenarios: one where only condensation and partitioning occur (Eq. 2-3 above), shown in black; and the other where condensation, partitioning, and particle-phase reaction all occur (Eq. 2-4), shown in red. These plots illustrate how the particle-phase reaction can increase the growth rate over what would be obtained by condensation and partitioning alone.

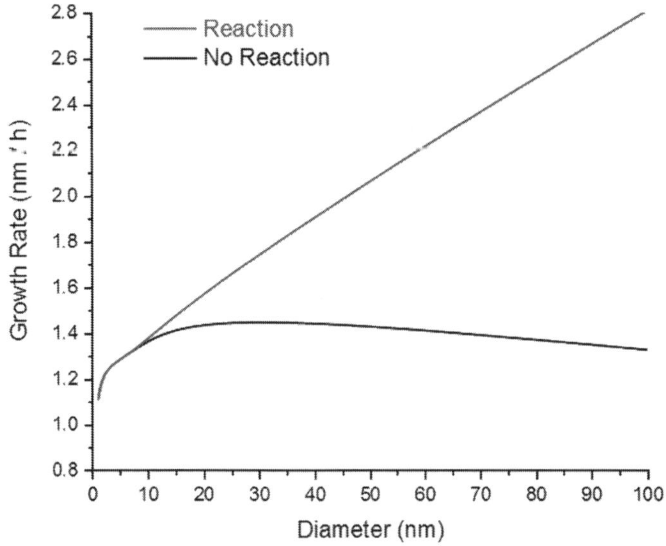

Figure 2. Growth rate vs. particle diameter for calculations with (red; $k_{DIMER} = 10^{-2}$ $M^{-1}s^{-1}$) or without (black; $k_{DIMER} = 0$ $M^{-1}s^{-1}$) accretion reaction in the particle phase. (see color insert)

The black curve in Figure 2 shows how particle growth proceeds by a combination of condensation and partitioning ($k_{DIMER} = 0$ $M^{-1}s^{-1}$). Under conditions dominated by condensational growth, a situation that usually exists in the atmosphere, the particle growth rate dd/dt becomes independent of particle diameter above ~10 nm, as illustrated by the following equation:

$$(5)\quad \frac{\mathrm{d}d}{\mathrm{d}t} = \frac{c_i}{2}\gamma C_{i,g}\,\beta_d V_{M,i}\,.$$

Since γ is independent of d for a surface-limited process such as condensation (*44*, *60*), the growth rate is also independent of particle diameter to a first approximation. In practice, the slight decrease of the growth rate with increasing particle diameter is due to the limitation of gas-phase diffusion on molecular transport to the particle surface. Partitioning of SVOC to the particle phase does increase with increasing particle diameter, but its overall contribution to the growth rate is small. In the same way, the composition of a particle at any size is dominated by the condensing species (NVOC and sulfuric acid along with neutralizing base), and the relative amounts of these species are also independent of particle diameter.

The red plot in Figure 2 shows how particle growth proceeds when condensation, partitioning, and particle-phase reaction are all included ($k_{DIMER} = 10^{-2}$ $M^{-1}s^{-1}$). Particle-phase reaction causes the growth rate to increase linearly with increasing particle diameter. This size dependence can be understood by realizing that γ in Eq. 5 is proportional to d, which is characteristic of a volume-limited process (*61*, *62*). Since γ is proportional to d, the growth rate due to multiphase chemistry is also linearly proportional to *d*, while the growth rate due to condensation of NVOC and sulfuric acid remain independent of *d*.

The mass ratio of non-volatile DIMER to NVOC monomer is shown in Figure 3. Below 10 nm in diameter, DIMER mass fraction is small. This is due to the Kelvin effect (Eq. 1), which greatly limits the uptake of SVOC. At particle sizes larger than 10 nm, DIMER production is significant and the DIMER to NVOC mass ratio is proportional to *d*. Condensation of NVOC is still independent of particle diameter, so the composition change in Figure 3 arises from the production of DIMER in the particle phase.

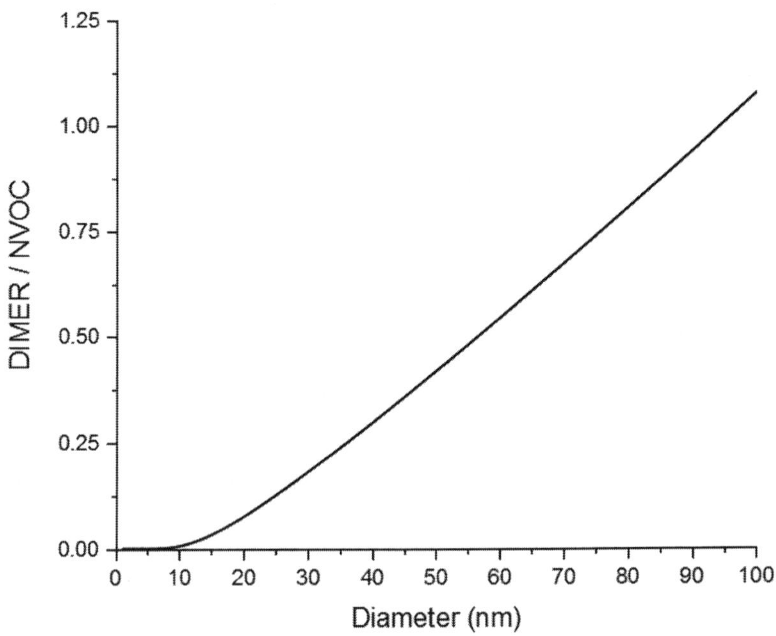

Figure 3. Mass ratio vs. particle diameter for DIMER to NVOC under conditions where both partitioning and particle-phase chemistry (SVOC-SVOC dimer formation) are included. Reprinted from reference (59). Copyright 2018.

In practice, accretion chemistry is not necessarily restricted to SVOC molecules. If one or both reactants are NVOC, then different results are obtained. Figure 4 shows growth rate plots for three scenarios: SVOC-SVOC reaction (red; same calculation and plot as in Figure 2), NVOC-NVOC reaction (black), and SVOC-NVOC reaction (blue). When two NVOC molecules react to form DIMER, then the diameter growth rate *will not* change with increasing particle diameter since NVOC uptake remains unaffected. Therefore, the black plot in Figure 4 is equivalent to the black plot in Figure 2. If one SVOC molecule reacts with one NVOC molecule to form DIMER, the growth rate increases with increasing particle diameter, but the dependencies are different from the reaction of two SVOC molecules. The SVOC-NVOC reaction enhances the particle growth rate to a greater extent than the SVOC-SVOC does at small particle diameters. This enhancement arises from the impact the Kelvin effect has on SVOC volatility, which favors formation of the SVOC-NVOC product over the SVOC-SVOC product at small particle diameters. Because of the greater potential of the SVOC-SVOC reaction to transform semi-volatile matter into non-volatile matter, its contribution to growth rate eventually overcomes that of the SVOC-NVOC reaction as the particle diameter increases.

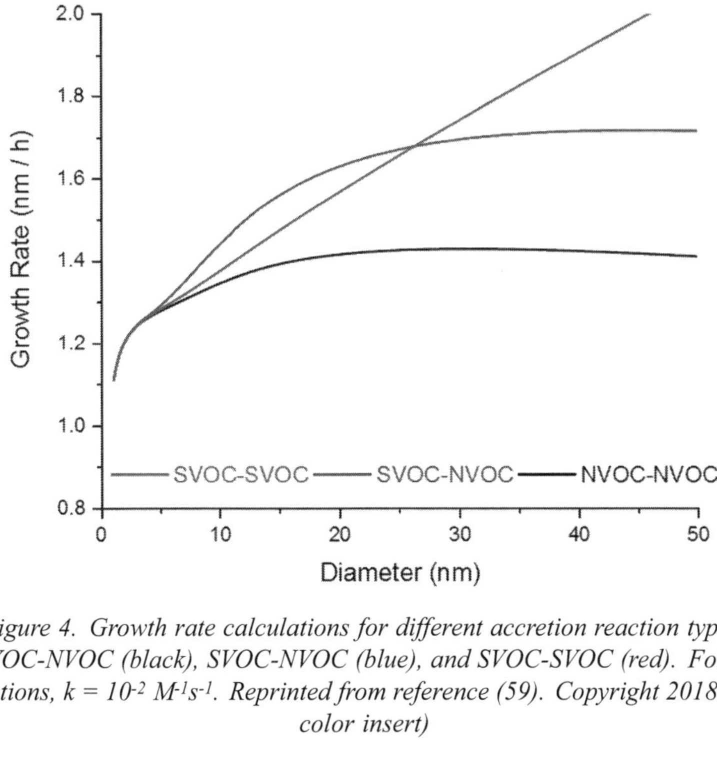

Figure 4. Growth rate calculations for different accretion reaction types: NVOC-NVOC (black), SVOC-NVOC (blue), and SVOC-SVOC (red). For all reactions, $k = 10^{-2}$ $M^{-1}s^{-1}$. Reprinted from reference (59). Copyright 2018. (see color insert)

The calculations in Figures 2-4 assume a dimerization rate constant of 10^{-2} $M^{-1}s^{-1}$, though a rate constant as low as 10^{-3} $M^{-1}s^{-1}$ could enhance the growth rate over that of condensation and partitioning alone (*59*). What types of accretion reactions have rate constants in this range? Ziemann and Atkinson (*63*) have reviewed thermodynamic and kinetic data for several types of reactions relevant to biogenic SOA. The reaction of a hydroperoxide with a carbonyl to give a peroxyhemiacetal, and the reaction of a peroxyacid with a carbonyl to form an acyl peroxyhemiacetal, have reported rate constants in the 10^{-4} to 10^{-2} $M^{-1}s^{-1}$ range depending on reaction conditions and are fast enough to be atmospherically relevant. Other reactions, such as aldol condensation of carbonyls and ester formation from an acid and alcohol, are much slower and unlikely to be atmospherically relevant for enhancing particle growth.

Multiphase Chemistry Viewed through Molecular Composition Measurements of Laboratory SOA

The modeling results give insight into the potential impact of multiphase chemistry on nanoparticle growth rate and chemical composition, and they provide context for interpreting experimental measurements. Here we discuss molecular composition measurements from two systems where multiphase chemistry assists secondary aerosol formation.

The first system is SOA from β-pinene ozonolysis. Biogenically-derived SOA is produced from a variety of mono- and sesquiterpene precursors that are emitted primarily from evergreens. β-Pinene is a particularly interesting precursor in that the yield of NVOC-like molecules (often referred to in the literature as highly oxidized molecules, or HOMs, because of their high oxygen content) from the ozonolysis process is 1-2 orders of magnitude lower than those of other monoterpene precursors, yet the overall aerosol yield is within about a factor of 2 of those of other precursors. This disparity suggests that nanoparticulate SOA from β-pinene ozonolysis grows substantially through multiphase chemistry rather than condensational growth of HOMs. (This disparity does not preclude the possibility that nanoparticulate SOA growth from other monoterpene precursors is also influenced by multiphase chemistry.) Specifically, aerosol yields for β-pinene ozonolysis under conditions similar to the experiments described here have been reported to be on the order of 30%, with products spanning a wide range of volatilities (*65*). Most products of β-pinene ozonolysis are volatile or semi-volatile molecules that partition between the gas and particle phases (*66*). HOMs are inefficiently produced from β-pinene ozonolysis owing to its exocyclic double bond, which results in an estimated yield that is less than 0.1% (*67*). In the experiments described here, β-pinene and ozone vapors are mixed in a flow tube reactor to produce SOA. Particles exiting the reactor span a size range of about 30 to 150 nm in diameter. Larger particles were produced earlier in the reactor and had a longer time to grow, while smaller particles were produced later in the reactor and had less time to grow. Thus, changes in molecular composition as a function of particle size indicate changes in the growth mechanism with time.

Figure 5 compares the mass spectra of 35 nm vs. 100 nm diameter particles based on positive and negative electrospray ionization (ESI) mass spectra. Overall, the mass spectra show families of ions corresponding to monomers (individual β-pinene molecules that have reacted to give a low molecular weight product), dimers (combinations of two β-pinene molecules that have reacted), and higher order oligomers (combinations of three or more β-pinene molecules that have reacted to give high molecular weight products). Monomers are necessarily highly oxidized since this is needed to achieve very low volatility and enter the particle phase, whereas dimers and higher order oligomers can be somewhat less oxidized owing to their larger molecular size. Inspection of the mass spectra in Figure 5 shows that monomers (defined as ions having molecular formulas with 9 or fewer C atoms, typically 220 m/z or smaller) are more highly represented at the smaller particle size, while dimers, (defined as ions having molecular formulas with 10-18 C atoms, typically 200-400 m/z) and especially higher order oligomers (defined as ions having molecular formulas with greater than 18 C atoms, typically 400 m/z or greater), are more dominant at the larger particle size. The abundance of oligomers in larger particles is deduced from their appearance in mass spectra at the higher molecular weight regions. These peaks are either non-existent or yield lower signal intensities in the mass spectra for 35 nm particles. This particle size dependence suggests that many dimers, and most higher order oligomers, are produced directly in the particle phase by an accretion reaction.

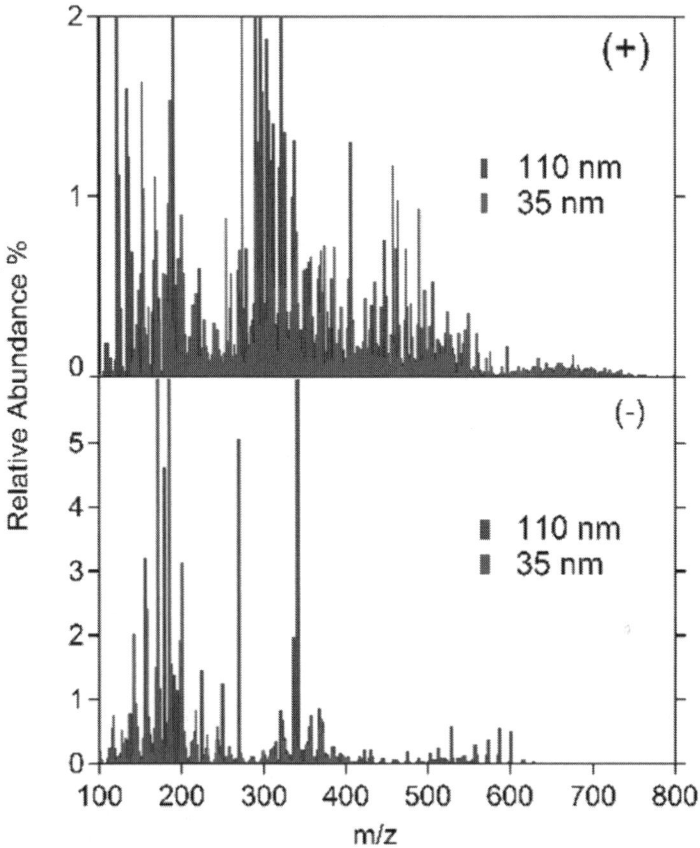

Figure 5. Positive (+) and negative (-) ion ESI mass spectra of 35 nm (red) and 110 nm (blue) monodisperse SOA samples. Reprinted from reference (64). Copyright 2017. (see color insert)

Further insight can be gained from the ion signal intensities of these ions. Figure 6 shows the fraction of the total signal intensity due to higher order oligomers as a function of particle diameter. Both ion polarities show an approximate linear increase of oligomer intensity with increasing particle diameter. The plot in Figure 6 is similar in concept to that in Figure 3 for the modeling study and is expected for a volume-limited process, such as accretion chemistry, relative to a surface-limited process, such as condensation. The total oligomer signal intensity in this experiment is comparable to our previous study of α-pinene SOA formed under similar reaction conditions, where the oligomer content was experimentally determined to be about 50% of the total SOA mass (*58*). Oligomerization has been considered for many years to be a significant contributor to SOA formation (*39, 44, 46, 68, 69*). Figures 5 and 6 show, through chemical measurement, that this contribution strongly depends on particle size.

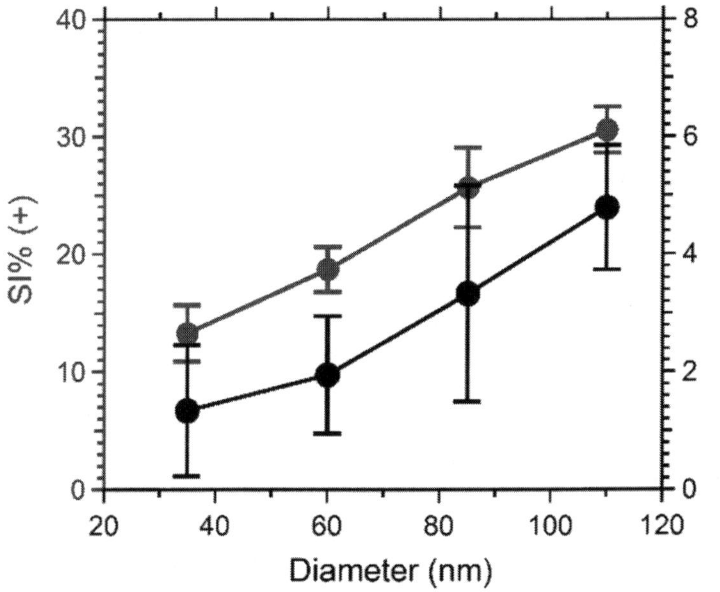

Figure 6. Percentage of total signal intensity (SI %) in positive (+) and negative (-) ion ESI mass spectra from higher order oligomers vs. particle diameter. Reprinted from reference (64). Copyright 2017. (see color insert)

The model described above was applied to the β-pinene ozonolysis experiment to better constrain the molecular processes responsible for particle growth. The NVOC mixing ratio is estimated to be approximately 2 ppbv based on the yield estimate of HOMs for β-pinene ozonolysis by Ehn et al. (*29*) Mixing ratios for SVOC having saturation concentrations in the 10 μg m^{-3} and 100 μg m^{-3} bins are estimated to be approximately 30 ppbv and 40 ppbv respectively, based on the volatility basis set (VBS) parameterization of Donahue and coworkers for α-pinene ozonolysis (*69, 70*). Using these inputs, we found that condensational growth and/or partitioning of molecular species alone are able to explain the overall growth of particles to about 80-100 nm in diameter during the ~20 s residence time of the flow tube reactor, though the modeling results are extremely sensitive to the mixing ratios and VBS parameterization used. Condensational growth and partitioning alone cannot explain the particle size dependence and high abundance of accretion reaction products that were measured. We estimate that the reaction rate constants needed to form dimers and higher order oligomers in the amounts detected had to be at least on the order of 10^{-3} $M^{-1}s^{-1}$. (The required rate constant is dependent on SVOC volatility, with a larger rate constant needed for saturation concentrations above 10 μg m^{-3}). Rate constants of this magnitude are consistent with reactions of hydroperoxides and/or peroxyacids with carbonyls to give accretion products.

The second system discussed here involves secondary aerosol produced by the reaction of decamethylcyclopentasiloxane (D_5) with OH. Cyclic volatile methylsiloxanes (cVMS) are widely used in personal care products and can have very high mixing ratios in indoor air. For example, Tang et al. examined the sources of volatile organic compounds (VOCs) in a university classroom and reported a D_5 concentration of 60 ± 32 μg m^{-3}, which represented >30% of all VOCs detected (*71*). Silicon is routinely observed in our ambient nanoparticle measurements, and gas-phase oxidation of cVMS, particularly D_5, has been suggested as a possible source (*9*). The reaction of D_5 with OH produces three main types of products: ring-opened species, oxidized D_5 molecules containing multiple OH and/or CH_2OH functionalities in place of CH_3, and dimers of D_5 and/or its oxidation products (*72*). In the experiments described below, D_5 and an OH precursor were mixed in a photochemical flow through chamber (*73*). The amount of aerosol produced was determined by the mixing ratio of D_5 added to the chamber. Lower D_5 mixing ratios produced smaller aerosol mass loadings, while higher D_5 mixing ratios produced higher mass loadings. Lower mass loadings generally contained smaller particles and a lower aerosol volume-to-surface area ratio, while higher mass loadings generally contained larger particles and a higher aerosol volume-to-surface area ratio.

Figure 7 shows the positive ESI mass spectra for three aerosol mass loadings. The oxidation products observed in these spectra are color coded into three main types. First are monomer products, defined as those containing exactly 5 Si atoms, at least 5 O atoms, and a molecular formula (and MS/MS spectra) consistent with one siloxane ring but no other sites of unsaturation. Relative to the D_5 precursor molecule, monomer products contain various combinations of functionalization where CH_3 groups are replaced by OH and/or CH_2OH groups. In Figure 7, monomer products are coded red.

Second are dimer products, defined here as those containing exactly 10 Si atoms, at least 10 O atoms, and a molecular formula (and MS/MS spectra) consistent with two siloxane rings but no other sites of unsaturation. Dimers also contain various combinations of functionalization where CH_3 groups are replaced by OH and/or CH_2OH, and the two siloxane rings can be linked by O, CH_2, or CH_2CH_2 groups. In Figure 7, dimer products are coded black.

Third are ring-opened products, defined here as those containing neither 5 nor 10 Si atoms. Formation of these products necessarily requires fragmentation of the original D_5 siloxane ring structure. Since most of these products contain more than 5 Si atoms, it is likely that they are formed by coupling of a D_5 molecule (or one of its monomer oxidation products) with the fragmentation product of another D_5 molecule. Molecular formulas and MS/MS spectra of these products are consistent with CH_3 replacement by OH and/or CH_2OH, and sometimes indicate an additional site of unsaturation, for example, Si=O. In Figure 7, ring-opened products are coded blue. As shown in Figure 7, ring-opened products dominate the mass spectra at low aerosol mass loading, while monomers and dimers become more prominent as the mass loading increases.

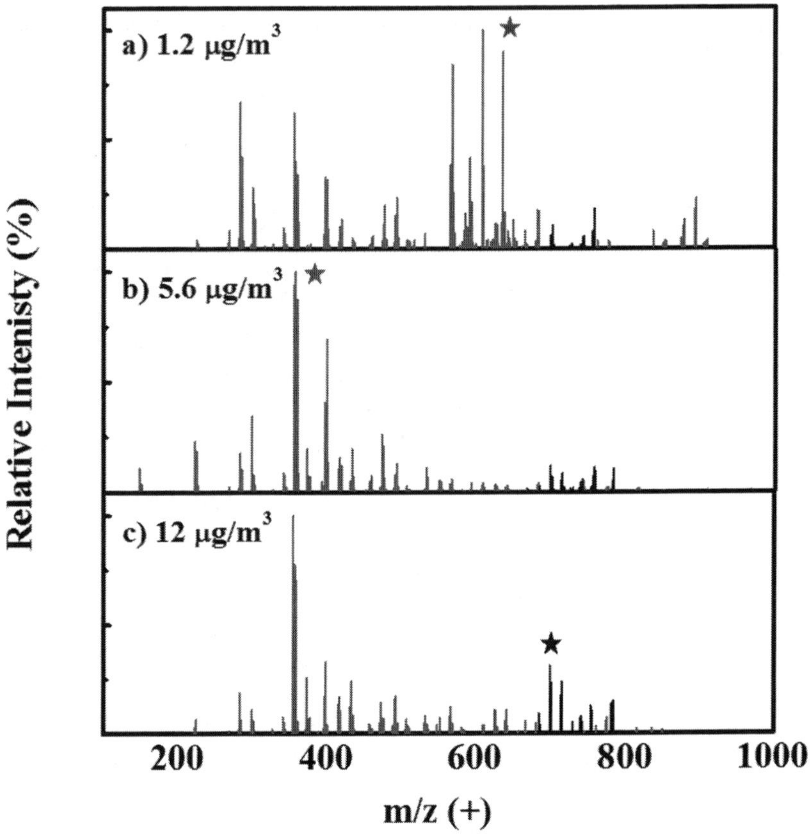

Figure 7. ESI mass spectra of secondary aerosol from D_5 oxidation under various aerosol mass loadings: a) 1.2 μg m⁻³, b) 5.6 μg m⁻³, c) 12 g m⁻³. Peaks considered as ring-opened products are coded blue, dimers are coded black, and monomers are coded red. Stars highlight prominent ions of each type. Reprinted from reference (73). Copyright 2017, the American Chemical Society. (see color insert)

The mass loading dependence is explored further in Figure 8 where the ratio of the summed signal intensities of dimers to ring-opened products is plotted as a function of aerosol volume-to-surface area ratio. Although all dimers and most ring-opened products are expected to be extremely non-volatile, with predicted saturation concentrations (C*) below10^{-2} μg m^{-3}, Figure 8 shows that their relative abundances are strongly dependent on aerosol loading. The approximately linear relationship suggests that ring-opened products are formed in the gas phase and condensationally grow particles, while dimers are formed directly in the particle phase through accretion reactions. As indicated by the modeling results in Figure 3, a linear dependence is expected for a volume-limited process, such as accretion reaction, relative to a surface-limited process, such as condensation. Since the volume-to-surface area ratio for a particle of diameter *d* is 2*d*/3, the plots in Figures 6 and 8 span similar ranges of volume-to-surface area ratio for these two secondary aerosol systems.

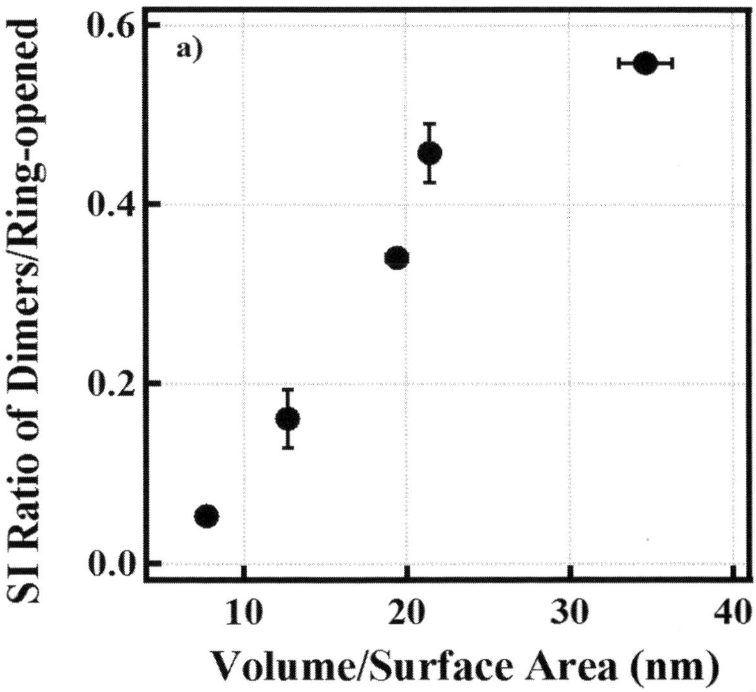

Figure 8. Signal intensity ratio of dimer to ring-opened products. Reprinted from reference (73). Copyright 2017, the American Chemical Society.

The model described above was applied to the D_5 experiment to better constrain the molecular processes responsible for particle growth. In these experiments, the mixing ratios of both D_5 and its monomer and ring-opened oxidation products are all expected to be in the low ppbv range. Using these values, we estimate the dimerization rate constant had to be within the 10^{-3} to 10^{-1} $M^{-1}s^{-1}$ range to achieve the level of dimerization observed, given that most of the monomers expected to participate in the reaction had estimated saturation concentrations within the 10^2 to 10^3 μg m^{-3} range. As with the β-pinene ozonolysis experiments, the accretion reaction rate constants needed to explain the D_5 secondary aerosol results are in the range expected for reactions of hydroperoxides and/or peroxyacids. The MS and MS/MS measurements performed in the D_5 experiments described here were unable to distinguish multiple OH and/or CH_2OH functionalities from e.g. OOH and/or CH_2OOH functionalities. Our earlier work with D_4 oxidation by OH did provide unequivocal evidence for the existence of peroxyl functionalities (*72*).

Atmospheric Implications: Connecting the Dots with Other Studies

This chapter has reviewed some of our recent work to study the impact of multiphase chemistry on nanoparticle formation and growth, specifically accretion reactions of semi-volatile molecules that have partitioned into the particle phase (*59*, *64*, *73*). These reactions can induce systematic changes in molecular composition as a function of particle size, or more broadly, aerosol volume-to-surface area ratio. For relevant atmospheric and laboratory conditions, we find that reaction rate constants on the order of 10^{-3} to 10^{-2} $M^{-1}s^{-1}$ are needed for particle growth by multiphase chemistry to compete favorably with condensational growth by non-volatile molecules formed directly in the gas phase. Modeling results here require oligomers to be formed in the particle phase. Recent ambient measurements have found that dimers produced from *gas-phase* monoterpene oxidation are present at much higher levels than previously thought (*74*). While this discovery has significant implications for those attempting to model SOA yields on regional or global scales from this gas-phase reaction chemistry, *gas-phase* dimers would be expected to follow a condensational growth mechanism (Eq. 2), but not cause the growth rate to increase with increasing particle diameter.

Recent measurements by Kourtchev et al. (*75*) of accretion oligomers in ambient aerosol from a boreal forest suggest that the oligomer content increases with increasing SOA mass concentration, an observation that is consistent with the particle size and aerosol volume-to-surface area ratio measurements summarized here. More importantly, the authors noted that aerosols enriched with oligomers were strongly correlated with higher CCN activity, and they suggested that this correlation could indicate that oligomers may speed up particle growth. The modeling results presented here suggest that multiphase chemistry, specifically accretion reactions of hydroperoxides and/or peroxyacids, is indeed capable of enhancing growth rates in a size range relevant to CCN activity.

Paasonen et al. (*76*) recently evaluated a 20-year record of ambient particle size measurements in a boreal forest. They found that particle growth rates generally increased with increasing particle diameter and attributed the effect to accretion chemistry, similar to that discussed in this chapter. Similarly, Burkart et al. (*77*) studied ambient particle growth over the summertime Arctic Ocean and found that large pre-existing particles (~90 nm dia.) grew faster than nucleation mode particles (~20 nm dia.). The reason for this size dependence is not known, but multiphase chemistry similar to that discussed in this chapter provides a reasonable explanation.

Although a variety of accretion reactions in the particle phase can be envisioned, the only types with known rate constants fast enough to compete with condensational growth are those involving reactions of hydroperoxides and/or peroxy acids with carbonyls (*63*). Molecules with these functionalities are readily produced in the gas phase, especially by oxidation of biogenic precursors (*36*, *67*, *78–80*), and not surprisingly, these functionalities are prevalent in freshly produced SOA (*63*, *81*). Furthermore, hydroperoxides and peroxyacids have been shown to exist in laboratory SOA almost exclusively within monomers rather than dimers or higher order oligomers (*80*). This distribution of molecular functionalities is consistent with our work showing that accretion reactions leading to particle growth require hydroperoxides and/or peroxyacids in the reacting monomer(s), which are subsequently lost when oligomers are formed.

Gas-phase organic hydroperoxide formation relevant to multiphase chemistry is likely to be important on a global scale. In addition to the oxidation of monoterpenes from pine trees, organic hydroperoxides can also originate from isoprene (*82–85*), which is emitted from deciduous trees and represents approximately 50% of all biogenic emissions annually (*86*), as well as alkanes, alkenes, and aromatic molecules emitted from fossil fuel combustion (*63, 87*).

Organic peroxides are photochemically active (*88–91*), and their formation is potentially reversible (*69*). They may also decompose to give products that do not participate in accretion chemistry, or to give additional reactive oxygen species (ROS) (*84*) that may lead to a different set of reactions. For these reasons, the modeling results presented here represent an upper limit for the impact of multiphase chemistry on nanoparticle growth.

Finally, it should be noted that the molecular functionalities of interest here, hydroperoxides and peroxyacids, are ROS themselves that may induce oxidative stress when nanoparticulate SOA is inhaled (*13, 14*). Based on the results presented here, this effect on a per-unit-mass basis should be greatest for small particles that are dominated by condensation of non-volatile monomers from the gas phase and then decrease as the particles grow and multiphase chemistry consumes the ROS functionalities.

Acknowledgments

The research summarized in this chapter was supported by the National Science Foundation under grant numbers CHE-1408455 and AGS-1649719.

References

1. Charlson, R. J.; Schwartz, S. E.; Hales, J. M.; Cess, R. D.; Coakley, J. A.; Hansen, J. E.; Hofmann, D. J. Climate Forcing by Anthropogenic Aerosols. *Science* **1992**, *255*, 423–430.
2. Kerminen, V. M.; Paramonov, M.; Anttila, T.; Riipinen, I.; Fountoukis, C.; Korhonen, H.; Asmi, E.; Laakso, L.; Lihavainen, H.; Swietlicki, E.; Svenningsson, B.; Asmi, A.; Pandis, S. N.; Kulmala, M.; Petäjä, T. Cloud Condensation Nuclei Production Associated with Atmospheric Nucleation: A Synthesis Based on Existing Literature and New Results. *Atmos. Chem. Phys.* **2012**, *12*, 12037–12059.
3. Lohmann, U.; Feichter, J. Global Indirect Aerosol Effects: A Review. *Atmos. Chem. Phys. Discuss.* **2004**, *4*, 7561–7614.
4. Bzdek, B. R.; Johnston, M. V. New Particle Formation and Growth in the Troposhere. *Anal. Chem.* **2010**, *82*, 7871–7878.
5. Dusek, U.; Frank, G. P.; Hildebrandt, L.; Curtius, J.; Schneider, J.; Walter, S.; Chand, D.; Drewnick, F.; Hings, S.; Jung, D.; Borrmann, S.; Andreae, M. O.

Size Matters More Than Chemistry for Cloud-Nucleating Ability of Aerosol Particles. *Science* **2006**, *312*, 1375–1378.

6. Kulmala, M.; Kontkanen, J.; Junninen, H.; Lehtipalo, K.; Manninen, H. E.; Nieminen, T.; Petäjä, T.; Sipilä, M.; Schobesberger, S.; Rantala, P.; Franchin, A.; Jokinen, T.; Järvinen, E.; Äijälä, M.; Kangasluoma, J.; Hakala, J.; Aalto, P. P.; Paasonen, P.; Mikkilä, J.; Vanhanen, J.; Aalto, J.; Hakola, H.; Makkonen, U.; Ruuskanen, T.; Mauldin, R. L.; Duplissy, J.; Vehkamäki, H.; Bäck, J.; Kortelainen, A.; Riipinen, I.; Kurtén, T.; Johnston, M. V; Smith, J. N.; Ehn, M.; Mentel, T. F.; Lehtinen, K. E. J.; Laaksonen, A.; Kerminen, V.-M.; Worsnop, D. R. Direct Observations of Atmospheric Aerosol Nucleation. *Science* **2013**, *339*, 943–946.
7. Zhang, R.; Khalizov, A.; Wang, L.; Hu, M.; Xu, W. Nucleation and Growth of Nanoparticles in the Atmosphere. *Chem. Rev.* **2012**, *112*, 1957–2011.
8. Kuang, C.; Riipinen, I.; Sihto, S. L.; Kulmala, M.; McCormick, A. V.; McMurry, P. H. An Improved Criterion for New Particle Formation in Diverse Atmospheric Environments. *Atmos. Chem. Phys.* **2010**, *10*, 8469–8480.
9. Carslaw, K. S.; Lee, L. a; Reddington, C. L.; Mann, G. W.; Pringle, K. J. The Magnitude and Sources of Uncertainty in Global Aerosol. *Faraday Discuss.* **2013**, *165*, 495–512.
10. Landrigan, P. J.; Fuller, R.; Acosta, N. J. R.; Adeyi, O.; Arnold, R.; Basu, N.; Baldé, A. B.; Bertollini, R.; Bose-O'Reilly, S.; Boufford, J. I.; Breysse, P. N.; Chiles, T.; Mahidol, C.; Coll-Seck, A. M.; Cropper, M. L.; Fobil, J.; Fuster, V.; Greenstone, M.; Haines, A.; Hanrahan, D.; Hunter, D.; Khare, M.; Krupnick, A.; Lanphear, B.; Lohani, B.; Martin, K.; Mathiasen, K. V; McTeer, M. A.; Murray, C. J. L.; Ndahimananjara, J. D.; Perera, F.; Potočnik, J.; Preker, A. S.; Ramesh, J.; Rockström, J.; Salinas, C.; Samson, L. D.; Sandilya, K.; Sly, P. D.; Smith, K. R.; Steiner, A.; Stewart, R. B.; Suk, W. A.; van Schayck, O. C. P.; Yadama, G. N.; Yumkella, K.; Zhong, M. The Lancet Commission on Pollution and Health. *Lancet* **2018**, *391*, 462–512.
11. Cohen, A. J.; Brauer, M.; Burnett, R.; Anderson, H. R.; Frostad, J.; Estep, K.; Balakrishnan, K.; Brunekreef, B.; Dandona, L.; Dandona, R.; Feigin, V.; Freedman, G.; Hubbell, B.; Jobling, A.; Kan, H.; Knibbs, L.; Liu, Y.; Martin, R.; Morawska, L.; Pope, C. A.; Shin, H.; Straif, K.; Shaddick, G.; Thomas, M.; van Dingenen, R.; van Donkelaar, A.; Vos, T.; Murray, C. J. L.; Forouzanfar, M. H. Estimates and 25-Year Trends of the Global Burden of Disease Attributable to Ambient Air Pollution: An Analysis of Data from the Global Burden of Diseases Study 2015. *Lancet* **2017**, *389*, 1907–1918.
12. Lelieveld, J. Clean Air in the Anthropocene. *Faraday Discuss.* **2017**, *200*, 693–703.
13. Lelieveld, J.; Evans, J. S.; Fnais, M.; Giannadaki, D.; Pozzer, A. The Contribution of Outdoor Air Pollution Sources to Premature Mortality on a Global Scale. *Nature* **2015**, *525*, 367–371.
14. Gaschen, A.; Lang, D.; Kalberer, M.; Savi, M.; Geiser, T.; Gazdhar, A.; Lehr, C.-M.; Bur, M.; Dommen, J.; Baltensperger, U.; Geiser, M. Cellular

Responses after Exposure of Lung Cell Cultures to Secondary Organic Aerosol Particles. *Environ. Sci. Technol.* **2010**, *44*, 1424–1430.

15. Shiraiwa, M.; Ueda, K.; Pozzer, A.; Lammel, G.; Kampf, C. J.; Fushimi, A.; Enami, S.; Arangio, A. M.; Fröhlich-Nowoisky, J.; Fujitani, Y.; Furuyama, A.; Lakey, P. S. J.; Lelieveld, J.; Lucas, K.; Morino, Y.; Pöschl, U.; Takahama, S.; Takami, A.; Tong, H.; Weber, B.; Yoshino, A.; Sato, K. Aerosol Health Effects from Molecular to Global Scales. *Environ. Sci. Technol.* **2017**, *51*, 13545–13567.
16. Bzdek, B. R.; Horan, A. J.; Pennington, M. R.; DePalma, J. W.; Zhao, J.; Jen, C. N.; Hanson, D. R.; Smith, J. N.; McMurry, P. H.; Johnston, M. V. Quantitative and Time-Resolved Nanoparticle Composition Measurements during New Particle Formation. *Faraday Discuss.* **2013**, *165*, 25–43.
17. Pennington, M. R.; Bzdek, B. R.; Depalma, J. W.; Smith, J. N.; Kortelainen, A. M.; Hildebrandt Ruiz, L.; Petäjä, T.; Kulmala, M.; Worsnop, D. R.; Johnston, M. V. Identification and Quantification of Particle Growth Channels during New Particle Formation. *Atmos. Chem. Phys.* **2013**, *13*, 10215–10225.
18. Smith, J. N.; Dunn, M. J.; VanReken, T. M.; Iida, K.; Stolzenburg, M. R.; McMurry, P. H.; Huey, L. G. Chemical Composition of Atmospheric Nanoparticles Formed from Nucleation in Tecamac, Mexico: Evidence for an Important Role for Organic Species in Nanoparticle Growth. *Geophys. Res. Lett.* **2008**, *35*, 2–6.
19. Stolzenburg, M. R.; McMurry, P. H.; Sakurai, H.; Smith, J. N.; Mauldin, R. L.; Eisele, F. L.; Clement, C. F. Growth Rates of Freshly Nucleated Atmospheric Particles in Atlanta. *J. Geophys. Res. Atmos.* **2005**, *110*, 1–10.
20. Weber, R. J.; Marti, J. J.; McMurry, P. H. Measured Atmospheric New Particle Formation Rates: Implications for Nucleation Mechanisms. *Chem. Eng. Commun.* **1996**, *151*, 53–64.
21. Kuang, C.; Chen, M.; Zhao, J.; Smith, J.; McMurry, P. H.; Wang, J. Size and Time-Resolved Growth Rate Measurements of 1 to 5 nm Freshly Formed Atmospheric Nuclei. *Atmos. Chem. Phys.* **2012**, *12*, 3573–3589.
22. Wehner, B.; Petäjä, T.; Boy, M.; Engler, C.; Birmili, W.; Tuch, T.; Wiedensohler, A.; Kulmala, M. The Contribution of Sulfuric Acid and Non-Volatile Compounds on the Growth of Freshly Formed Atmospheric Aerosols. *Geophys. Res. Lett.* **2005**, *32*, 3–6.
23. Bzdek, B. R.; Zordan, C. A.; Pennington, M. R.; Luther, G. W., III; Johnston, M. V. Quantitative Assessment of the Sulfuric Acid Contribution to New Particle Growth. *Environ. Sci. Technol.* **2012**, *46*, 4365–4373.
24. Bzdek, B. R.; Zordan, C. A.; Luther, G. W.; Johnston, M. V. Nanoparticle Chemical Composition During New Particle Formation. *Aerosol Sci. Technol.* **2011**, *45*, 1041–1048.
25. Bzdek, B. R.; Horan, A. J.; Pennington, M. R.; Janechek, N. J.; Baek, J.; Stanier, C. O.; Johnston, M. V. Silicon Is a Frequent Component of Atmospheric Nanoparticles. *Environ. Sci. Technol.* **2014**, *48*, 11137–11145.
26. Bzdek, B. R.; Lawler, M. J.; Horan, A. J.; Pennington, M. R.; DePalma, J. W.; Zhao, J.; Smith, J. N.; Johnston, M. V. Molecular Constraints on Particle

Growth during New Particle Formation. *Geophys. Res. Lett.* **2014**, *41*, 6045–6054.

27. Riipinen, I.; Yli-Juuti, T.; Pierce, J. R.; Petäjä, T.; Worsnop, D. R.; Kulmala, M.; Donahue, N. M. The Contribution of Organics to Atmospheric Nanoparticle Growth. *Nat. Geosci.* **2012**, *5*, 453–458.
28. Bianchi, F.; Junninen, H.; Frege, C.; Henne, S.; Hoyle, C. R.; Molteni, U.; Herrmann, E.; Adamov, A.; Bukowiecki, N.; Chen, X.; Duplissy, J.; Gysel, M.; Hutterli, M.; Kangasluoma, J.; Kontkanen, J.; Manninen, H. E.; Rondo, L.; Williamson, C.; Curtius, J.; Worsnop, D. R.; Kulmala, M.; Dommen, J.; Baltensperger, U. New Particle Formation in the Free Troposphere: A Question of Chemistry and Timing. *Science* **2016**, *352*, 1109–1112.
29. Ehn, M.; Thornton, J. A.; Kleist, E.; Sipila, M.; Junninen, H.; Pullinen, I.; Springer, M.; Rubach, F.; Tillmann, R.; Lee, B.; Lopez-Hilfiker, F.; Andres, S.; Acir, I. H.; Rissanen, M.; Jokinen, T.; Schobesberger, S.; Kangasluoma, J.; Kontkanen, J.; Nieminen, T.; Kurten, T.; Nielsen, L. B.; Jorgensen, S.; Kjaergaard, H. G.; Canagaratna, M.; Dal Maso, M.; Berndt, T.; Petaja, T.; Wahner, A.; Kerminen, V. M.; Kulmala, M.; Worsnop, D. R.; Wildt, J.; Mentel, T. F. A Large Source of Low-Volatility Secondary Organic Aerosol. *Nature* **2014**, *506*, 476–479.
30. Riccobono, F.; Schobesberger, S.; Scott, C. E.; Dommen, J.; Ortega, I. K.; Rondo, L.; Almeida, J.; Amorim, A.; Bianchi, F.; Breitenlechner, M.; David, A.; Downard, A.; Dunne, E. M.; Duplissy, J.; Ehrhart, S.; Flagan, R. C.; Franchin, A.; Hansel, A.; Junninen, H.; Kajos, M.; Keskinen, H.; Kupc, A.; Kurten, A.; Kvashin, A. N.; Laaksonen, A.; Lehtipalo, K.; Makhmutov, V.; Mathot, S.; Nieminen, T.; Onnela, A.; Petaja, T.; Praplan, A. P.; Santos, F. D.; Schallhart, S.; Seinfeld, J. H.; Sipila, M.; Spracklen, D. V.; Stozhkov, Y.; Stratmann, F.; Tome, A.; Tsagkogeorgas, G.; Vaattovaara, P.; Viisanen, Y.; Vrtala, A.; Wagner, P. E.; Weingartner, E.; Wex, H.; Wimmer, D.; Carslaw, K. S.; Curtius, J.; Donahue, N. M.; Kirkby, J.; Kulmala, M.; Worsnop, D. R.; Baltensperger, U.; Kürten, A.; Kvashin, A. N.; Laaksonen, A.; Lehtipalo, K.; Makhmutov, V.; Mathot, S.; Nieminen, T.; Onnela, A.; Petäjä, T.; Praplan, A. P.; Santos, F. D.; Schallhart, S.; Seinfeld, J. H.; Sipilä, M.; Spracklen, D. V.; Stozhkov, Y.; Stratmann, F.; Tomé, A.; Tsagkogeorgas, G.; Vaattovaara, P.; Viisanen, Y.; Vrtala, A.; Wagner, P. E.; Weingartner, E.; Wex, H.; Wimmer, D.; Carslaw, K. S.; Curtius, J.; Donahue, N. M.; Kirkby, J.; Kulmala, M.; Worsnop, D. R.; Baltensperger, U. Oxidation Products of Biogenic Emissions Contribute to Nucleation of Atmospheric Particles. *Science* **2014**, *344*, 717–721.
31. Hallquist, M.; Wenger, J. C.; Baltensperger, U.; Rudich, Y.; Simpson, D.; Claeys, M.; Dommen, J. The Formation, Properties and Impact of Secondary Organic Aerosol : Current and Emerging Issues. *Atmos. Chem. Phys.* **2009**, *9*, 5155–5236.
32. Tröstl, J.; Chuang, W. K.; Gordon, H.; Heinritzi, M.; Yan, C.; Molteni, U.; Ahlm, L.; Frege, C.; Bianchi, F.; Wagner, R.; Simon, M.; Lehtipalo, K.; Williamson, C.; Craven, J. S.; Duplissy, J.; Adamov, A.; Almeida, J.; Bernhammer, A.-K.; Breitenlechner, M.; Brilke, S.; Dias, A.; Ehrhart, S.;

Flagan, R. C.; Franchin, A.; Fuchs, C.; Guida, R.; Gysel, M.; Hansel, A.; Hoyle, C. R.; Jokinen, T.; Junninen, H.; Kangasluoma, J.; Keskinen, H.; Kim, J.; Krapf, M.; Kürten, A.; Laaksonen, A.; Lawler, M.; Leiminger, M.; Mathot, S.; Möhler, O.; Nieminen, T.; Onnela, A.; Petäjä, T.; Piel, F. M.; Miettinen, P.; Rissanen, M. P.; Rondo, L.; Sarnela, N.; Schobesberger, S.; Sengupta, K.; Sipilä, M.; Smith, J. N.; Steiner, G.; Tomè, A.; Virtanen, A.; Wagner, A. C.; Weingartner, E.; Wimmer, D.; Winkler, P. M.; Ye, P.; Carslaw, K. S.; Curtius, J.; Dommen, J.; Kirkby, J.; Kulmala, M.; Riipinen, I.; Worsnop, D. R.; Donahue, N. M.; Baltensperger, U. The Role of Low-Volatility Organic Compounds in Initial Particle Growth in the Atmosphere. *Nature* **2016**, *533*, 527–531.

33. Reinhardt, A.; Emmenegger, C.; Gerrits, B.; Panse, C.; Dommen, J.; Baltensperger, U.; Zenobi, R.; Kalberer, M. Ultrahigh Mass Resolution and Accurate Mass Measurements as a Tool to Characterize Oligomers in Secondary Organic Aerosols. *Anal. Chem.* **2007**, *79*, 4074–4082.
34. Bateman, A. P.; Walser, M. L.; Desyaterik, Y.; Laskin, J.; Laskin, A.; Nizkorodov, S. A.; Bateman, A. P. The Effect of Solvent on the Analysis of Secondary Organic Aerosol Using Electrospray Ionization Mass Spectrometry. *Environ. Sci. Technol.* **2008**, *42*, 7341–7346.
35. Tu, P.; Hall, W. A.; Johnston, M. V. Characterization of Highly Oxidized Molecules in Fresh and Aged Biogenic Secondary Organic Aerosol. *Anal. Chem.* **2016**, *88*, 4495–4501.
36. Mentel, T. F.; Springer, M.; Ehn, M.; Kleist, E.; Pullinen, I.; Kurtén, T.; Rissanen, M.; Wahner, A.; Wildt, J. Formation of Highly Oxidized Multifunctional Compounds: Autoxidation of Peroxy Radicals Formed in the Ozonolysis of Alkenes – Deduced from Structure–product Relationships. *Atmos. Chem. Phys.* **2015**, *15*, 6745–6765.
37. Pankow, J. F. An Absorption Model of Gas / Particle Partitioning of Organic Compounds in the Atmosphere. *Atmos. Environ.* **1994**, *28*, 185–188.
38. Barsanti, K. C.; Kroll, J. H.; Thornton, J. A. Formation of Low-Volatility Organic Compounds in the Atmosphere: Recent Advancements and Insights. *J. Phys. Chem. Lett.* **2017**, *8*, 1503–1511.
39. Kroll, J. H.; Seinfeld, J. H. Chemistry of Secondary Organic Aerosol: Formation and Evolution of Low-Volatility Organics in the Atmosphere. *Atmos. Environ.* **2008**, *42*, 3593–3624.
40. Kulmala, M.; Riipinen, I.; Sipilä, M.; Manninen, H. E.; Petäjä, T.; Junninen, H.; Maso, M. D.; Mordas, G.; Mirme, A.; Vana, M.; Hirsikko, A.; Laakso, L.; Harrison, R. M.; Hanson, I.; Leung, C.; Lehtinen, K. E. J.; Kerminen, V.-M. Toward Direct Measurement of Atmospheric Nucleation. *Science* **2007**, *318*, 89–92.
41. Bzdek, B. R.; Reid, J. P. Perspective: Aerosol Microphysics: From Molecules to the Chemical Physics of Aerosols. *J. Chem. Phys.* **2017**, *147*, 220901.
42. Bzdek, B. R.; DePalma, J. W.; Johnston, M. V. Mechanisms of Atmospherically Relevant Cluster Growth. *Acc. Chem. Res.* **2017**, *50*, 1965–1975.

43. Seinfeld, J. H.; Pandis, S. N. *Atmospheric Chemistry and Physics: From Air Pollution to Climate Change*, 3rd ed.; John Wiley & Sons: New York, NY, 2016.
44. Tolocka, M. P.; Jang, M.; Ginter, J. M.; Cox, J.; Kamens, R. M.; Johnston, M. V; Tolocka, M. P. Formation of Oligomers in Secondary Organic Aerosol. *Environ. Sci. Technol.* **2004**, *38*, 1428–1434.
45. Barsanti, K. C.; Pankow, J. F. Thermodynamics of the Formation of Atmospheric Organic Particulate Matter by Accretion Reactions-Part 1: Aldehydes and Ketones. *Atmos. Environ.* **2004**, *38*, 4371–4382.
46. Kalberer, M.; Paulsen, D.; Sax, M.; Steinbacher, M.; Dommen, J.; Prevot, A. S. H.; Fisseha, R.; Weingartner, E.; Frankevich, V.; Zenobi, R.; Baltensperger, U. Identification of Polymers as Major Components of Atmospheric Organic Aerosols. *Science* **2004**, *303*, 1659–1662.
47. De Haan, D. O.; Hawkins, L. N.; Kononenko, J. A.; Turley, J. J.; Jimenez, J. L. Formation of Nitrogen-Containing Oligomers by Methylglyoxal and Amines in Simulated Evaporating Cloud Droplets. *Environ. Sci. Technol.* **2011**, *45*, 984–991.
48. Lee, A. K. Y.; Zhao, R.; Li, R.; Liggio, J.; Li, S. M.; Abbatt, J. P. D. Formation of Light Absorbing Organo-Nitrogen Species from Evaporation of Droplets Containing Glyoxal and Ammonium Sulfate. *Environ. Sci. Technol.* **2013**, *47*, 12819–12826.
49. Galloway, M. M.; Powelson, M. H.; Sedehi, N.; Wood, S. E.; Millage, K. D.; Kononenko, J. A.; Rynaski, A. D.; De Haan, D. O. Secondary Organic Aerosol Formation during Evaporation of Droplets Containing Atmospheric Aldehydes, Amines, and Ammonium Sulfate. *Environ. Sci. Technol.* **2014**, *48*, 14417–14425.
50. Stangl, C. M.; Johnston, M. V. Aqueous Reaction of Dicarbonyls with Ammonia as a Potential Source of Organic Nitrogen in Airborne Nanoparticles. *J. Phys. Chem. A* **2017**, *121*, 3720–3727.
51. Surratt, J. D.; Lewandowski, M.; Offenberg, J. H.; Kleindienst, T. E.; Edney, E. O.; Seinfeld, J. H.; Surratt, J. D. Effect of Acidity on Secondary Organic Aerosol Formation from Isoprene Effect of Acidity on Secondary Organic Aerosol Formation from Isoprene. *Environ. Sci. Technol.* **2007**, *41*, 5363–5369.
52. Riva, M.; Budisulistiorini, S. H.; Chen, Y.; Zhang, Z.; D'Ambro, E. L.; Zhang, X.; Gold, A.; Turpin, B. J.; Thornton, J. A.; Canagaratna, M. R.; Surratt, J. D. Chemical Characterization of Secondary Organic Aerosol from Oxidation of Isoprene Hydroxyhydroperoxides. *Environ. Sci. Technol.* **2016**, *50*, 9889–9899.
53. Wong, J. P. S.; Lee, A. K. Y.; Abbatt, J. P. D. Impacts of Sulfate Seed Acidity and Water Content on Isoprene Secondary Organic Aerosol Formation. *Environ. Sci. Technol.* **2015**, *49*, 13215–13221.
54. Xu, L.; Guo, H.; Boyd, C. M.; Bougiatioti, A.; Cerully, K. M.; Hite, J. R.; Isaacman-vanwertz, G.; Kreisberg, N. M.; Olson, K.; Koss, A.; Goldstein, A. H.; Susanne, V.; Gouw, J. De; Baumann, K.; Lee, S.; Nenes, A.; Weber, R. J.; Ng, N. L. Effects of Anthropogenic Emissions on Aerosol Formation from

Isoprene and Monoterpenes in the Southeastern United States. *Proc. Natl. Acad. Sci. U.S.A.* **2015**, *112*, 37–42.

55. Surratt, J. D.; Kroll, J. H.; Kleindienst, T. E.; Edney, E. O.; Claeys, M.; Sorooshian, A.; Ng, N. L.; Offenberg, J. H.; Lewandowski, M.; Jaoui, M.; Flagan, R. C.; Seinfeld, J. H. Evidence for Organosulfates in Secondary Organic Aerosol. *Environ. Sci. Technol.* **2007**, *41*, 517–527.
56. Lopez-Hilfiker, F. D.; Mohr, C.; D'Ambro, E. L.; Lutz, A.; Riedel, T. P.; Gaston, C. J.; Iyer, S.; Zhang, Z.; Gold, A.; Surratt, J. D.; Lee, B. H.; Kurten, T.; Hu, W. W.; Jimenez, J.; Hallquist, M.; Thornton, J. A. Molecular Composition and Volatility of Organic Aerosol in the Southeastern U.S.: Implications for IEPOX Derived SOA. *Environ. Sci. Technol.* **2016**, *50*, 2200–2209.
57. Shiraiwa, M.; Berkemeier, T.; Schilling-Fahnestock, K. A.; Seinfeld, J. H.; Pöschl, U. Molecular Corridors and Kinetic Regimes in the Multiphase Chemical Evolution of Secondary Organic Aerosol. *Atmos. Chem. Phys.* **2014**, *14*, 8323–8341.
58. Hall, W. A., IV; Johnston, M. V. Oligomer Content of α -Pinene Secondary Organic Aerosol. *Aerosol Sci. Technol.* **2011**, *45*, 37–45.
59. Apsokardu, M. J.; Johnston, M. V. Nanoparticle Growth by Particle-Phase Chemistry. *Atmos. Chem. Phys* **2018**, *18*, 1895–1907.
60. Smith, G. D.; Woods, E., III; Baer, T.; Miller, R. E. Aerosol Uptake Described by Numerically Solving the Diffusion-Reaction Equations in the Particle. *J. Phys. Chem. A* **2003**, *107*, 9582–9587.
61. Tolocka, M. P.; Saul, T. D.; Johnston, M. V. Reactive Uptake of Nitric Acid into Aqueous Sodium Chloride Particles Using Real-Time Single-Particle Mass Spectrometry. *J. Phys. Chem. A* **2004**, *108*, 2659–2665.
62. Saul, T. D.; Tolocka, M. P.; Johnston, M. V. Reactive Uptake of Nitric Acid onto Sodium Chloride Aerosols across a Wide Range of Relative Humidities. *J. Phys. Chem. A* **2006**, *110*, 7614–7620.
63. Ziemann, P. J.; Atkinson, R. Kinetics, Products, and Mechanisms of Secondary Organic Aerosol Formation. *Chem. Soc. Rev.* **2012**, *41*, 6582.
64. Tu, P.; Johnston, M. V. Particle Size Dependence of Biogenic Secondary Organic Aerosol Molecular Composition. *Atmos. Chem. Phys.* **2017**, *17*, 7593–7603.
65. von Hessberg, C.; von Hessberg, P.; Pöschl, U.; Bilde, M.; Nielsen, O. J.; Moortgat, G. K. Temperature and Humidity Dependence of Secondary Organic Aerosol Yield from the Ozonolysis of β-Pinene. *Atmos. Chem. Phys.* **2009**, *9*, 3583–3599.
66. Jaoui, M.; Kamens, R. M. Gaseous and Particulate Oxidation Products Analysis of a Mixture of α-Pinene + β-pinene/O_3/Air in the Absence of Light and α-Pinene + β-pinene/NO_x/Air in the Presence of Natural Sunlight. *J. Atmos. Chem.* **2003**, *44*, 259–297.
67. Ehn, M.; Thornton, J. A.; Kleist, E.; Sipilä, M.; Junninen, H.; Pullinen, I.; Springer, M.; Rubach, F.; Tillmann, R.; Lee, B.; Lopez-Hilfiker, F.; Andres, S.; Acir, I.-H.; Rissanen, M.; Jokinen, T.; Schobesberger, S.; Kangasluoma, J.; Kontkanen, J.; Nieminen, T.; Kurtén, T.; Nielsen, L. B.; Jørgensen, S.; Kjaergaard, H. G.; Canagaratna, M.; Maso, M. D.; Berndt, T.;

Petäjä, T.; Wahner, A.; Kerminen, V.-M.; Kulmala, M.; Worsnop, D. R.; Wildt, J.; Mentel, T. F. A Large Source of Low-Volatility Secondary Organic Aerosol. *Nature* **2014**, *506*, 476–479.

68. Hall IV, W. A.; Johnston, M. V. Oligomer Content of α-Pinene Secondary Organic Aerosol. *Aerosol Sci. Technol.* **2011**, *45*, 37–45.
69. Trump, E. R.; Donahue, N. M. Oligomer Formation within Secondary Organic Aerosols: Equilibrium and Dynamic Considerations. *Atmos. Chem. Phys.* **2014**, *14*, 3691–3701.
70. Donahue, N. M.; Kroll, J. H.; Pandis, S. N.; Robinson, A. L. A Two-Dimensional Volatility Basis Set-Part 2: Diagnostics of Organic-Aerosol Evolution. *Atmos. Chem. Phys.* **2012**, *12*, 615–634.
71. Tang, X.; Misztal, P. K.; Nazaroff, W. W.; Goldstein, A. H. Siloxanes Are the Most Abundant Volatile Organic Compound Emitted from Engineering Students in a Classroom. *Environ. Sci. Technol. Lett.* **2015**, *2*, 303–307.
72. Wu, Y.; Johnston, M. V. Molecular Characterization of Secondary Aerosol from Oxidation of Cyclic Methylsiloxanes. *J. Am. Soc. Mass Spectrom.* **2016**, *27*, 402–409.
73. Wu, Y.; Johnston, M. V. Aerosol Formation from OH Oxidation of the Volatile Cyclic Methyl Siloxane (cVMS) Decamethylcyclopentasiloxane. *Environ. Sci. Technol.* **2017**, *51*, 4445–4451.
74. Mohr, C.; Lopez-Hilfiker, F. D.; Yli-Juuti, T.; Heitto, A.; Lutz, A.; Hallquist, M.; D'Ambro, E. L.; Rissanen, M. P.; Hao, L.; Schobesberger, S.; Kulmala, M.; Mauldin, R. L.; Makkonen, U.; Sipilä, M.; Petäjä, T.; Thornton, J. A. Ambient Observations of Dimers from Terpene Oxidation in the Gas Phase: Implications for New Particle Formation and Growth. *Geophys. Res. Lett.* **2017**, *44*, 2958–2966.
75. Kourtchev, I.; Giorio, C.; Manninen, A.; Wilson, E.; Mahon, B.; Aalto, J.; Kajos, M.; Venables, D.; Ruuskanen, T.; Levula, J.; Loponen, M.; Connors, S.; Harris, N.; Zhao, D.; Kiendler-Scharr, A.; Mentel, T.; Rudich, Y.; Hallquist, M.; Doussin, J.-F.; Maenhaut, W.; Bäck, J.; Petäjä, T.; Wenger, J.; Kulmala, M.; Kalberer, M. Enhanced Volatile Organic Compounds Emissions and Organic Aerosol Mass Increase the Oligomer Content of Atmospheric Aerosols. *Sci. Rep.* **2016**, *6*, 35038/9.
76. Paasonen, P.; Kajos, M. K.; Rantala, P.; Junninen, H.; Petaja, T.; Kulmala, M. Comprehensive Analysis of Particle Growth Rates from Nucleation Mode to Cloud Condensation Nuclei in Boreal Forest. *Atmos. Chem. Phys. Discuss.* **2018**, *18*, 12085–12103.
77. Burkart, J.; Hodshire, A. L.; Mungall, E. L.; Pierce, J. R.; Collins, D. B.; Ladino, L. A.; Lee, A. K. Y.; Irish, V.; Wentzell, J. J. B.; Liggio, J.; Papakyriakou, T.; Murphy, J.; Abbatt, J. Organic Condensation and Particle Growth to CCN Sizes in the Summertime Marine Arctic Is Driven by Materials More Semivolatile Than at Continental Sites. *Geophys. Res. Lett.* **2017**, *44*, 10725–10734.
78. Rissanen, M. P.; Kurtén, T.; Sipilä, M.; Thornton, J. A.; Kangasluoma, J.; Sarnela, N.; Junninen, H.; Jørgensen, S.; Schallhart, S.; Kajos, M. K.; Taipale, R.; Springer, M.; Mentel, T. F.; Ruuskanen, T.; Petäjä, T.; Worsnop, D. R.; Kjaergaard, H. G.; Ehn, M. The Formation of Highly

Oxidized Multifunctional Products in the Ozonolysis of Cyclohexene. *J. Am. Chem. Soc.* **2014**, *136*, 15596–15606.

79. Jokinen, T.; Berndt, T.; Makkonen, R.; Kerminen, V.-M.; Junninen, H.; Paasonen, P.; Stratmann, F.; Herrmann, H.; Guenther, A. B.; Worsnop, D. R.; Kulmala, M.; Ehn, M.; Sipilä, M. Production of Extremely Low Volatile Organic Compounds from Biogenic Emissions: Measured Yields and Atmospheric Implications. *Proc. Natl. Acad. Sci. U.S.A.* **2015**, *112*, 7123–7128.
80. Zhou, S.; Rivera-Rios, J. C.; Keutsch, F. N.; Abbatt, J. P. D. Identification of Organic Hydroperoxides and Peroxy Acids Using Atmospheric Pressure Chemical Ionization – Tandem Mass Spectrometry (APCI-MS/MS): Application to Secondary Organic Aerosol. *Aerosol Meas. Tech. Discuss.* **2018**, *11*, 3081–3089.
81. Docherty, K. S.; Wu, W.; Lim, Y. B.; Ziemann, P. J. Contributions of Organic Peroxides to Secondary Aerosol Formed from Reactions of Monoterpenes with O_3. *Environ. Sci. Technol.* **2005**, *39*, 4049–4059.
82. Liu, J.; D'Ambro, E. L.; Lee, B. H.; Lopez-Hilfiker, F. D.; Zaveri, R. A.; Rivera-Rios, J. C.; Keutsch, F. N.; Iyer, S.; Kurten, T.; Zhang, Z.; Gold, A.; Surratt, J. D.; Shilling, J. E.; Thornton, J. A. Efficient Isoprene Secondary Organic Aerosol Formation from a Non-IEPOX Pathway. *Environ. Sci. Technol.* **2016**, *50*, 9872–9880.
83. Riva, M.; Bell, D. M.; Hansen, A. M. K.; Drozd, G. T.; Zhang, Z.; Gold, A.; Imre, D.; Surratt, J. D.; Glasius, M.; Zelenyuk, A. Effect of Organic Coatings, Humidity and Aerosol Acidity on Multiphase Chemistry of Isoprene Epoxydiols. *Environ. Sci. Technol.* **2016**, *50*, 5580–5588.
84. Liu, Y. J.; Kuwata, M.; McKinney, K. A.; Martin, S. T. Uptake and Release of Gaseous Species Accompanying the Reactions of Isoprene Photo-Oxidation Products with Sulfate Particles. *Phys. Chem. Chem. Phys.* **2016**, *18*, 1595–1600.
85. Surratt, J. D.; Chan, A. W.; Eddingsaas, N. C.; Chan, M.; Loza, C. L.; Kwan, a J.; Hersey, S. P.; Flagan, R. C.; Wennberg, P. O.; Seinfeld, J. H. Reactive Intermediates Revealed in Secondary Organic Aerosol Formation from Isoprene. *Proc. Natl. Acad. Sci. U.S.A.* **2010**, *107*, 6640–6645.
86. Guenther, A. B.; Jiang, X.; Heald, C. L.; Sakulyanontvittaya, T.; Duhl, T.; Emmons, L. K.; Wang, X. The Model of Emissions of Gases and Aerosols from Nature Version 2.1 (MEGAN2.1): An Extended and Updated Framework for Modeling Biogenic Emissions. *Geosci. Model Dev.* **2012**, *5*, 1471–1492.
87. Calvert, J. G.; Atkinson, R.; Kerr, J. A.; Madronich, S.; Moortgat, G. K.; Wallington, T. J.; Yarwood, G. *The Mechanisms of Atmospheric Oxidation of Alkenes*, 1st Ed.; Oxford University: New York, NY, 2000.
88. Epstein, S. A.; Blair, S. L.; Nizkorodov, S. A. Direct Photolysis of α-Pinene Ozonolysis Secondary Organic Aerosol: Effect on Particle Mass and Peroxide Content. *Environ. Sci. Technol.* **2014**, *48*, 11251–11258.
89. Zhao, R.; Aljawhary, D.; Lee, A. K. Y.; Abbatt, J. P. D. Rapid Aqueous-Phase Photooxidation of Dimers in the α-Pinene Secondary Organic Aerosol. *Environ. Sci. Technol. Lett.* **2017**, *4*, 205–210.

90. Romonosky, D. E.; Li, Y.; Shiraiwa, M.; Laskin, A.; Laskin, J.; Nizkorodov, S. A. Aqueous Photochemistry of Secondary Organic Aerosol of α-Pinene and α-Humulene Oxidized with Ozone, Hydroxyl Radical, and Nitrate Radical. *J. Phys. Chem. A* **2017**, *121*, 1298–1309.
91. Krapf, M.; El Haddad, I.; Bruns, E. A.; Molteni, U.; Daellenbach, K. R.; Prévôt, A. S. H.; Baltensperger, U.; Dommen, J. Labile Peroxides in Secondary Organic Aerosol. *Chem* **2016**, *1*, 603–616.

Chapter 3

Interfacial Criegee Chemistry

Shinichi Enami*

National Institute for Environmental Studies, 16-2 Onogawa, Tsukuba 305-8506, Japan
***E-mail: enami.shinichi@nies.go.jp; phone: +81-29-850-2770**

Criegee intermediates (CIs) in the gas-phase play key roles in the atmospheric HOx cycles and particle formation from organic precursors. The roles at the gas-aerosol interface, however, remain unknown. Recently, interface-specific mass spectrometric study reveals the mechanisms for reactions of CIs with a series of atmospherically relevant compounds including carboxylic acids, alcohols, saccharides and water molecules on the water/acetonitrile mixture surface. CIs generated from ozonolysis of β-caryophyllene or α-humulene on the liquid surface preferentially react with surface-active and acidic species (e.g., *cis*-pinonic acid) to form larger mass, less volatile products. Surprisingly, levoglucosan and other saccharides efficiently react with CIs at the gas-liquid interface. Large gas-phase acidities of the multiple OH-groups of saccharides underlie their exceptional reactivities toward CIs. Based on these new findings, the roles of interfacial CI reactions in the atmosphere are discussed.

Criegee intermediates (CIs), the carbonyl oxides (RR′COO, biradicals/zwitterions) produced in the ozonation of unsaturated hydrocarbons, are key intermediates that play important roles in atmospheric chemistry impacting both the HOx cycles and the formation of organic particles. CIs rapidly react with water dimers $(H_2O)_2$ at typical water vapor concentrations in the lower troposphere (*1*,

2). CIs may also react with acids, alcohols, and SO_2 or unimolecularly decompose in the atmosphere (*3–11*), while the reactivity of CIs is largely structure-dependent (*1, 12*). Because CIs have short lifetimes, studying their reactions is challenging and little was known about how these intermediates behave on and in an aqueous droplet or water film; both of these phases might play important roles in the atmosphere.

Recently, interface-specific mass spectrometric study reveals the mechanisms for reactions of CIs with a series of atmospherically relevant compounds including carboxylic acids, alcohols, saccharides and water molecules on the liquid surface. Whether the CIs on the surface of aqueous aerosols will react exclusively with water therein is an important issue. In contrast with extensive studies on CIs chemistry in the gas-phase (*2, 11–28*), the chemistry of CIs at the air-water interface relevant to that taking place on fog droplets, aqueous aerosol and thin water films (e.g., covering leaves and soil) was unexplored (*14*). Recent theoretical calculations predict that the reaction of CH_2OO (the simplest CI) with water at the air-water interface takes place about 2−3 orders of magnitude faster than in the gas phase, via interface-specific reaction pathways (*14*). In contrasting with CH_2OO, theory also predicts that some larger-size CIs may be inert toward water molecules on aqueous interfaces (*29*). The presence of a hydrophobic methyl substituent on the CI's C-atom would lower the proton transfer ability and inhibit the formation of a pre-reaction complex for the CIs + water reaction (*29*).

In the laboratory, solutions of water(W):acetonitrile(AN) can be used as a surrogate of atmospheric aqueous organic media (*30–33*). Direct detection of products from reactions of CIs with H_2O, D_2O, $H_2{}^{18}O$, alkyl carboxylic acids, *cis*-pinonic acid, alkyl alcohols, and saccharides (levoglucosan, glucose, arabitol, and mannitol) on fresh surfaces of β-caryophyllene (β-C) or α-humulene (α-H) solutions in W:AN exposed to $O_3(g)$ for ~ 10 μs is reported (*34–37*). The relatively small ozone exposures (i.e., $E = [O_3(g)] \times \tau_R \leq 2.4 \times 10^{11}$ molecules cm^{-3} s) used in these experiments enables them to probe the hitherto inaccessible early stages of alkene ozonations on liquid surfaces. By adding NaCl in the solution, some neutral products such as hydroxy-hydroperoxides are detected as the chloride-adducts (*25, 38, 39*). The composition of its interfacial layers of W:AN mixture solvent is well characterized (*30–33*). The experiments take place in the spraying chamber of the electrospray mass spectrometer that is continuously flushed with $O_3(g)/O_2(g)/N_2(g)$ mixtures under ambient conditions (Figure 1) (*40*). This design allows direct detection and measurement of genuinely early stages of the heterogeneous oxidation process which can be used to determine elemental composition of the reaction products.

CIs produced by prompt ozonolysis of β-C/α-H are found to preferentially react with millimolar longer-chain carboxylic acids (e.g., octanoic acid) and *cis*-pinonic acid (CPA) rather than water molecules $(H_2O)_m$ in the interfacial layers of model aqueous organic aerosols. Note that the bulk concentration of water in W:AN (1:4 = vol:vol) is ~23 M, much larger than those of acids. It is also found that smaller-chain carboxylic acids $R_{n\leq3}COOH$, due to the larger hydrophilicities, cannot react with CIs at the interface. These results imply that the interfacial availability of the reactant (*41*) is an important controlling factor.

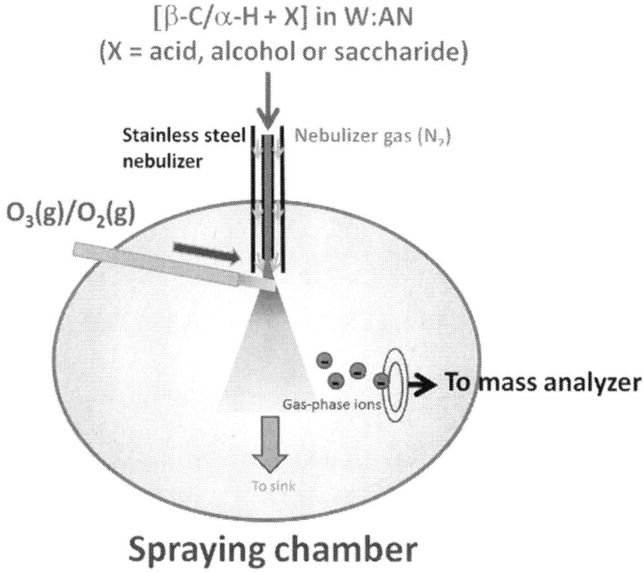

Figure 1. Schematic diagram of the experimental setup for the reaction of Criegee intermediates at the air-aqueous interface (34–37). β-C, α-H, W, and AN stands for β-caryophyllene, α-humulene, water and acetonitrile, respectively.

There is no evidence for the OH radical formation via unimolecular decomposition in these experiments (*37*), implying that fully stabilized CIs in collisions with the solvent (i.e., W:AN) would preferentially react with reactants rather than undergo isomerization via H-atom migration, that is required to precede OH radical emission. These results are in interesting contrast with the CI reactions in the gas-phase (*3*, *21*, *42*). Unimolecular decompositions to yield OH radicals are particularly important for larger CIs in the gas-phase (*42*). For example, the observed yield of OH radical from gas-phase ozonolysis of β-C and α-H is reported to be (10.4 ± 2.3) and (10.5 ± 0.7) %, respectively (*43*, *44*).

Among tested acids, CPA has an exceptional reactivity toward both β-C's and α-H's CIs on the aqueous surface (Figure 2). This result would be explained by the unique molecular geometry of CPA that positions the reactive –C(O)OH group close to the interface (*45*, *46*). A previous molecular dynamics calculation predicted that all 144 CPA molecules partition to the interfacial layers of a $(H_2O)_{2000}$ droplet, showing the exceptionally large affinity of CPA for the air-water interface (*46*). Previous experiments employing interface-specific mass spectrometry also showed that *cis*-pinonate has a significantly larger affinity than *n*-octanoate for the air–water interface (*45*). Importantly, since C_{25} products from CIs + CPA (MW 436, detected as m/z 471; 473, see Figure 2) have one (for β-C) and two (for α-H) remaining C=C double bond(s), they will undergo further ozonolysis at the interface leading to high O/C products (*37*). The observed products also have hydroperoxide -OOH groups, that could propagate oligomerizations (*24*, *25*, *47*). Such processes provide hitherto unrecognized routes for the formation of the extremely low-volatility organic compounds

(ELVOCs) found in field studies (*48–51*). Since CPA is accumulated in ambient particles, an efficient scavenger of the CIs produced on aqueous organic surfaces, and a precursor of ELVOCs, it could be an important contributor to the formation and growth of atmospheric particles. These results underscore the important fact that competitive reactions at the air–liquid interface depend on interfacial availability rather than bulk reactant concentrations.

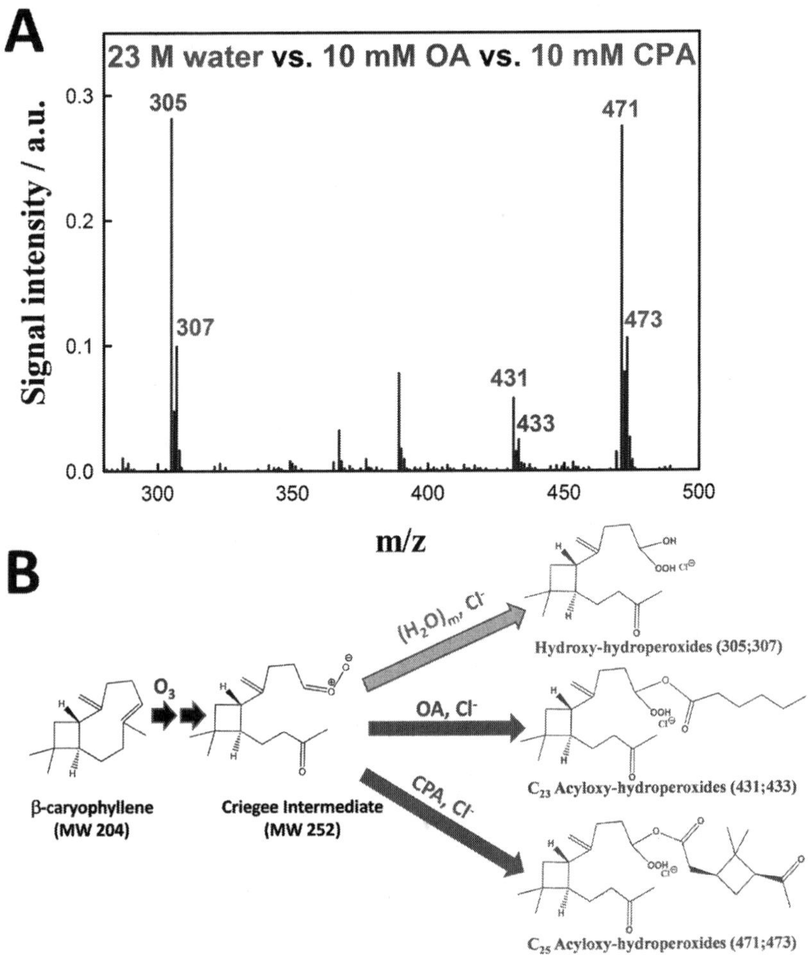

Figure 2. A) Negative ion mass spectra of 1 mM β-caryophyllene + 0.2 mM NaCl + 10 mM cis-pinonic acid (CPA) + 10 mM octanoic acid (OA) in W:AN (1 : 4 = vol : vol) solution microjets exposed to O_3(g) at exposure (= concentration x time) = 2.3 x 10^{11} molecules cm^{-3} s under 1 atm and 298 K. The m/z 305;307, 431;433, and 471;473 signals correspond to chloride-adducts of C_{15} hydroxy-hydroperoxide, C_{23} acyloxy-hydroperoxides, and C_{25} acyloxy-hydroperoxides, respectively. B) Reaction schemes for CI + water, CI + OA and CI + CPA (36). Note that only representative structures among possible isomers are shown.

Among hydroxylic species, alcohols R_n-OH are found to be significantly less reactive than carboxylic acids R_n-COOH of similar size at the same molar concentration towards CIs, suggesting that their reactivities depend on not only interfacial affinities but also their acidities (*35*). Furthermore, the reactivities of R_n-OH toward CIs are found to be correlated with their gas-phase acidity $\Delta G_{acidity}$ (*35*). These results are consistent with the report by Tobias and Ziemann, who showed that rate constants of gas-phase C_{13} CIs reactions with various species increase in the order: water < methanol < 2-propanol < formic acid < heptanoic acid (*52*).

Surprisingly, CIs produced by ozonolysis of α-H/β-C react with atmospherically relevant saccharides (C_5 or C_6 polyols) including levoglucosan (Levo), the most abundant saccharide found in typical biomass burning aerosol (*53*), in the interfacial layers of model aqueous organic aerosols (*34*). Mass-specific identification of the products generated in very short reaction times on W:AN surfaces proves that sesquiterpene's CIs react with Levo, glucose, arabitol, and mannitol rather than water molecules. At the interface, Levo is found to be ~20 times more reactive than 1-octanol, and ~2 times more reactive than *n*-octanoic acid towards CIs. The C_{21} ether product is tentatively identified as a major product from the reaction of CIs with Levo at the interface (Figure 3) (*34*). The larger gas-phase acidities of the multiple OH-groups of Levo and other saccharides, relative to monoalkanols, underlie their exceptionally large reactivities toward CIs. In fact, reported $\Delta G_{acidity}$ for the Levo's three -OH groups: $\Delta G_{acidity}$ = 1454-1486 kJ mol^{-1} (*54*), that are significantly lower (i.e., -OH groups are more acidic) than those of *n*-alkanols, which range from 1563 kJ mol^{-1} (methanol) to 1525 kJ mol^{-1} (1-octanol) (*55*). As described above, the reactivities of OH-species toward CIs are correlated with their $\Delta G_{acidity}$. The availability of reactive OH groups of saccharides at the gas-liquid interface may also contribute to the enhanced reactivities (*36*). A theoretical study on molecular geometries and orientations of saccharides at the gas-liquid interface would be desirable.

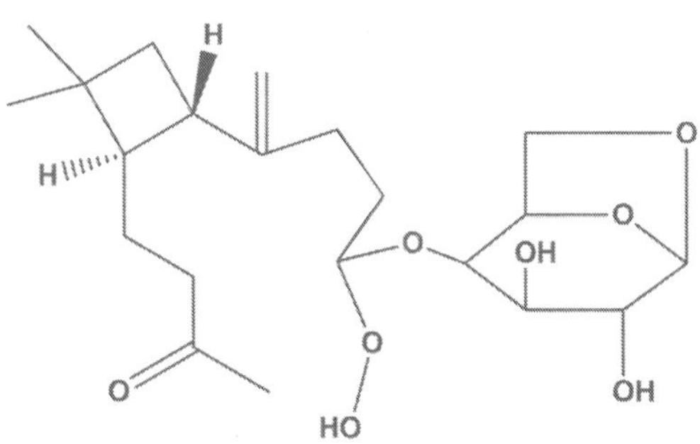

C_{21} **Ether product (MW 414)**

Figure 3. A structure of the observed product from the reaction of β-caryophyllene's CIs with levoglucosan at the gas-liquid interface.

These results suggest that saccharides could be important, hitherto unrecognized, contributors to the growth/augmentation of atmospheric particles and play critical roles in the troposphere. Recent experimental studies have shown that Levo may transform into oxidized products during the daytime OH-radical oxidations, casting doubts about its role as a stable biomass burning tracer (*56–63*). Notably, the -C=C- + O_3 + sugar reactions on aqueous organic aerosol occur during not only daytime but also nighttime, in contrast with OH-radical oxidations (*34*).

Atmospheric implications of interfacial CI chemistry are briefly discussed. It has been demonstrated that the ozonolysis of biogenic sesquiterpenes in the presence of atmospherically relevant OH-species at the interface potentially produce large-mass and less-volatile products via CIs on aqueous organic particles. This is in striking contrast with the gas-phase fate of CIs, that are mostly controlled by the reactions of water vapors, and unimolecular decomposition (*42*). Unsaturated organic species (e.g., terpenes), which are taken up and transformed into oligomers on acidic surfaces (*64–69*), will be present as such on the acidic atmospheric particles (*70–75*). Under such conditions, the CIs produced via interfacial ozonolysis are expected to react with ubiquitous saccharides (*76–79*) or surface-active *cis*-pinonic acid (*80–84*), rather than water and hydrophilic acids. It is noted that O_3 is rather insoluble in water (Henry's law constant H = 0.01 M atm^{-1}), hence the interfacial ozonolysis would play the key role. Recently, high molecular mass compounds presumably originating from the reaction of CIs with pinonic/pinic acids were observed as major products in aerosols produced in the chamber experiment via α-pinene ozonolysis, as well as in fresh aerosols collected over boreal forests (*51*). Since terpenes and the derivatives are naturally surface-active, and saccharides have multiple, acidic -OH groups, their interfacial CI reactions will contribute to the growth and mass loading of secondary organic aerosol (SOA). Atmospheric modeling including the interfacial CI reactions will be required to elucidate the impact of recent findings on tropospheric chemistry, which may narrow the gap between field observations and model calculations that persistently underestimate SOA yields.

References

1. Jr-Min Lin, J.; Chao, W. Structure-Dependent Reactivity of Criegee Intermediates Studied with Spectroscopic Methods. *Chem. Soc. Rev.* **2017**, *46*, 7483–7497.
2. Chao, W.; Hsieh, J. T.; Chang, C. H.; Lin, J. J. M. Direct Kinetic Measurement of the Reaction of the Simplest Criegee Intermediate with Water Vapor. *Science* **2015**, *347*, 751–754.
3. Kidwell, N. M.; Li, H. W.; Wang, X. H.; Bowman, J. M.; Lester, M. I. Unimolecular Dissociation Dynamics of Vibrationally Activated CH_3CHOO Criegee Intermediates to OH Radical Products. *Nat. Chem.* **2016**, *8*, 509–514.

4. Stone, D.; Whalley, L. K.; Heard, D. E. Tropospheric OH and HO_2 Radicals: Field Measurements and Model Comparisons. *Chem. Soc. Rev.* **2012**, *41*, 6348–6404.
5. Osborn, D. L.; Taatjes, C. A. The Physical Chemistry of Criegee Intermediates in the Gas Phase. *Int. Rev. Phys. Chem.* **2015**, *34*, 309–360.
6. Stone, D.; Blitz, M.; Daubney, L.; Howes, N. U. M.; Seakins, P. Kinetics of CH_2OO Reactions with SO_2, NO_2, NO, H_2O and CH_3CHO as a Function of Pressure. *Phys. Chem. Chem. Phys.* **2014**, *16*, 1139–1149.
7. Lee, Y. P. Perspective: Spectroscopy and Kinetics of Small Gaseous Criegee Intermediates. *J. Chem. Phys.* **2015**, *143*, 020901.
8. Newland, M. J.; Rickard, A. R.; Alam, M. S.; Vereecken, L.; Munoz, A.; Rodenas, M.; Bloss, W. J. Kinetics of Stabilised Criegee Intermediates Derived from Alkene Ozonolysis: Reactions with SO_2, H_2O and Decomposition under Boundary Layer Conditions. *Phys. Chem. Chem. Phys.* **2015**, *17*, 4076–4088.
9. McGillen, M. R.; Curchod, B. F. E.; Chhantyal-Pun, R.; Beames, J. M.; Watson, N.; Khan, M. A. H.; McMahon, L.; Shallcross, D. E.; Orr-Ewing, A. J. Criegee Intermediate–Alcohol Reactions, A Potential Source of Functionalized Hydroperoxides in the Atmosphere. *ACS Earth Space Chem.* **2017**, *1*, 664–672.
10. Tadayon, S. V.; Foreman, E. S.; Murray, C. Kinetics of the Reactions between the Criegee Intermediate CH_2OO and Alcohols. *J. Phys. Chem. A* **2018**, *122*, 258–268.
11. Mauldin, R. L.; Berndt, T.; Sipila, M.; Paasonen, P.; Petaja, T.; Kim, S.; Kurten, T.; Stratmann, F.; Kerminen, V. M.; Kulmala, M. A New Atmospherically Relevant Oxidant of Sulphur Dioxide. *Nature* **2012**, *488*, 193–196.
12. Huang, H. L.; Chao, W.; Lin, J. J. M. Kinetics of a Criegee Intermediate That Would Survive High Humidity and May Oxidize Atmospheric SO_2. *Proc. Natl. Acad. Sci. U. S. A.* **2015**, *112*, 10857–10862.
13. Long, B.; Bao, J. L.; Truhlar, D. G. Atmospheric Chemistry of Criegee Intermediates: Unimolecular Reactions and Reactions with Water. *J. Am. Chem. Soc.* **2016**, *138*, 14409–14422.
14. Zhu, C. Q.; Kumar, M.; Zhong, J.; Li, L.; Francisco, J. S.; Zeng, X. C. New Mechanistic Pathways for Criegee-Water Chemistry at the Air/Water Interface. *J. Am. Chem. Soc.* **2016**, *138*, 11164–11169.
15. Kjaergaard, H. G.; Kurten, T.; Nielsen, L. B.; Jorgensen, S.; Wennberg, P. O. Criegee Intermediates React with Ozone. *J. Phys. Chem. Lett.* **2013**, *4*, 2525–2529.
16. Vereecken, L.; Harder, H.; Novelli, A. The Reaction of Criegee Intermediates with NO, RO_2, and SO_2, and Their Fate in the Atmosphere. *Phys. Chem. Chem. Phys.* **2012**, *14*, 14682–14695.
17. Vereecken, L.; Harder, H.; Novelli, A. The Reactions of Criegee Intermediates with Alkenes, Ozone, and Carbonyl Oxides. *Phys. Chem. Chem. Phys.* **2014**, *16*, 4039–4049.

18. Vereecken, L.; Rickard, A. R.; Newland, M. J.; Bloss, W. J. Theoretical Study of the Reactions of Criegee Intermediates with Ozone, Alkylhydroperoxides, and Carbon Monoxide. *Phys. Chem. Chem. Phys.* **2015**, *17*, 23847–23858.
19. Aplincourt, P.; Anglada, J. M. Theoretical Studies of the Isoprene Ozonolysis under Tropospheric Conditions. 2. Unimolecular and Water-Assisted Decomposition of the A-Hydroxy Hydroperoxides. *J. Phys. Chem. A* **2003**, *107*, 5812–5820.
20. Taatjes, C. A.; Shallcross, D. E.; Percival, C. J. Research Frontiers in the Chemistry of Criegee Intermediates and Tropospheric Ozonolysis. *Phys. Chem. Chem. Phys.* **2014**, *16*, 1704–1718.
21. Nguyen, T. B.; Tyndall, G. S.; Crounse, J. D.; Teng, A. P.; Bates, K. H.; Schwantes, R. H.; Coggon, M. M.; Zhang, L.; Feiner, P.; Milller, D. O.; Skog, K. M.; Rivera-Rios, J. C.; Dorris, M.; Olson, K. F.; Koss, A.; Wild, R. J.; Brown, S. S.; Goldstein, A. H.; de Gouw, J. A.; Brune, W. H.; Keutsch, F. N.; Seinfeld, J. H.; Wennberg, P. O. Atmospheric Fates of Criegee Intermediates in the Ozonolysis of Isoprene. *Phys. Chem. Chem. Phys.* **2016**, *18*, 10241–10254.
22. Welz, O.; Eskola, A. J.; Sheps, L.; Rotavera, B.; Savee, J. D.; Scheer, A. M.; Osborn, D. L.; Lowe, D.; Booth, A. M.; Xiao, P.; Khan, M. A. H.; Percival, C. J.; Shallcross, D. E.; Taatjes, C. A. Rate Coefficients of C1 and C2 Criegee Intermediate Reactions with Formic and Acetic Acid near the Collision Limit: Direct Kinetics Measurements and Atmospheric Implications. *Angew. Chem., Int. Ed.* **2014**, *53*, 4547–4550.
23. Lin, L. C.; Chang, H. T.; Chang, C. H.; Chao, W.; Smith, M. C.; Chang, C. H.; Lin, J. J. M.; Takahashi, K. Competition between H_2O and $(H_2O)_2$ Reactions with CH_2OO/CH_3CHOO. *Phys. Chem. Chem. Phys.* **2016**, *18*, 4557–4568.
24. Sakamoto, Y.; Yajima, R.; Inomata, S.; Hirokawa, J. Water Vapour Effects on Secondary Organic Aerosol Formation in Isoprene Ozonolysis. *Phys. Chem. Chem. Phys.* **2017**, *19*, 3165–3175.
25. Sakamoto, Y.; Inomata, S.; Hirokawa, J. Oligomerization Reaction of the Criegee Intermediate Leads to Secondary Organic Aerosol Formation in Ethylene Ozonolysis. *J. Phys. Chem. A* **2013**, *117*, 12912–12921.
26. Hatakeyama, S.; Akimoto, H. Reactions of Criegee Intermediates in the Gas-Phase. *Res. Chem. Intermed.* **1994**, *20*, 503–524.
27. Smith, M. C.; Chang, C. H.; Chao, W.; Lin, L. C.; Takahashi, K.; Boering, K. A.; Lin, J. J. M. Strong Negative Temperature Dependence of the Simplest Criegee Intermediate CH_2OO Reaction with Water Dimer. *J. Phys. Chem. Lett.* **2015**, *6*, 2708–2713.
28. Lewis, T. R.; Blitz, M. A.; Heard, D. E.; Seakins, P. W. Direct Evidence for a Substantive Reaction between the Criegee Intermediate, CH_2OO, and the Water Vapour Dimer. *Phys. Chem. Chem. Phys.* **2015**, *17*, 4859–4863.
29. Zhong, J.; Kumar, M.; Zhu, C. Q.; Francisco, J. S.; Zeng, X. C. Surprising Stability of Larger Criegee Intermediates on Aqueous Interfaces. *Angew. Chem., Int. Ed.* **2017**, *56*, 7740–7744.
30. Perrine, K. A.; Van Spyk, M. H.; Margarella, A. M.; Winter, B.; Faubel, M.; Bluhm, H.; Hemminger, J. C. Characterization of the Acetonitrile Aqueous

Solution/Vapor Interface by Liquid-Jet X-Ray Photoelectron Spectroscopy. *J. Phys. Chem. C* **2014**, *118*, 29378–29388.

31. Makowski, M. J.; Stern, A. C.; Hemminger, J. C.; Tobias, D. J. Orientation and Structure of Acetonitrile in Water at the Liquid–Vapor Interface: A Molecular Dynamics Simulation Study. *J. Phys. Chem. C* **2016**, *120*, 17555–17563.
32. Rao, Y.; Turro, N. J.; Eisenthal, K. B. Water Structure at Air/Acetonitrile Aqueous Solution Interfaces. *J. Phys. Chem. C* **2009**, *113*, 14384–14389.
33. Paul, S.; Chandra, A. Liquid-Vapor Interfaces of Water-Acetonitrile Mixtures of Varying Composition. *J. Chem. Phys.* **2005**, *123*, 184706.
34. Enami, S.; Hoffmann, M. R.; Colussi, A. J. Criegee Intermediates React with Levoglucosan on Water. *J. Phys. Chem. Lett.* **2017**, *8*, 3888–3894.
35. Enami, S.; Colussi, A. J. Reactions of Criegee Intermediates with Alcohols at Air-Aqueous Interfaces. *J. Phys. Chem. A* **2017**, *121*, 5175–5182.
36. Enami, S.; Colussi, A. J. Efficient Scavenging of Criegee Intermediates on Water by Surface-Active *cis*-Pinonic Acid. *Phys. Chem. Chem. Phys.* **2017**, *19*, 17044–17051.
37. Enami, S.; Colussi, A. J. Criegee Chemistry on Aqueous Organic Surfaces. *J. Phys. Chem. Lett.* **2017**, *8*, 1615–1623.
38. Larson, J. W.; McMahon, T. B. Fluoride and Chloride Affinities of Main Group Oxides, Fluorides, Oxofluorides, and Alkyls. Quantitative Scales of Lewis Acidities from Ion Cyclotron Resonance Halide-Exchange Equilibria. *J. Am. Chem. Soc.* **1985**, *107*, 766–773.
39. Bohringer, H.; Fahey, D. W.; Fehsenfeld, F. C.; Ferguson, E. E. Bond-Energies of the Molecules H_2O, SO_2, H_2O_2, and HCl to Various Atmospheric Negative-Ions. *J. Chem. Phys.* **1984**, *81*, 2805–2810.
40. Enami, S.; Sakamoto, Y.; Colussi, A. J. Fenton Chemistry at Aqueous Interfaces. *Proc. Natl. Acad. Sci. U. S. A.* **2014**, *111*, 623–628.
41. Enami, S.; Fujii, T.; Sakamoto, Y.; Hama, T.; Kajii, Y. Carboxylate Ion Availability at the Air-Water Interface. *J. Phys. Chem. A* **2016**, *120*, 9224–9234.
42. Vereecken, L.; Novelli, A.; Taraborrelli, D. Unimolecular Decay Strongly Limits the Atmospheric Impact of Criegee Intermediates. *Phys. Chem. Chem. Phys.* **2017**, *19*, 31599–31612.
43. Beck, M.; Winterhalter, R.; Herrmann, F.; Moortgat, G. K. The Gas-Phase Ozonolysis of Alpha-Humulene. *Phys. Chem. Chem. Phys.* **2011**, *13*, 10970–11001.
44. Winterhalter, R.; Herrmann, F.; Kanawati, B.; Nguyen, T. L.; Peeters, J.; Vereecken, L.; Moortgat, G. K. The Gas-Phase Ozonolysis of Beta-Caryophyllene ($C_{15}H_{24}$). Part I: An Experimental Study. *Phys. Chem. Chem. Phys.* **2009**, *11*, 4152–4172.
45. Enami, S.; Sakamoto, Y. OH-Radical Oxidation of Surface-Active *cis*-Pinonic Acid at the Air–Water Interface. *J. Phys. Chem. A* **2016**, *120*, 3578–3587.
46. Li, X.; Hede, T.; Tu, T.; Leck, C.; Agren, H. Surface-Active *cis*-Pinonic Acid in Atmospheric Droplets: A Molecular Dynamics Study. *J. Phys. Chem. Lett.* **2010**, *1*, 769–773.

47. Riva, M.; Budisulistiorini, S. H.; Zhang, Z.; Gold, A.; Thornton, J. A.; Turpin, B. J.; Surratt, J. D. Multiphase Reactivity of Gaseous Hydroperoxide Oligomers Produced from Isoprene Ozonolysis in the Presence of Acidified Aerosols. *Atmos. Environ.* **2017**, *152*, 314–322.
48. Barsanti, K. C.; Kroll, J. H.; Thornton, J. A. Formation of Low-Volatility Organic Compounds in the Atmosphere: Recent Advancements and Insights. *J. Phys. Chem. Lett.* **2017**, *8*, 1503–1511.
49. Ehn, M.; Thornton, J. A.; Kleist, E.; Sipila, M.; Junninen, H.; Pullinen, I.; Springer, M.; Rubach, F.; Tillmann, R.; Lee, B.; Lopez-Hilfiker, F.; Andres, S.; Acir, I. H.; Rissanen, M.; Jokinen, T.; Schobesberger, S.; Kangasluoma, J.; Kontkanen, J.; Nieminen, T.; Kurten, T.; Nielsen, L. B.; Jorgensen, S.; Kjaergaard, H. G.; Canagaratna, M.; Dal Maso, M.; Berndt, T.; Petaja, T.; Wahner, A.; Kerminen, V. M.; Kulmala, M.; Worsnop, D. R.; Wildt, J.; Mentel, T. F. A Large Source of Low-Volatility Secondary Organic Aerosol. *Nature* **2014**, *506*, 476–479.
50. Riccobono, F.; Schobesberger, S.; Scott, C. E.; Dommen, J.; Ortega, I. K.; Rondo, L.; Almeida, J.; Amorim, A.; Bianchi, F.; Breitenlechner, M.; David, A.; Downard, A.; Dunne, E. M.; Duplissy, J.; Ehrhart, S.; Flagan, R. C.; Franchin, A.; Hansel, A.; Junninen, H.; Kajos, M.; Keskinen, H.; Kupc, A.; Kurten, A.; Kvashin, A. N.; Laaksonen, A.; Lehtipalo, K.; Makhmutov, V.; Mathot, S.; Nieminen, T.; Onnela, A.; Petaja, T.; Praplan, A. P.; Santos, F. D.; Schallhart, S.; Seinfeld, J. H.; Sipila, M.; Spracklen, D. V.; Stozhkov, Y.; Stratmann, F.; Tome, A.; Tsagkogeorgas, G.; Vaattovaara, P.; Viisanen, Y.; Vrtala, A.; Wagner, P. E.; Weingartner, E.; Wex, H.; Wimmer, D.; Carslaw, K. S.; Curtius, J.; Donahue, N. M.; Kirkby, J.; Kulmala, M.; Worsnop, D. R.; Baltensperger, U. Oxidation Products of Biogenic Emissions Contribute to Nucleation of Atmospheric Particles. *Science* **2014**, *344*, 717–721.
51. Kristensen, K.; Watne, Å. K.; Hammes, J.; Lutz, A.; Petäjä, T.; Hallquist, M.; Bilde, M.; Glasius, M. High-Molecular Weight Dimer Esters Are Major Products in Aerosols from Alpha-Pinene Ozonolysis and the Boreal Forest. *Environ. Sci. Technol. Lett.* **2016**, *3*, 280–285.
52. Tobias, H. J.; Ziemann, P. J. Kinetics of the Gas-Phase Reactions of Alcohols, Aldehydes, Carboxylic Acids, and Water with the C13 Stabilized Criegee Intermediate Formed from Ozonolysis of 1-Tetradecene. *J. Phys. Chem. A* **2001**, *105*, 6129–6135.
53. Fu, P. Q.; Kawamura, K.; Okuzawa, K.; Aggarwal, S. G.; Wang, G. H.; Kanaya, Y.; Wang, Z. F. Organic Molecular Compositions and Temporal Variations of Summertime Mountain Aerosols over Mt. Tai, North China Plain. *J. Geophys. Res. Atmos.* **2008**, *113*, D19107.
54. Rocha, I. M.; Galvão, T. L. P.; Sapei, E.; Ribeiro da Silva, M. D. M. C.; Ribeiro da Silva, M. A. V. Levoglucosan: A Calorimetric, Thermodynamic, Spectroscopic, and Computational Investigation. *J. Chem. Eng. Data* **2013**, *58*, 1813–1821.
55. Boand, G.; Houriet, R.; Gaumann, T. Gas-Phase Acidity of Aliphatic-Alcohols. *J. Am. Chem Soc.* **1983**, *105*, 2203–2206.

56. Hennigan, C. J.; Miracolo, M. A.; Engelhart, G. J.; May, A. A.; Presto, A. A.; Lee, T.; Sullivan, A. P.; McMeeking, G. R.; Coe, H.; Wold, C. E.; Hao, W. M.; Gilman, J. B.; Kuster, W. C.; de Gouw, J.; Schichtel, B. A.; Collett, J. L., Jr.; Kreidenweis, S. M.; Robinson, A. L. Chemical and Physical Transformations of Organic Aerosol from the Photo-Oxidation of Open Biomass Burning Emissions in an Environmental Chamber. *Atmos. Chem. Phys.* **2011**, *11*, 7669–7686.
57. Kessler, S. H.; Smith, J. D.; Che, D. L.; Worsnop, D. R.; Wilson, K. R.; Kroll, J. H. Chemical Sinks of Organic Aerosol: Kinetics and Products of the Heterogeneous Oxidation of Erythritol and Levoglucosan. *Environ. Sci. Technol.* **2010**, *44*, 7005–7010.
58. Hoffmann, D.; Tilgner, A.; Iinuma, Y.; Herrmann, H. Atmospheric Stability of Levoglucosan: A Detailed Laboratory and Modeling Study. *Environ. Sci. Technol.* **2010**, *44*, 694–699.
59. Hennigan, C. J.; Sullivan, A. P.; Collett, J. L., Jr.; Robinson, A. L. Levoglucosan Stability in Biomass Burning Particles Exposed to Hydroxyl Radicals. *Geophys. Res. Lett.* **2010**, *37*, L09806.
60. Zhao, R.; Mungall, E. L.; Lee, A. K. Y.; Aljawhary, D.; Abbatt, J. P. D. Aqueous-Phase Photooxidation of Levoglucosan - A Mechanistic Study Using Aerosol Time-of-Flight Chemical Ionization Mass Spectrometry (Aerosol TOF-CIMS). *Atmos. Chem. Phys.* **2014**, *14*, 9695–9706.
61. Lai, C.; Liu, Y.; Ma, J.; Ma, Q.; He, H. Degradation Kinetics of Levoglucosan Initiated by Hydroxyl Radical under Different Environmental Conditions. *Atmos. Environ.* **2014**, *91*, 32–39.
62. Slade, J. H.; Knopf, D. A. Heterogeneous OH Oxidation of Biomass Burning Organic Aerosol Surrogate Compounds: Assessment of Volatilisation Products and the Role of OH Concentration on the Reactive Uptake Kinetics. *Phys. Chem. Chem. Phys.* **2013**, *15*, 5898–5915.
63. Holmes, B. J.; Petrucci, G. A. Oligomerization of Levoglucosan by Fenton Chemistry in Proxies of Biomass Burning Aerosols. *J. Atmos. Chem.* **2007**, *58*, 151–166.
64. Limbeck, A.; Kulmala, M.; Puxbaum, H. Secondary Organic Aerosol Formation in the Atmosphere via Heterogeneous Reaction of Gaseous Isoprene on Acidic Particles. *Geophys. Res. Lett.* **2003**, *30*, 1996.
65. Liggio, J.; Li, S. M.; Brook, J. R.; Mihele, C. Direct Polymerization of Isoprene and Alpha-Pinene on Acidic Aerosols. *Geophys. Res. Lett.* **2007**, *34*, L05814.
66. Connelly, B. M.; Tolbert, M. A. Reaction of Isoprene on Thin Sulfuric Acid Films: Kinetics, Uptake, and Product Analysis. *Environ. Sci. Technol.* **2010**, *44*, 4603–4608.
67. Matsuoka, K.; Sakamoto, Y.; Hama, T.; Kajii, Y.; Enami, S. Reactive Uptake of Gaseous Sesquiterpenes on Aqueous Surfaces. *J. Phys. Chem. A* **2017**, *121*, 810–818.
68. Enami, S.; Mishra, H.; Hoffmann, M. R.; Colussi, A. J. Protonation and Oligomerization of Gaseous Isoprene on Mildly Acidic Surfaces: Implications for Atmospheric Chemistry. *J. Phys. Chem. A* **2012**, *116*, 6027–6032.

69. Enami, S.; Hoffmann, M. R.; Colussi, A. J. Dry Deposition of Biogenic Terpenes Via Cationic Oligomerization on Environmental Aqueous Surfaces. *J. Phys. Chem. Lett.* **2012**, *3*, 3102–3108.
70. Pye, H. O.; Murphy, B. N.; Xu, L.; Ng, N. L.; Carlton, A. G.; Guo, H.; Weber, R.; Vasilakos, P.; Appel, K. W.; Budisulistiorini, S. H. On the Implications of Aerosol Liquid Water and Phase Separation for Organic Aerosol Mass. *Atmos. Chem. Phys.* **2017**, *17*, 343–369.
71. Guo, H.; Sullivan, A. P.; Campuzano-Jost, P.; Schroder, J. C.; Lopez-Hilfiker, F. D.; Dibb, J. E.; Jimenez, J. L.; Thornton, J. A.; Brown, S. S.; Nenes, A.; Weber, R. J. Fine Particle pH and the Partitioning of Nitric Acid During Winter in the Northeastern United States. *J. Geophys. Res. Atmos.* **2016**, *121*, 10355–10376.
72. Bougiatioti, A.; Nikolaou, P.; Stavroulas, I.; Kouvarakis, G.; Weber, R.; Nenes, A.; Kanakidou, M.; Mihalopoulos, N. Particle Water and Ph in the Eastern Mediterranean: Source Variability and Implications for Nutrient Availability. *Atmos. Chem. Phys.* **2016**, *16*, 4579–4591.
73. Guo, H.; Xu, L.; Bougiatioti, A.; Cerully, K. M.; Capps, S. L.; Hite, J. R.; Carlton, A. G.; Lee, S. H.; Bergin, M. H.; Ng, N. L.; Nenes, A.; Weber, R. J. Fine-Particle Water and pH in the Southeastern United States. *Atmos. Chem. Phys.* **2015**, *15*, 5211–5228.
74. Guo, H.; Liu, J.; Froyd, K. D.; Roberts, J. M.; Veres, P. R.; Hayes, P. L.; Jimenez, J. L.; Nenes, A.; Weber, R. J. Fine Particle pH and Gas-Particle Phase Partitioning of Inorganic Species in Pasadena, California, During the 2010 Calnex Campaign. *Atmos. Chem. Phys.* **2017**, *17*, 5703–5719.
75. Fang, T.; Guo, H.; Zeng, L.; Verma, V.; Nenes, A.; Weber, R. J. Highly Acidic Ambient Particles, Soluble Metals, and Oxidative Potential: A Link between Sulfate and Aerosol Toxicity. *Environ. Sci. Technol.* **2017**, *51*, 2611–2620.
76. Kumar, S.; Aggarwal, S. G.; Fu, P. Q.; Kang, M.; Sarangi, B.; Sinha, D.; Kotnala, R. K. Size-Segregated Sugar Composition of Transported Dust Aerosols from Middle-East over Delhi During March 2012. *Atmos. Res.* **2017**, *189*, 24–32.
77. Liang, L. L.; Engling, G.; Du, Z. Y.; Cheng, Y.; Duan, F. K.; Liu, X. Y.; He, K. B. Seasonal Variations and Source Estimation of Saccharides in Atmospheric Particulate Matter in Beijing, China. *Chemosphere* **2016**, *150*, 365–377.
78. Li, X.; Jiang, L.; Hoa, L. P.; Lyu, Y.; Xu, T. T.; Yang, X.; Iinuma, Y.; Chen, J. M.; Herrmann, H. Size Distribution of Particle-Phase Sugar and Nitrophenol Tracers During Severe Urban Haze Episodes in Shanghai. *Atmos. Environ.* **2016**, *145*, 115–127.
79. Li, X.; Chen, M. X.; Le, H. P.; Wang, F. W.; Guo, Z. G.; Iinuma, Y.; Chen, J. M.; Herrmann, H. Atmospheric Outflow of PM2.5 Saccharides from Megacity Shanghai to East China Sea: Impact of Biological and Biomass Burning Sources. *Atmos. Environ.* **2016**, *143*, 1–14.
80. Vestenius, M.; Hellen, H.; Levula, J.; Kuronen, P.; Helminen, K. J.; Nieminen, T.; Kulmala, M.; Hakola, H. Acidic Reaction Products of Monoterpenes and Sesquiterpenes in Atmospheric Fine Particles in a Boreal Forest. *Atmos. Chem. Phys.* **2014**, *14*, 7883–7893.

81. Müller, L.; Reinnig, M. C.; Naumann, K. H.; Saathoff, H.; Mentel, T. F.; Donahue, N. M.; Hoffmann, T. Formation of 3-Methyl-1,2,3-Butanetricarboxylic Acid via Gas Phase Oxidation of Pinonic Acid - A Mass Spectrometric Study of Soa Aging. *Atmos. Chem. Phys.* **2012**, *12*, 1483–1496.
82. Cheng, Y.; Brook, J. R.; Li, S. M.; Leithead, A. Seasonal Variation in the Biogenic Secondary Organic Aerosol Tracer *cis*-Pinonic Acid: Enhancement Due to Emissions from Regional and Local Biomass Burning. *Atmos. Environ.* **2011**, *45*, 7105–7112.
83. Zhang, Y. Y.; Muller, L.; Winterhalter, R.; Moortgat, G. K.; Hoffmann, T.; Poschl, U. Seasonal Cycle and Temperature Dependence of Pinene Oxidation Products, Dicarboxylic Acids and Nitrophenols in Fine and Coarse Air Particulate Matter. *Atmos. Chem. Phys.* **2010**, *10*, 7859–7873.
84. Kavouras, I. G.; Mihalopoulos, N.; Stephanou, E. G. Formation of Atmospheric Particles from Organic Acids Produced by Forests. *Nature* **1998**, *395*, 683–686.

Chapter 4

Tropospheric Aqueous-Phase OH Oxidation Chemistry: Current Understanding, Uptake of Highly Oxidized Organics and Its Effects

Andreas Tilgner and Hartmut Herrmann*

Atmospheric Chemistry Department (ACD), Leibniz Institute for Tropospheric Research (TROPOS), Permoserstr. 15, 04318 Leipzig, Germany

***E-mail: herrmann@tropos.de**

The oxidation budget of the tropospheric gas phase is relatively well known. However, the concentrations and turnover rates of important oxidants such as OH that are present in the tropospheric aqueous phase are considerably more uncertain, as a result of the higher complexity of associated multiphase chemical interactions involved. This chapter outlines (i) the current understanding of the aqueous-phase OH oxidation budget, (ii) recent progress within the field with a focus on the uptake of highly oxidized organics as well as their effects on the oxidation capacity in aqueous aerosols, and (iii) future research objectives. In detail, the first part presents an overview on photochemical OH sinks and sources within the tropospheric aqueous phase, focusing on modelled and measured in-situ formation rates of OH. It also discusses current discrepancies between models and measurements as well as, finally, the limitations of both. In this part, model simulations using CAPRAM are presented. They demonstrate that utilizing a more detailed organic chemistry leads to substantially lowered aqueous-phase OH concentrations, more closely aligning modelled and measured OH concentrations. In the second part, a summary of current state-of-the-art knowledge on the role and fate of organic peroxides, labile hydroperoxides, and other organic peroxy species as potential OH sources is given. Furthermore, recent results of a model case study utilizing

CAPRAM are presented, allowing an examination of the uptake and chemistry of highly oxidized organics from the gas phase and its effects. The case study demonstrates that the uptake and subsequent decomposition of labile hydroperoxides may lead to a mean increase of 19% of the aqueous OH sink and source rates. Finally, a brief perspective is given, including an outline of current gaps within the knowledge on the oxidation budget in aqueous aerosols as well as research objectives for future laboratory, field and model investigations.

Introduction

The troposphere is a multiphase oxidizing environment where emitted volatile gaseous and particulate compounds are generally oxidized in (i) the gas phase, (ii) on available interfaces and (iii) in the bulk organic and aqueous phase of aerosol particles through a large variety of chemical processes (*1–8*). The tropospheric aqueous phase comprises cloud and fog droplets as well as deliquesced water-containing particles characterized by huge variations in microphysical and chemical parameters, which results in partially different chemical environments (*6*). The various chemical processes occurring in the aqueous phase and their interaction with the surrounding gas phase, including the uptake of trace gases, have the potential to change both the composition and, hence, the chemical and physical properties of tropospheric aerosols as well as the resulting effects on health and climate. Moreover, chemical processes occurring within the bulk of aqueous aerosol particles and cloud/fog droplets can alter the composition and oxidizing capacity of the troposphere (*1, 5, 9–11*). Thus, aqueous-phase processes are important when regarding environmental and societal issues such as climate change, air pollution, the tropospheric oxidation capacity and its related effects on, e.g., human health (*12*). Therefore, improved knowledge on the processing and fate of inorganic and, particularly, organic compounds is necessary in order to clarify the role of aqueous-phase processes and their impact on present environmental and the related societal issues.

In general, the oxidation and fate of various organic and inorganic compounds in the tropospheric aqueous phase of cloud droplets and deliquesced particles are strongly dependent on the concentration of oxidants, i.e., their chemical source and sink fluxes (rates). Besides other oxidants, the hydroxyl radical (OH) represents the most powerful oxidant due to its high oxidation potential. Therefore, OH is often referred to as the "detergent" of the troposphere, being able to alter the aerosol composition. However, current knowledge of aqueous-phase oxidant sources and sinks, including clouds and aqueous aerosols, is less detailed than that for the gas phase. Therefore, our understanding of chemical transformations occurring in the tropospheric aqueous phase is still fairly limited, and investigations on sources and sinks of aqueous-phase oxidants pose a key topic in atmospheric chemistry (*6*).

The aim of this chapter is to outline the state of the art as well as recent advances in the field of atmospheric multiphase processes by focusing on the oxidation capacity in tropospheric aqueous systems. We will not offer a full survey of present knowledge but rather focus on potential radical sources in aqueous particles and cloud droplets as well as new directions of research in the field. In the first part, a brief overview of photochemical OH sinks and sources in the tropospheric aqueous phase is provided, with a focus on modelled and measured in-situ formation rates of OH, current discrepancies between models and measurements, and, finally, the limitations of both. In the second part, we focus on the current state-of-the-art knowledge on the role and fate of organic peroxides, labile hydroperoxides, and other organic peroxy species as potential OH sources. Furthermore, recent results of a model case study using the Chemical Aqueous Phase RAdical Mechanism (CAPRAM (*5*, *11*, *13*, *14*)) are presented in order to study the uptake and aqueous-phase chemistry of highly oxidized organics (HOMs), and the possible effects. Finally, an outlook on research objectives of future laboratory, field and model investigations is given.

Photochemical OH Sources and Sinks: Discrepancies between Models and Measurements

Aqueous-Phase Sources and Sinks of Hydroxyl Radicals

OH radicals are formed by photochemical processes in both the tropospheric gas phase and the aqueous phase. For the aqueous-phase budget of the reactive daytime radical OH, both phase transfer from the gas phase and in-situ aqueous-phase formation sources are important (*5*, *13*, *15*). The most significant chemical in-situ sources of OH in the aqueous phase are directly linked to photochemical processes and/or transition metal ion (TMI) related reactions. The formation of OH results from, e.g., the photo-Fenton (or "Fenton-like") reactions (R-1a/b (*16*)), the decomposition of ozone initiated by superoxide radical anions (R-2a-c (*17*)) as well as the photolysis of nitrate (R-3 (*18*)), nitric acid/nitrite (R-4 (*18*)), Fe(III)-hydroxy complexes (R-5/6 (*19*)), hydrogen peroxide (R-7 (*18*)) and other organic peroxides (R-8 (*18*, *20*, *21*)). Furthermore, the photochemistry of CDOM ("Colored" Dissolved Organic Matter) constituents acting as an OH radical source in atmospheric waters (*21–23*) is discussed. Overall, it should be noted that the complexities of the interacting formation pathways of OH radicals have not yet been fully characterized and, therefore, still represent an on-going research topic.

$$OH_{gas} \rightleftharpoons OH_{gas} \qquad \text{(U-1)}$$

$$Fe^{2+} + H_2O_2 \longrightarrow Fe^{3+} + OH_{aq} + OH^- \qquad \text{(R-1a)}$$

$$Cu^{+} + H_2O_2 \longrightarrow Cu^{2+} + OH_{aq} + OH^- \qquad \text{(R-1b)}$$

$$O_{3,aq} + O_2^- \longrightarrow O_{2,aq} + O_3^- \qquad \text{(R-2a)}$$

$$O_3^- + H^+ \rightleftharpoons HO_3 \qquad \text{(R-2b)}$$

$$HO_3 \longrightarrow OH_{aq} + O_{2,aq} \quad \text{(R-2c)}$$

$$NO_3^- + h\nu \overset{H^+}{\rightarrow} NO_2 + OH_{aq} \quad \text{(R-3)}$$

$$NO_2^- + h\nu \overset{H^+}{\rightarrow} NO + OH_{aq} \quad \text{(R-4)}$$

$$[Fe(OH)]^{2+} + h\nu \longrightarrow Fe^{2+} + OH_{aq} \quad \text{(R-5)}$$

$$[Fe(OH)_2]^+ + h\nu \longrightarrow Fe^{2+} + OH_{aq} + OH^- \quad \text{(R-6)}$$

$$H_2O_2 + h\nu \longrightarrow OH_{aq} + OH_{aq} \quad \text{(R-7)}$$

$$ROOH + h\nu \longrightarrow RO \cdot + OH_{aq} \quad \text{(R-8)}$$

$$CDOM + h\nu \longrightarrow\longrightarrow products + OH_{aq} \quad \text{(R-9)}$$

The most important OH sinks in tropospheric aqueous solutions include reactions with halogen anions (e.g. Cl^- and Br^- (*24*); R-10a/b), in particular with dissolved organic matter (DOM; R-11). The first reaction pathway initiates the formation of halogen radicals and their follow-up chemistry. The oxidation of DOM leads to the formation of carbon centered organic radicals, further reacting with oxygen to produce organic peroxy radicals (RO_2). The RO_2 chemistry is subsequently leading to the formation of stable organic compounds (RH*) and secondary reactive oxygen species (ROS), such as hydroperoxyl/superoxide radicals (HO_2/O_2^-) and hydrogen peroxide (H_2O_2) (*25*). Secondary ROS species can recycle OH radicals via, for example, photo-Fenton chemistry. Therefore, the DOM photooxidation does not necessarily lead to a reduced or interrupted OH processing, as long as OH recycling processes are effective.

$$OH_{aq} + Cl^-/Br^- \rightleftharpoons ClOH^-/BrOH^- \quad \text{(R-10a)}$$

$$ClOH^-/BrOH^- + H^+ \rightleftharpoons Cl_{aq}/Br_{aq} + H_2O \quad \text{(R-10b)}$$

$$OH_{aq} + DOM \overset{+O2}{\longrightarrow} H_2O + RO_2 \cdot \longrightarrow\longrightarrow RH^* + \alpha \cdot HO_2 + \beta \cdot H_2O_2$$

$$(\alpha, \beta = \text{branching ratios}) \quad \text{(R-11)}$$

Due to the significant role of OH in the tropospheric oxidation reactions, the aqueous steady-state OH concentration ($[OH_{aq}])_{SS}$ is an important parameter to estimate the tropospheric lifetime of pollutants in and the overall oxidation capacity of the tropospheric aqueous phase. The tropospheric lifetime of a compound X with regard to a second-order aqueous OH oxidation reaction is inversely proportional to the product of the ($[OH_{aq}])_{SS}$ and the specific reaction rate constant ($k_{OH,X}$).

$$\tau_{1/2} = 1/k_{OH,X} \cdot [OH_{aq}]$$

Assuming an equilibrium of chemical sink rates (loss fluxes) and source rates (formation fluxes) of OH (steady-state conditions), the $([OH_{aq}])_{SS}$ is defined as the ratio of the OH formation rate (R_{OH}) and the sum of the first order loss rates $(\sum_i k_{OH,X_i} \cdot [X_i])$.

$$[OH_{aq}]_{SS} = \frac{R_{OH_{aq}}}{\sum_i k_{OH,X_i} \cdot [X_i]}$$

The aqueous-phase OH concentration $([OH_{aq}])_{SS}$ as well as the OH formation rate (R_{OH}) are crucial parameters that can be determined by means of both field measurements and models in order to quantify the oxidation capacity of the aqueous phase and, thus, the fate of pollutants in the tropospheric multiphase system. Additionally, the formation potential of oxidants in ambient aerosol particles, such as OH, is also suggested to be a potential pathway that causes adverse health effects, such as pulmonary and cardiovascular diseases, in regions of high air pollution. In the following paragraphs, the findings of both field and model studies are outlined and subsequently compared.

Measured in-Situ Aqueous-Phase OH Formation Rates and OH Concentrations

Since the early 1990s, the OH radical formation rate and the corresponding $([OH_{aq}])_{SS}$ have been measured in cloud droplet and rain samples as well as aqueous extracts of aerosol particles (see Arakaki et al. (*26*) and references therein). The OH radical formation is usually investigated by means of a chemical probe technique. In bulk solutions, the OH photoformation is quantified by means of an added OH radical scavenger with an OH reaction rate constant (e.g., benzene or benzoic acid) that is known to be high. Within the examined solutions, the concentrations of the added OH radical scavenger and the formed stable product (usually less reactive with OH) are monitored as functions of time by high-performance liquid chromatography (HPLC). The initial formation rate $R_{product}$ (M s^{-1}) of the product measured using HPLC is proportional to the OH radical formation R_{OH} (M s^{-1}). In brief, the OH radical formation R_{OH} (M s^{-1}) can be derived from the initial formation rate $R_{product}$ (M s^{-1}) of the product as well as the chemical yield of the scavenging reaction. Further details on the applied determination procedures of R_{OH} (M s^{-1}) and $([OH_{aq}])_{SS}$ are given in the literature (see, e.g., (*26*, *27*)).

A comprehensive overview of in-situ OH formation rates and derived $([OH_{aq}])_{SS}$ in different aqueous solutions is given in Table 1. Where possible, the formation rates are indicated in two different units: (i) mol $L^{-1}_{(water)}$ s^{-1} and (ii) mol $m^{-3}_{(air)}$ s^{-1}. The latter unit allows for better comparability of in-situ OH formation rates under different aqueous solution types. It must, however, be taken into account that the measured OH formation rates represent aqueous photochemical sources in atmospheric waters, which do not reflect any contributions related to the gas-phase partitioning of OH and its chemical precursors (e.g. HO_2, H_2O_2,

O_3, etc.). Therefore, most likely, the values can only be said to represent a lower limit of OH formation rates in ambient particles.

In Table 1, it is demonstrated that the obtained OH formation rates (mol $m^{-3}{}_{(air)}$ s^{-1}) in aerosol particle extracts include a broad range of values, ranging from $2.2 \cdot 10^{-15}$ to $2.3 \cdot 10^{-12}$. OH photoformation rates of more than $1 \cdot 10^{-13}$ mol $m^{-3}{}_{(air)}$ s^{-1} are measured at urban sites, while lower values are usually obtained at marine coastal or continental remote locations. The reason might be because higher concentrations of soluble TMI and organic matter are present under continental urban conditions. For instance, Arakaki et al. (*26*) found that the OH photoformation rates are strongly correlated with soluble TMI ($R = 0.88$) and organic matter ($R = 0.69$), respectively.

Furthermore, the results of other studies investigating natural waters suggest that HULIS (humic-like substances), representing a large fraction of the DOM, may act, either directly (via photolysis) or indirectly (through photo-Fenton chemistry), as sources of OH (White et al. (*28*) and references therein). Likewise, Badali et al. (*29*) demonstrate that the photolysis of secondary organic aerosol material, that is formed in terpene ozonolysis chamber experiments, can lead to the formation of OH radicals. The OH radicals likely originate from ROOH compounds, when the dissolved chamber samples are exposed to ultraviolet light in a photochemical reactor. In addition, Anastasio et al. (*30–32*) have shown that OH formation rates (mol $m^{-3}{}_{(air)}$ s^{-1}) are generally higher in fine mode aerosol compared to coarse mode aerosol.

In contrast to aqueous aerosols, the OH photoformation rates of clouds show a much more narrow range, approximately $5 \cdot 10^{-14}$ to $2.1 \cdot 10^{-13}$ mol $m^{-3}{}_{(air)}$ s^{-1} (see Table 1). This difference is a result of higher organic mass amounts present in the droplets due to uptake of soluble trace gases into the larger water volume of droplets. The aqueous-phase partitioning of soluble trace gases is much higher under cloud conditions. Consequently, the dissolved organic matter contribution is much smaller in the aqueous aerosol extracts. Therefore, very low OH photoformation rates may not be observed under diluted cloud conditions.

A comparison of OH photoformation (mol $L^{-1}{}_{(water)}$ s^{-1}) in aerosol and cloud waters to OH formation in other non-atmospheric aqueous environments (see White et al. (*28*)) shows that natural water samples, e.g., from lakes, rivers, wetlands, seawater, etc., exhibit OH production rates within a similar range of approximately $0.003\text{-}1.7 \cdot 10^{-9}$ mol L^{-1} s^{-1}.

Finally, Table 1 summarizes the obtained OH concentrations ($[OH_{aq}])_{SS}$ in different aqueous solution types. As opposed to OH formation rates, including a rather broad range of values of about four orders of magnitude, ($[OH_{aq}])_{SS}$ values measured in aerosol particle, cloud and rain water samples show a relatively narrow range. For ($[OH_{aq}])_{SS}$, data ranges of approximately $0.1\text{-}6 \cdot 10^{-15}$ mol L^{-1}, $0.5\text{-}7 \cdot 10^{-15}$ mol L^{-1} and $1\text{-}2 \cdot 10^{-15}$ mol L^{-1} for aerosol particle, cloud and rain waters, respectively, have been reported (see Table 1 and references therein). Furthermore, the rather sparse values exhibit no significant difference between different environmental regimes.

Table 1. Overview of measured in-situ OH formation rates (in mol $L^{-1}{}_{water}$ s^{-1} / mol $m^{-3}{}_{(air)}$ s^{-1}) and steady-state OH concentrations (mol $L^{-1}{}_{water}$) in aqueous aerosol particle extracts and cloud/rain water samples.

Study	*$[OH]_{SS}$ [mol L^{-1}]*	*In-situ OH formation rate [mol L^{-1} s^{-1}]*	*In-situ OH formation rate [mol m^{-3} s^{-1}]*	*Conditions and other remarks*
Aerosol particle extracts				
Anastasio and Jorden (2004) (*33*)	—	$1.0 \cdot 10^{-9}$ [c] ($0.09\text{-}3.0 \cdot 10^{-9}$)	$2.8 \cdot 10^{-15}$ [c] ($2.2\text{-}3.6 \cdot 10^{-15}$)[a]	Bulk aerosol particles (Alert, Nunavut, Canada)
Arakaki et al. (2006) (*34*)	—	$1.8 \cdot 10^{-10}$ [c]	$3.2 \cdot 10^{-14}$	Marine aerosol particles (Okinawa Island, Japan), bulk filter measurements
Anastasio and Newberg (2007) (*30*)	$3.8 \cdot 10^{-16}$ [c] ($1\text{-}57 \cdot 10^{-16}$)	$9.4 \cdot 10^{-8}$ [c] ($0.01\text{-}2.3 \cdot 10^{-6}$)	$2.6 \cdot 10^{-14}$ [a]	Sea salt aerosol particles (Bodega Bay, CA, USA), average of stage 2 and 3
Zhou et al. (2008) (*35*)	—	$1.1 \cdot 10^{-8}$ [c]	—	Marine aerosol particles (Sargasso Sea)
Kondo et al. (2009) (*36*)	—	—	$9.2 \cdot 10^{-14}$ [c] ($0.06\text{-}5.3 \cdot 10^{-13}$) [a]	Bulk aerosol particles (Higashi-Hiroshima, Japan)
Shen and Anastasio (2011) (*31*)	—	—	$1\text{-}3 \cdot 10^{-13}$	Fine aerosol, urban site (Fresno)
Shen and Anastasio (2012) (*32*)	—	—	$0.3\text{-}3.6 \cdot 10^{-13}$	Fine and coarse aerosol particles, urban site (Fresno)
Nomi et al. (2012) (*37*)	—	—	$3.9 \cdot 10^{-13}$ [c] ($0.3\text{-}23.3 \cdot 10^{-13}$)	Bulk aerosol particles (Higashi-Hiroshima, Japan)

Continued on next page.

Table 1. (Continued). Overview of measured in-situ OH formation rates (in mol L^{-1}_{water} s^{-1} / mol $m^{-3}_{(air)}$ s^{-1}) and steady-state OH concentrations (mol L^{-1}_{water}) in aqueous aerosol particle extracts and cloud/rain water samples.

Study	*$[OH]_{SS}$ [mol L^{-1}]*	*In-situ OH formation rate [mol L^{-1} s^{-1}]*	*In-situ OH formation rate [mol m^{-3} s^{-1}]*	*Conditions and other remarks*
Arakaki et al. (2013) (*26*)	$1.4 \cdot 10^{-15}$ [c]	—	—	Marine aerosol particles
Badali et al. (2015) (*29*)		$1.0 \cdot 10^{-9}$ [c] (0.9-$1.1 \cdot 10^{-9}$)		Secondary organic aerosol (SOA) samples from chamber experiments (terpene ozonolysis)
Cloud/fog water				
Faust and Allen (1993) (*38*)	—	$4.4 \cdot 10^{-10}$ [c]	($1.3 \cdot 10^{-13}$)	Cloud remote
Arakaki and Faust (1998) (*16*)	$7.2 \cdot 10^{-15}$ [cb]	$1.8 \cdot 10^{-10}$ [c]	($5.4 \cdot 10^{-14}$) [d]	Cloud remote
Anastasio and McGregor (2001) (*39*)	$4.7 \cdot 10^{-16}$ [c]	$5.0 \cdot 10^{-10}$ [c]	($1.5 \cdot 10^{-13}$) [d]	Cloud marine (Tenerife, Canary Islands, Spain), size-segregated cloud sample
Anastasio and McGregor (2001) (*39*)	$5.9 \cdot 10^{-16}$ [c]	$9.2 \cdot 10^{-10}$ [c]	$9.7 \cdot 10^{-14}$ [c]	Continental fog (Davis, CA, USA), bulk samples
Bianco et al. (2015) (*40*)	—	$7.0 \cdot 10^{-11}$	($2.1 \cdot 10^{-13}$) [d]	Cloud marine (Puy de Dôme station, France)

Study	*$[OH]_{SS}$ [mol L^{-1}]*	*In-situ OH formation rate [mol L^{-1} s^{-1}]*	*In-situ OH formation rate [mol m^{-3} s^{-1}]*	*Conditions and other remarks*
Bianco et al. (2015) (*40*)	—	$1.5 \cdot 10^{-10}$	$(4.5 \cdot 10^{-14})$ [d]	Clouds with continental influence (Puy de Dôme station, France)
Kaur and Anastasio (2017) (*27*)	$4.9 \cdot 10^{-16}$ [c] (0.26-$1.1 \cdot 10^{-15}$)	$3.3 \cdot 10^{-10}$ [c] (1.3-$7.0 \cdot 10^{-10}$)	$(9.9 \cdot 10^{-14})$ [d]	Continental fog (Davis, CA, USA; Baton Rouge, LA, USA)
Rain water				
Arakaki and Faust (1998) (*16*)	$1.1 \cdot 10^{-15}$ [cb]	$7.7 \cdot 10^{-11}$ [cb]	—	Rain water sample
Alibinet et al. (2010) (*22*)	$1.2 \cdot 10^{-15}$ [c] (0.9-$1.5 \cdot 10^{-15}$)	$3.3 \cdot 10^{-11}$ [c]	—	Rain water sample

[a] Value taken from Nomi et al. (2012). [b] Value taken from Arakaki et al. (2013). [c] Average value. [d] Calculated with an estimated cloud/fog LWC of 0.3 g $m^{-3}_{(air)}$.

Modeled in-Situ OH Formation Rates and OH Concentrations

In the last 30 years, several aqueous-phase chemical mechanisms and models have been developed and applied to investigate the atmospheric multiphase chemistry of deliquesced particles and cloud droplets (see Table 2 and references therein). Based on growing kinetic and mechanistic knowledge gained in laboratory experiments (see Herrmann et al. (*6*) and references therein), chemical mechanisms have been continuously extended. Nevertheless, the current mechanisms are still not able to reflect all the chemical reactions occurring in atmospheric waters as well as the uptake of important ROS from the gas-phase that acts as a potential aqueous-phase OH source.

Two key parameters usually provided by multiphase models that characterize the aqueous-phase oxidation budget are aqueous-phase OH processing rates (chemical sinks and sources fluxes) and modeled OH concentrations. Unlike the values obtained from measurements, these values do not reflect steady-state conditions. Furthermore, it should be noted that the predicted model rates and concentrations are based on considering (i) contributions of the direct phase transfer of OH and its precursors from the gas phase, (ii) contributions of in-situ formation pathways (such as the Fenton-type reaction) and (iii) the chemical sink reactions of OH and its precursors.

In Table 2, the modeled in-situ OH formation rates and OH concentrations from different aqueous-phase model studies are listed. Due to the stronger focus on in-cloud chemistry processing within model investigations, several datasets have been reported for cloud conditions and only few for concentrated aerosol solutions.

The reported OH concentrations in concentrated aerosol solutions show a broad data interval, ranging from $1.4\cdot10^{-16}$ to $8.0\cdot10^{-12}$ mol $L^{-1}{}_{(water)}$. The few OH formation rates in concentrated aerosol solutions that have been reported show a rather narrow range of $1.0\cdot10^{-13}$ to $3.5\cdot10^{-12}$ mol $m^{-3}{}_{(air)}$ s^{-1} with no substantial differences between remote and urban conditions. For cloud droplet conditions, the predicted OH concentrations and OH formation rates range from approximately $3.0\cdot10^{-15}$ to $8.4\cdot10^{-12}$ mol $L^{-1}{}_{(water)}$ and $1.0\cdot10^{-14}$ to $1.6\cdot10^{-11}$ mol $m^{-3}{}_{(air)}$ s^{-1}, respectively. Based on the dataset in Table 2, it can be seen that studies (*5*, *44*) that apply a more comprehensive description of the organic chemistry tend to identify somewhat lower OH concentrations, compared to model studies (*52*, *54*) considering chemistry of organic compounds with one and two carbon atoms (C_1/C_2 chemistry) only. The tendency of lower radical concentrations being predicted in the model, as a result of more organic sinks considered, has already been reported, e.g., by Herrmann et al. (*13*). Under remote conditions, taking into account additional C_2-C_4 chemistry pathways led to about 40% lower diurnal peak droplet concentrations of OH being predicted by the model (*60*). Under urban conditions featuring higher amounts of dissolved organic matter, even larger OH peak concentration reductions (of up to a factor of four) have been modeled using CAPRAM 3.0i compared to CAPRAM 2.4 (see Figure 1). Both studies anticipated a notable impact on the aqueous OH budget of the dissolved organic compounds considered in the mechanism. It was also anticipated that the more precise CAPRAM 3.0 mechanism may still overestimate OH concentrations as a result of not yet considered or hitherto unknown sinks. Nevertheless, such studies reveal the importance of treating more detailed organic chemistry in models in order to better predict the aqueous-phase OH radical budget as well as a need for continuous improvement of multiphase mechanisms.

Gaps between Measured and Modeled OH Concentrations and Formation Rates

A comparison of measured and modeled OH concentrations and formation rates as presented in the two former subsections partially reveals substantial differences (see Figure 2). Both measured OH concentrations and formation rates tend to be lower than the values predicted within models, and the gap is slightly larger for the OH concentrations. Arakaki et al. (*26*), for the first time, have systematically investigated the existing gaps and outlined possible root causes. However, as mentioned by Ervens et al. (*44*), the comparison of the measured and modeled data is not straightforward. Hence, in the following section, we will discuss possible reasons for the existing gaps, with a focus on limitations of both the measurements and models including their chemical mechanisms.

Limitations of Measurements

The study presented by Arakaki et al. (*26*) tries to overcome some of the limitations related to measurements as well as close the gap between field and model data. Due to the fact that aerosol samples are usually collected on a filter and subsequently diluted for analysis, while cloud water, instead, is sampled in bulk, important OH formation rates related to gas-phase uptake of OH and its precursors cannot be reflected by the measured rates. Therefore, Arakaki et al. (*26*) applied corrections to the measured OH formation rates in order to consider possible contributions of the direct OH gas-phase uptake (U-1) and the reaction of ozone (from the gas phase) with superoxide (R-2). However, other possibly significant OH sources (not directly related to photolytic processes), such as Fenton type reactions (R-1a/b) or indirect contributions via the uptake of important precursors (HO_2, H_2O_2), were not considered in this study. Arakaki et al. (*26*) suggested that models generally overestimate OH formation rates in clouds by a factor of 6 for continental remote clouds and a factor of 5 for marine clouds. Additionally, the study revealed that the calculated OH concentrations of models are substantially higher than the values derived from field samples. Compared to the relatively narrow data range of OH concentrations obtained in the field samples ($0.5\text{-}7 \cdot 10^{-15}$ M, see (*26*)), the modeled average values are found to be about 70 times higher for remote clouds and about 1000 times higher for marine clouds.

OH photoformation measurements using the chemical probe technique with simulated sunlight illumination or monochromatic light (e.g., $\lambda = 313$ nm) intend to reproduce tropospheric conditions in the laboratory as far as possible. However, there may still be essential processes (see the following discussion) that are not yet adequately reflected in the experiments.

Table 2. Overview of modeled in-situ OH formation rates (in mol L^{-1}_{water} s^{-1} / mol $m^{-3}_{(air)}$ s^{-1}) and OH concentrations (mol L^{-1}_{water}) under aqueous aerosol particle (APC), cloud (CC), fog (FC) and rain (RC) water conditions.

Study	*[OH]* *[mol L^{-1}]*	*OH formation rate* *[mol L^{-1} s^{-1}]*	*OH formation rate* *[mol m^{-3} s^{-1}]*	*Conditions and other remarks*
Aerosol particles				
Warneck (2005) (*41*)	$4.0\cdot10^{-16}$ (noon)	—	—	Marine APC
Herrmann et al. (2010) (*42*) / Tilgner et al. (2013) (*5*)	$4.4\cdot10^{-13}$ [a] ($1.4\cdot10^{-16}$-$1.9\cdot10^{-12}$)	0.1-$1.4\cdot10^{-4}$	0.25-$3.5\cdot10^{-12}$	Urban APC
	$3.0\cdot10^{-12}$ [a] ($5.5\cdot10^{-14}$-$8.0\cdot10^{-12}$)	0.4-$8\cdot10^{-5}$	0.1-$2\cdot10^{-12}$	Remote APC
	$1.0\cdot10^{-13}$ [a] ($4.6\cdot10^{-15}$-$3.3\cdot10^{-12}$)	—	—	Marine APC
Ervens and Volkamer (2010) (*43*)	$3\cdot10^{-12}$	—	—	—
Bräuer et al. (2013) (*24*)	$1\cdot10^{-13}$ (max.)	—	—	Marine APC
Ervens et al. (2014) (*44*)	$3.2\cdot10^{-13}$	0.3-$5.0\cdot10^{-3}$	0.3-$5.0\cdot10^{-11}$	—
Hoffmann et al. (2016) (*14*)	$< 1.9\cdot10^{-14}$	—	—	Marine APC
Cloud/fog droplets				
Chameides and Davis (1982) (*45*)	0.1-$3.5\cdot10^{-12}$	0.3-$2.9\cdot10^{-9}$	0.1-$4.4\cdot10^{-13}$	Remote CC

Study	*[OH] [mol L⁻¹]*	*OH formation rate [mol L⁻¹ s⁻¹]*	*OH formation rate [mol m⁻³ s⁻¹]*	*Conditions and other remarks*
Jacob (1986) (*46*)	$2.3 \cdot 10^{-13}$ (bulk) $1.8 \cdot 10^{-12}$ (surface)	$5.9 \cdot 10^{-9}$ (total) $3.4 \cdot 10^{-9}$ (in situ)	$2.5 \cdot 10^{-12}$ (total) $1.7 \cdot 10^{-12}$ (in situ)	Remote tropical CC
Pandis and Seinfeld (1989) (*47*)	$5.6 \cdot 10^{-14}$	—	—	Remote CC
Lelieveld and Crutzen (1990) (*48*)	$5.8 \cdot 10^{-13}$	$5.0 \cdot 10^{-9}$	$2 \cdot 10^{-12}$	Remote CC
Matthijsen et al. (1995) (*49*)	$1.0\text{-}2.0 \cdot 10^{-13}$ $1.5\text{-}5.2 \cdot 10^{-13}$ $8.4 \cdot 10^{-12}$	—	—	Marine CC Continental CC Urban CC
Monod and Carlier (1999) (*50*)	$0.2\text{-}1.8 \cdot 10^{-12}$	$0.8\text{-}3.3 \cdot 10^{-8}$	$0.4\text{-}1.6 \cdot 10^{-11}$	Rural CC with different pH conditions
Warneck (1999) (*51*)	$2.6 \cdot 10^{-14}$ $5.0 \cdot 10^{-14}$	$8.4 \cdot 10^{-9}$ $4.3 \cdot 10^{-9}$	$1.4 \cdot 10^{-12}$ $7.3 \cdot 10^{-13}$	Continental CC (no TMIs) Continental CC (with TMIs)
Herrmann et al. (2000) (*52*)	$< 1.4 \cdot 10^{-12}$ $< 1.7 \cdot 10^{-12}$ $< 2.0 \cdot 10^{-12}$	$2.4 \cdot 10^{-8}$ (noon) $1.7 \cdot 10^{-8}$ (noon) $1.3 \cdot 10^{-8}$ (noon)	$7.2 \cdot 10^{-12}$ $5.1 \cdot 10^{-12}$ $3.9 \cdot 10^{-12}$	Urban CC Remote CC Marine CC
Warneck (2003) (*53*)	$1.8 \cdot 10^{-13}$	—	—	Marine CC
Ervens et al. (2003) (*15*)	$< 1.0 \cdot 10^{-13}$ $< 2.0 \cdot 10^{-13}$ $< 4.5 \cdot 10^{-13}$	—	—	Urban CC Remote CC Marine CC

Continued on next page.

Table 2. (Continued). Overview of modeled in-situ OH formation rates (in mol L^{-1}_{water} s^{-1} / mol $m^{-3}_{(air)}$ s^{-1}) and OH concentrations (mol L^{-1}_{water}) under aqueous aerosol particle (APC), cloud (CC), fog (FC) and rain (RC) water conditions.

Study	*[OH]* *[mol L^{-1}]*	*OH formation rate* *[mol L^{-1} s^{-1}]*	*OH formation rate* *[mol m^{-3} s^{-1}]*	*Conditions and other remarks*
Barth et al. (2003) (*54*)	$2.4\cdot10^{-12}$	—	—	Remote CC
Deguillaume et al. (2004) (*55*)	$< 1.2\cdot10^{-12}$ $< 1.5\cdot10^{-12}$ $< 4.8\cdot10^{-12}$	$1.1\cdot10^{-8}$ (noon) $9.5\cdot10^{-9}$ (noon) $5.5\cdot10^{-9}$ (noon)	$3.3\cdot10^{-12}$ (noon) $2.9\cdot10^{-12}$ (noon) $1.7\cdot10^{-12}$ (noon)	Urban CC Remote CC Marine CC
Ervens et al. (2004) (*56*)	$1.5\cdot10^{-13}$ $3.8\cdot10^{-13}$	—	—	Urban CC Remote CC
Warneck (2005) (*41*)	$3.9\cdot10^{-13}$ (noon)	—	—	Marine CC
Herrmann et al. (2005) (*13*)	$0.3\text{-}1.5\cdot10^{-13}$	$4.6\cdot10^{-9}$	$1.4\cdot10^{-12}$ (noon)	Remote CC, permanent cloud scenario
Lim et al. (2005) (*57*)	$0.3\text{-}2.4\cdot10^{-14}$	—	—	Remote tropical CC
Deguillaume et al. (2010) (*58*)	$2.2\text{-}4.4\cdot10^{-13}$	$1.1\text{-}4.2\cdot10^{-9}$	$0.3\text{-}1.2\cdot10^{-12}$	Remote CC, permanent cloud scenario
Herrmann et al. (2010) / Tilgner et al. (2013) (*5*)	$3.5\cdot10^{-15}$ [a] $(0.03\text{-}1.6\cdot10^{-14})$ $2.2\cdot10^{-14}$ [a] $(0.5\text{-}6.9\cdot10^{-14})$ $2.0\cdot10^{-12}$ [a] $(0.04\text{-}5.3\cdot10^{-12})$	$6.0\cdot10^{-9}$ $8.0\cdot10^{-9}$ —	$3.0\cdot10^{-12}$ $4.0\cdot10^{-12}$ —	Urban CC, non-permanent cloud scenario Remote CC, non-permanent cloud scenario Marine CC, non-permanent cloud scenario

Study	*[OH] [mol L⁻¹]*	*OH formation rate [mol L⁻¹ s⁻¹]*	*OH formation rate [mol m⁻³ s⁻¹]*	*Conditions and other remarks*
Bräuer et al. (2013) (*24*)	$1.4\text{-}2.8 \cdot 10^{-12}$	—	—	Marine CC, non-permanent cloud scenario
Ervens et al. (2014) (*44*)	$1.4 \cdot 10^{-14}$	$0.1\text{-}8.0 \cdot 10^{-8}$	$0.3\text{-}8.0 \cdot 10^{-12}$	—
Hoffmann et al. (2016) (*14*)	$1.8 \cdot 10^{-13}$ [a] ($0.1\text{-}3.0 \cdot 10^{-13}$)	—	—	Marine CC
Rain				
Graedel and Goldberg (1983) (*59*)	$3.1\text{-}6.8 \cdot 10^{-14}$	$1.0\text{-}2.2 \cdot 10^{-12}$		

[a] Average value.

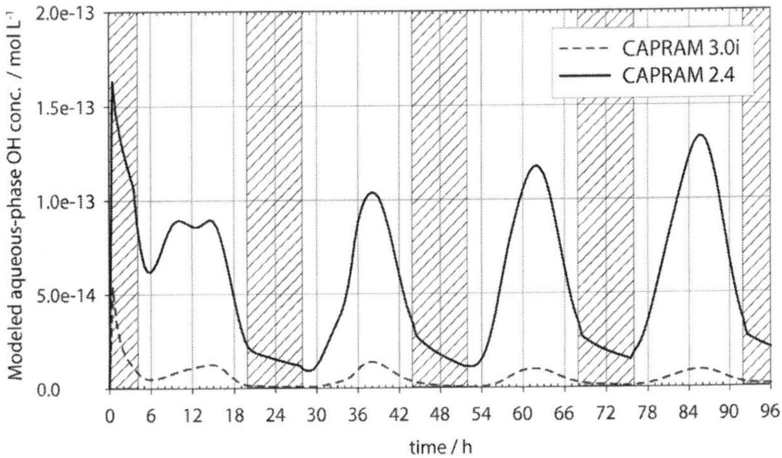

Figure 1. Aqueous-phase OH radical concentrations modeled with CAPRAM 3.0i (dashed line) and CAPRAM 2.4 (solid line) over the simulation time of four days under urban environmental conditions using a permanent cloud scenario (60).

One of the issues to be considered may be the sampling and treatment durations (hours), storage in the dark as well as the missing link to the gas-phase oxidant budget. Important oxidants and potential OH precursors found in particles and cloud droplets (such as H_2O_2, HO_2 and ROOH) are most likely reacted, and no replenishment from the gas phase can occur in the bulk experiments. Additionally, concentrations of reduced TMIs (Fe(II), etc.) can be substantially lowered in comparison to real single particles or cloud droplets as a result of performed dilutions. Altogether, this may cause a lower rate of OH recycling (e.g., via R-1a/b, R-7, or HO_2/O_2^- reactions with TMIs) in the OH photoformation experiments when compared to real aerosols linked to the gas phase. Related to this, Nomi et al. (*37*) have demonstrated that Fenton-type reactions may be potentially significant to total OH photoformation. In total, the above-mentioned issues may be the cause of the underestimation of the OH formation rates.

A further issue that may introduce differences between measured and modeled OH photoformation rates in aerosol particles is the dissolution of aerosol samples (see Arakaki et al. (*34*)), and the corresponding alteration of aerosol water pH and ionic strength. In the study presented by Arakaki et al. (*34*), for example, the bulk aerosol was added to a beaker together with 200 mL Milli-Q water. Based on the mean volume of the sampled air, a liquid water content of 0.17 g $m^{-3}{}_{air}$ can be calculated. Due to the added water, the water content is about four orders of magnitude higher than what is typical for aerosol water (see Herrmann et al. (*6*)). Thus, the liquid water content is more similar to typical cloud water samples rather than to the liquid water content of an aerosol particle. Compared to cloud water samples, the formed artificial solution does not include gaseous compounds taken up from the gas phase. Therefore, the solution represents a fairly sparse aqueous solution that is not comparable with ambient cloud water samples. Moreover, the performed dilution leads to a less concentrated solution with an altered acidity compared to the concentrated aerosol water solution. Therefore, the chemical processes and turnovers occurring in these different solutions, including the OH formation, can be affected considerably. As a result of pH and ionic strength being affected, the TMI speciation is also likely different. The chemical turnovers, e.g., from R-5/R-6, may be found to be quite different. Additionally, the chemical turnover of reactions under dilute (low-ionic-strength) and concentrated (high-ionic-strength) conditions can differ substantially, while the formation rates cannot be scaled linearly with the dilution. The performed dilution of the aerosol phase can result in considerably lower OH formation rates (mol $L^{-1}{}_{water}$ s^{-1}) as compared to modeled rates due to the fact that the liquid water content is four orders of magnitude higher.

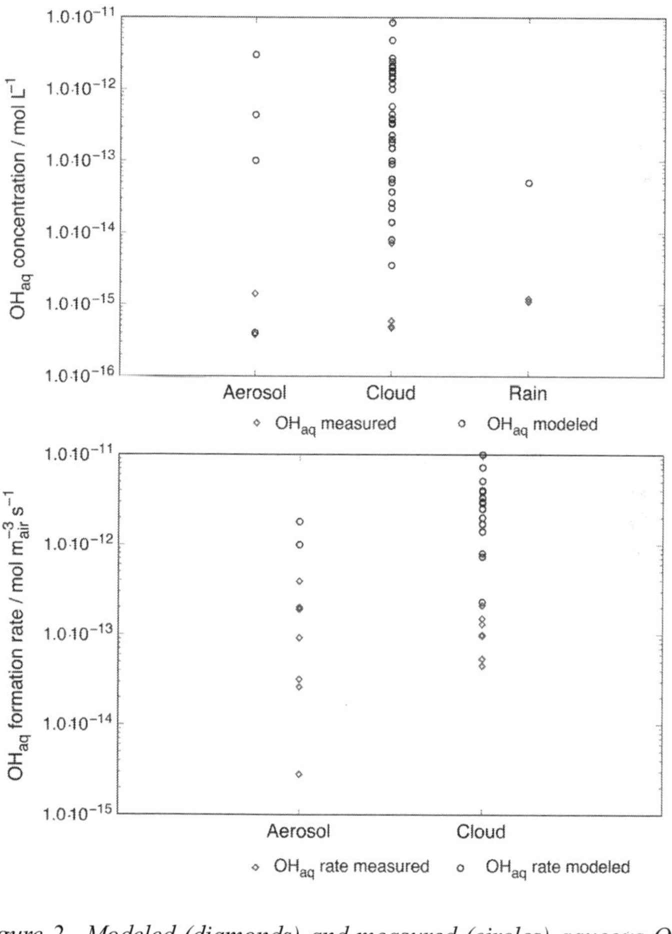

Figure 2. Modeled (diamonds) and measured (circles) aqueous OH concentrations (mol L^{-1}) in deliquesced aerosol, cloud and rain water (top), and modeled (diamonds) and measured (circles) aqueous OH formation rates (mol m^{-3} s^{-1}) in deliquesced aerosol and cloud water (down). Data based on Table 1 and 2, and references therein.

Another issue, possibly explaining why correctly measuring bulk formation rates is not straightforward, could lie in composition differences (acidity, etc.) of various particles/droplets of differing sizes. This issue can lead to different preferred chemical pathways in various particles/droplets. It will be difficult to correct such differences in chemical composition within the particle/droplet spectra by means of bulk chemical information only. On the other hand, size effects can be addressed within models and may, thereby, lead to a more efficient formation, compared to a pure bulk treatment under externally mixed aerosol conditions.

Limitations of Current Models and Mechanisms

The limitations of multiphase models and their corresponding chemical mechanisms when predicting OH formation rates and OH concentrations are mainly related to the fact that they only partially take into account the complex organic aqueous-phase chemistry. The tropospheric aqueous phases contain a wide variety of organic and inorganic compounds that can take part in chemical processing, including OH cycling. Atmospheric waters can contain a range of individual organic compounds, on the order of approximately 10^4 (*61*, *62*). However, only a relatively small mass fraction can usually be identified; typically, more than 50 % of the organic matter measured in fog and cloud droplets remains unknown (see Herckes et al. (*63*) and references therein). Due to the still limited knowledge on its organic composition, chemical oxidation schemes considered in current models can only describe the chemistry of organic compounds with a smaller carbon skeleton, since these compounds have been characterized better than organic compounds containing larger carbon numbers. At present, the most sophisticated aqueous-phase mechanisms describe the chemistry of compounds with up to six carbon atoms and about 10^3 individual organic compounds (including individual hydrated and dissociated compound forms) (*14*, *64*). Accordingly, the chemistry of larger organic compounds that are, e.g., taken up from the gas phase during cloud conditions, is not yet reflected by state-of-the-art multiphase mechanisms. These uncharacterized organic compounds can not only be expected to form sinks for OH radicals, but also to act as sources of OH or its precursors (e.g.., in case of ROOHs (*21*, *29*)). Arakaki et al. (*26*) suggested that current models underestimate OH sink rates and, therefore, overestimate the predicted OH radical concentrations. Consequently, it was proposed that improving the level of sink consideration in the models would, firstly, lower the predicted OH radical concentrations and, secondly, reduce the significance of OH in oxidizing individual organic compounds. The suggestion that a feedback of the reduced OH radical concentrations occurs is surely correct, as studies applying a more comprehensive description of the organic chemistry involved tend to predict lower OH concentrations than model studies that merely consider more limited organic chemistry, such as only C_1-C_2 chemistry.

Present model studies applying the CAPRAM4.0 mechanism with a complex description of C_1-C_4 chemistry, as well as an additional reaction module considering OH and NO_3 sinks of higher water soluble organic aerosol carbon (WSOC) and the organic complex formation of HULIS with iron, show a substantial reduction of the radical concentrations (see Figure 3). Due to the consideration of additional sinks as well as the lowered redox-cycling, OH concentration levels in the aqueous aerosols are predominantly reduced. Under cloud conditions, the soluble gaseous compounds taken up from the gas phase into droplets mainly influence OH levels and the OH sink strengths. Figure 3 shows OH concentrations (full model run) within a range of $1\text{-}2\cdot10^{-14}$ M under cloud conditions and $< 1.5\cdot10^{-14}$ M in the deliquesced particles. Uptake of OH presents the main source under cloud conditions (about 90%, see (*5*)), while other sources, as a consequence, are smaller by at least one order of magnitude. The gas-phase uptake is not reflected in bulk experiments. Without this key source, the measured OH concentrations obtained in bulk experiments should be substantially smaller (by about one order of magnitude) when compared to modeled concentrations. Thus, the modeled concentrations of approximately one order of magnitude higher seem to be a reasonable prediction of OH conditions in real droplets.

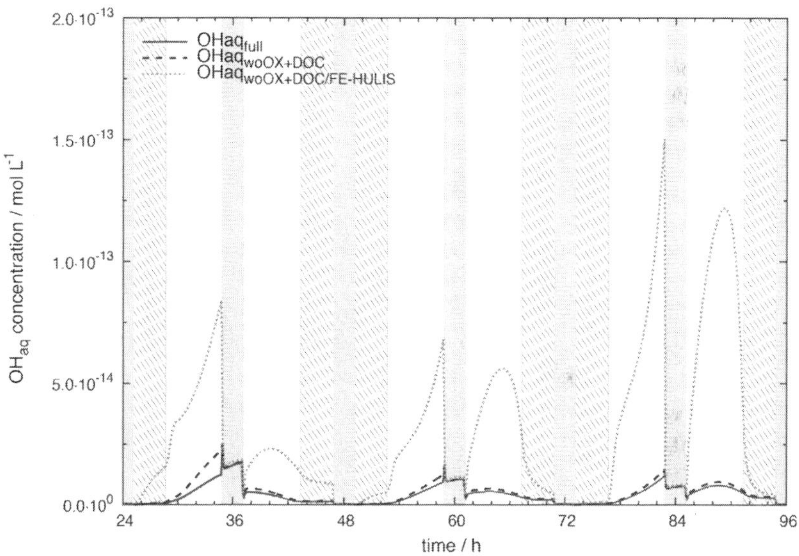

Figure 3. Modeled OH concentrations with CAPRAM 4.0, without/with consideration of WSOC radical sink reactions (dotted green/dashed blue line) and with consideration of both WSOC radical sink reactions and HULIS complex formations with iron (solid red line) simulated for an urban summer scenario.

The second statement of Arakaki et al. (*26*), suggesting a reduced processing of individual organic compounds by OH as well as lowered concentrations of secondarily formed radicals, may only be partially true. It is ignored that the HO_x/ HO_y processing and possible recycling of OH can occur within the aqueous phase. Therefore, the consideration of additional OH sinks (as a result of unidentified dissolved organic matter, such as HULIS or amino acids (*65*)) may lead not only to higher OH sink rates, but likely also slightly higher OH in-situ formation rates.

Overall, the combination of high sink and high source rates has the potential to characterize a reactive chemical system with low steady-state OH concentrations but high chemical turnovers. In this regard, it should be noted that chemical rates pose an even better parameter than steady-state OH concentrations when characterizing the chemical reactivity of a system, the latter merely resulting from the corresponding sink and source rates. Systems with both low and high chemical sink and source rates may exhibit identical steady-state OH concentrations.

It should be noted that current models/mechanisms are not only restricted in terms of missing OH sinks. Some potential OH sources may not yet have been considered within present models. Such OH sources may be, e.g., direct photolysis of HULIS or its indirect photo-Fenton chemistry (R-9, White et al. (*28*) and references therein), photolysis of α-hydroxyhydroperoxides (α-HHPs, R-9 (*21*, *66*)), or also, following up on a recent development in atmospheric chemistry, the decay of highly oxidized multifunctional molecules (HOMs) (*20*) formed in the gas phase, containing numerous organic hydroperoxide groups. Further information on this topic is given in the following sub-section.

Apart from restrictions in the applied chemical mechanism, predictions of models simulating aerosol particle chemistry are often limited as a result of the handling of non-ideality. Without consideration of non-ideality effects, OH formation rates resulting from, e.g., Fenton chemistry, could be largely overestimated. The activity of triply charged ions can be substantially lower than their corresponding molar concentration (*67*). Thus, present models that do not include adequate handling of non-ideal chemistry most likely overestimate chemical OH processing (see Herrmann et al. (*6*) and references therein).

In brief, it can be concluded that state-of-the-art models and mechanisms have not yet been sufficiently developed to reliably predict the OH concentration and processing rates in aqueous aerosols and cloud droplets. Both further mechanistic improvements and more advanced model developments are necessary in order to provide more sophisticated predictions. On the other hand, it has been demonstrated that field measurements of OH formation rates and derived OH concentrations are also limited. In order to gain a better understanding of OH processing, combined field and model investigations may be a first essential step. The direct combination of both means of investigation may help to interactively overcome the limitations of either approach. Furthermore, first combined comparisons of modeled with measured OH formation rates and OH concentrations that consider the same aerosol and meteorological conditions will be possible. A perspective on needs and research objectives of future laboratory, field and model investigations is given in the last sub-section of this chapter.

Role and Fate of Organic Peroxides, Labile Hydroperoxides, Other Organic Peroxy Species as Potential OH Source

Reactive oxygen species (ROS) are defined as chemically reactive species containing oxygen, such as peroxides, superoxide, hydroxyl radical as well as singlet oxygen (*68–71*). Aside from their effects on human health (*72–75*), ROS can potentially act as precursors of OH radicals in particles. Organic hydroperoxides (ROOH compounds) in particular are currently suggested to act as important precursors of OH in the aqueous phase, hence affecting the particle oxidant budget (*29*). OH radicals may be released by either thermal or photochemical cleavage of labile hydroperoxide bonds, according to the following reactions:

$$\mathrm{ROOH} \longrightarrow \mathrm{RO} \cdot + \mathrm{OH}_{aq} \qquad \text{(R-12)}$$

or

$$\mathrm{ROOH} + \mathrm{h\nu} \longrightarrow \mathrm{RO} \cdot + \mathrm{OH}_{aq} \qquad \text{(R-8)}$$

Overview on HOMs Related ROOHs

Labile peroxides, i.e., so-called HOMs or 'highly oxidized multifunctional molecules', have recently been discussed by Krapf et al. (*76*), following a first systematic investigation of HOMs phase partitioning and conversion in a combined chamber and field study conducted by Mutzel et al. (*77*). The uptake of HOMs has the potential to significantly improve the prediction of the budget of tropospheric particle SOA. However, since it is expected that only some HOMs will stay intact after their transfer from the gas phase, there are a multitude of open research questions following up on the HOMs formation in the gas phase. The following are some of the key questions:

- What are decomposition and functionalization products of HOMs following their uptake? How large are respective contributions to the budgets of important particle phase constituent groups, such as organosulphates (OS)?
- Which of the HOM species stay intact? Do they undergo particle phase reactions?
- If HOMs decompose along the peroxide bond, how is this going to affect the particle phase OH budget?
- If HOMs decompose based on other mechanisms, e.g., the so-called Korchev-mechanism (*77*, *78*), what are the resulting products? Are they going to remain present in the particle phase or are smaller organic products, possibly, going to partition back into the gas phase?

In order to answer these questions, clearly, a great effort of multiphase chemistry investigation is necessary.

The Fate of HOMs in the Particle Phase

The work of Mutzel et al. (*77*) has demonstrated that, following the phase transfer of HOMs, a multitude of particle phase reaction products may be observed. They can be grouped as follows: (i) HOMs that remain intact, (ii) HOMs undergoing functionalization towards highly oxidized organosulphates or 'HOOS', (iii) smaller carbonyl compounds resulting from the decomposition of HOMs, and finally, (iv) HOMs, including multiple organic hydroperoxides, undergo O,O cleavage and release OH into the particle phase. All of these possibilities need to be properly considered when discussing the contributions of HOMs that are formed through gas-phase reactions leading to particle phase organic mass (or, in short, SOA). The exact nature of the HOOS formation mechanism as well as the decomposition yielding to smaller carbonyl compounds requires further intensive study.

Do HOMs Remain Intact after Phase Transfer?

In regard to this question, the current state of knowledge may be summarized as follows: Only a minority of HOMs that are formed in the gas phase remain chemically unaltered while undergoing transfer into particles. In fact, in the study conducted by Mutzel et al. (*77*), only one HOM could be identified in the particle phase that had previously been present in the gas phase. Hence, pathways (ii), (iii) and (iv), as defined above, are expected to govern the fate of HOMs following uptake into the particle phase. The relative yields for each pathway, however, are still unclear. In the following, the effect of a particle-phase OH release is examined by means of simulations with CAPRAM.

Case Study Using CAPRAM: HOMs as Aqueous Particle-Phase OH Sources

Recently, model studies have been performed that apply the MCM3.2/ CAPRAM4.0 mechanism and an additional reaction module in order to investigate the formation, uptake and aqueous-phase fate of HOMs from α-pinene and β-pinene oxidation via OH and O_3. The developed reaction module takes into account a HOMs yield of 5% for both the OH and O_3 oxidation pathways in the MCM3.2. The Henry's Law uptake coefficients of gaseous HOMs have been calculated by means of the prediction method HENRYWIN (*79*), which is based on an OH-initiated HOM compound structure (SMILES: C12CC(OO)(CC(O)C1(C)OO)C(C)(C)OO2) (*80*). Accordingly, a Henry's Law coefficient of $8 \cdot 10^{11}$ has been utilized for all four formed HOMs. In the aqueous phase, the chemical fate of HOMs is based on (i) photolysis (estimated to be identical to methyl hydroperoxide in CAPRAM [13]), (ii) thermal decay (estimated with k_{1st} of $1.0 \cdot 10^{-3}$ s^{-1}), (iii) the oxidation of the HOM and its reaction products (estimated with k_{2nd} of $3.8 \cdot 10^{8}$ M^{-1} s^{-1} (*26*)) and (iv) the reaction of HOMs with sulfur(IV) (estimated to be identical to methyl hydroperoxide in CAPRAM (*13*)). In total, the reaction mechanism includes four uptake processes and 72 reactions of HOMs and its reaction products. The additional reaction module is schematically summarized in Figure 4. The chemistry of the final reaction products is not considered within this reaction scheme, and therefore, these compounds accumulate in the particle phase.

Phase Transfer:

$HOMs_{(gas)\text{-}x\text{-}PIN,y} \rightleftharpoons HOMs_{(aq)\text{-}x\text{-}PIN,y}$ (x=α,β; y=OH,O_3)

Photolyis (P):

$HOMs_{(aq)\text{-}x\text{-}PIN,y} + h\nu \rightarrow OH + HOM_{(aq)\text{-}x\text{-}PIN,y,Pi}$ (i=1)

$HOMs_{(aq)\text{-}x\text{-}PIN,y,Pi} + h\nu \rightarrow OH + HOM_{(aq)\text{-}x\text{-}PIN,y,Pi+1}$ (i=1,2)

Thermal Decomposition (TD):

$HOMs_{(aq)\text{-}x\text{-}PIN,y} \rightarrow OH + HOM_{(aq)\text{-}x\text{-}PIN,y,TDi}$ (i=1)

$HOMs_{(aq)\text{-}x\text{-}PIN,y,TDi} \rightarrow OH + HOM_{(aq)\text{-}x\text{-}PIN,y,Pi+1}$ (i=1,2)

Oxidation (OX) by OH:

$HOMs_{(aq)\text{-}x\text{-}PIN,y} + OH \rightarrow HOM_{(aq)\text{-}x\text{-}PIN,y,OXi} + 0.5\ HO_2$ (i=1)

$HOMs_{(aq)\text{-}x\text{-}PIN,Pi} + OH \rightarrow HOM_{(aq)\text{-}x\text{-}PIN,y,P|OXi} + 0.5\ HO_2$ (i=1,2,3)

$HOMs_{(aq)\text{-}x\text{-}PIN,TDi} + OH \rightarrow HOM_{(aq)\text{-}x\text{-}PIN,y,TD|OXi} + 0.5\ HO_2$ (i=1,2,3)

$HOMs_{(aq)\text{-}x\text{-}PIN,TDi} + OH \rightarrow HOM_{(aq)\text{-}x\text{-}PIN,y,TD|OXi} + 0.5\ HO_2$ (i=1,2,3)

$HOMs_{(aq)\text{-}x\text{-}PIN,Si} + OH \rightarrow HOM_{(aq)\text{-}x\text{-}PIN,y,S|OXi} + 0.5\ HO_2$ (i=1,2,3)

S(IV) Oxidation (S):

$HOMs_{(aq)\text{-}x\text{-}PIN,y} + HSO_3^- + H^+ \rightarrow SO_4^{2-} + HOM_{(aq)\text{-}x\text{-}PIN,y,Si} + 2\ H^+$ (i=1)

$HOMs_{(aq)\text{-}x\text{-}PIN,y,Si} + HSO_3^- + H^+ \rightarrow SO_4^{2-} + HOM_{(aq)\text{-}x\text{-}PIN,y,Si+1} + 2\ H^+$ (i=1,2)

$HOMs_{(aq)\text{-}x\text{-}PIN,y,Zi} + HSO_3^- + H^+ \rightarrow SO_4^{2-} + HOM_{(aq)\text{-}x\text{-}PIN,y,Zi+1} + 2\ H^+$ (i=1,2,3; Z=P, TD,OX)

Figure 4. Schematic representation of the applied uptake and chemical reaction scheme of HOMs products from α-pinene and β-pinene oxidation.

First simulations were carried out using a non-permanent cloud scenario under remote environmental conditions (see Tilgner et al. (*5*)). In order to study the effects of HOMs on the aqueous-phase chemical processing, model runs with and without utilization of the HOMs reaction module have been performed.

Expectedly, the model simulations show a relatively small effect on the aqueous concentrations of the very reactive oxidant OH. However, the flux analysis reveals a considerable increase in the radical sink and source fluxes, particularly under aqueous particle conditions (see Figure 5). In contrast, the differences in sink and source fluxes under cloud conditions are quite small. Due to the considered HOMs chemistry, the OH sink and source fluxes in the aqueous aerosols are increased by 7% throughout the whole model run when considering both deliquesced particle and cloud periods. Observing non-cloud periods only, OH sink and source fluxes are increased by about 19% compared to the model run without consideration of HOMs chemistry. However, it should be noted that not only direct OH formation via HOMs thermal decay and photolysis plays an important role, but also indirect feedbacks are significant due to their effect on the HO_2/O_2^- budget and related chemistry influencing aqueous OH budget.

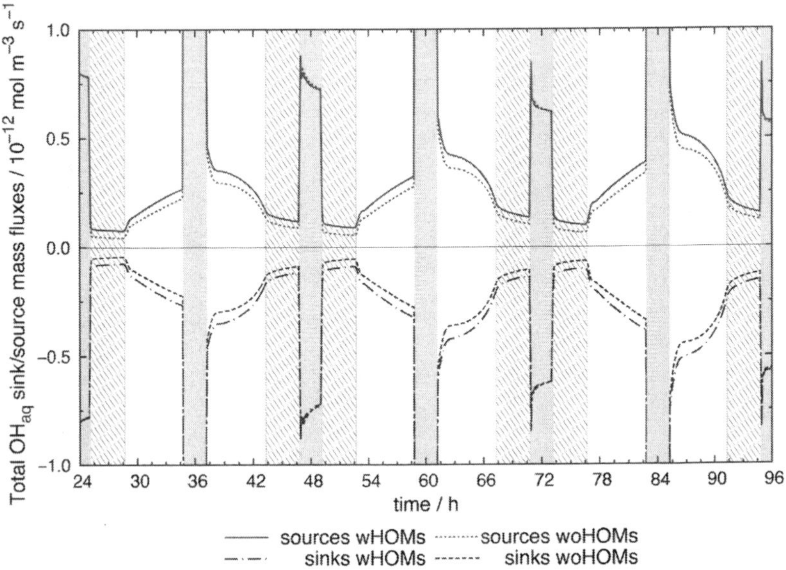

Figure 5. Modeled total aqueous-phase OH sink (blue) and source (red) fluxes using MCM3.2/CAPRAM4.0 with (wHOMs) and without (woHOMs) consideration of formation, uptake and aqueous-phase chemical processing HOMs from gas-phase α-pinene oxidation under remote environmental conditions. Blue dashed-dotted line: sink with HOM consideration, blue dashed line: sink without HOM consideration, red solid line: source with HOM consideration, red dotted line: sink without HOM consideration. Blue solid and striped bars mark cloud and nighttime conditions, respectively.

Moreover, the aqueous uptake and oxidation of the water-soluble HOMs can represent a considerable source of aqSOA (see Figure 6). The present model simulations show that OM formation is increased as a result of considering HOMs. Overall, the presented model simulations demonstrate the relevance of HOMs, potentially acting both as additional HOx and aqSOA sources; however, further HOMs sources (e.g., ISOPOOHs from the isoprene oxidation) need to be considered within further mechanisms in order to comprehensively explore the significance of HOMs for the aqueous-phase chemistry.

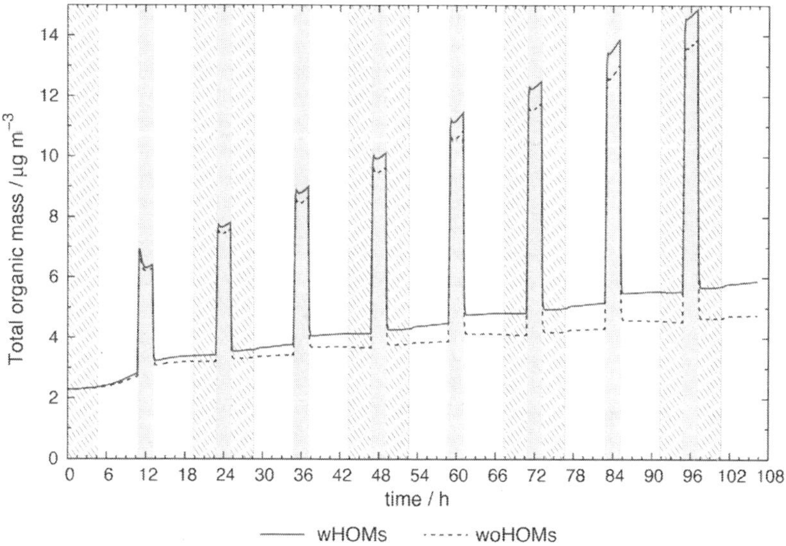

Figure 6. Total aqueous-phase organic mass concentration (µg m-3) modeled using MCM3.2/CAPRAM4.0 with (wHOMs, red solid line) and without (woHOMs, blue dashed line) consideration of formation, uptake and aqueous-phase chemical processing HOMs from gas-phase α-pinene oxidation under remote environmental conditions. Solid blue and striped bars mark cloud and nighttime conditions, respectively.

Need for More Advanced Laboratory, Field and Model Investigations Focusing on Tropospheric Aqueous-Phase Reactivity

The current state of knowledge on the tropospheric aqueous-phase reactivity features gaps resulting from limitations in field measurements, and restrictions within the mechanisms and models (*8*) that are currently available, as well as the available kinetic laboratory data (*6*). Radical reactivity measurements are only known for OH and NO_3 in the gas phase (*81, 82*); however, aqueous-phase measurements are rare or even non-existent for NO_3. A small number of measurements of OH in cloud droplets have been reported, while even less have been performed for OH present in aerosol particles. Overall, there is a definite need to experimentally characterize aqueous-phase budgets and concentrations of several radical and non-radical oxidants beyond OH, H_2O_2, and simple ROOHs. In this line of thought, the following subsection addresses needs and objectives for future investigations in further detail.

More Advanced Field Investigations and Analytical Methods

At present, knowledge on the budget of particle oxidants (concentrations and turnover rates) is very limited. Since the early 1990s, the OH radical formation rate (R_{OH}), reactivity (k'_{OH}) and corresponding steady-state OH concentration ($OH_{aq,ss}$) have been measured in cloud and rain samples (see subsections above). However, very few studies have investigated k'_{OH} and $OH_{aq,ss}$ in aerosol particles. The three studies on aerosol particles available are focused mainly on marine systems (*26, 30*). Arakaki et al. (*26*) have shown that the obtained reactivities (k'_{OH}) in marine aerosol particle extracts lie within a narrow range of values, ranging from $2.6 \cdot 10^{+8}$ to $3.8 \cdot 10^{+8}$ s^{-1}. However, k'_{OH} values in continental aerosols may be increased due to higher contents of soluble transition metal ions (TMIs) and organic matter. Both are known to strongly correlate with OH photoformation rates (*26, 27*). The only investigation of continental aerosols, that is available, has been performed by Shen and Anastasio (*32*). They have shown substantially higher OH formation rates of $0.3\text{-}3.6 \cdot 10^{-13}$ mol L^{-1} s^{-1} (cp. Table 1) in comparison to marine aerosol particles. Hence, it can be concluded that more particle reactivity measurements under different environmental conditions need to be performed, under consideration of different radical and non-radical oxidants. Those measurements should at the same time investigate the role of important parameters, such as TMI concentrations, organic matter content, aerosol liquid water content, aerosol acidity (pH) and ionic strength. It should be noted that former field studies often examine the OH radical oxidation budget only. However with more detailed data on particle and linked gas-phase composition, future field studies could examine important dependencies and correlations of oxidant concentrations and turnover rates in aerosols, e.g., from water-soluble metal content, organic matter, aerosol acidity (pH), aerosol liquid water content, and ionic strength.

The two parameters of (i) aerosol acidity (pH) and (ii) ionic strength are rather important for aerosol particles, as concentrated aqueous aerosol solutions are most likely to differ substantially from well-diluted cloud/fog solutions. As the aerosol samples are usually collected on filters and subsequently diluted for analysis, this procedure may introduce differences between measured and modelled OH photoformation rates in aerosol particles. Due to the sample preparation (see Arakaki et al. (*34*)), the aerosol water pH and ionic strength is altered and, thus, as mentioned above, chemical conditions are changed. Again, this has the potential to substantially modify the chemical processing and the oxidant budget. Therefore, based on the current measurement limitations, it is suggested that the methods currently applied need to be improved in order to gain a more advanced knowledge of the aqueous particle oxidation capacity. One possible improvement, for example, could be the miniaturization of photoreactor and the application of miniaturized analytical techniques, such as chip-based capillary electrophoresis (CE) techniques, in order to minimize the difficulties accompanied by diluting aerosol particle samples. Furthermore, adjusting both particle acidity and ionic strength to aerosol values could present a valuable improvement.

Combined Field and Modeling Investigations

As outlined in the first part of this chapter, the subsequent offline analysis of field samples by means of a chemical probe experimental technique, e.g., for OH, is only appropriate when attempting to determine in-situ oxidant formation rates that are not dependent on the gas-phase uptake of the oxidant itself and its precursors (e.g. HO_2, H_2O_2, O_3, ROOHs, etc.). Implemented corrections (*26*) of the measured OH formation rates as a result of possible uptake contributions (e.g. OH gas-phase uptake) can possibly be applied but are most likely too simple. Other indirect OH sources or related reaction sequences, such as Fenton type reactions, TMI redox processes and ROOH reactions (see above), affecting the radical budget and depending on the uptake of ROS compounds, are not considered. Furthermore, ROS species present in particles have most likely been reacted and have not been replenished from the gas phase, as well as reduced TMIs (Fe(II), etc.) have been lowered compared to ambient single particles. To overcome these limitations, combined field and kinetic model investigations need to be performed in order to more comprehensively understand the tropospheric aerosol oxidation capacity. Multiphase models provide time-resolved oxidant concentrations and formation/ sink rates. Thus, modelling should be used to determine contributions related to the above-mentioned indirect source strengths; e.g., contributions of gas-phase oxidant precursors. This could contribute to a more realistic characterization of the aqueous oxidant formation fluxes related to gas-phase uptake as well as immediate follow-up reactions under ambient conditions. Additionally, related modelling should also enable an advanced interpretation of measured oxidant concentrations and formation rates, as well as better explanations of the gaps between current measured and modelled concentrations/chemical rates.

Field, Chamber and Laboratory Investigations on the Aqueous Processing of Hydroperoxides (H_2O_2, ROOHs)

In studies of Paulson et al. (*83–87*), the concentrations of peroxides have been determined almost exclusively for urban aerosol particles. Here, H_2O_2 concentrations have been mainly determined (*83–87*), as well as, in some systems, organic hydroperoxides (*84*) - in almost all cases for urban particles collected in western US cities. The measured H_2O_2 aerosol concentrations are high and lie mainly within the range of 10^{-4}-10^{-2} M (*83, 84*). The studies available have shown that H_2O_2 represents the major hydroperoxide that is present in aerosols. Organic hydroperoxides such as CH_3OOH have only occasionally been detected in trace amounts (*84*). Moreover, several studies (*83, 84, 87*) have demonstrated that measured H_2O_2 concentrations exceed the levels predicted by Henry's law by two to three orders of magnitude, indicating an effective in-situ formation of H_2O_2 in aerosols. These reported H_2O_2 levels have the potential to cause cellular lung damage (*83*). Finally, within more recent studies, initial H_2O_2 formation rates have also been reported, e.g., $9.8 \cdot 10^{-11}$ - $1.3 \cdot 10^{-9}$ M s^{-1} (*85, 86*). The obtained rates were found to lie within similar ranges as OH radical formation rates as a result of TMI chemistry. Moreover, the majority of H_2O_2 generation is found to be related to TMI chemistry or redox cycling of quinoid compounds (*85–87*).

The kinetic reaction data of such processes are still incomplete and subject to ongoing research. TMIs are known to form inorganic and organic complexes (*19*, *88*), while TMI-complexes are able to react similarly to their corresponding not complexed metal ions – however, reaction kinetics can change considerably. An important example of such types of reactions are Fenton-like reactions, i.e., reactions of H_2O_2 with TMIs other than iron(II) or TMI-complexes, which may contribute to the formation of OH radicals (see R-13). Overall, the kinetic data of processes related to TMI-complexes are largely not yet well known and can therefore not be adequately considered within current chemical tropospheric aqueous-phase chemistry mechanisms. In atmospheric aqueous solutions, TMIs are often present in the form of complexes (*89*). Therefore, in order to be able to consider photochemical processes of any reactions of TMI-complexes including photochemical processes in future models, further laboratory investigations are required.

$$[\text{TMIcomplex}]^{n+} + H_2O_2 \longrightarrow [\text{TMIcomplex}]^{(n+1)+} + OH_{aq} + OH^- \quad \text{(R-13)}$$

Recently, studies of chamber aerosols conducted by Badali et al. (*29*) implied that photolysis of SOA generates OH in aqueous solution, likely originating from the photochemical decay of ROOH aerosol species. Due to the importance of both H_2O_2 and ROOHs (including HOMs), potentially acting as OH precursors, future investigations need to determine both oxidants in both field and chamber studies simultaneously. However, it should be noted that both H_2O_2 and oxygenated ROOHs may be strongly linked to gas-phase processes and subsequent aqueous-phase processes. In case of HOMs with hydroperoxide function, further chamber and laboratory investigations are needed in order to examine their potential to act as aqueous-phase ROS and potential OH precursors. Much more kinetic data on their thermal and photochemical decay than is currently available (*76*, *77*, *90*) need to be obtained in order to assess their significance and to be able to consider those processes in future kinetic mechanisms.

It should be noted that the highly oxidized molecules (HOMs) containing one or more organic hydroperoxide groups can also chemically interact with transition metal ions (TMIs) present in tropospheric aqueous solutions (*19*). The TMI interactions may contribute to both the formation of organic radicals and, therefore, the processing of HOMs in the aqueous phase and the redox cycling of metal ions. Formed RO· or RO_2· radicals can undergo further chemical processing, thus contributing to the formation of OH radical precursors such as HO_2 and H_2O_2. In order to examine the relevance of such TMI-HOM interaction processes, kinetic laboratory investigations must be performed using available synthesized HOM compounds such as isoprene hydroxyhydroperoxides (ISOPOOHs) or terpene-derived larger HOMs. However, there is a great need of chemical synthesis of HOMs-like compounds to further elucidate this chemistry. Only in cases where this is realistically impossible, surrogate compounds should be used.

Another pathway discussed in the literature that may be relevant to ROOH and, hence, OH budget in the aqueous phase, is the reaction of HO_2 with organic peroxy radicals (RO_2) (*21*). The resulting ROOHs from radical-induced oxidations are expected to be able to recycle the OH radical by subsequent aqueous-phase photochemical pathways. However, the HO_2 + RO_2 reaction competes with RO_2 + RO_2 recombination reactions, so that ROOH formation might be suppressed. The ROOH yield strongly depends on the decomposition rate of the tetroxide intermediate, which has not been constrained by many kinetic laboratory experiments. This issue may hence represent a limitation within current chemical mechanisms/models.

Singlet Oxygen ($^1O_2^*$): An Important Tropospheric Aqueous-Phase Oxidant?

It has long been known that photochemical processes in natural surface waters can produce singlet oxygen ($^1O_2^*$) (*18*, *91*). Singlet oxygen ($^1O_2^*$) is a photooxidant that represents the electronically excited form of molecular oxygen. It can be formed, e.g., from the photochemical interaction of photosensitizers, after quenching of exited organic species by molecular ground state oxygen. The main reaction of the photoexcited DOC in natural aqueous systems is an energy transfer to the dissolved molecular oxygen in its ground state ($^3O_2$). In the aqueous phase, $^1O_2^*$ can react with electron-rich organic compounds and, therefore, represents a more selective oxidant than the hydroxyl radical. Recently, atmospheric chemistry laboratory and field studies have focused on photosensitization processes and the linked formation of singlet oxygen ($^1O_2^*$) (*27*, *92*, *93*). Recent field measurements have observed singlet oxygen ($^1O_2^*$) in fog droplets (*27*) and aerosol particles (*93*). The observed steady-state concentrations in fog droplets lie within the range of $0.1\text{-}3.0 \cdot 10^{-13}$ M and, thus, show values of about 2 orders of magnitude higher than measured OH concentrations. Similar $^1O_2^*$ values have been observed in road dust aerosols by Cote et al. (*93*), showing steady-state concentrations of approximately $1 \cdot 10^{-13}$ M. Kaur and Anastasio (*27*) used a simple reactivity calculation to conclude that the selective oxidant $^1O_2^*$ may generally be less important than OH. Nevertheless, $^1O_2^*$ can play a role in the oxidation of electron-rich organic compounds found in atmospheric aqueous solutions. Because the number of available measurements for ambient aerosols is still rather small, further field measurements need to be performed under different atmospheric conditions.

Mechanism and Model Improvements

The limitations of multiphase models and their corresponding chemical mechanisms in the attempt to predict OH formation rates and OH concentrations are mainly related to the still insufficient consideration of the complex organic chemistry involved, including the issues addressed above. The chemistry of higher organic compounds (>C_5) and other possible oxidants is not yet reflected by state-of-the-art multiphase mechanisms. These hitherto unconsidered organic compounds can surely be expected to act as an important sink for OH radicals and other oxidants, indicating the need for further efforts in mechanism development.

Apart from the limits of the chemical mechanism, predictions of models simulating aqueous particle chemistry are often limited due to the handling of non-ideality. Without consideration of non-ideal effects, the predicted OH formation rates, e.g., based on the Fenton chemistry, can be inadequate. Further model developments should aim at developing comprehensive chemical mechanisms that describe the key organic chemistry pathways of different oxidants as well as an adequate handling of non-ideal aerosol chemistry in order to predict the aqueous aerosol oxidation capacity more realistically. However, an advanced handling of non-ideality in future models is definitely going to rely on appropriate interaction parameters based on laboratory investigations, which are not yet available for all treated compounds within the mechanisms.

Finally, it should be noted that studies on the aerosol oxidation capacity are closely related to questions regarding the health effects of tropospheric particulate matter. It has been demonstrated clearly that oxidative stress can cause negative effects on human health that are triggered by the inhalation of particulate matter (*94*). Particles have the potential to introduce transition metal ions (*95*), redox-active organic compounds such as quinones (*96*) or reactive oxygen species (ROS) into the human body. Accordingly, the intake of ROS and other oxidative substances into human bodies as well as their respective formation potential are of vast importance within current studies regarding human health (*97*). Therefore, the aforementioned issues, questions and limitations related to the current state of knowledge on aqueous-phase oxidation capacity should be addressed by future research studies in the field of atmospheric chemistry and related disciplines. Overall, a more detailed knowledge of oxidant speciation and reactivity is going to present a crucial step in the attempt to gain an improved understanding of the properties and effects of tropospheric aqueous systems as they occur in aerosol particles, fog and clouds.

Acknowledgments

Support of some of the work described as well as the preparation of this chapter by DFG project MISOX2 (grant number HE 3086/13-1), DFG project PhotoSOA (grant number HE 3086/32-1) as well as HORIZON 2020 projects MARSU as project 690958 in RISE and EUROCHAMP-2020 under project 730997 is gratefully acknowledged. Many aspects of the work of TROPOS-ACD (Leipzig) are supported by funding of the European Union through the European Regional Development Fund (EFRE) and the Free State of Saxony, Germany.

References

1. Ravishankara, A. R. Heterogeneous and multiphase chemistry in the troposphere. *Science* **1997**, *276*, 1058–1065.
2. Herrmann, H. Kinetics of aqueous phase reactions relevant for atmospheric chemistry. *Chem. Rev.* **2003**, *103*, 4691–4716.

3. Hallquist, M.; Wenger, J. C.; Baltensperger, U.; Rudich, Y.; Simpson, D.; Claeys, M.; Dommen, J.; Donahue, N. M.; George, C.; Goldstein, A. H.; Hamilton, J. F.; Herrmann, H.; Hoffmann, T.; Iinuma, Y.; Jang, M.; Jenkin, M. E.; Jimenez, J. L.; Kiendler-Scharr, A.; Maenhaut, W.; McFiggans, G.; Mentel, T. F.; Monod, A.; Prevot, A. S. H.; Seinfeld, J. H.; Surratt, J. D.; Szmigielski, R.; Wildt, J. The formation, properties and impact of secondary organic aerosol: current and emerging issues. *Atmos. Chem. Phys.* **2009**, *9*, 5155–5236.
4. Ervens, B.; Turpin, B. J.; Weber, R. J. Secondary organic aerosol formation in cloud droplets and aqueous particles (aqSOA): a review of laboratory, field and model studies. *Atmos. Chem. Phys.* **2011**, *11*, 11069–11102.
5. Tilgner, A.; Brauer, P.; Wolke, R.; Herrmann, H. Modelling multiphase chemistry in deliquescent aerosols and clouds using CAPRAM3.0i. *J. Atmos. Chem.* **2013**, *70*, 221–256.
6. Herrmann, H.; Schaefer, T.; Tilgner, A.; Styler, S. A.; Weller, C.; Teich, M.; Otto, T. Tropospheric aqueous-phase chemistry: kinetics, mechanisms, and its coupling to a changing gas phase. *Chem. Rev.* **2015**, *115*, 4259–4334.
7. Pöschl, U.; Shiraiwa, M. Multiphase chemistry at the atmosphere–biosphere interface influencing climate and public health in the anthropocene. *Chem. Rev.* **2015**, *115*, 4440–4475.
8. Ervens, B. Modeling the processing of aerosol and trace gases in clouds and fogs. *Chem. Rev.* **2015**, *115*, 4157–4198.
9. Jacob, D. J. Heterogeneous chemistry and tropospheric ozone. *Atmos. Environ.* **2000**, *34*, 2131–2159.
10. Kreidenweis, S. M.; Walcek, C. J.; Feingold, G.; Gong, W. M.; Jacobson, M. Z.; Kim, C. H.; Liu, X. H.; Penner, J. E.; Nenes, A.; Seinfeld, J. H. Modification of aerosol mass and size distribution due to aqueous-phase SO2 oxidation in clouds: Comparisons of several models. *J. Geophys. Res.-Atmos.* **2003**, *108*, 4213.
11. Tilgner, A.; Herrmann, H. Radical-driven carbonyl-to-acid conversion and acid degradation in tropospheric aqueous systems studied by CAPRAM. *Atmos. Environ.* **2010**, *44*, 5415–5422.
12. Lelieveld, J.; Evans, J. S.; Fnais, M.; Giannadaki, D.; Pozzer, A. The contribution of outdoor air pollution sources to premature mortality on a global scale. *Nature* **2015**, *525*, 367–371.
13. Herrmann, H.; Tilgner, A.; Barzaghi, P.; Majdik, Z.; Gligorovski, S.; Poulain, L.; Monod, A. Towards a more detailed description of tropospheric aqueous phase organic chemistry: CAPRAM 3.0. *Atmos. Environ.* **2005**, *39*, 4351–4363.
14. Hoffmann, E. H.; Tilgner, A.; Schrodner, R.; Brauer, P.; Wolke, R.; Herrmann, H. An advanced modeling study on the impacts and atmospheric implications of multiphase dimethyl sulfide chemistry. *Proc. Natl. Acad. Sci. U.S.A.* **2016**, *113*, 11776–11781.
15. Ervens, B.; George, C.; Williams, J. E.; Buxton, G. V.; Salmon, G. A.; Bydder, M.; Wilkinson, F.; Dentener, F.; Mirabel, P.; Wolke, R.; Herrmann, H. CAPRAM 2.4 (MODAC mechanism): An extended and

condensed tropospheric aqueous phase mechanism and its application. *J. Geophys. Res.-Atmos.* **2003**, *108*, 4426.

16. Arakaki, T.; Faust, B. C. Sources, sinks, and mechanisms of hydroxyl radical ((OH)-O-center dot) photoproduction and consumption in authentic acidic continental cloud waters from Whiteface Mountain, New York: The role of the Fe(r) (r=II, III) photochemical cycle. *J. Geophys. Res.-Atmos.* **1998**, *103*, 3487–3504.
17. Forni, L.; Bahnemann, D.; Hart, E. J. Mechanism of the Hydroxide Ion Initiated Decomposition of Ozone in Aqueous-Solution. *J. Phys. Chem.* **1982**, *86*, 255–259.
18. Zellner, R.; Exner, M.; Herrmann, H. Absolute Oh Quantum Yields in the Laser Photolysis of Nitrate, Nitrite and Dissolved H2o2 at 308 and 351 Nm in the Temperature-Range 278-353 K. *J. Atmos. Chem.* **1990**, *10*, 411–425.
19. Deguillaume, L.; Leriche, M.; Desboeufs, K.; Mailhot, G.; George, C.; Chaumerliac, N. Transition metals in atmospheric liquid phases: sources, reactivity, and sensitive parameters. *Chem. Rev.* **2005**, *105*, 3388–3431.
20. Tong, H. J.; Arangio, A. M.; Lakey, P. S. J.; Berkemeier, T.; Liu, F. B.; Kampf, C. J.; Brune, W. H.; Poschl, U.; Shiraiwa, M. Hydroxyl radicals from secondary organic aerosol decomposition in water. *Atmos. Chem. Phys.* **2016**, *16*, 1761–1771.
21. Zhao, R.; Lee, A. K. Y.; Wang, C.; Wania, F.; Wong, J. P. S.; Zhou, S.; Abbatt, J. P. D. The Role of Water in Organic Aerosol Multiphase Chemistry: Focus on Partitioning and Reactivity In *Advances in Atmospheric Chemistry*; World Scientific: London (UK), 2016; Vol. 1, pp 95−184.
22. Albinet, A.; Minero, C.; Vione, D. Photochemical generation of reactive species upon irradiation of rainwater: Negligible photoactivity of dissolved organic matter. *Sci. Total Environ.* **2010**, *408*, 3367–3373.
23. Arakaki, T.; Saito, K.; Okada, K.; Nakajima, H.; Hitomi, Y. Contribution of fulvic acid to the photochemical formation of Fe(II) in acidic Suwannee River fulvic acid solutions. *Chemosphere* **2010**, *78*, 1023–1027.
24. Bräuer, P.; Tilgner, A.; Wolke, R.; Herrmann, H. Mechanism development and modelling of tropospheric multiphase halogen chemistry: The CAPRAM Halogen Module 2.0 (HM2). *J. Atmos. Chem.* **2013**, *70*, 19–52.
25. Alfassi, Z. B.; Khaikin, G. I.; Johnson, R. D.; Neta, P. Formylmethyl and formylmethylperoxyl radicals and their chlorinated derivatives. Formation and reactions in irradiated aqueous solutions. *J. Phys. Chem.* **1996**, *100*, 15961–15967.
26. Arakaki, T.; Anastasio, C.; Kuroki, Y.; Nakajima, H.; Okada, K.; Kotani, Y.; Handa, D.; Azechi, S.; Kimura, T.; Tsuhako, A.; Miyagi, Y. A general scavenging rate constant for reaction of hydroxyl radical with organic carbon in atmospheric waters. *Environ. Sci. Technol.* **2013**, *47*, 8196–8203.
27. Kaur, R.; Anastasio, C. Light absorption and the photoformation of hydroxyl radical and singlet oxygen in fog waters. *Atmos. Environ.* **2017**, *164*, 387–397.
28. White, E.; Vaughan, P.; Zepp, R. *Abstracts of papers of the American Chemical Society*, 2003; U833−U833.

29. Badali, K. M.; Zhou, S.; Aljawhary, D.; Antinolo, M.; Chen, W. J.; Lok, A.; Mungall, E.; Wong, J. P. S.; Zhao, R.; Abbatt, J. P. D. Formation of hydroxyl radicals from photolysis of secondary organic aerosol material. *Atmos. Chem. Phys.* **2015**, *15*, 7831–7840.
30. Anastasio, C.; Newberg, J. T. Sources and sinks of hydroxyl radical in sea-salt particles. *J. Geophys. Res.-Atmos.* **2007**, *112*, D10306.
31. Shen, H.; Anastasio, C. Formation of hydroxyl radical from San Joaquin Valley particles extracted in a cell-free surrogate lung fluid. *Atmos. Chem. Phys.* **2011**, *11*, 9671–9682.
32. Shen, H.; Anastasio, C. A Comparison of Hydroxyl Radical and Hydrogen Peroxide Generation in Ambient Particle Extracts and Laboratory Metal Solutions. *Atmos. Environ.* **2012**, *46*, 665–668.
33. Anastasio, C.; Jordan, A. L. Photoformation of hydroxyl radical and hydrogen peroxide in aerosol particles from Alert, Nunavut: implications for aerosol and snowpack chemistry in the Arctic. *Atmos. Environ.* **2004**, *38*, 1153–1166.
34. Arakaki, T.; Kuroki, Y.; Okada, K.; Nakama, Y.; Ikota, H.; Kinjo, M.; Higuchi, T.; Uehara, M.; Tanahara, A. Chemical composition and photochemical formation of hydroxyl radicals in aqueous extracts of aerosol particles collected in Okinawa, Japan. *Atmos. Environ.* **2006**, *40*, 4764–4774.
35. Zhou, X. L.; Davis, A. J.; Kieber, D. J.; Keene, W. C.; Maben, J. R.; Maring, H.; Dahl, E. E.; Izaguirre, M. A.; Sander, R.; Smoydzyn, L. Photochemical production of hydroxyl radical and hydroperoxides in water extracts of nascent marine aerosols produced by bursting bubbles from Sargasso seawater. *Geophys. Res. Lett.* **2008**, *35*, L20803.
36. Kondo, H.; Chiwa, M.; Sakugawa, H. Photochemical formation and scavenging mechanisms of hydroxyl radical in water-extracts of atmospheric aerosol collected in Higashi-Hiroshima, Japan. *Geochemistry* **2009**, *43*, 15–25.
37. Nomi, S. N.; Kondo, H.; Sakugawa, H. Photoformation of OH radical in water-extract of atmospheric aerosols and aqueous solution of water-soluble gases collected in Higashi-Hiroshima, Japan. *Geochem. J.* **2012**, *46*, 21–29.
38. Faust, B. C.; Allen, J. M. Aqueous-Phase Photochemical Formation of Hydroxyl Radical in Authentic Cloudwaters and Fogwaters. *Environ. Sci. Technol.* **1993**, *27*, 1221–1224.
39. Anastasio, C.; McGregor, K. G. Chemistry of fog waters in California's Central Valley: 1. In situ photoformation of hydroxyl radical and singlet molecular oxygen. *Atmos. Environ.* **2001**, *35*, 1079–1089.
40. Bianco, A.; Passananti, M.; Perroux, H.; Voyard, G.; Mouchel-Vallon, C.; Chaumerliac, N.; Mailhot, G.; Deguillaume, L.; Brigante, M. A better understanding of hydroxyl radical photochemical sources in cloud waters collected at the puy de Dome station - experimental versus modelled formation rates. *Atmos. Chem. Phys.* **2015**, *15*, 9191–9202.
41. Warneck, P. Multi-phase chemistry of C2 and C3 organic compounds in the marine atmosphere. *J. Atmos. Chem.* **2005**, *51*, 119–159.

42. Herrmann, H.; Hoffmann, D.; Schaefer, T.; Brauer, P.; Tilgner, A. Tropospheric aqueous-phase free-radical chemistry: radical sources, spectra, reaction kinetics and prediction tools. *ChemPhysChem* **2010**, *11*, 3796–3822.
43. Ervens, B.; Volkamer, R. Glyoxal processing by aerosol multiphase chemistry: towards a kinetic modeling framework of secondary organic aerosol formation in aqueous particles. *Atmos. Chem. Phys.* **2010**, *10*, 8219–8244.
44. Ervens, B.; Sorooshian, A.; Lim, Y. B.; Turpin, B. J. Key parameters controlling OH-initiated formation of secondary organic aerosol in the aqueous phase (aqSOA). *J. Geophys. Res.-Atmos.* **2014**, *119*, 3997–4016.
45. Chameides, W. L.; Davis, D. D. The Free-Radical Chemistry of Cloud Droplets and Its Impact Upon the Composition of Rain. *J. Geophys. Res.-Oceans* **1982**, *87*, 4863–4877.
46. Jacob, D. J. Chemistry of Oh in Remote Clouds and Its Role in the Production of Formic-Acid and Peroxymonosulfate. *J. Geophys. Res.-Atmos.* **1986**, *91*, 9807–9826.
47. Pandis, S. N.; Seinfeld, J. H. Sensitivity Analysis of a Chemical Mechanism for Aqueous-Phase Atmospheric Chemistry. *J. Geophys. Res.-Atmos.* **1989**, *94*, 1105–1126.
48. Lelieveld, J.; Crutzen, P. J. Influences of Cloud Photochemical Processes on Tropospheric Ozone. *Nature* **1990**, *343*, 227–233.
49. Matthijsen, J.; Builtjes, P. J. H.; Sedlak, D. L. Cloud Model Experiments of the Effect of Iron and Copper on Tropospheric Ozone under Marine and Continental Conditions. *Meteorol. Atmos. Phys.* **1995**, *57*, 43–60.
50. Monod, A.; Carlier, P. Impact of clouds on the tropospheric ozone budget: Direct effect of multiphase photochemistry of soluble organic compounds. *Atmos. Environ.* **1999**, *33*, 4431–4446.
51. Warneck, P. The relative importance of various pathways for the oxidation of sulfur dioxide and nitrogen dioxide in sunlit continental fair weather clouds. *Phys. Chem. Chem. Phys.* **1999**, *1*, 5471–5483.
52. Herrmann, H.; Ervens, B.; Jacobi, H. W.; Wolke, R.; Nowacki, P.; Zellner, R. CAPRAM2.3: A chemical aqueous phase radical mechanism for tropospheric chemistry. *J. Atmos. Chem.* **2000**, *36*, 231–284.
53. Warneck, P. In-cloud chemistry opens pathway to the formation of oxalic acid in the marine atmosphere. *Atmos. Environ.* **2003**, *37*, 2423–2427.
54. Barth, M. C.; Sillman, S.; Hudman, R.; Jacobson, M. Z.; Kim, C. H.; Monod, A.; Liang, J. Summary of the cloud chemistry modeling intercomparison: Photochemical box model simulation. *J. Geophys. Res.-Atmos.* **2003**, *108*, 4214.
55. Deguillaume, L.; Leriche, M.; Monod, A.; Chaumerliac, N. The role of transition metal ions on HO x radicals in clouds: a numerical evaluation of its impact on multiphase chemistry. *Atmos. Chem. Phys.* **2004**, *4*, 95–110.
56. Ervens, B.; Feingold, G.; Frost, G. J.; Kreidenweis, S. M. A modeling study of aqueous production of dicarboxylic acids: 1. Chemical pathways and speciated organic mass production. *J. Geophys. Res.-Atmos.* **2004**, *109*, D15205.

57. Lim, H. J.; Carlton, A. G.; Turpin, B. J. Isoprene forms secondary organic aerosol through cloud processing: Model simulations. *Environ. Sci. Technol.* **2005**, *39*, 4441–4446.
58. Deguillaume, L.; Desboeufs, K. V.; Leriche, M.; Long, Y.; Chaumerliac, N. Effect of iron dissolution on cloud chemistry: from laboratory measurements to model results. *Atmos. Pollut. Res.* **2010**, *1*, 220–228.
59. Graedel, T. E.; Goldberg, K. I. Kinetic-Studies of Raindrop Chemistry .1. Inorganic and Organic Processes. *J. Geophys. Res.-Oceans* **1983**, *88*, 865–882.
60. Tilgner, A. Modelling of the physico-chemical multiphase processing of tropospheric aerosols. Ph.D. thesis, University of Leipzig, Leipzig, Germany, 2009.
61. Goldstein, A. H.; Galbally, I. E. Known and unexplored organic constituents in the earth's atmosphere. *Environ. Sci. Technol.* **2007**, *41*, 1514–1521.
62. Mouchel-Vallon, C.; Brauer, P.; Camredon, M.; Valorso, R.; Madronich, S.; Herrmann, H.; Aumont, B. Explicit modeling of volatile organic compounds partitioning in the atmospheric aqueous phase. *Atmos. Chem. Phys.* **2013**, *13*, 1023–1037.
63. Herckes, P.; Valsaraj, K. T.; Collett, J. L. A review of observations of organic matter in fogs and clouds: Origin, processing and fate. *Atmos. Res.* **2013**, *132*, 434–449.
64. Hoffmann, E. H.; Tilgner, A.; Wolke, R.; Boge, O.; Walter, A.; Herrmann, H. Oxidation of substituted aromatic hydrocarbons in the tropospheric aqueous phase: kinetic mechanism development and modelling. *Phys. Chem. Chem. Phys.* **2018**, *20*, 10960–10977.
65. Bianco, A.; Voyard, G.; Deguillaume, L.; Mailhot, G.; Brigante, M. Improving the characterization of dissolved organic carbon in cloud water: Amino acids and their impact on the oxidant capacity. *Sci. Rep.* **2016**, *6*, 37420.
66. Zhao, R.; Lee, A. K. Y.; Soong, R.; Simpson, A. J.; Abbatt, J. P. D. Formation of aqueous-phase alpha-hydroxyhydroperoxides (alpha-HHP): potential atmospheric impacts. *Atmos. Chem. Phys.* **2013**, *13*, 5857–5872.
67. Rusumdar, A. J.; Wolke, R.; Tilgner, A.; Herrmann, H. Treatment of non-ideality in the SPACCIM multiphase model - Part 1: Model development. *Geosci. Model. Dev.* **2016**, *9*, 247–281.
68. Halliwell, B. Oxidative stress and neurodegeneration: where are we now? *J. Neurochem.* **2006**, *97*, 1634–1658.
69. Hayyan, M.; Hashim, M. A.; AlNashef, I. M. Superoxide Ion: Generation and Chemical Implications. *Chem. Rev.* **2016**, *116*, 3029–3085.
70. Liou, G. Y.; Storz, P. Reactive oxygen species in cancer. *Free Radical Res.* **2010**, *44*, 479–496.
71. Turrens, J. F. Reactive Oxygen Species. In *Encyclopedia of Biophysics*; Springer: Berlin, 2013; pp 2198–2200.
72. Brown, D. I.; Griendling, K. K. Regulation of signal transduction by reactive oxygen species in the cardiovascular system. *Circ. Res.* **2015**, *116*, 531–549.
73. Durand, N.; Storz, P. Targeting reactive oxygen species in development and progression of pancreatic cancer. *Expert Rev. Anticanc.* **2017**, *17*, 19–31.

74. Kovac, S.; Dinkova-Kostova, A. T.; Abramov, A. Y. The role of reactive oxygen species in epilepsy. *Reactive Oxygen Species* **2016**, *1*, 38–52.
75. Schumacker, P. T. Reactive oxygen species in cancer: a dance with the devil. *Cancer Cell* **2015**, *27*, 156–157.
76. Krapf, M.; El Haddad, I.; Bruns, E. A.; Molteni, U.; Daellenbach, K. R.; Prevot, A. S. H.; Baltensperger, U.; Dommen, J. Labile Peroxides in Secondary Organic Aerosol. *Chem.* **2016**, *1*, 603–616.
77. Mutzel, A.; Poulain, L.; Berndt, T.; Iinuma, Y.; Rodigast, M.; Boge, O.; Richters, S.; Spindler, G.; Sipila, M.; Jokinen, T.; Kulmala, M.; Herrmann, H. Highly Oxidized Multifunctional Organic Compounds Observed in Tropospheric Particles: A Field and Laboratory Study. *Environ. Sci. Technol.* **2015**, *49*, 7754–7761.
78. Jalan, A.; Alecu, I. M.; Meana-Paneda, R.; Aguilera-Iparraguirre, J.; Yang, K. R.; Merchant, S. S.; Truhlar, D. G.; Green, W. H. New pathways for formation of acids and carbonyl products in low-temperature oxidation: the Korcek decomposition of gamma-ketohydroperoxides. *J. Am. Chem. Soc.* **2013**, *135*, 11100–11114.
79. US-EPA *Estimation Programs Interface Suite™ for Microsoft® Windows, v 4.11*; United States Environmental Protection Agency: Washington, DC, U.S.A., 2012.
80. Schindelka, J.; Böge, O.; Mutzel, A.; Poulain, L.; Herrmann, H. Highly Oxidized Multifunctional Organic Compounds from α-pinene OH oxidation – First insights into phase partitioning parameters and particle phase chemistry; Leibniz Institute for Tropospheric Research (TROPOS), 2018 (unpublished).
81. Ng, N. L.; Brown, S. S.; Archibald, A. T.; Atlas, E.; Cohen, R. C.; Crowley, J. N.; Day, D. A.; Donahue, N. M.; Fry, J. L.; Fuchs, H.; Griffin, R. J.; Guzman, M. I.; Herrmann, H.; Hodzic, A.; Iinuma, Y.; Jimenez, J. L.; Kiendler-Scharr, A.; Lee, B. H.; Luecken, D. J.; Mao, J.; McLaren, R.; Mutzel, A.; Osthoff, H. D.; Ouyang, B.; Picquet-Varrault, B.; Platt, U.; Pye, H. O. T.; Rudich, Y.; Schwantes, R. H.; Shiraiwa, M.; Stutz, J.; Thornton, J. A.; Tilgner, A.; Williams, B. J.; Zaveri, R. A. Nitrate radicals and biogenic volatile organic compounds: oxidation, mechanisms, and organic aerosol. *Atmos. Chem. Phys.* **2017**, *17*, 2103–2162.
82. Whalley, L. K.; Stone, D.; George, I. J.; Mertes, S.; van Pinxteren, D.; Tilgner, A.; Herrmann, H.; Evans, M. J.; Heard, D. E. The influence of clouds on radical concentrations: observations and modelling studies of HO_x during the Hill Cap Cloud Thuringia (HCCT) campaign in 2010. *Atmos. Chem. Phys.* **2015**, *15*, 3289–3301.
83. Arellanes, C.; Paulson, S. E.; Fine, P. M.; Sioutas, C. Exceeding of Henry's law by hydrogen peroxide associated with urban aerosols. *Environ. Sci. Technol.* **2006**, *40*, 4859–4866.
84. Hasson, A. S.; Paulson, S. E. An investigation of the relationship between gas-phase and aerosol-borne hydroperoxides in urban air. *J. Aerosol Sci.* **2003**, *34*, 459–468.

85. Wang, Y.; Arellanes, C.; Curtis, D. B.; Paulson, S. E. Probing the source of hydrogen peroxide associated with coarse mode aerosol particles in southern California. *Environ. Sci. Technol.* **2010**, *44*, 4070–4075.
86. Wang, Y.; Arellanes, C.; Paulson, S. E. Hydrogen Peroxide Associated with Ambient Fine-Mode, Diesel, and Biodiesel Aerosol Particles in Southern California. *Aerosol Sci. Technol.* **2012**, *46*, 394–402.
87. Wang, Y.; Kim, H.; Paulson, S. E. Hydrogen peroxide generation from alpha- and beta-pinene and toluene secondary organic aerosols. *Atmos. Environ.* **2011**, *45*, 3149–3156.
88. Daugherty, E. E.; Gilbert, B.; Nico, P. S.; Borch, T. Complexation and Redox Buffering of Iron(II) by Dissolved Organic Matter. *Environ. Sci. Technol.* **2017**, *51*, 11096–11104.
89. Scheinhardt, S.; Müller, K.; Spindler, G.; Herrmann, H. Complexation of trace metals in size-segregated aerosol particles at nine sites in Germany. *Atmos. Environ.* **2013**, *74*, 102–109.
90. Riva, M. Multiphase Chemistry of Highly Oxidized Molecules: The Case of Organic Hydroperoxides. *Chem* **2016**, *1*, 526–528.
91. Haag, W. R.; Hoigne, J. Singlet oxygen in surface waters. 3. Photochemical formation and steady-state concentrations in various types of waters. *Environ. Sci. Technol.* **1986**, *20*, 341–348.
92. Aregahegn, K. Z.; Noziere, B.; George, C. Organic aerosol formation photo-enhanced by the formation of secondary photosensitizers in aerosols. *Faraday Discuss.* **2013**, *165*, 123–134.
93. Cote, C. D.; Schneider, S. R.; Lyu, M.; Gao, S.; Gan, L.; Holod, A. J.; Chou, T. H. H.; Styler, S. A. Photochemical Production of Singlet Oxygen by Urban Road Dust. *Environ. Sci. Technol. Lett.* **2018**, *5*, 92–97.
94. Kelly, F. J. Oxidative stress: its role in air pollution and adverse health effects. *Occup. Environ. Med.* **2003**, *60*, 612–616.
95. Charrier, J. G.; Anastasio, C. Rates of Hydroxyl Radical Production from Transition Metals and Quinones in a Surrogate Lung Fluid. *Environ. Sci. Technol.* **2015**, *49*, 9317–9325.
96. Charrier, J. G.; McFall, A. S.; Richards-Henderson, N. K.; Anastasio, C. Hydrogen peroxide formation in a surrogate lung fluid by transition metals and quinones present in particulate matter. *Environ. Sci. Technol.* **2014**, *48*, 7010–7017.
97. Poschl, U.; Shiraiwa, M. Multiphase chemistry at the atmosphere-biosphere interface influencing climate and public health in the anthropocene. *Chem. Rev.* **2015**, *115*, 4440–4475.

Chapter 5

Photochemistry in Model Aqueous-Organic Atmospheric Condensed Phases

Tara F. Kahan,* Philip P. A. Malley, Jarod N. Grossman,[1] and Alexa A. Stathis

Department of Chemistry, Syracuse University, Syracuse, New York 13244, United States
***E-mail: tfkahan@syr.edu; phone: (315) 443-3285**
[1]Currently at Agilent Technologies, Inc., Santa Clara, California 95051, United States

Photolysis in condensed phases is an important transformation process for aromatic pollutants such as pesticides and aromatic species associated with fossil fuel combustion. Photolysis kinetics and mechanisms (and therefore ultimate effects on human and environmental health) can depend strongly on the physical and chemical nature of these complex reaction media. While numerous studies have investigated photolysis kinetics and mechanisms in liquid deionized water, little is known about kinetics in other condensed phases such as organic and aqueous-organic aerosols and ice. We have measured photolysis rate constants of aromatic pollutants including substituted benzenes and polycyclic aromatic hydrocarbons (PAHs) in simple models for atmospheric condensed phases. By systematically varying the physical and chemical properties of the reaction media, we have improved our understanding of how factors such as polarity and state of matter affect photolysis kinetics. These results can be incorporated into atmospheric models to improve predictions of pollutant fate and to better understand how the physicochemical properties of atmospheric reaction media affect reactivity.

Introduction

Photochemistry in atmospheric condensed phases has received attention in recent years due to its potential contribution to aerosol mass and to direct and indirect climate effects. For example, there is a growing body of evidence that photochemistry of organic molecules in aerosols and cloud droplets can be an important source of secondary organic aerosol mass (*1*, *2*). Photochemistry at the interface between the gas and condensed phases (heterogeneous photochemistry) has also received attention due to its effects on atmospheric composition and pollutant fate. For example, photochemistry in polar snowpacks initiates catalytic ozone loss in the polar boundary layer (the layer of the atmosphere nearest to the Earth's surface) (e.g. Reference (*3*)). While important, the factors that determine photochemical reaction kinetics of organic species in and at the surface of atmospheric condensed phases remain uncertain. This uncertainty is largely due to the technical challenges associated with measuring reaction kinetics in complex environmental media.

We have measured photolysis rate constants of a range of aromatic pollutants in laboratory-prepared media designed to model physical characteristics of condensed phases such as aerosols, cloud droplets, surface waters, and snowpacks (*4–7*). The analytes chosen include several substituted benzenes and polycyclic aromatic hydrocarbons (PAHs). These pollutants are associated with the extraction and combustion of fossil fuels, and several are EPA priority pollutants, due in part to their toxicity after oxidation in the environment (*8*). These pollutants are often partitioned to atmospheric condensed phases, and multiphase and heterogeneous reactions can determine their environmental fate. By investigating the photolysis kinetics of these pollutants in model atmospheric condensed phases we can better predict their environmental fates and improve our understanding of the role of the physical properties of atmospheric condensed phases on reactivity.

Experimental

We measured photolysis rate constants of several substituted benzenes and PAHs in aqueous solution, organic solvents, and in aqueous-organic mixtures over a range of temperatures above and below water's freezing point (*4–7*). Liquid samples were placed in a quartz cuvette or a flat-bottomed dish for photolysis studies, while ice samples (in the form of either ~20 mm diameter ice cubes or ~2 mm diameter ice granules formed by crushing ice cubes with a mortar and pestle) were placed in a temperature-controlled stainless steel reaction vessel. Kinetics measured in ice cubes are expected to reflect reactivity within bulk ice, while those in ice granules are expected to reflect reactivity at air-ice interfaces (*9*). A xenon arc lamp filtered to remove light at wavelengths shorter than 295 nm was used as the illumination source for these experiments. Each sample was removed from the light after known irradiation times and the analyte concentration was measured by fluorescence spectroscopy. The time-dependent decrease in fluorescence intensity of the analyte was fit to both first- and second-order rate constants for all experiments. Control ("dark") experiments were performed for each study. Complete descriptions of experimental procedures are provided in References (*4–7*).

Discussion

Liquid Phases

Liquid atmospheric condensed phases including aerosols, surface waters, and cloud and fog droplets can contain varying amounts of water and organic matter. Figure 1 demonstrates some possible mixing states for aqueous-organic aerosols and other condensed phases. As shown, particles may be homogeneously mixed or phase-separated.

Figure 1. Representation of aqueous (top left) and organic (top right) aerosols, as well as various possible mixtures: Homogeneous (bottom left), phase-separated core-shell (bottom middle), and phase-separated partially-engulfed (bottom right).

Effects of Non-Chromophoric Organic Matter

Organic molecules are common constituents of many environmental condensed phases. Many organic species can absorb sunlight and participate in photolysis (in ways that both hinder and enhance reactivity of other species). Organic matter that does not absorb sunlight can also affect the photolysis rates of other species. For example, a few studies have reported much more rapid photolysis of aromatic pollutants in aqueous solution than in organic solvents (*4, 10–13*). We measured PAH photolysis kinetics in water and in organic solvents that do not absorb sunlight to determine what factors lead to the different reactivity (*4*). As shown in Figure 2, anthracene and pyrene photolyzed more rapidly in aqueous solution than in organic solvents, and rate constants in the different organic solvents generally increased with increasing solvent polarity. The one exception was that pyrene photolyzed very slowly in aprotic solvents, perhaps suggesting a role for hydrogen bonding in pyrene's photolysis mechanism.

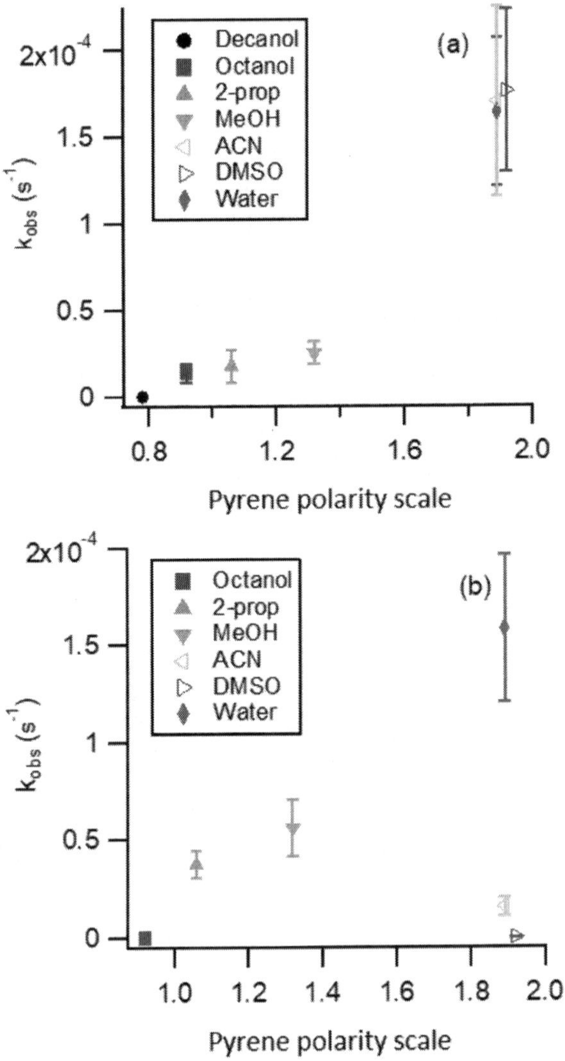

Figure 2. First-order photolysis rate constants of (a) 1.5 × 10^{-7} M anthracene and (b) 3.0 × 10^{-7} M pyrene in water and organic solvents of varying polarity (decanol, octanol, 2-propanol (2-prop), methanol (MeOH), acetonitrile (ACN), and dimethylsulfoxide (DMSO)). The x-axis represents polarity as determined by the pyrene polarity scale; increasing values reflect increasing polarity (14). Rate constants for anthracene in decanol and for pyrene in octanol and DMSO were below our detection limit (~1 × 10^{-5} s^{-1}). Rate constants in the aprotic solvents (ACN and DMSO) are represented by open symbols. Error bars represent the standard deviation about the mean of at least three trials. Reproduced with permission from Reference (4). Copyright 2017 Elsevier.

We investigated the contribution of factors other than polarity to the different rate constants in water and organic solvents. The following factors were determined *not* to significantly affect PAH reactivity under our experimental conditions: different PAH molar absorptivities, different molecular oxygen or singlet oxygen concentrations, and PAH-PAH interactions (e.g. complex formation). These results suggest that polarity will be the major determinant of photochemical reactivity (of at least some aromatic pollutants) in inert aerosol matrices (*4*). To further investigate the role of polarity in atmospheric condensed phases, we measured anthracene and pyrene photolysis kinetics in aqueous solutions containing various concentrations of methanol and DMSO. As shown in Figure 3, both PAHs photolyzed more slowly as methanol concentrations increased. A stronger negative dependence was observed for pyrene photolysis in aqueous DMSO solutions. Anthracene photolysis was insensitive to DMSO concentration, as expected based on the similar kinetics in pure water and pure DMSO (Figure 2) (*4*). In agreement with our observations, anthracene and benzo[*a*]anthracene photolysis rate constants have been reported to depend negatively on acetonitrile mole fraction in aqueous solutions (*12*). These results suggest that we can predict direct photolysis kinetics of PAHs in atmospheric and surface waters as long as the solution polarity – based on the relative fractions of water and organic matter – is known.

Photolysis kinetics in phase-separated mixtures have not been widely studied, despite their prevalence in the environment. We measured anthracene photolysis rate constants in water-octanol mixtures, which are phase-separated at most aqueous-organic ratios (*4*). As shown in Figure 4, when solutions were irradiated without stirring the sample ("stagnant conditions"), anthracene's photolysis rate constant depended positively but non-linearly on water content. When solutions were stirred during irradiation ("turbulent conditions"), the observed rate constant increased rapidly with increasing water content, reaching values similar to those in pure water at aqueous fractions as low as 25% v/v. Anthracene partitions preferentially to the organic phase, which explains the negative deviation from linearity under stagnant conditions. We hypothesize that the much stronger dependence of the rate constant on water content under turbulent conditions is due to more rapid partitioning of anthracene between the two phases; this would rapidly replenish anthracene in the aqueous phase where photolysis is much faster. An important take-away from this result is that atmospheric condensed phases are often not at equilibrium, and that treating all systems as being at equilibrium could result in inaccurate predictions of atmospheric lifetimes (*4*).

Effects of Chromophoric Organic Matter

Many organic species in atmospheric and environmental waters absorb sunlight, and can affect the photolysis of other species by participating in photochemistry. In surface waters, the complex mixture of organic species that absorb sunlight are collectively referred to as chromophoric dissolved organic matter (CDOM), while sunlight-absorbing organic species in atmospheric waters (aerosols and cloud water) are referred to as humic-like substances (HULIS). These compounds can enhance photolysis rates of other species by acting as photosensitizers and generating reactive species such as hydroxyl radicals (OH), singlet oxygen ($^1O_2$), and superoxide (O_2^-). They can also suppress photolysis by blocking available photons (the "inner filter effect") and scavenging reactive species (*15*, *16*).

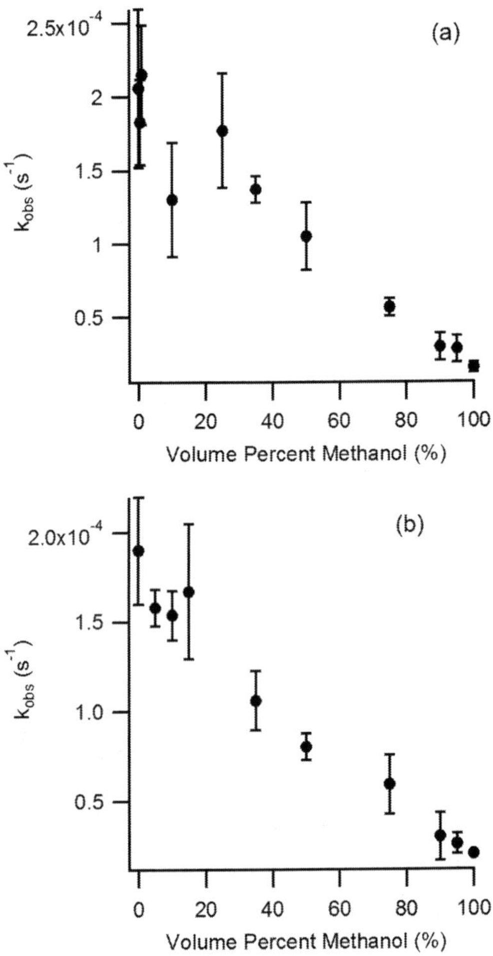

Figure 3. Photolysis rate constants of (a) anthracene and (b) pyrene in aqueous solutions containing methanol. Error bars represent the standard deviation about the mean of three trials. Reproduced with permission from Reference (4). Copyright 2017 Elsevier.

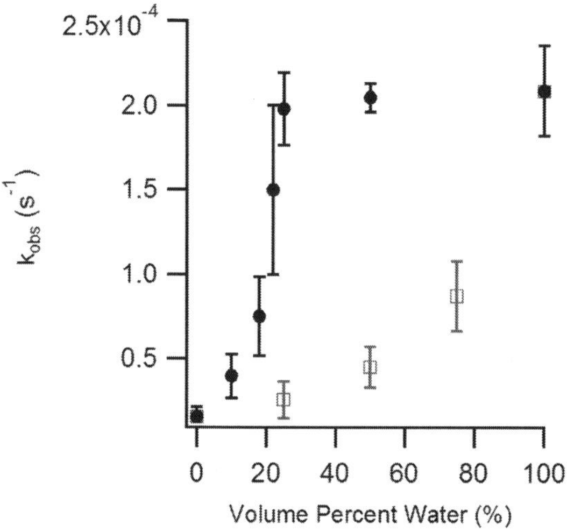

Figure 4. Effect of water on anthracene photolysis kinetics in octanol solutions with constant stirring (solid circles), and stirring during analysis only (open squares). Error bars represent the standard deviation about the mean of three trials. Reproduced with permission from Reference (4). Copyright 2017 Elsevier.

We measured anthracene photolysis kinetics in aqueous solutions containing up to 30 mg L^{-1} CDOM. The photolysis rate constants were the same as those in the absence of CDOM within our experimental uncertainty (*7*). This is consistent with previous reports of small or no effects of CDOM on the photolysis kinetics of numerous PAHs, including anthracene, in aqueous solution (*17*, *18*). Although CDOM and HULIS can increase pollutant photolysis rates via photosensitization and the production of reactive species, this has been suggested to be most important to species that absorb sunlight only weakly (*16*).

Solid Phases

Photolysis in Ice and at Air-Ice Interfaces

Photolysis kinetics in ice and at air-ice interfaces can be very different from those in liquid water (e.g. References (*19*, *20*)). Different reaction kinetics in ice compared to in liquid water can often be explained by enhanced solute concentrations. "Freeze exclusion" or "salting out" occurs when solutes are excluded from the ice matrix during freezing, resulting in regions (at the ice surface and in veins and pockets within the ice bulk) with very high solute concentrations. Freeze exclusion has been familiar to food chemists for approximately 80 years, prompted by observations of greater enzymatic activity in frozen solutions compared to super cooled solutions at the same temperature due to higher local enzyme concentrations (e.g. Reference (*21*)). Many bimolecular reactions of atmospheric importance also occur more rapidly in ice or at ice surfaces due to freeze exclusion. For example, heterogeneous reactions between bromide and gas-phase oxidants such as OH and ozone occur much more rapidly at air-ice interfaces than at air-water interfaces due to enhanced bromide concentrations at the air-ice interface (*22*).

Freeze exclusion is not expected to affect the direct photolysis kinetics of aromatic pollutants in ice, since direct photolysis is a unimolecular process with rate constants that do not depend on reactant concentrations. Indeed, photolysis rate constants of several aromatic species have been reported to be similar in ice and in liquid water, despite the much higher reactant concentrations in the ice samples (*5, 9, 23*). This suggests that direct photolysis kinetics of aromatic pollutants in ice are well described by kinetics in liquid water. However, this only applies to the photolysis of pollutants that are *within* the ice. Direct photolysis of several aromatic species at air-ice interfaces has been reported to occur much more rapidly than photolysis in liquid water or within bulk ice (*5, 24, 25*). This enhanced reactivity at air-ice interfaces is not due to freeze exclusion; rapid photolysis has been reported at ice surfaces when low surface coverages of PAHs were deposited from the gas phase (*24*).

Several possible reasons for enhanced PAH photolysis rates at air-ice interfaces have been investigated, as discussed in References (*7*) and (*24*). Scattering of photons within ice and water particles, which can increase the availability of light (the photon flux), has been suggested to be at least partially responsible for faster photolysis at air-ice interfaces (*26*). In our experiments, photon fluxes were approximately 66% greater in ice granules than in liquid water. Given that anthracene photolyzed ~5× more rapidly at air-ice interfaces than in liquid water, the increased local photon flux was determined to account for no more than 20% of anthracene's enhanced reactivity at air-ice interfaces (*7*).

The majority of the rate enhancement of PAHs observed at air-ice interfaces cannot be explained by increased photon fluxes. Several factors, including temperature, the availability of molecular oxygen, and changes in pH have also been determined not to contribute to this enhancement (*24*). We have recently provided evidence that PAH-PAH interactions at air-ice interfaces may be responsible for faster photolysis there (*7*). It is believed that aromatic species such as PAHs self-associate (aggregate) in island-like formations at air-ice interfaces (e.g. Reference (*19*)). This is often associated with red-shifts in excitation and / or emission spectra (*24, 27–30*). We have shown that anthracene photolysis rate constants increase in liquid water at very high concentrations where spectral evidence of self-association is observed (*7*). Figure 5 shows that anthracene's photolysis rate constant remained invariant at concentrations below ~8 × 10^{-6} M, but that a strong positive concentration dependence was observed at higher concentrations. At the concentrations where more rapid photolysis was observed, anthracene emission spectra displayed evidence of self-association. Further, the reaction order switched from first order to second order at these same concentrations, suggesting that the reaction shifted from unimolecular to bimolecular (*7*). This mechanism may also take place on atmospheric surfaces other than ice. Self-association has been reported for anthracene and other aromatic species on surfaces such as silica, and faster photolysis at these surfaces compared to in aqueous solution has been observed in some cases (*31–33*). It is possible that photolysis of some aromatic pollutants will be faster at the surfaces of a range of atmospheric solids, including various forms of particulate matter.

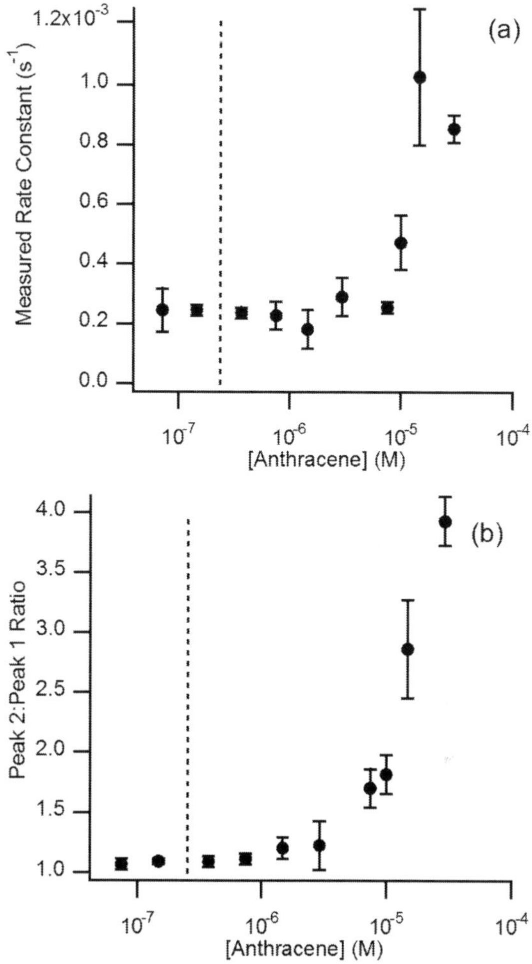

Figure 5. Effect of anthracene concentration on (a) first-order photolysis rate constants, and (b) relative fraction of self-associated molecules in aqueous solutions containing 1% methanol (v/v) (higher Peak 2:Peak 1 ratios indicate more extensive self-association). The vertical dashed traces denote anthracene's saturated concentration in aqueous solution (34). The error bars represent the standard deviation about the mean of at least three trials. Reproduced with permission from Reference (7). Copyright American Chemical Society 2017.

Even more striking than the enhancement of PAH photolysis rates at air-ice interfaces is the photolysis of benzene and substituted benzenes there. Benzene, toluene, ethylbenzene, and xylenes (collectively referred to as BTEX) do not undergo photolysis in liquid water because their absorption spectra do not overlap with sunlight that reaches Earth's surface (λ > 290 nm). They do, however, undergo photolysis at air-ice interfaces (*6*, *28*). As shown in Figure 6, toluene, ethylbenzene, and xylene concentrations decreased only slightly after 50 minutes of irradiation by simulated sunlight. In contrast, concentrations decreased rapidly under irradiation at air-ice interfaces (*6*). Benzene's absorption spectrum has been shown to undergo a large red shift at air-ice interfaces. This enables it to absorb sunlight and undergo photolysis (*28*). Ethylbenzene photolyzed when irradiated with photons at wavelengths as long as 305 nm (*6*). This suggests that, like benzene, photolysis of toluene, ethylbenzene, and xylene at ice surfaces is made possible by red-shifted absorption spectra.

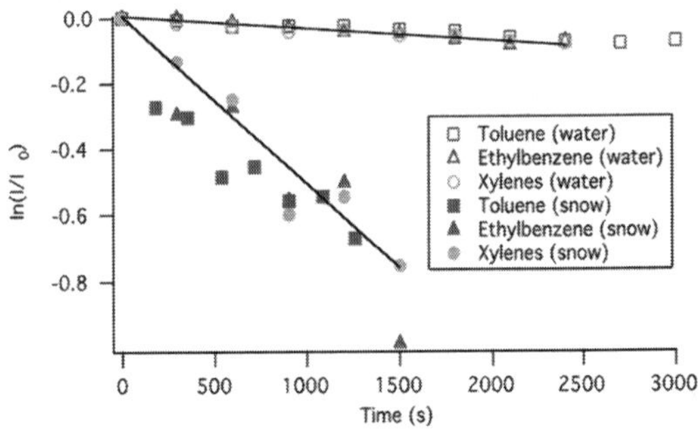

Figure 6. Time-dependent decay of fluorescence intensity of toluene, ethylbenzene, and xylenes in aqueous solution (~24 °C) and in ice granules (−15 °C). Solid traces are fits to the averaged data in each medium. Reproduced with permission from Reference (6). Copyright American Chemical Society 2016.

Some photochemical reactions are bimolecular, and so can be greatly affected by freeze exclusion. For example, "indirect photolysis" occurs when a molecule other than the analyte absorbs a photon and transfers energy or an electron to the analyte (photosensitization) or photolyzes to form a reactive species such as OH or $^1O_2$ that reacts with the analyte. Enhanced local concentrations of reactive species caused by freeze exclusion increase analyte indirect photolysis rates (*35*, *36*). Similarly, the photochemically-initiated reaction between *p*-nitroanisole and pyridine is faster in ice than in liquid water due to enhanced local reactant concentrations (*37*). This effect is expected to be important both at air-ice interfaces and in liquid regions within bulk ice.

Solid and Liquid Aqueous-Organic Mixtures

As discussed above, freeze exclusion can result in high local solute concentrations in frozen solutions. We have reported that octanol and decanol suppress PAH photolysis kinetics in and at the surface of ice (*5*). At ice surfaces, the presence of 7.5 mM octanol or decanol reduced anthracene photolysis rate constants to 50 – 80% of those measured in the absence of organics. Even greater suppression was observed in bulk ice: no anthracene photolysis was observed in the presence of 250 μM octanol, and suppression was observed at octanol concentrations as low as 25 μM. These low concentrations of octanol and decanol had no effect on anthracene photolysis kinetics in liquid aqueous solutions (*5*).

Organic matter and PAHs may be co-excluded during freezing, causing PAHs to reside in environments with characteristics more similar to octanol than to ice. In support of this, fluorescence spectra of the PAH naphthalene in ice granules show less self-association as octanol concentrations increase (Figure 7) (*5*). Further, molecular dynamics simulations suggest that naphthalene will interact with both octanol and ice at octanol-coated ice surfaces, resulting in a less polar environment than that presented by bare ice surfaces (*38*).

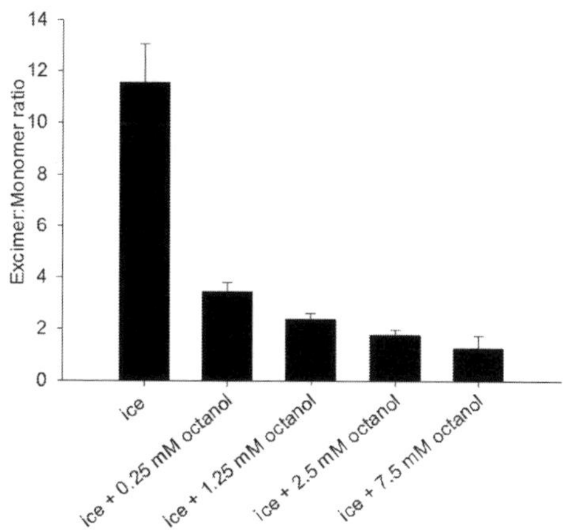

Figure 7. Effects of octanol on the fraction of self-associated naphthalene molecules in ice granules at -15 °C. Larger excimer:monomer ratios indicate more extensive self-association. The error represents the standard deviation about the mean for three samples. Reproduced with permission from Reference (5). Copyright American Chemical Society 2014.

While the physical state of the major component of a solution or particle (in this case water) plays a large role in determining photolysis kinetics, the state of minor (in this case organic) components can also be important. We noted that decanol suppressed anthracene photolysis to a slightly greater extent than octanol in experiments performed at −15 °C. At this temperature, octanol is a liquid (m.p. = −16 °C), while decanol is a solid (m.p. = 6.4 °C). When we reduced the temperature to −25 °C, anthracene photolysis rate constants in pure ice samples and in samples containing decanol were the same as at −15 °C, but photolysis was too slow to detect in samples containing octanol. This dramatic reduction in reactivity appears to be due to the change in octanol's state from liquid to solid (*5*).

Finally, the identity of the organic species is important to reactivity. As discussed above, different organic solutes may exist at different states at tropospherically-relevant temperatures. Chromophoric and non-chromophoric organic matter may also affect reactivity differently in ice. Figure 8 shows the effects of fulvic acid (a common form of CDOM) on anthracene photolysis kinetics in aqueous solution, in bulk ice, and at air-ice interfaces (*7*). As discussed above, CDOM did not affect anthracene photolysis in liquid water. It did, however, suppress photolysis in ice and at ice surfaces. Suppression was observed both when the fulvic acid was dissolved in the sample with anthracene and when an aqueous fulvic acid solution was suspended above the solution. In this second case, CDOM could only affect kinetics by competitively absorbing photons. The observed suppression suggests that the inner filter effect is important in ice, despite being negligible at atmospherically-relevant CDOM concentrations in the liquid phase. The observation of greater suppression when CDOM was incorporated into the ice sample compared to when it was suspended above it suggests the influence of freeze exclusion. The enhanced suppression could be due to either reduced polarity or a stronger inner filter effect (or a combination of the two). Either way, freeze exclusion appears to increase CDOM's suppressing effects on anthracene photolysis (*7*).

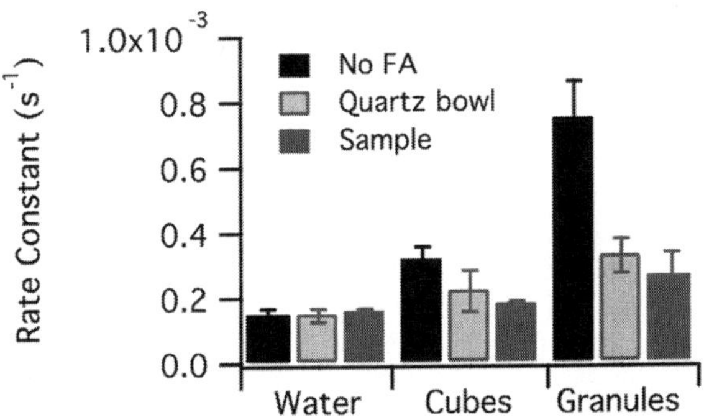

Figure 8. Effects of 5 mg L^{-1} fulvic acid (FA) on anthracene photolysis rate constants in water, ice cubes (reflective of reactions primarily occurring in bulk ice), and ice granules (reflective of reactions primarily occurring at air-ice interfaces). Error bars indicate the standard deviation about the mean of at least three trials. "Quartz bowl" indicates that aqueous FA solutions were suspended above the sample, and "sample" indicates that the FA was incorporated into the anthracene solution. Reproduced with permission from Reference (7). Copyright American Chemical Society 2017.

Conclusions

The fate of aromatic pollutants and other species may strongly depend on the nature of the condensed phase to which they are partitioned. Factors such as polarity and state (solid or liquid) can influence photolysis kinetics in aerosols and other atmospheric condensed phases. In some cases, previously unconsidered reaction mechanisms – such as the photolysis of BTEX at air-ice interfaces – may occur. Understanding the role of individual variables on reactivity in condensed phases will improve predictions of pollutant lifetimes. The factors examined in this work are summarized in Figure 9. As shown in the figure, PAH photolysis in liquid condensed phases will slow down as polarity decreases (for example due to increased organic content). Environmentally-relevant loadings of CDOM will likely have little effect on photolysis kinetics. Pollutant photolysis rate constants in liquid regions within bulk ice are expected to be similar to those in liquid water, but those at air-ice interfaces will be larger due to self-association. Low concentrations of organic matter that do not affect photolysis kinetics in liquid water can suppress PAH photolysis in bulk ice and at ice surfaces due to freeze exclusion.

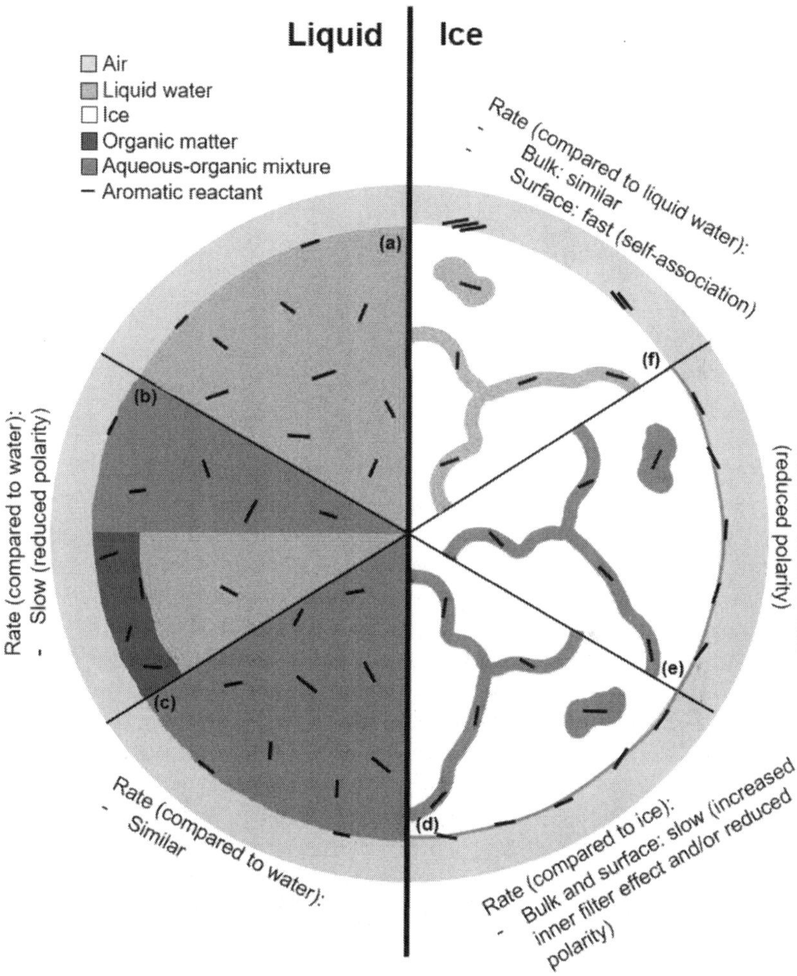

Figure 9. Representation of the reactive environments discussed in this work: (a) liquid water; (b) aqueous solutions containing non-chromophoric organic matter (top: homogeneously mixed, bottom: phase separated); (c) aqueous solutions containing CDOM; (d) ice containing CDOM; (e) ice containing non-chromophoric organic matter; (f) ice.

Incorporating condensed phase reaction media other than liquid water into environmental models could improve predictions of pollutant fate under some conditions. The greatest improvements will be obtained when pollutants are in media in which reactivity differs greatly from that in liquid water. In the context of this work, these are organic phases and air-ice interfaces. Aromatic pollutants in organic phases will photolyze much less rapidly than predicted by models based on reactivity in liquid water. In some cases, photolysis rate constants may be less than 1% of the currently used values (*4*, *10*, *11*). At air-ice interfaces, PAHs may photolyze more than 10× faster than currently predicted (*5*, *24*). Further, photolysis is not currently included as a reaction mechanism for pollutants such as BTEX in environmental condensed phases. Including this novel reaction pathway at air-ice interfaces would greatly alter predictions of their fate in snow-covered regions.

While large differences in predicted pollutant reactivity are expected in the environments discussed above, rate constants measured in liquid water will likely accurately capture reactivity in many other condensed phases. For example, at atmospherically-relevant loadings of water-soluble chromophoric and non-chromophoric organic matter (e.g. in surface waters, aerosols, and fog and cloud water), photolysis kinetics should be well-described by those measured in deionized water (*4*, *7*, *17*, *18*). It should be noted that other aromatic pollutants such as some pesticides and pharmaceuticals have been reported to photolyze more rapidly in the presence of CDOM (*16*). Accounting for the presence of CDOM may be important to accurately predicting the environmental fate of these compounds.

Several aromatic pollutants photolyze much more rapidly at air-ice interfaces than in liquid water (*5*, *6*, *24*, *28*). However, not all aromatic pollutants associated with snow or ice are expected to reside at the air-ice interface. Some will be in liquid regions within the ice bulk, and others will be partitioned to phases that are distinct from ice (e.g. organic phases, particulate matter within the ice, or liquid regions at the air-ice interface formed by solute-induced surface melting). Photolysis kinetics of aromatic pollutants in these environments will be determined by the specific phase to which they are partitioned. Rates may be faster or slower than those in liquid water, depending on the properties of the local environment.

Photolysis rate constants of aromatic pollutants in complex condensed phases are difficult to predict because the partitioning of reactants between different phases is often not well understood. In addition to kinetics studies such as those described in this work, investigations of the partitioning of aromatic pollutants within multi-phase environments such as core-shell aerosols and solute-containing ice are necessary to improve our understanding of pollutant fate.

Acknowledgments

Some of the studies discussed in this work were funded by NSF award 1454959.

References

1. Ervens, B. Modeling the Processing of Aerosol and Trace Gases in Clouds and Fogs. *Chem. Rev.* **2015**, *115*, 4157–4198.

2. McNeill, V. F. Aqueous Organic Chemistry in the Atmosphere: Sources and Chemical Processing of Organic Aerosols. *Environ. Sci. Technol.* **2015**, *49*, 1237–1244.
3. Simpson, W. R.; von Glasow, R.; Riedel, K.; Anderson, P.; Ariya, P.; Bottenheim, J.; Burrows, J.; Carpenter, L. J.; Friess, U.; Goodsite, M. E.; Heard, D.; Hutterli, M.; Jacobi, H. W.; Kaleschke, L.; Neff, B.; Plane, J.; Platt, U.; Richter, A.; Roscoe, H.; Sander, R.; Shepson, P.; Sodeau, J.; Steffen, A.; Wagner, T.; Wolff, E. Halogens and Their Role in Polar Boundary-Layer Ozone Depletion. *Atmos. Chem. Phys.* **2007**, *7*, 4375–4418.
4. Grossman, J. N.; Stern, A. P.; Kirich, M. L.; Kahan, T. F. Anthracene and Pyrene Photolysis Kinetics in Aqueous, Organic, and Mixed Aqueous-Organic Phases. *Atmos. Environ.* **2016**, *128*, 158–164.
5. Malley, P. P. A.; Kahan, T. F. Nonchromophoric Organic Matter Suppresses Polycyclic Aromatic Hydrocarbon Photolysis in Ice and at Ice Surfaces. *J. Phys. Chem. A* **2014**, *118*, 1638–1643.
6. Stathis, A. A.; Hendrickson-Stives, A. K.; Kahan, T. F. Photolysis Kinetics of Toluene, Ethylbenzene, and Xylenes at Ice Surfaces. *J. Phys. Chem. A* **2016**, *120*, 6693–6697.
7. Malley, P. P. A.; Grossman, J. N.; Kahan, T. F. Effects of Chromophoric Dissolved Organic Matter on Anthracene Photolysis Kinetics in Aqueous Solution and Ice. *J. Phys. Chem. A* **2017**, *121*, 7619–7626.
8. United States Environmental Protection Agency. Initial List of Hazardous Air Pollutants with Modifications. https://www.epa.gov/haps/initial-list-hazardous-air-pollutants-modifications (acessed May 14, 2018).
9. Kahan, T. F.; Zhao, R.; Jumaa, K. B.; Donaldson, D. J. Anthracene Photolysis in Aqueous Solution and Ice: Photon Flux Dependence and Comparison of Kinetics in Bulk Ice and at the Air-Ice Interface. *Environ. Sci. Technol.* **2010**, *44*, 1302–1306.
10. Zepp, R. G.; Schlotzhauer, P. F. Photoreactivity of Selected Aromatic Hydrocarbons in Water. In *Polynuclear Aromatic Hydrocarbons; Third International Symposium on Chemistry and Biology Carcinogenesis and Mutagenesis*; Jones, P. W., Leber, P., Eds.; Ann Arbor Science Publishers, Inc.: Ann Arbor, MI, 1979; pp 141−158.
11. Donaldson, D. J.; Kahan, T. F.; Kwamena, N.-O. A.; Handley, S. R.; Barbier, C., Atmospheric Chemistry of Urban Surface Films. In *Atmospheric Aerosols: Characterization, Chemistry, Modeling, and Climate*; Valsaraj, K. T., Kommalapati, R. R., Eds.; American Chemical Society: Washington, DC, USA, 2009; pp 79−89.
12. Librando, V.; Bracchitta, G.; de Guidi, G.; Minniti, Z.; Perrini, G.; Catalfo, A. Photodegradation of Anthracene and Benzo a Anthracene in Polar and Apolar Media: New Pathways of Photodegradation. *Polycycl. Aromat. Comp.* **2014**, *34*, 263–279.
13. Fioressi, S.; Arce, R. Photochemical Transformations of Benzo[E]Pyrene in Solution and Adsorbed on Silica Gel and Alumina Surfaces. *Environ. Sci. Technol.* **2005**, *39*, 3646–3655.

14. Dong, D. C.; Winnik, M. A. The Py Scale of Solvent Polarities - Solvent Effects on the Vibronic Fine-Structure of Pyrene Fluorescence and Empirical Correlations with Et-Value and Y-Value. *Photochem. Photobiol.* **1982**, *35*, 17–21.
15. Hassett, J. P. Chemistry - Dissolved Natural Organic Matter as a Microreactor. *Science* **2006**, *311*, 1723–1724.
16. Lam, M. W.; Tantuco, K.; Mabury, S. A. Photofate: A New Approach in Accounting for the Contribution of Indirect Photolysis of Pesticides and Pharmaceuticals in Surface Waters. *Environ. Sci. Technol.* **2003**, *37*, 899–907.
17. Fasnacht, M. P.; Blough, N. V. Aqueous Photodegradation of Polycyclic Aromatic Hydrocarbons. *Environ. Sci. Technol.* **2002**, *36*, 4364–4369.
18. Bertilsson, S.; Widenfalk, A. Photochemical Degradation of Pahs in Freshwaters and Their Impact on Bacterial Growth - Influence of Water Chemistry. *Hydrobiologia* **2002**, *469*, 23–32.
19. Kahan, T. F.; Wren, S. N.; Donaldson, D. J. A Pinch of Salt Is All It Takes: Chemistry at the Frozen Water Surface. *Acc. Chem. Res.* **2014**, *47*, 1587–1594.
20. Bartels-Rausch, T.; Jacobi, H. W.; Kahan, T. F.; Thomas, J. L.; Thomson, E. S.; Abbatt, J. P. D.; Ammann, M.; Blackford, J. R.; Bluhm, H.; Boxe, C.; Domine, F.; Frey, M. M.; Gladich, I.; Guzman, M. I.; Heger, D.; Huthwelker, T.; Klan, P.; Kuhs, W. F.; Kuo, M. H.; Maus, S.; Moussa, S. G.; McNeill, V. F.; Newberg, J. T.; Pettersson, J. B. C.; Roeselova, M.; Sodeau, J. R. A Review of Air-Ice Chemical and Physical Interactions (AICI): Liquids, Quasi-Liquids, and Solids in Snow. *Atmos. Chem. Phys.* **2014**, *14*, 1587–1633.
21. Pincock, R. E. Reactions in Frozen Systems. *Accounts Chem. Res.* **1969**, *2*, 97–103.
22. Wren, S. N.; Kahan, T. F.; Jumaa, K. B.; Donaldson, D. J. Spectroscopic Studies of the Heterogeneous Reaction between O-3(G) and Halides at the Surface of Frozen Salt Solutions. *J. Geophys. Res.-Atmos.* **2010**, *115*, 8.
23. Ram, K.; Anastasio, C. Photochemistry of Phenanthrene, Pyrene, and Fluoranthene in Ice and Snow. *Atmos. Environ.* **2009**, *43*, 2252–2259.
24. Kahan, T. F.; Donaldson, D. J. Photolysis of Polycyclic Aromatic Hydrocarbons on Water and Ice Surfaces. *J. Phys. Chem. A* **2007**, *111*, 1277–1285.
25. Kahan, T. F.; Kwamena, N. O. A.; Donaldson, D. J. Different Photolysis Kinetics at the Surface of Frozen Freshwater Vs. Frozen Salt Solutions. *Atmos. Chem. Phys.* **2010**, *10*, 10917–10922.
26. McFall, A. S.; Anastasio, C. Photon Flux Dependence on Solute Environment in Water Ices. *Environ. Chem.* **2016**, *13*, 682–687.
27. Kahan, T. F.; Donaldson, D. J. Heterogeneous Ozonation Kinetics of Phenanthrene at the Air-Ice Interface. *Environ. Res. Lett.* **2008**, *3*, 045006.
28. Kahan, T. F.; Donaldson, D. J. Benzene Photolysis on Ice: Implications for the Fate of Organic Contaminants in the Winter. *Environ. Sci. Technol.* **2010**, *44*, 3819–3824.

29. Kahan, T. F.; Reid, J. P.; Donaldson, D. J. Spectroscopic Probes of the Quasi-Liquid Layer on Ice. *J. Phys. Chem. A* **2007**, *111*, 11006–11012.
30. Kahan, T. F.; Zhao, R.; Donaldson, D. J. Hydroxyl Radical Reactivity at the Air-Ice Interface. *Atmos. Chem. Phys.* **2010**, *10*, 843–854.
31. Barbas, J. T.; Sigman, M. E.; Dabestani, R. Photochemical Oxidation of Phenanthrene Sorbed on Silica Gel. *Environ. Sci. Technol.* **1996**, *30*, 1776–1780.
32. Dabestani, R.; Ellis, K. J.; Sigman, M. E. Photodecomposition of Anthracene on Dry Surfaces: Products and Mechanism. *J. Photochem. Photobiol., A* **1995**, *86*, 231–239.
33. Wilkinson, F.; Worrall, D. R.; Williams, S. L. Primary Photochemical Processes of Anthracene Adsorbed on Silica Gel. *J. Phys. Chem.* **1995**, *99*, 6689–6696.
34. Pearlman, R. S.; Yalkowsky, S. H.; Banerjee, S. Water Solubilities of Polynuclear Aromatic and Heteroaromatic Compounds. *J. Phys. Chem. Ref. Data* **1984**, *13*, 555–562.
35. Bower, J. P.; Anastasio, C. Measuring a 10,000-Fold Enhancement of Singlet Molecular Oxygen (O-1*(2)) Concentration on Illuminated Ice Relative to the Corresponding Liquid Solution. *Atmos. Environ.* **2013**, *75*, 188–195.
36. Bower, J. P.; Anastasio, C. Degradation of Organic Pollutants in/on Snow and Ice by Singlet Molecular Oxygen (O-1(2)*) and an Organic Triplet Excited State. *Environ. Sci.: Processes Impacts* **2014**, *16*, 748–756.
37. Grannas, A. M.; Bausch, A. R.; Mahanna, K. M. Enhanced Aqueous Photochemical Reaction Rates after Freezing. *J. Phys. Chem. A* **2007**, *111*, 11043–11049.
38. Liyana-Arachchi, T. P.; Valsaraj, K. T.; Hung, F. R. Adsorption of Naphthalene and Ozone on Atmospheric Air/Ice Interfaces Coated with Surfactants: A Molecular Simulation Study. *J. Phys. Chem. A* **2012**, *116*, 2519–2528.

Chapter 6

Organic Nitrates and Secondary Organic Aerosol (SOA) Formation from Oxidation of Biogenic Volatile Organic Compounds

M. Takeuchi[1] and N. L. Ng[2,3,*]

[1]School of Civil and Environmental Engineering, Georgia Institute of Technology, 790 Atlantic Dr. NW, Atlanta, Georgia 30332, United States
[2]School of Chemical and Biomolecular Engineering, Georgia Institute of Technology, 311 Ferst Dr. NW, Atlanta, Georgia 30332, United States
[3]School of Earth and Atmospheric Sciences, Georgia Institute of Technology, 311 Ferst Dr. NW, Atlanta, Georgia 30332, United States
[*]E-mail: ng@chbe.gatech.edu

Organic nitrates (ON) originating from biogenic volatile organic compounds (BVOC) are ubiquitous in the atmosphere and play an important role in NO_x recycling, ozone, and secondary organic aerosol (SOA) formation, though a fundamental understanding on their formation and fates remains largely unexplored. This chapter reviews the current knowledge of BVOC-derived ON chemistry. Topics range from ON formation mechanisms, yields, fates, measurement techniques, implication on SOA formation, field observations, and recommendations for future research.

Introduction

Organic aerosols (OA) constitute a substantial fraction of fine particulate matter ($PM_{2.5}$) in the atmosphere and have important impacts on climate, visibility, and human health (*1*, *2*). Primary OA (POA) are directly emitted into the atmosphere as PM; secondary OA (SOA) are formed in the atmosphere from the oxidation of gas-phase compounds followed by gas-particle partitioning. Ambient field studies have found that SOA contribute a significant fraction of OA and often dominate over POA even in urban areas (*3*–*7*). A detailed understanding of the formation and evolution of SOA is vital to assess the effects of PM on climate and health.

Owing to their high emissions and high reactivity with the major atmospheric oxidants, such as ozone, hydroxyl radical (OH) and nitrate radical (NO_3), the oxidation of biogenic volatile organic compounds (BVOC) emitted by vegetation, such as isoprene (C_5H_8), monoterpenes ($C_{10}H_{16}$), and sesquiterpenes ($C_{15}H_{24}$), is a dominant contributor to global SOA budget (*1*). In recent years, it has been increasingly recognized that anthropogenic emissions play a critical role in biogenic SOA formation in ambient environments (*8–10*). A key mechanism that couples anthropogenic emissions with biogenic emissions is the effect of nitrogen oxides (NO_x = NO + NO_2) on biogenic SOA production. In clean or moderately polluted environments, higher NO_x will result in more oxidant formation which can enhance SOA production from BVOC. In the presence of NO_x, OH oxidation and NO_3 oxidation of BVOC can lead to the formation of organic nitrates (ON), a major component of reactive oxidized nitrogen. ON formation represents a large instantaneous sink of NO_x that effectively couples the HO_x and NO_x cycles and impacts ozone and SOA production. The formation of ON removes NO_x from the atmosphere and directly suppresses ozone production near source regions (*11*). Transportation and subsequent chemical reactions of ON can release NO_x and promote ozone formation in locations far away from the initial NO_x emissions. Owing to their semi-volatile/low-volatility nature, ON can also undergo gas-particle partitioning and contribute to SOA. Results from ambient field measurements have revealed that particulate ON contribute a large fraction of OA at multiple sites around the world (*12*). The ubiquitous presence of ON highlights the importance of understanding their formation and fates to accurately evaluate their roles in NO_x recycling, ozone, and SOA production.

This chapter serves an overview of recently published reviews by Perring et al. (*11*), Ng et al. (*12*), and Wennberg et al. (*13*) and specifically reviews ON chemistry from oxidation of BVOC, with a focus on the role of ON in SOA formation. We refer the readers to the aforementioned review articles for comprehensive discussions related to this chemistry. Here, we outline the formation and fates of ON, methods for measuring gas-phase and particle-phase ON, the relationships between ON and SOA production, ambient observations of particulate ON, and conclude with recommendations for future research.

Sources of Organic Nitrates: Formation Mechanism and Yield

Atmospheric ON are typically formed via secondary reactions in the ambient air as direct emissions account for a minor fraction. There are two major gas-phase sources of ON from the oxidations of BVOC: OH oxidation of BVOC in the presence of nitric oxide (NO) and NO_3 oxidation of BVOC.

OH Oxidation in the Presence of NO

OH oxidation of BVOC is one of the major oxidation pathways in the ambient air during daytime. In general, OH initiates the chain of radical reactions by either abstracting a hydrogen atom or attacking a double bond (if present). Owing to the olefinic nature of most BVOC, the majority of OH oxidation proceeds with the attack of OH on a double bond. The intermediate radical immediately reacts with an oxygen molecule to form a peroxy radical (RO_2) which can follow various chemical pathways depending on the conditions of the ambient environment. In the presence of higher levels of NO_x, such as in urban areas, RO_2 preferentially reacts with NO, leading to the formation of hydroxy ON or decomposed products (alkoxy radical and NO_2), as shown in Figure 1. Although the formation of ON is a minor channel of the two, the relative importance of the ON channel varies depending on the identity of the parent BVOC, temperature, and pressure (*11*, *14*).

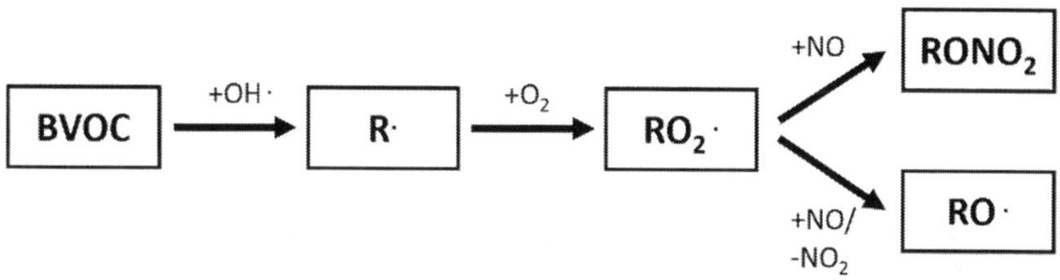

Figure 1. Generic formation mechanism of ON from OH oxidation of BVOC in the presence of NO.

The detailed and simplified formation mechanisms of major BVOC-derived ON from OH oxidation in the presence of NO have been reported in a number of prior studies (*15*, *16*). Shown in Figure 2 is the schematic of the simplified ON formation mechanism from OH oxidation of isoprene in the presence of NO that is currently implemented in the global chemical transport model (GEOS-Chem) (*17*). For more detailed isoprene oxidation chemistry, readers are directed to the recently published review by Wennberg et al. (*13*).

In many studies, the yield of ON is defined as the amount of either gaseous or total ON produced per parent VOC reacted on a molar basis and, thus, the ON yield can be also considered as the branching ratio of the RO_2+NO reaction leading to the formation of ON, as shown in Figure 1. While there has been an extensive number of studies on isoprene hydroxy nitrate yields reporting a range of 0.04-0.15, Wennberg et al. (*13*) discussed that the reported lower yields could be inaccurate due to differences in experimental conditions (i.e., pressure) and issues in instrument sensitivities, providing a recommended yield of 0.13. Nonetheless, due to the high sensitivity of ozone to changes in the isoprene ON yield (*18*), further investigations are needed to reduce uncertainty in the isoprene ON yield. On the other hand, studies on other BVOC are scarce or not in agreement with each other when available (*11*, *13*). For instance, the measured gaseous ON yields from the OH oxidation of α-pinene greatly vary, ranging from 0.01 to 0.26 (*19–21*). To our best knowledge, there are no experimental data on the ON yields from OH oxidation of other monoterpenes or sesquiterpenes to date.

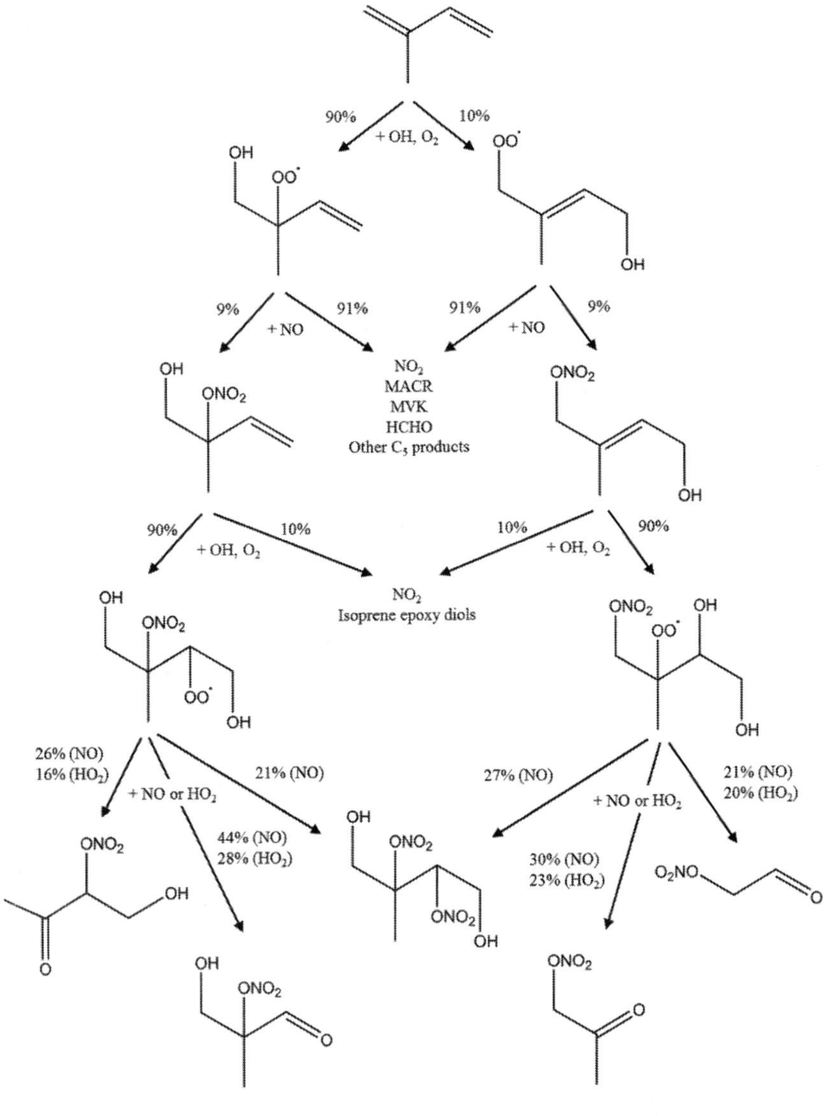

Figure 2. Schematic of simplified ON formation mechanism from isoprene OH oxidation in the presence of NO. (Reproduced with permission from reference (17). Copyright 2016, the authors.)

NO_3 Oxidation

The initial oxidation of BVOC at night can be quite distinct from that of daytime. With the absence of solar radiation at night, the concentration of NO_3 is sufficient to be one of the dominant oxidation pathways along with the oxidation by ozone. Different from OH, NO_3 has low H abstraction efficiency such that NO_3 oxidation dominantly proceeds by attacking a double bond. Due to ubiquitous presence of double bonds in BVOC, NO_3 oxidation is therefore considered an important oxidation pathway at night.

Detailed formation mechanisms of major BVOC-derived ON from NO_3 oxidation have been reported only for a few BVOC (*12, 22–24*). Shown in Figure 3 are the schematics of the isoprene and β-pinene ON formation mechanisms. As evident from Figure 3, a nitrate functional group is formed in the initial oxidation step and it either remains or leaves depending on the fate of RO_2 and further chemistry.

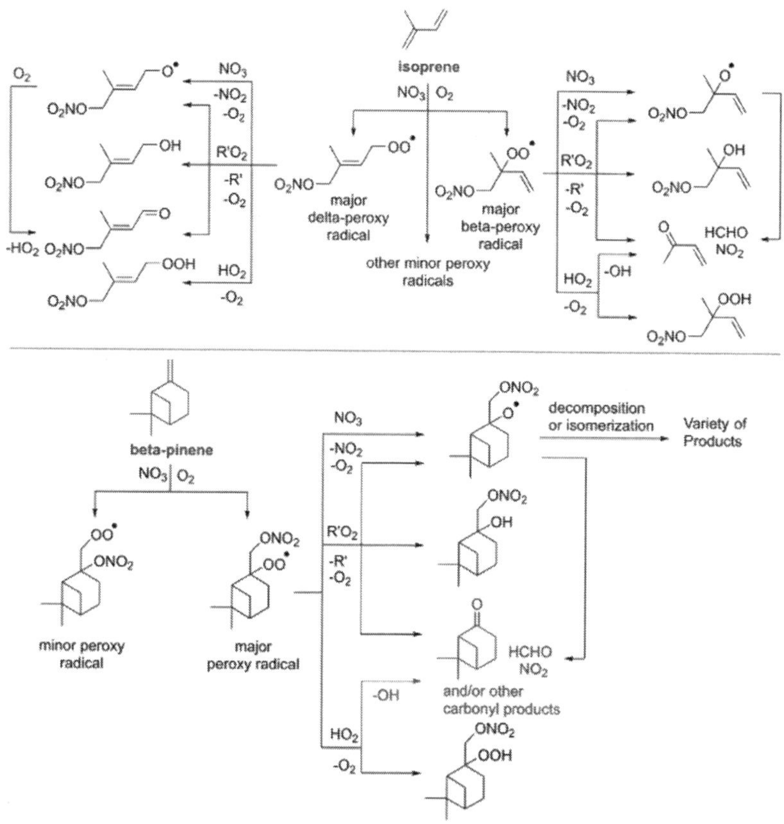

Figure 3. Schematics of ON formation mechanisms from isoprene and β-pinene NO_3 oxidation. (Adapted with permission from reference (12). Copyright 2017, the authors.)

The higher yield of ON with NO_3 oxidation (0.62-0.8; Table 1) than with OH oxidation in the presence of NO (0.13) is reported for isoprene owing to the direct addition of a nitrate functional group to a double bond in the initial step (*25*). The ON yields from NO_3 oxidation for other BVOC are also high (except for α-pinene) (*12*). Table 1 summarizes the ON yields as well as other important results from NO_3 oxidation of various BVOC (*12*).

Table 1. ON and SOA mass yields from NO_3 oxidation of various BVOC

BVOC	*ON yield*	*SOA mass yield*
isoprene	0.62-0.80	0.02-0.24
α-pinene	0.10-0.29	0-0.16
β-pinene	0.22-0.74	0.07-1.04
Δ-carene	0.68-0.77	0.12-0.65
d-limonene	0.30-0.72	0.14-1.74
β-caryophyllene	N/A	0.86-1.46

SOURCE: Reproduced with permission from reference (*12*). Copyright 2017, the authors.

Ozonolysis is another important oxidation process in the atmosphere. Studies on the ON formation and yields from ozonolysis of BVOC are even more scarce due in part to the difficulty in controlling the fate of RO_2 in laboratory experiments because NO can be quickly consumed by ozone. To our knowledge, there do not exist studies on ON yields from ozonolysis of BVOC. A recent study investigated the production of ON from α-pinene ozonolysis in the presence of NO_x but concluded that the majority of observed ON stemmed from RO_2 formed by α-pinene reacting with secondary OH generated from ozonolysis (*26*). In the atmosphere, since ozone and NO coexist in non-negligible amounts, further studies are required to explore the potential formation of ON from BVOC ozonolysis.

Fates of Organic Nitrates

Owing to their semi-/low-volatile nature, BVOC-derived ON exist in both gas and particle phases in the atmosphere. The loss mechanisms vary from further oxidation, photolysis, particle-phase chemical processing, deposition, and thermal loss. Figure 4 illustrates the various fates of atmospheric ON. The following two sub-sections will discuss major fates of BVOC-derived ON in gas and particle phases.

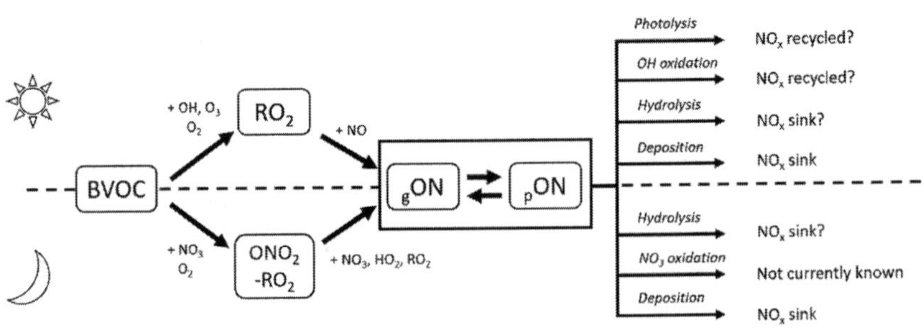

Figure 4. Potentially important fates of atmospheric ON derived from BVOC.

Fates of Gas-Phase Organic Nitrates

Gaseous ON can undergo a variety of chemical and physical processes: partitioning to particle phase, dissolution in aerosol water, further oxidation by oxidants (such as OH and NO_3), photolysis reaction, and deposition to the ground. The availability of synthetic standards has allowed for measurements of rate constants for the reactions of OH, ozone, and NO_3 with various isoprene hydroxy nitrates, carbonyl nitrates, and hydroperoxy nitrates. In general, the reaction rate constants are on the order of 1×10^{-11}, 1×10^{-14}, and 1×10^{-19} cm^3 $molecule^{-1}$ s^{-1}, for OH, ozone, and NO_3 reactions, respectively (*13*). Recent studies shed important insights on the role of photolysis as a sink of gaseous ON. Carbonyl ON derived from isoprene is found to undergo much faster photolysis than previously expected (*27*, *28*). However, the dependence on the structure and proximity of other functional groups to a nitrate group warrants further studies. In terms of deposition, rapid deposition fluxes of both isoprene nitrates and monoterpene nitrates have been observed (*29*). The photolysis and photochemical oxidation of ON formed from monoterpenes and sesquiterpenes are largely unknown.

Fates of Particle-Phase Organic Nitrates

As with gaseous ON, particulate ON can undergo a variety of chemical and physical processes: evaporation due to dilution and/or temperature change, aqueous reactions, such as hydrolysis, and deposition to the ground.

Hydrolysis in aerosol water has been proposed as a dominant particle-phase ON fate from field and laboratory studies (*30*, *31*). To date, there have been several laboratory experiments exploring the rate of ON hydrolysis from BVOC oxidation systems. Representative α-pinene derived ON via OH oxidation in the presence of NO has been shown to undergo fast hydrolysis on the order of minutes to hours depending on the aerosol pH (*32*). Another study has also found that the decay rate of α-pinene derived ON via OH oxidation in the presence of NO, attributed to hydrolysis, is as fast as 3.4 h^{-1} (*33*). On the other hand, little hydrolysis has been observed for ON formed via NO_3 oxidation of β-pinene and limonene (*23*, *34*). The difference in overall hydrolysis rates of particulate ON is attributed to the structural difference of ON formed via OH oxidation and NO_3 oxidation. Previous studies using bulk solutions reported the dependence of ON hydrolysis rate on the site to which a nitrate group is attached (*35*, *36*). Tertiary ON, in which a nitrate functional group is attached to a tertiary carbon, undergoes fast hydrolysis, while primary ON does not. However, unsaturated ON, such as first-generation isoprene hydroxy nitrates, are found to undergo hydrolysis on the order of minutes to hours regardless of the site of a nitrate functional group due to allylic stabilization of the cationic intermediates (*37*). Thus, the representativeness of results from these recent studies for other monoterpene and sesquiterpene systems requires further work.

A striking difference in the behaviors of α- and β-pinene derived particulate ON upon photochemical aging (i.e., photolysis + OH oxidation) has been observed, as shown in Figure 5 (*34*). While α-pinene derived particulate ON from NO_3 oxidation evaporated upon photochemical aging, β-pinene derived particle-phase ON from NO_3 oxidation did not undergo such changes. The distinct behaviors are attributed to the difference in ON products formed from the different BVOC precursors, though the effect of OH oxidation and photolysis cannot be separated. Another study observed similar degradation of α-pinene derived particulate ON from NO_3 oxidation upon photochemical aging in the aqueous solution (*38*). However, the relative importance of photolysis and OH aging in releasing NO_2 remains unknown.

Online Measurement Methods for Organic Nitrates

An endeavor to develop analytical instruments capable of detecting and quantifying gaseous and particulate ON had commenced in order to resolve the unanswered question of "missing NO_y". This missing NO_y problem had been observed during multiple field measurements in which the total NO_y concentration, measured as NO via catalytic conversion of all the NO_y species to NO, always appeared greater than the sum of individually measured NO_y species, such as NO, NO_2, HNO_3, HONO, particulate nitrate, peroxyacetyl nitrate (PAN) (*39–41*). A wide range of analytical instruments has been applied in field measurements and laboratory experiments. Several decades of work by various research groups have enabled detection and quantification of gaseous and particulate ON, though analytical challenges still remain to date. Here, the measurement approaches are divided into two major categories: gas-phase ON and particle-phase ON measurements. A comprehensive discussion of the current available ON measurement techniques has been presented in a recent review on NO_3 oxidation of BVOC by Ng et al. (*12*)

Gas-Phase ON Measurement

Early attempts to identify and quantify individual gaseous ON relied on chromatographic techniques, such as gas chromatography (GC) and liquid chromatography (LC), coupled with various detectors, such as electron capture detection and electron impact mass spectrometry. Although these techniques allow both detecting individual gaseous ON and quantifying them, the major disadvantage is in the measurement of multi-functional ON, which are traditionally difficult to efficiently sample and separate. Mass spectrometry (MS) without an initial separation stage, including chemical ionization MS (CIMS) and proton transfer reaction MS (PTR-MS), provides capabilities of measuring multi-functional ON, though the challenge of quantification of such individual ON remains (*42*, *43*).

Another distinct approach is to measure the sum of all ON without speciated information utilizing the thermal properties of nitrate functional groups attached to organic compounds. Initiated by the development and successful implementation of this technique by thermal dissociation laser induced fluorescence (TD-LIF) (*44*), the TD inlet has been coupled to other NO_2 detectors, such as cavity ring-down spectroscopy (TD-CRDS) (*45*) and cavity attenuated phase shift spectroscopy (TD-CAPS) (*46*).

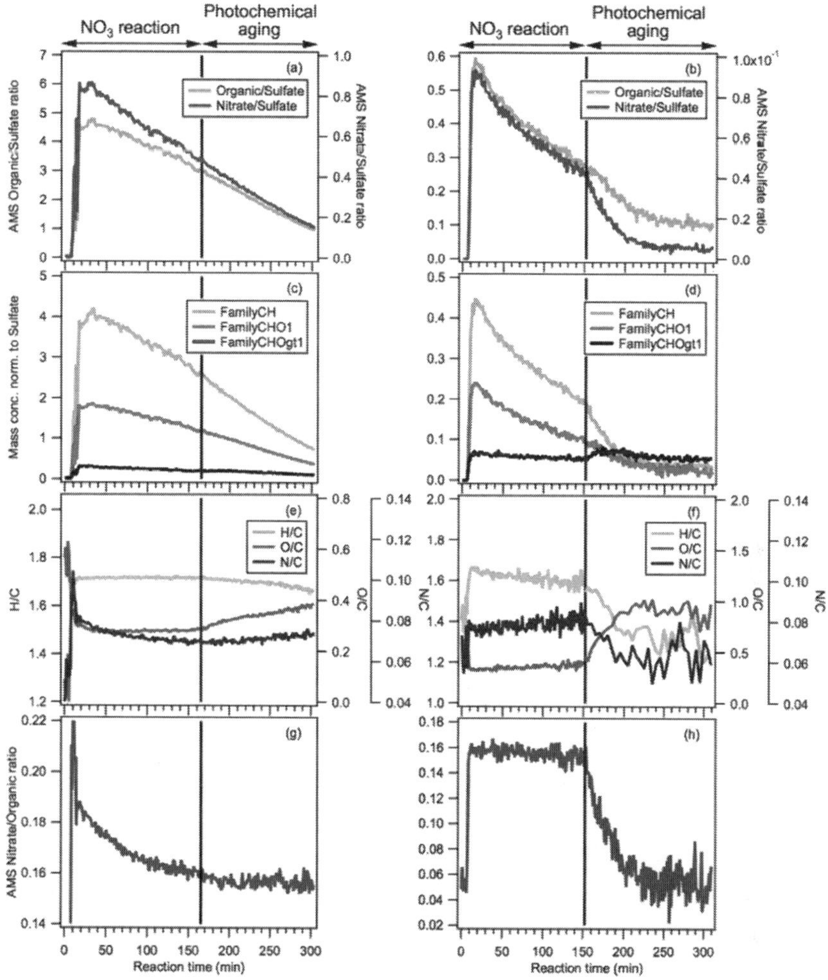

Figure 5. High resolution time-of-flight aerosol mass spectrometer (HR-ToF-AMS) measurements of the β-pinene (panels a, c, e, and g) and α-pinene (panels b, d, f, and h) reactions. Panels (a) and (b): Time series of AMS organic and nitrate mass concentrations normalized to the sulfate mass concentration of the SOA. Panels (c) and (d): Time series of major AMS organic families (CH, CHO1, and CHOgt1) normalized to the sulfate mass concentration of the SOA. Panels (e) and (f): Time series of H/C, O/C, and N/C ratios of the SOA. Panels (g) and (h): Time series of AMS nitrate mass concentration normalized to the organic mass concentration of the SOA. (Adapted with permission from reference (34). Copyright 2016 American Chemical Society.)

Particle-Phase ON Measurement

Some approaches presented in the gas-phase ON measurement section above are also capable of measuring particulate ON with modifications in the inlet. TD-LIF equipped with an activated charcoal denuder that selectively scrubs gas-phase ON has been demonstrated to measure particulate ON (*47*, *48*). A special filter inlet to collect and subsequently desorb aerosols has been coupled to high resolution time-of-flight CIMS (FIGAERO-HR-ToF-CIMS) and has demonstrated specifically the capability of measuring particle-phase ON in field and laboratory studies (*34*, *43*).

Aside from the aforementioned instruments, the high-resolution time-of-flight aerosol mass spectrometer (HR-ToF-AMS) that has been extensively utilized for field and laboratory studies is capable of estimating ON concentration based on three approaches (*49*, *50*). First, owing to significant fragmentations of analytes due to flash evaporation and hard ionization, nitrate groups are mainly detected as NO^+ or NO_2^+. The ratio of the two ions in the HR-ToF-AMS mass spectra varies depending on the origin of nitrates (i.e., inorganic or organic nitrates), which allows apportioning the amount of nitrate signals to inorganic and organic origins (*51*). Second, by co-locating an instrument that is capable of measuring inorganic nitrate, such as an ion chromatograph, one can also estimate the concentration of particulate ON by subtracting the inorganic nitrate concentration from the total nitrate concentration measured by HR-ToF-AMS. Finally, one can conduct positive matrix factorization (PMF) analysis on organic mass spectra together with NO^+ and NO_2^+ ions to obtain insights into the relative contribution of organic and inorganic nitrates.

Organic Nitrate and Secondary Organic Aerosol (SOA) Formation

OH Oxidation in the Presence of NO

The presence and abundance of NO not only affect the yields of ON but also influence the properties and yields of bulk SOA. Traditionally, the effect of NO_x level on the SOA yield was studied as a binary approach, namely, by comparing high and low NO_x conditions (where RO_2+NO and RO_2+HO_2 reactions, respectively, dominate). There is a general understanding that OH oxidation under high-NO_x condition produces lower SOA yields for isoprene and monoterpenes, while the opposite has been observed for sesquiterpenes (*52*, *53*). For isoprene, multiple studies have demonstrated the non-linear nature of NO effect on SOA yields (*52*, *54–59*). Shown in Figure 6 is the dependence of SOA yield on the NO level parametrized in terms of a ratio of NO to isoprene. The non-linear NO effect is typically attributed to difference in RO_2 fates or oxidant levels (*52*). At a very low and high concentration of NO, the fate of RO_2 is dominated by the reaction with HO_2 and NO, respectively, whereas RO_2 chemistry can be rather complex in a condition where RO_2+HO_2 and RO_2+NO compete with each other. Also, the level of NO_x affects the OH concentration in a non-linear manner owing to the close coupling of HO_x and NO_x cycling. At a low NO_x level, HO_2 preferentially undergoes a radical termination process by reacting with RO_2 or HO_2, leading to less OH production via the reaction HO_2 with NO. As the NO_x level increases, the HO_2+NO reaction becomes more important, leading to a higher concentration of OH. However, at a high NO_x level, the formation of HNO_3 via the reaction of OH with NO_2 accelerates such that the loss of OH becomes faster than its production, leading to a low concentration of OH. A recent study suggested that the non-linear effects of NO_x on the SOA yield mainly stems from the variation in OH concentration (*60*). Further studies are needed to systematically investigate the role of NO_x on SOA formation both in terms of the impact on RO_2 fates as well as on the level of oxidants.

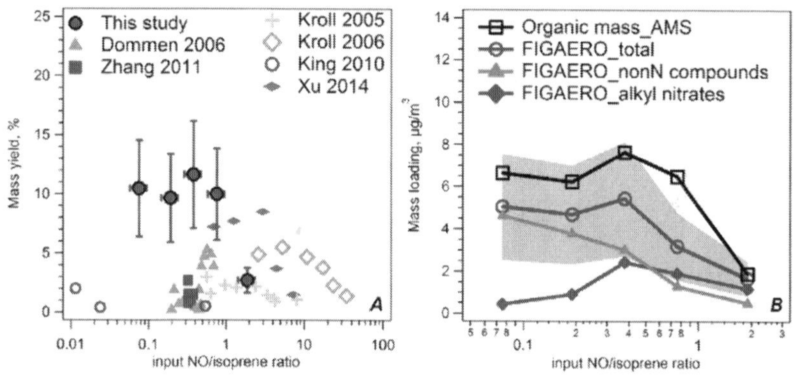

Figure 6. (A) Effect of NO_x level on SOA yield from isoprene OH oxidation experiments. (B) Approximate concentrations of the sum of non-N-containing C5 compounds and of alkyl nitrate compounds plotted with the sum of both quantities and the AMS-measured total OA mass concentration. (Reproduced with permission from reference (59). Copyright 2016 American Chemical Society.)

Volatility of bulk SOA, one of the important physical properties of SOA, can also be influenced by the level of NO. It has been shown that isoprene SOA formed in the presence of NO is less volatile than that formed in the absence of NO in laboratory chamber experiments up to certain NO level (*58, 61*). As with the SOA yield, the bulk SOA property also behaves in a non-linear manner in response to the level of NO as shown in Figure 7.

While the particulate ON and total SOA masses formed from the OH oxidation of isoprene also showed a non-linear trend with increasing NO, the fraction of particulate ON to total SOA mass has been found to monotonically increase as the level of NO increases, though the rate of increase varies for different NO levels (*61*). In other words, even though there is less total SOA mass at a high NO level, more of the mass is in the form of ON. It is interesting to observe such dependence on NO and indicates the importance of considering the yield of total ON not just as a function of parent BVOC identity but also as a function of both BVOC concentration and NO level.

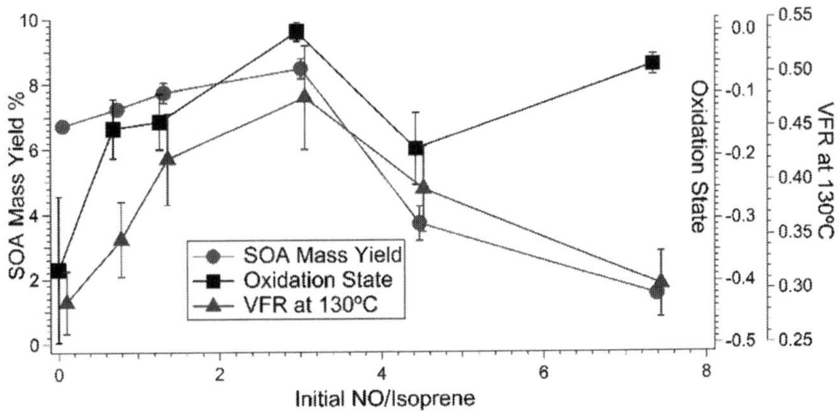

Figure 7. Non-linear behavior of the bulk SOA volatility expressed as volume fraction remaining (VFR) formed via isoprene OH oxidation at varying NO levels. (Reproduced with permission from reference (58). Copyright 2014 American Chemical Society.)

NO_3 Oxidation of BVOC

Recent studies have revealed high SOA formation efficiency from NO_3 oxidation of BVOC. Table 1 summarizes the SOA mass yield from various BVOC in NO_3 oxidation. Except for α-pinene, other monoterpenes and sesquiterpenes have high SOA mass yield up to 1.74, highlighting the importance of NO_3 oxidation of BVOC not only in the formation of ON but also in the bulk SOA (*12*, *62*). Note that SOA mass yields above 1 are possible because of the addition of oxygen and nitrogen to the parent BVOC precursors.

Figure 8 illustrates the SOA mass yield as a function of organic mass loading for NO_3 oxidation of β-pinene (*23*). The SOA mass yield remains high even at a low organic mass loading, implying the low-volatility nature of SOA formed via NO_3 oxidation of β-pinene. The same study also investigated the effect of relative humidity (RH) and RO_2 fates on the SOA yields from NO_3 oxidation of β-pinene but found no significant effects.

Field Observations of Particulate Organic Nitrates

Results from various field studies have demonstrated that particulate ON are ubiquitous across the U.S. and Europe and contribute to a substantial fraction of submicron organic aerosols by up to 77%, as shown in Figure 9 (*12*). Online, advanced analytical techniques in measuring particulate ON, such as TD-LIF, HR-ToF-AMS, and FIGAERO-CIMS, have provided new insights into the importance of NO_3 oxidation of BVOC, whereas the co-location of such instruments capable of independently measuring ON confirmed the substantial contribution of ON aerosols.

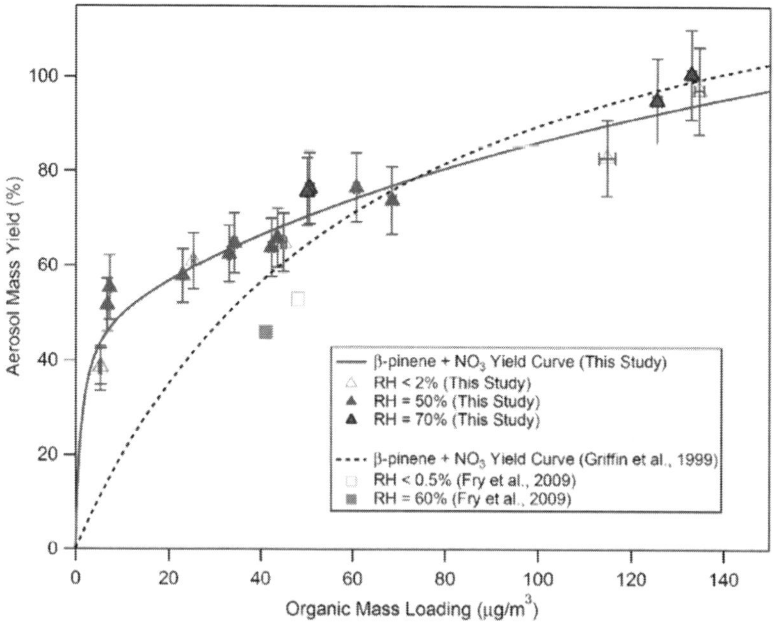

Figure 8. SOA mass yield as a function of organic mass loading for NO_3 oxidation of β-pinene. (Reproduced with permission from reference (23). Copyright 2015, the authors.)

A prominent example of successful ON measurements is the Southern Oxidant and Aerosol Study (SOAS) that took place in the southeastern U.S. in 2013. A total of five independent approaches to quantify particulate ON from TD-LIF, HR-ToF-AMS, and FIGAERO-CIMS have revealed the important contribution of NO_3 oxidation of monoterpenes to aerosol ON (*43*, *49*). As with the SOAS campaign, many field studies conducted in BVOC-rich environments have observed the particulate ON peaking after midnight, confirming the important role of monoterpene-derived ON via nighttime chemistry (*63*).

Spatially extensive AMS data sets from two field measurements across various countries in Europe (i.e., EUCAARI and EMEP) also provided evidence of substantial particle-phase ON formation via oxidations of BVOC (*50*). High concentrations of particulate ON are usually observed in areas that are in proximity to NO_x emission sources during the night, while remote regions with less anthropogenic impacts during daytime have experienced low levels of particulate ON.

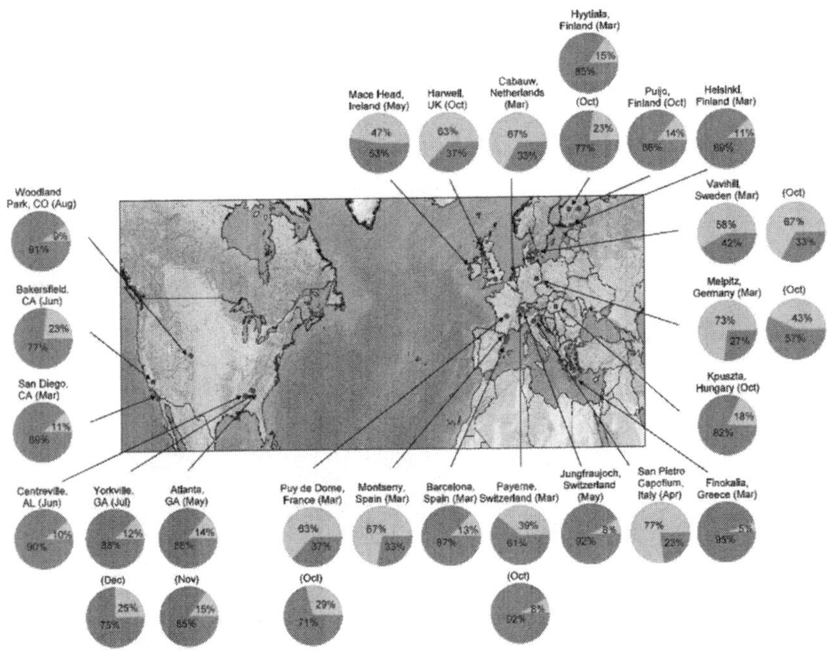

Figure 9. Mass-based fraction of particle-phase ON to submicron total organic aerosols, indicated by light blue (print in color) or light gray (print in grayscale) in pie charts. (Adapted with permission from reference (12). Copyright 2017, the authors.)

Future Research Needs

Recent advances in measurements of gaseous and particulate ON have demonstrated their ubiquitous presence in the atmosphere. One of the largest uncertainties in our understanding of ON chemistry is the extent to which ON act as a permanent versus temporary sink of NO_x. This will depend on their formation yields and fates as they can either retain or release NO_x upon further reactions. Such knowledge is imperative in improving regional and global model simulations of NO_x budget, ozone, and SOA formation.

The synthesis of ON formed from first-generation isoprene chemistry has led to substantial advancement in the quantification of their formation yields and gas-phase oxidation chemistry. Nevertheless, more studies are needed to further constrain the ON yields from isoprene chemistry. Much less is known regarding organic nitrogen chemistry of monoterpenes and sesquiterpenes. There is a critical need to measure the formation yields of ON from monoterpenes and sesquiterpenes oxidations under different reaction conditions, including the dependence on NO_x levels (RO_2 fates), temperature, and pressure. The photolysis and photochemical oxidation of ON formed from BVOC are poorly understood. The rates and products of these reactions for larger multifunctional ON (containing more than five carbon atoms) and whether NO_2 is released in photochemical aging of gas and particulate ON are generally not known. Specifically, whether

ON act as a permanent or temporary sink for NO_x can be highly dependent on the BVOC precursor. It is imperative to not only conduct experiments to investigate ON formation from different BVOC oxidation pathways, but also how gas and particulate ON evolve over their lifetimes and whether they return or retain NO_x upon photochemical aging by photolysis and/or OH reactions. Beyond gas-phase chemistry, future research needs to investigate the loss ON via interactions with aqueous aerosols and cloud droplets. The solubility of multifunctional ON and the extent and rate to which hydrolysis proceeds in aerosol water warrant future studies. The effect of particle acidity on ON hydrolysis should also be considered.

Owing to a lack of fundamental laboratory data, monoterpene and sesquiterpene organic nitrogen chemistry are largely incomplete and missing in current atmospheric models. Future laboratory and field studies and advancement in measurement techniques will allow for better constraints on the formation and fates of ON. A fundamental and quantitative understanding of the formation mechanisms, yields, gas-particle partitioning, and fates of ON from BVOC oxidations is critical to accurately predict their impacts on NO_x cycling, oxidation capacity, and ozone formation as NO_x emissions continue to decrease throughout the U.S.

Acknowledgments

The authors acknowledged support from US Environmental Protection Agency STAR RD-83540301 and National Science Foundation AGS-1555034 (CAREER). This publication's contents are solely the responsibility of the grantee and do not necessarily represent the official views of the US EPA. Further, US EPA does not endorse the purchase of any commercial products or services mentioned in the publication.

References

1. Kanakidou, M.; Seinfeld, J. H.; Pandis, S. N.; Barnes, I.; Dentener, F. J.; Facchini, M. C.; Van Dingenen, R.; Ervens, B.; Nenes, A.; Nielsen, C. J.; Swietlicki, E.; Putaud, J. P.; Balkanski, Y.; Fuzzi, S.; Horth, J.; Moortgat, G. K.; Winterhalter, R.; Myhre, C. E. L.; Tsigaridis, K.; Vignati, E.; Stephanou, E. G.; Wilson, J. Organic aerosol and global climate modelling: a review. *Atmos. Chem. Phys.* **2005**, *5*, 1053–1123.
2. Lim, S. S.; Vos, T.; Flaxman, A. D. A comparative risk assessment of burden of disease and injury attributable to 67 risk factors and risk factor clusters in 21 regions, 1990-2010: a systematic analysis for the Global Burden of Disease Study 2010 (vol 380, pg 2224, 2012). *Lancet* **2013**, *381*, 1276–1276.
3. Zhang, Q.; Alfarra, M. R.; Worsnop, D. R.; Allan, J. D.; Coe, H.; Canagaratna, M. R.; Jimenez, J. L. Deconvolution and quantification of hydrocarbon-like and oxygenated organic aerosols based on aerosol mass spectrometry. *Environ. Sci. Technol.* **2005**, *39*, 4938–4952.
4. de Gouw, J. A.; Middlebrook, A. M.; Warneke, C.; Goldan, P. D.; Kuster, W. C.; Roberts, J. M.; Fehsenfeld, F. C.; Worsnop, D. R.; Canagaratna, M. R.;

Pszenny, A. A. P.; Keene, W. C.; Marchewka, M.; Bertman, S. B.; Bates, T. S. Budget of organic carbon in a polluted atmosphere: Results from the New England Air Quality Study in 2002. *J. Geophys. Res.-Atmos.* **2005**, *110*, D16305.

5. Volkamer, R.; Jimenez, J. L.; San Martini, F.; Dzepina, K.; Zhang, Q.; Salcedo, D.; Molina, L. T.; Worsnop, D. R.; Molina, M. J. Secondary organic aerosol formation from anthropogenic air pollution: Rapid and higher than expected. *Geophys. Res. Lett.* **2006**, *33*, L17811.
6. Jimenez, J. L.; Canagaratna, M. R.; Donahue, N. M.; Prevot, A. S. H.; Zhang, Q.; Kroll, J. H.; DeCarlo, P. F.; Allan, J. D.; Coe, H.; Ng, N. L.; Aiken, A. C.; Docherty, K. S.; Ulbrich, I. M.; Grieshop, A. P.; Robinson, A. L.; Duplissy, J.; Smith, J. D.; Wilson, K. R.; Lanz, V. A.; Hueglin, C.; Sun, Y. L.; Tian, J.; Laaksonen, A.; Raatikainen, T.; Rautiainen, J.; Vaattovaara, P.; Ehn, M.; Kulmala, M.; Tomlinson, J. M.; Collins, D. R.; Cubison, M. J.; Dunlea, E. J.; Huffman, J. A.; Onasch, T. B.; Alfarra, M. R.; Williams, P. I.; Bower, K.; Kondo, Y.; Schneider, J.; Drewnick, F.; Borrmann, S.; Weimer, S.; Demerjian, K.; Salcedo, D.; Cottrell, L.; Griffin, R.; Takami, A.; Miyoshi, T.; Hatakeyama, S.; Shimono, A.; Sun, J. Y.; Zhang, Y. M.; Dzepina, K.; Kimmel, J. R.; Sueper, D.; Jayne, J. T.; Herndon, S. C.; Trimborn, A. M.; Williams, L. R.; Wood, E. C.; Middlebrook, A. M.; Kolb, C. E.; Baltensperger, U.; Worsnop, D. R. Evolution of Organic Aerosols in the Atmosphere. *Science* **2009**, *326*, 1525–1529.
7. Ng, N. L.; Canagaratna, M. R.; Zhang, Q.; Jimenez, J. L.; Tian, J.; Ulbrich, I. M.; Kroll, J. H.; Docherty, K. S.; Chhabra, P. S.; Bahreini, R.; Murphy, S. M.; Seinfeld, J. H.; Donahue, N. M.; Hildebrandt, L.; Decarlo, P. F.; Lanz, V. A.; Prevot, A. S. H.; Dinar, E.; Rudich, Y.; Worsnop, D. R. Organic aerosol components observed in Northern Hemispheric datasets from Aerosol Mass Spectrometry. *Atmos. Chem. Phys.* **2010**, *10*, 4625–4641.
8. Weber, R. J.; Sullivan, A. P.; Peltier, R. E.; Russell, A.; Yan, B.; Zheng, M.; de Gouw, J.; Warneke, C.; Brock, C.; Holloway, J. S.; Atlas, E. L.; Edgerton, E. A study of secondary organic aerosol formation in the anthropogenic-influenced southeastern United States. *J. Geophys. Res.-Atmos.* **2007**, *112*, D13302.
9. Carlton, A. G.; Pinder, R. W.; Bhave, P. V.; Pouliot, G. A. To What Extent Can Biogenic SOA be Controlled? *Environ. Sci. Technol.* **2010**, *44*, 3376–3380.
10. Xu, L.; Guo, H. Y.; Boyd, C. M.; Klein, M.; Bougiatioti, A.; Cerully, K. M.; Hite, J. R.; Isaacman-VanWertz, G.; Kreisberg, N. M.; Knote, C.; Olson, K.; Koss, A.; Goldstein, A. H.; Hering, S. V.; de Gouw, J.; Baumann, K.; Lee, S. H.; Nenes, A.; Weber, R. J.; Ng, N. L. Effects of anthropogenic emissions on aerosol formation from isoprene and monoterpenes in the southeastern United States. *Proc. Natl. Acad. Sci. U. S. A.* **2015**, *112*, 37–42.
11. Perring, A. E.; Pusede, S. E.; Cohen, R. C. An Observational Perspective on the Atmospheric Impacts of Alkyl and Multifunctional Nitrates on Ozone and Secondary Organic Aerosol. *Chem. Rev.* **2013**, *113*, 5848–5870.
12. Ng, N. L.; Brown, S. S.; Archibald, A. T.; Atlas, E.; Cohen, R. C.; Crowley, J. N.; Day, D. A.; Donahue, N. M.; Fry, J. L.; Fuchs, H.; Griffin, R. J.; Guzman, M. I.; Herrmann, H.; Hodzic, A.; Iinuma, Y.; Jimenez, J. L.;

Kiendler-Scharr, A.; Lee, B. H.; Luecken, D. J.; Mao, J. Q.; McLaren, R.; Mutzel, A.; Osthoff, H. D.; Ouyang, B.; Picquet-Varrault, B.; Platt, U.; Pye, H. O. T.; Rudich, Y.; Schwantes, R. H.; Shiraiwa, M.; Stutz, J.; Thornton, J. A.; Tilgner, A.; Williams, B. J.; Zaveri, R. A. Nitrate radicals and biogenic volatile organic compounds: oxidation, mechanisms, and organic aerosol. *Atmos. Chem. Phys.* **2017**, *17*, 2103–2162.

13. Wennberg, P. O.; Bates, K. H.; Crounse, J. D.; Dodson, L. G.; McVay, R. C.; Mertens, L. A.; Nguyen, T. B.; Praske, E.; Schwantes, R. H.; Smarte, M. D.; St Clair, J. M.; Teng, A. P.; Zhang, X.; Seinfeld, J. H. Gas-Phase Reactions of Isoprene and Its Major Oxidation Products. *Chem. Rev.* **2018**, *118*, 3337–3390.
14. Atkinson, R.; Carter, W. P. L.; Winer, A. M. Effects of Temperature and Pressure on Alkyl Nitrate Yields in the Nox Photooxidations of Normal-Pentane and Normal-Heptane. *J. Phys. Chem.* **1983**, *87*, 2012–2018.
15. Eddingsaas, N. C.; Loza, C. L.; Yee, L. D.; Seinfeld, J. H.; Wennberg, P. O. alpha-pinene photooxidation under controlled chemical conditions - Part 1: Gas-phase composition in low- and high-NOx environments. *Atmos. Chem. Phys.* **2012**, *12*, 6489–6504.
16. Paulot, F.; Crounse, J. D.; Kjaergaard, H. G.; Kroll, J. H.; Seinfeld, J. H.; Wennberg, P. O. Isoprene photooxidation: new insights into the production of acids and organic nitrates. *Atmos. Chem. Phys.* **2009**, *9*, 1479–1501.
17. Fisher, J. A.; Jacob, D. J.; Travis, K. R.; Kim, P. S.; Marais, E. A.; Miller, C. C.; Yu, K. R.; Zhu, L.; Yantosca, R. M.; Sulprizio, M. P.; Mao, J. Q.; Wennberg, P. O.; Crounse, J. D.; Teng, A. P.; Nguyen, T. B.; St Clair, J. M.; Cohen, R. C.; Romer, P.; Nault, B. A.; Wooldridge, P. J.; Jimenez, J. L.; Campuzano-Jost, P.; Day, D. A.; Hu, W. W.; Shepson, P. B.; Xiong, F. L. Z.; Blake, D. R.; Goldstein, A. H.; Misztal, P. K.; Hanisco, T. F.; Wolfe, G. M.; Ryerson, T. B.; Wisthaler, A.; Mikoviny, T. Organic nitrate chemistry and its implications for nitrogen budgets in an isoprene- and monoterpene-rich atmosphere: constraints from aircraft (SEAC(4)RS) and ground-based (SOAS) observations in the Southeast US. *Atmos. Chem. Phys.* **2016**, *16*, 5969–5991.
18. Paulot, F.; Henze, D. K.; Wennberg, P. O. Impact of the isoprene photochemical cascade on tropical ozone. *Atmos. Chem. Phys.* **2012**, *12*, 1307–1325.
19. Aschmann, S. M.; Atkinson, R.; Arey, J. Products of reaction of OH radicals with alpha-pinene. *J. Geophys. Res.-Atmos.* **2002**, *107*, ACH 6-1–ACH 6-7.
20. Noziere, B.; Barnes, I.; Becker, K. H. Product study and mechanisms of the reactions of alpha-pinene and of pinonaldehyde with OH radicals. *J. Geophys. Res.-Atmos.* **1999**, *104*, 23645–23656.
21. Rindelaub, J. D.; McAvey, K. M.; Shepson, P. B. The photochemical production of organic nitrates from alpha-pinene and loss via acid-dependent particle phase hydrolysis. *Atmos. Environ.* **2015**, *100*, 193–201.
22. Schwantes, R. H.; Teng, A. P.; Nguyen, T. B.; Coggon, M. M.; Crounse, J. D.; St Clair, J. M.; Zhang, X.; Schilling, K. A.; Seinfeld, J. H.; Wennberg, P. O. Isoprene NO3 Oxidation Products from the RO2 + HO2 Pathway. *J. Phys. Chem. A* **2015**, *119*, 10158–10171.

23. Boyd, C. M.; Sanchez, J.; Xu, L.; Eugene, A. J.; Nah, T.; Tuet, W. Y.; Guzman, M. I.; Ng, N. L. Secondary organic aerosol formation from the beta-pinene+NO3 system: effect of humidity and peroxy radical fate. *Atmos. Chem. Phys.* **2015**, *15*, 7497–7522.
24. Claflin, M. S.; Ziemann, P. J. Identification and Quantitation of Aerosol Products of the Reaction of beta-Pinene with NO3 Radicals and Implications for Gas- and Particle-Phase Reaction Mechanisms. *J. Phys. Chem. A* **2018**, *122*, 3640–3652.
25. Rollins, A. W.; Kiendler-Scharr, A.; Fry, J. L.; Brauers, T.; Brown, S. S.; Dorn, H. P.; Dube, W. P.; Fuchs, H.; Mensah, A.; Mentel, T. F.; Rohrer, F.; Tillmann, R.; Wegener, R.; Wooldridge, P. J.; Cohen, R. C. Isoprene oxidation by nitrate radical: alkyl nitrate and secondary organic aerosol yields. *Atmos. Chem. Phys.* **2009**, *9*, 6685–6703.
26. Berkemeier, T.; Ammann, M.; Mentel, T. F.; Poschl, U.; Shiraiwa, M. Organic Nitrate Contribution to New Particle Formation and Growth in Secondary Organic Aerosols from alpha-Pinene Ozonolysis. *Environ. Sci. Technol.* **2016**, *50*, 6334–6342.
27. Suarez-Bertoa, R.; Picquet-Varrault, B.; Tamas, W.; Pangui, E.; Doussin, J. F. Atmospheric Fate of a Series of Carbonyl Nitrates: Photolysis Frequencies and OH-Oxidation Rate Constants. *Environ. Sci. Technol.* **2012**, *46*, 12502–12509.
28. Muller, J. F.; Peeters, J.; Stavrakou, T. Fast photolysis of carbonyl nitrates from isoprene. *Atmos. Chem. Phys.* **2014**, *14*, 2497–2508.
29. Nguyen, T. B.; Crounse, J. D.; Teng, A. P.; St. Clair, J. M.; Paulot, F.; Wolfe, G. M.; Wennberg, P. O. Rapid deposition of oxidized biogenic compounds to a temperate forest. *Proc. Natl. Acad. Sci. U.S.A.* **2015**, *112*, E392–E401.
30. Russell, L. M.; Bahadur, R.; Ziemann, P. J. Identifying organic aerosol sources by comparing functional group composition in chamber and atmospheric particles. *Proc. Natl. Acad. Sci. U.S.A.* **2011**, *108*, 3516–3521.
31. Day, D. A.; Liu, S.; Russell, L. M.; Ziemann, P. J. Organonitrate group concentrations in submicron particles with high nitrate and organic fractions in coastal southern California. *Atmos. Environ.* **2010**, *44*, 1970–1979.
32. Rindelaub, J. D.; Borca, C. H.; Hostetler, M. A.; Slade, J. H.; Lipton, M. A.; Slipchenko, L. V.; Shepson, P. B. The acid-catalyzed hydrolysis of an alpha-pinene-derived organic nitrate: kinetics, products, reaction mechanisms, and atmospheric impact. *Atmos. Chem. Phys.* **2016**, *16*, 15425–15432.
33. Bean, J. K.; Hildebrandt Ruiz, L. Gas-particle partitioning and hydrolysis of organic nitrates formed from the oxidation of alpha-pinene in environmental chamber experiments. *Atmos. Chem. Phys.* **2016**, *16*, 2175–2184.
34. Nah, T.; Sanchez, J.; Boyd, C. M.; Ng, N. L. Photochemical Aging of alpha-pinene and beta-pinene Secondary Organic Aerosol formed from Nitrate Radical Oxidation. *Environ. Sci. Technol.* **2016**, *50*, 222–231.
35. Darer, A. I.; Cole-Filipiak, N. C.; O'Connor, A. E.; Elrod, M. J. Formation and Stability of Atmospherically Relevant Isoprene-Derived Organosulfates and Organonitrates. *Environ. Sci. Technol.* **2011**, *45*, 1895–1902.

36. Hu, K. S.; Darer, A. I.; Elrod, M. J. Thermodynamics and kinetics of the hydrolysis of atmospherically relevant organonitrates and organosulfates. *Atmos. Chem. Phys.* **2011**, *11*, 8307–8320.
37. Jacobs, M. I.; Burke, W. J.; Elrod, M. J. Kinetics of the reactions of isoprene-derived hydroxynitrates: gas phase epoxide formation and solution phase hydrolysis. *Atmos. Chem. Phys.* **2014**, *14*, 8933–8946.
38. Romonosky, D. E.; Li, Y.; Shiraiwa, M.; Laskin, A.; Laskin, J.; Nizkorodov, S. A. Aqueous Photochemistry of Secondary Organic Aerosol of alpha-Pinene and alpha-Humulene Oxidized with Ozone, Hydroxyl Radical, and Nitrate Radical. *J. Phys. Chem. A* **2017**, *121*, 1298–1309.
39. Fahey, D. W.; Hubler, G.; Parrish, D. D.; Williams, E. J.; Norton, R. B.; Ridley, B. A.; Singh, H. B.; Liu, S. C.; Fehsenfeld, F. C. Reactive Nitrogen Species in the Troposphere - Measurements of No, No2, Hno3, Particulate Nitrate, Peroxyacetyl Nitrate (Pan), O-3, and Total Reactive Odd Nitrogen (Noy) at Niwot Ridge, Colorado. *J. Geophys. Res.-Atmos.* **1986**, *91*, 9781–9793.
40. Ridley, B. A.; Shetter, J. D.; Walega, J. G.; Madronich, S.; Elsworth, C. M.; Grahek, F. E.; Fehsenfeld, F. C.; Norton, R. B.; Parrish, D. D.; Hubler, G.; Buhr, M.; Williams, E. J.; Allwine, E. J.; Westberg, H. H. The Behavior of Some Organic Nitrates at Boulder and Niwot Ridge, Colorado. *J. Geophys. Res.-Atmos.* **1990**, *95*, 13949–13961.
41. Williams, E. J.; Roberts, J. M.; Baumann, K.; Bertman, S. B.; Buhr, S.; Norton, R. B.; Fehsenfeld, F. C. Variations in NOy composition at Idaho Hill, Colorado. *J. Geophys. Res.-Atmos.* **1997**, *102*, 6297–6314.
42. Duncianu, M.; David, M.; Kartigueyane, S.; Cirtog, M.; Doussin, J. F.; Picquet-Varrault, B. Measurement of alkyl and multifunctional organic nitrates by proton-transfer-reaction mass spectrometry. *Atmos. Meas. Tech.* **2017**, *10*, 1445–1463.
43. Lopez-Hilfiker, F. D.; Iyer, S.; Mohr, C.; Lee, B. H.; D'Ambro, E. L.; Kurten, T.; Thornton, J. A. Constraining the sensitivity of iodide adduct chemical ionization mass spectrometry to multifunctional organic molecules using the collision limit and thermodynamic stability of iodide ion adducts. *Atmos. Meas. Tech.* **2016**, *9*, 1505–1512.
44. Day, D. A.; Wooldridge, P. J.; Dillon, M. B.; Thornton, J. A.; Cohen, R. C. A thermal dissociation laser-induced fluorescence instrument for in situ detection of NO2, peroxy nitrates, alkyl nitrates, and HNO3. *J. Geophys. Res.-Atmos.* **2002**, *107*, ACH 4-1–ACH 4-14.
45. Thieser, J.; Schuster, G.; Schuladen, J.; Phillips, G. J.; Reiffs, A.; Parchatka, U.; Pohler, D.; Lelieveld, J.; Crowley, J. N. A two-channel thermal dissociation cavity ring-down spectrometer for the detection of ambient NO2, RO2NO2 and RONO2. *Atmos. Meas. Tech.* **2016**, *9*, 553–576.
46. Sadanaga, Y.; Takaji, R.; Ishiyama, A.; Nakajima, K.; Matsuki, A.; Bandow, H. Thermal dissociation cavity attenuated phase shift spectroscopy for continuous measurement of total peroxy and organic nitrates in the clean atmosphere. *Rev. Sci. Instrum.* **2016**, *87*, 074102/1–074102/6.

47. Rollins, A. W.; Smith, J. D.; Wilson, K. R.; Cohen, R. C. Real Time In Situ Detection of Organic Nitrates in Atmospheric Aerosols. *Environ. Sci. Technol.* **2010**, *44*, 5540–5545.
48. Rollins, A. W.; Browne, E. C.; Min, K. E.; Pusede, S. E.; Wooldridge, P. J.; Gentner, D. R.; Goldstein, A. H.; Liu, S.; Day, D. A.; Russell, L. M.; Cohen, R. C. Evidence for NOx Control over Nighttime SOA Formation. *Science* **2012**, *337*, 1210–1212.
49. Xu, L.; Suresh, S.; Guo, H.; Weber, R. J.; Ng, N. L. Aerosol characterization over the southeastern United States using high-resolution aerosol mass spectrometry: spatial and seasonal variation of aerosol composition and sources with a focus on organic nitrates. *Atmos. Chem. Phys.* **2015**, *15*, 7307–7336.
50. Kiendler-Scharr, A.; Mensah, A. A.; Friese, E.; Topping, D.; Nemitz, E.; Prevot, A. S. H.; Aijala, M.; Allan, J.; Canonaco, F.; Canagaratna, M.; Carbone, S.; Crippa, M.; Dall Osto, M.; Day, D. A.; De Carlo, P.; Di Marco, C. F.; Elbern, H.; Eriksson, A.; Freney, E.; Hao, L.; Herrmann, H.; Hildebrandt, L.; Hillamo, R.; Jimenez, J. L.; Laaksonen, A.; McFiggans, G.; Mohr, C.; O'Dowd, C.; Otjes, R.; Ovadnevaite, J.; Pandis, S. N.; Poulain, L.; Schlag, P.; Sellegri, K.; Swietlicki, E.; Tiitta, P.; Vermeulen, A.; Wahner, A.; Worsnop, D.; Wu, H. C. Ubiquity of organic nitrates from nighttime chemistry in the European submicron aerosol. *Geophys. Res. Lett.* **2016**, *43*, 7735–7744.
51. Farmer, D. K.; Matsunaga, A.; Docherty, K. S.; Surratt, J. D.; Seinfeld, J. H.; Ziemann, P. J.; Jimenez, J. L. Response of an aerosol mass spectrometer to organonitrates and organosulfates and implications for atmospheric chemistry. *Proc. Natl. Acad. Sci. U.S.A.* **2010**, *107*, 6670–6675.
52. Kroll, J. H.; Ng, N. L.; Murphy, S. M.; Flagan, R. C.; Seinfeld, J. H. Secondary organic aerosol formation from isoprene photooxidation. *Environ. Sci. Technol.* **2006**, *40*, 1869–1877.
53. Ng, N. L.; Chhabra, P. S.; Chan, A. W. H.; Surratt, J. D.; Kroll, J. H.; Kwan, A. J.; McCabe, D. C.; Wennberg, P. O.; Sorooshian, A.; Murphy, S. M.; Dalleska, N. F.; Flagan, R. C.; Seinfeld, J. H. Effect of NOx level on secondary organic aerosol (SOA) formation from the photooxidation of terpenes. *Atmos. Chem. Phys.* **2007**, *7*, 5159–5174.
54. Kroll, J. H.; Ng, N. L.; Murphy, S. M.; Flagan, R. C.; Seinfeld, J. H. Secondary organic aerosol formation from isoprene photooxidation under high-NOx conditions. *Geophys. Res. Lett.* **2005**, *32*, L18808/1–L18808/4.
55. Dommen, J.; Metzger, A.; Duplissy, J.; Kalberer, M.; Alfarra, M. R.; Gascho, A.; Weingartner, E.; Prevot, A. S. H.; Verheggen, B.; Baltensperger, U. Laboratory observation of oligomers in the aerosol from isoprene/NOx photooxidation. *Geophys. Res. Lett.* **2006**, *33*, L13805/1–L13805/5.
56. King, S. M.; Rosenoern, T.; Shilling, J. E.; Chen, Q.; Wang, Z.; Biskos, G.; McKinney, K. A.; Poschl, U.; Martin, S. T. Cloud droplet activation of mixed organic-sulfate particles produced by the photooxidation of isoprene. *Atmos. Chem. Phys.* **2010**, *10*, 3953–3964.

57. Zhang, H.; Surratt, J. D.; Lin, Y. H.; Bapat, J.; Kamens, R. M. Effect of relative humidity on SOA formation from isoprene/NO photooxidation: enhancement of 2-methylglyceric acid and its corresponding oligoesters under dry conditions. *Atmos. Chem. Phys.* **2011**, *11*, 6411–6424.
58. Xu, L.; Kollman, M. S.; Song, C.; Shilling, J. E.; Ng, N. L. Effects of NOx on the Volatility of Secondary Organic Aerosol from Isoprene Photooxidation. *Environ. Sci. Technol.* **2014**, *48*, 2253–2262.
59. Liu, J. M.; D'Ambro, E. L.; Lee, B. H.; Lopez-Hilfiker, F. D.; Zaveri, R. A.; Rivera-Rios, J. C.; Keutsch, F. N.; Iyer, S.; Kurten, T.; Zhang, Z. F.; Gold, A.; Surratt, J. D.; Shilling, J. E.; Thornton, J. A. Efficient Isoprene Secondary Organic Aerosol Formation from a Non-IEPDX Pathway. *Environ. Sci. Technol.* **2016**, *50*, 9872–9880.
60. Sarrafzadeh, M.; Wildt, J.; Pullinen, I.; Springer, M.; Kleist, E.; Tillmann, R.; Schmitt, S. H.; Wu, C.; Mentel, T. F.; Zhao, D. F.; Hastie, D. R.; Kiendler-Scharr, A. Impact of NOx and OH on secondary organic aerosol formation from beta-pinene photooxidation. *Atmos. Chem. Phys.* **2016**, *16*, 11237–11248.
61. D'Ambro, E. L.; Lee, B. H.; Liu, J. M.; Shilling, J. E.; Gaston, C. J.; Lopez-Hilfiker, F. D.; Schobesberger, S.; Zaveri, R. A.; Mohr, C.; Lutz, A.; Zhang, Z. F.; Gold, A.; Surratt, J. D.; Rivera-Rios, J. C.; Keutsch, F. N.; Thornton, J. A. Molecular composition and volatility of isoprene photochemical oxidation secondary organic aerosol under low- and high-NOx conditions. *Atmos. Chem. Phys.* **2017**, *17*, 159–174.
62. Boyd, C. M.; Nah, T.; Xu, L.; Berkemeier, T.; Ng, N. L. Secondary Organic Aerosol (SOA) from Nitrate Radical Oxidation of Monoterpenes: Effects of Temperature, Dilution, and Humidity on Aerosol Formation, Mixing, and Evaporation. *Environ. Sci. Technol.* **2017**, *51*, 7831–7841.
63. Fry, J. L.; Draper, D. C.; Zarzana, K. J.; Campuzano-Jost, P.; Day, D. A.; Jimenez, J. L.; Brown, S. S.; Cohen, R. C.; Kaser, L.; Hansel, A.; Cappellin, L.; Karl, T.; Roux, A. H.; Turnipseed, A.; Cantrell, C.; Lefer, B. L.; Grossberg, N. Observations of gas- and aerosol-phase organic nitrates at BEACHON-RoMBAS 2011. *Atmos. Chem. Phys.* **2013**, *13*, 8585–8605.

Chapter 7

Reactive Uptake of Ammonia by Biogenic and Anthropogenic Organic Aerosols

Julia Montoya-Aguilera,[1] Mallory L. Hinks,[1] Paige K. Aiona,[1] Lisa M. Wingen,[1] Jeremy R. Horne,[2] Shupeng Zhu,[2] Donald Dabdub,[2] Alexander Laskin,[3] Julia Laskin,[3] Peng Lin,[3] and Sergey A. Nizkorodov[1,*]

[1]Department of Chemistry, University of California, Irvine, California 92697, United States
[2]Department of Mechanical and Aerospace Engineering, University of California, Irvine, California 92697, United States
[3]Department of Chemistry, Purdue University, West Lafayette, Indiana 47907, United States
[*]E-mail: nizkorod@uci.edu

Secondary organic aerosol (SOA) produced by photooxidation of biogenic and anthropogenic volatile organic compounds (VOCs) represent a large fraction of fine atmospheric particulate matter ($PM_{2.5}$). The chemical composition of SOA particles is continuously changing as a result of various chemical and physical aging processes. One of the recently discovered aging processes, observed in both laboratory and field experiments, is the reactive uptake of ammonia by carbonyl species in SOA, leading to the formation of nitrogen-containing organic compounds (NOC). These NOC have attracted a lot of attention because of their propensity to absorb visible radiation and increase the amount of solar energy trapped in the atmosphere. Another potentially important, but poorly explored, consequence of NOC formation is that these compounds are less efficient than ammonia at neutralizing acids in particles. This paper summarizes existing experimental evidence for the reactive uptake of ammonia by SOA particles and describes our recent efforts to model the effect of this complicated process on air quality at the regional and continental scale. The modeling results predict that the reactive uptake of ammonia by SOA particles can significantly reduce the concentration of gas-phase ammonia, thereby indirectly affecting particle acidity as well as the amount of ammonium sulfate and ammonium nitrate in $PM_{2.5}$. Since emissions of ammonia to the atmosphere are expected to increase due to the growing agricultural needs of the human population, these findings have important implications for future air pollution control strategies.

Sources and Sinks of Atmospheric Ammonia

Ammonia (NH_3) is an important atmospheric trace gas emitted into the atmosphere by a number of natural and anthropogenic sources including soils, oceans, animal husbandry, fertilizer use, automobiles, and biomass burning (*1–3*). The largest source of ammonia is agriculture, which accounts for roughly half of the global ammonia emissions, including those from animal waste and fertilizer use. Remote observations of ammonia from space have revealed hot spots of ammonia concentrations in areas known for intense agricultural activities such as the Central Valley of California, North China Plain, Indo-Gangetic Plain, and Fergana Valley in Uzbekistan (*4*).

Depending on the location, the sources of ammonia can deviate strongly from the global average. For example, in the South Coast Air Basin of California (SoCAB), the amount of ammonia emitted by automobiles is on the same order of magnitude as that from agricultural sources (*5*). This results in a more uniform distribution of ammonia throughout the SoCAB, but hot spots around agricultural areas are still present, for example in Chino, a major center for dairy farming.

Ammonia is a highly water-soluble molecule, therefore it is primarily lost by deposition on water-containing particles and aquatic surfaces (*6*). Oxidation of gaseous ammonia by OH is relatively slow, but the chemistry of the resulting NH_2 radical contributes to the sources of N_2O (*2*). Atmospheric NH_3 in the gas-phase and NH_4^+ in the condensed-phase are often considered together as NH_x in literature dealing with the global nitrogen cycle. Human activities not only dominate ammonia emissions, but also determine where reactive nitrogen is deposited, with terrestrial ecosystems now receiving relatively more NH_x than marine ecosystems (*7*).

Chemistry of Ammonia in Particulate Matter

Ammonia is the most important atmospheric basic species capable of neutralizing inorganic acids, such as sulfuric acid and nitric acid, commonly found in polluted air. The resulting inorganic ammonium salts have low volatility and condense as solids into fine particulate matter ($PM_{2.5}$, particles with sizes below 2.5 μm, which more easily penetrate the respiratory tract).

$$NH_3(g) + H_2SO_4(g) \rightleftharpoons NH_4HSO_4(s) \qquad (R1)$$

$$NH_3(g) + NH_4HSO_4(s) \rightleftharpoons (NH_4)_2SO_4(s) \quad \text{(R2)}$$

$$NH_3(g) + HNO_3(g) \rightleftharpoons NH_4NO_3(s) \quad \text{(R3)}$$

For example, almost half of the $PM_{2.5}$ mass in the SoCAB is due to ammonium sulfate and ammonium nitrate (*8, 9*). The equilibria R1-R3 are simplified representations of the actual reactions, which occur in multiple phases and involve water. In particles containing liquid water, ammonia and acids are scavenged by droplets resulting in coupled aqueous equilibria involving NH_4^+(aq), HSO_4^-(aq), SO_4^{2-}(aq), and NO_3^-(aq). We note that complete neutralization of sulfuric acid in (R2) requires unrealistically high concentrations of ammonia (*10, 11*), so ambient particles remain acidic under typical atmospheric conditions.

In addition to the reasonably well constrained inorganic acid-base chemistry of ammonia, recent experiments have shown that ammonia can react with certain organic compounds in primary organic aerosol (POA, particles containing organic compounds directly emitted into the atmosphere by various sources) and in secondary organic aerosol (SOA, a collection of particle-phase organic compounds formed by photooxidation of volatile organic compounds (VOCs)). The general mechanism involves the reaction of ammonia with carbonyls within the organic particles to produce relatively unstable imines and amines, which may be stabilized by intramolecular cyclization into heteroaromatic products based on imidazole, pyrrole, indole, etc. These reactions change not only the chemical composition of organic particles but also their physical properties. For example, they can produce chromophoric organic compounds capable of absorbing visible radiation, known as brown carbon (BrC) (*12*). Additionally, uptake of ammonia modifies particle viscosity and condensed-phase diffusivity of SOA components (*13–15*).

Early interest in this chemistry can be traced to a serendipitous observation by Mang et al. (2008) that SOA samples prepared by ozonolysis of d-limonene (LIM/O_3 SOA) reproducibly turned brown when they were exposed to ambient air for a few days (*16*). Figure 1 reproduces the absorption spectrum published by Mang et al. (2008) of LIM/O_3 SOA particles collected on an optical window as a function of storage time in open air. The SOA material became more light-absorbing throughout the visible spectrum, with the largest change appearing in the form of a characteristic absorption band at 510 nm.

The reason for the browning of LIM/O_3 SOA material was not known until Bones et al. (2010) demonstrated that this process involves a complex reaction between ammonium/ammonia and carbonyl compounds in LIM/O_3 SOA (*17*). There must have been enough ammonia emitted by the laboratory personnel to cause the LIM/O_3 SOA material to slowly change color over time in the original Mang et al. (2008) experiments. Follow up experiments have shown that browning chemistry is not limited to LIM/O_3 SOA; similar browning processes were observed in a broad range of other types of biogenic and anthropogenic SOA (*18*). The mechanism of the reaction was examined by high-resolution mass spectrometry (HRMS), which found that highly-conjugated light-absorbing nitrogen-containing organic compounds (NOC) were responsible for the brown color of the aged SOA material (*19–23*).

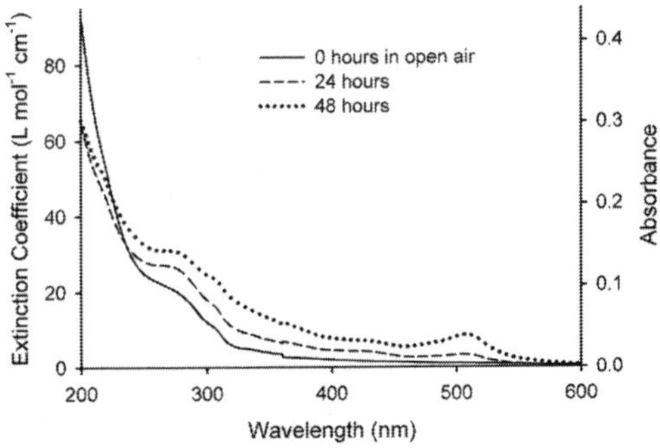

Figure 1. Absorption spectra of LIM/O_3 SOA material collected on a CaF_2 window as a function of storage time in open room air in darkness. The observed changes in the absorption profile were accompanied by a visible transition of the material from colorless to red-brown. The figure is reproduced with permission from ref. (16). Copyright 2017 American Chemical Society.

Even when SOA particles did not visibly change their color upon exposure to ammonia, NOC were still produced in the particles. For example, α-pinene ozonolysis SOA (APIN/O_3 SOA) remained colorless, meanwhile LIM/O_3 SOA browned after exposure to ammonia, but both samples had a number of relatively similar NOC products observed by HRMS methods (*20*). The stark contrast in the browning behavior of APIN/O_3 and LIM/O_3 SOA suggested that chromophoric NOC are highly conjugated oligomers of Schiff bases present at relatively low concentrations. The formation of these types of chromophoric compounds in APIN/O_3 SOA is hindered by the structural rigidity of the α-pinene oxidation products, compared to the much more flexible products of d-limonene oxidation (*20*).

Another important browning reaction identified in laboratory experiments is the aqueous reaction of 1,2-dicarbonyls, such as glyoxal and methyl glyoxal, with ammonia, amines, and amino acids (*24–54*). These reactions are capable of chemically trapping volatile glyoxal and methyl glyoxal in the form of oligomerized, light-absorbing products, thus providing an efficient pathway to secondary BrC formation. Their formation proceeds through imidazole by a mechanism described by Heinrich Debus in 1858 (*55*). The imidazole compounds attributed to this reaction have now been observed in ambient particles (*56, 57*) and shown to act as efficient photosensitizers (*40, 48, 58–60*). These types of compounds are believed to build up in aerosol particles at night but are quickly photolyzed during the day (*36, 45, 48, 50, 51*). The photolysis experiments cited here have been carried out only in the aqueous phase. The photochemical stability of these imidazole compounds in dry aerosol particles has not been studied yet.

The browning reactions between ammonia and 1,2-dicarbonyls are closely related to the Maillard chemistry responsible for the browning of foods during cooking driven by reactions of amino acids and reducing sugars (*49, 61–64*). Our current understanding of the SOA browning mechanism is that this reaction also requires a dicarbonyl species that can cyclize into a stable heterocyclic aromatic compound in a reaction with ammonia (*19, 65*). For example, Figure 2 shows the mechanism of the reaction of 4-oxopentanal, the simplest atmospherically relevant 1,4-dicarbonyl, with ammonia. The reaction involves a carbonyl-imine conversion of the aldehyde group, followed by intramolecular cyclization to form 2-methyl pyrrole (*19*). The amine intermediate is not very stable, but the cyclization drives the reaction downhill in free energy. The pyrrole can continue reacting with additional 4-oxopentanal molecules and ammonia to produce larger products capable of absorbing visible light. Similar chemical processes are expected to occur during reactions of ammonia with ketoaldehydes derived from oxidation of monoterpenes, for example in the case of browning of ketolimononaldehyde (*21*).

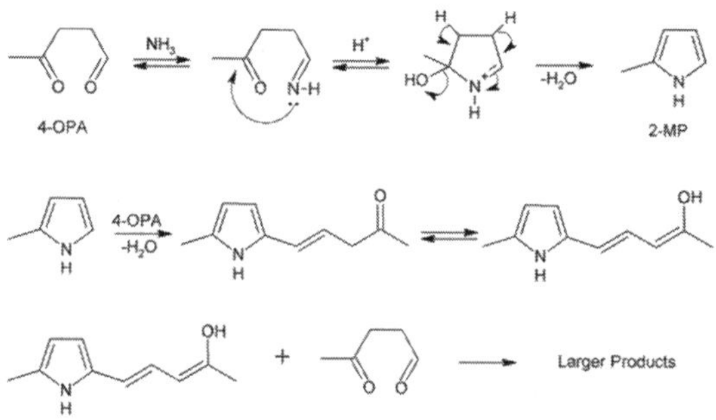

Figure 2. (Top) 4-oxopentanal reacts with NH_3 to form 2-methyl pyrrole, an intermediate to the production of BrC chromophores. (Bottom) 2-methyl pyrrole further reacts with 4-oxopentanal to form dimer products. A series of such reactions can produce larger, conjugated products potentially capable of absorbing visible light. The figure is reproduced with permission from ref. (19). Copyright 2017 American Chemical Society.

The majority of previous studies have described mechanistic aspects of reactions between ammonia and SOA compounds, but only a few studies have examined the kinetics of reactive uptake of ammonia by SOA particles. The most important study for this discussion is that by Liu et al. (2015) who reported uptake coefficients for ammonia onto APIN/O_3 SOA and m-xylene photooxidation SOA particles (XYL/OH SOA) (*66*). The uptake coefficients were obtained by measuring the rate of appearance of NOC in particles with a time-of-flight aerosol mass spectrometer (ToF-AMS) after their exposure to a pulse of gaseous ammonia. All experiments were carried out at a relative humidity (RH) of 50% (the critical role of RH in reactive uptake of ammonia is discussed below). Figure 3 reproduces a key result from that study, in which the uptake resulting in NOC formation is quantified for APIN/O_3 SOA grown on acidic or neutral seed particles. The observed initial uptake coefficients were quite high, on the order of $\gamma = 10^{-2}$ to 10^{-3} in both APIN/O_3 SOA and XYL/OH SOA, but after several hours of reaction they decreased to $\gamma \sim 10^{-5}$. The uptake coefficient was higher for particles that were acidified, with signals from both NOC and ammonium ion increasing with particle acidity. The weight fraction of NOC after the exposure was estimated as ~10 wt% for APIN/O_3 SOA and ~30 wt% for XYL/OH SOA. This suggested that a significant fraction of nitrogen could enter the particles not in the traditional form of organic and inorganic salts of ammonia, but also in the form of NOC. The decreasing uptake coefficients with time suggest a kinetic limitation to this uptake.

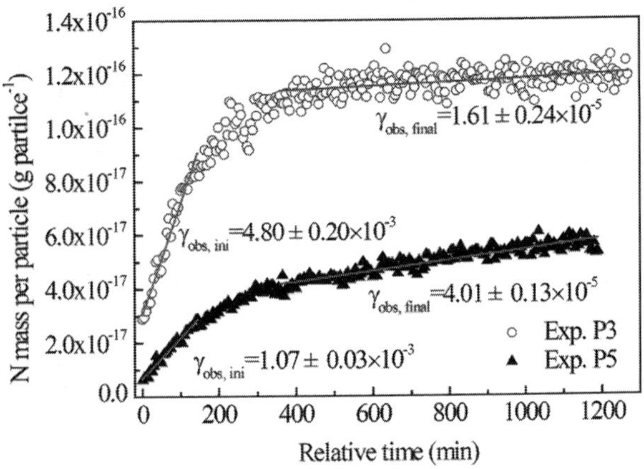

Figure 3. Reactive uptake coefficients for ammonia onto APIN/O_3 SOA particles at 50% RH. Experiment P3 corresponds to SOA produced using seed particles acidified with sulfuric acid, whereas experiment P5 corresponds to SOA produces using neutral seed particles. The figure is reproduced with permission from authors of Ref. (66). Copyright 2015 under creative commons attribution license.

A recent series of laboratory experiments highlighted the important role that RH can play in the reactive uptake of ammonia by SOA particles (*13–15*). Liu et al. (2018) examined the interactions between ammonia and SOA particles prepared by low-NO_x photooxidation of toluene (TOL/OH SOA) by tracing characteristic $C_xH_yN_z^+$ ions from ToF-AMS measurements that can only be produced by NOC (*15*). Figure 4 shows that the ratio of NOC mass concentration to total organic mass concentration is low under dry conditions, but increases at higher RH. This result was attributed to the high viscosity and slow molecular diffusion in TOL/OH SOA particles under dry conditions (*67*). The model accounting for the slow diffusion showed that the reaction was limited by the low diffusivity of the large organic molecules from the interior region of the particle to the surface. The diffusivity limitations no longer applied at higher RH because of the known ability of humid air to make organic materials less viscous (*68*). The observed transition from the lack of reactivity to full reactivity between 30-60 % RH is consistent with the independent result of Ye et al. (2016), who observed that the mass-transfer limitation for gas-particle partitioning in TOL/OH SOA was removed at RH above 40% (*69*). In another set of experiments (*14*), APIN/O_3 SOA particles exposed to ammonia under dry conditions developed a semi-solid coating on the surface (consisting of ammonium carboxylates) which prevented further reaction and also slowed down the particle coagulation. From these observations, we conclude that ammonia + SOA particle reactions leading to NOC are not efficient under dry conditions.

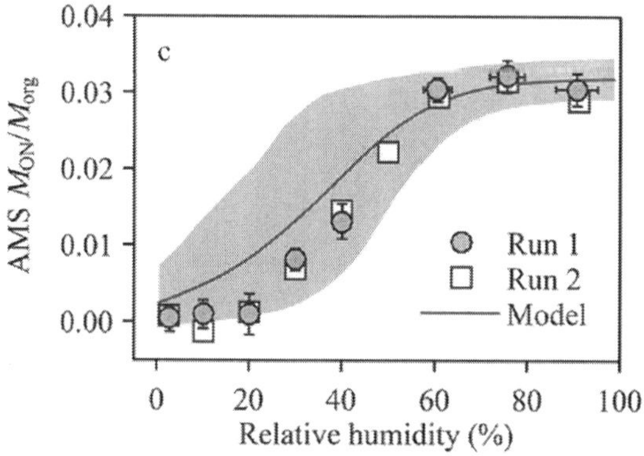

Figure 4. The ratio of NOC mass concentration to total organic mass concentration at different RH values. The reaction is suppressed at low RH by the high viscosity of TOL/OH SOA particles. The figure is reproduced with permission from ref. (15). Copyright 2018 American Chemical Society.

We carried out related experiments in our laboratory with LIM/O_3 SOA, as well as SOA particles prepared by low-NO_x and high-NO_x photooxidation of toluene (TOL/OH SOA and TOL/OH/NO_x SOA), n-hexadecane (HEX/OH SOA and HEX/OH/NO_x SOA), and d-limonene (LIM/OH SOA and LIM/OH/NO_x SOA) in a 5 m^3 Teflon chamber. Details of the experiments are provided in the Ph.D. dissertation of Dr. Mallory L. Hinks (*70*). All experiments were carried out at 50% RH to prevent the mass-transfer limitations described above. While oxidation of anthropogenic toluene and n-hexadecane usually happens under high-NO_x conditions in ambient air, the control experiments carried out under low NO_x conditions were helpful in distinguishing contributions to particulate nitrogen from ammonium nitrate, NOC produced during the photooxidation, and NOC produced by the subsequent reactive uptake of ammonia. Both high-NO_x and low-NO_x conditions are relevant for biogenic d-limonene, because it is found in both urban and remote environments.

Figure 5 shows how the mass concentration of TOL, HEX, and LIM SOA particles responds to a 50 ppb pulse of ammonia. In all cases, the low NO_x experiments showed no significant change in particle mass loading, whereas there was a clear increase in the particle mass loading in high-NO_x experiments. This increase in particle mass concentration in the high-NO_x experiments was accompanied by an increase in the average particle size, and it can be attributed to the reaction (R3) between ammonia and nitric acid, where the latter is produced by the normal oxidation of NO_x in the chamber. The lack of effect of ammonia on the particle mass concentration and size in the low NO_x experiments can be best understood by examining the reaction shown at the top of Figure 2. The uptake of ammonia is expected to convert carbonyls into imines and aromatic NOC by the loss of one or several water molecules. Since the molecular weight of ammonia reactant and water product is about the same, we do not expect a large change in the molecular weight of the organic reactants when they are converted to NOC. Indeed, the experiments by Liu et al. (2015) described above also found no significant change in the SOA particle mass concentration after exposure to ammonia (*66*).

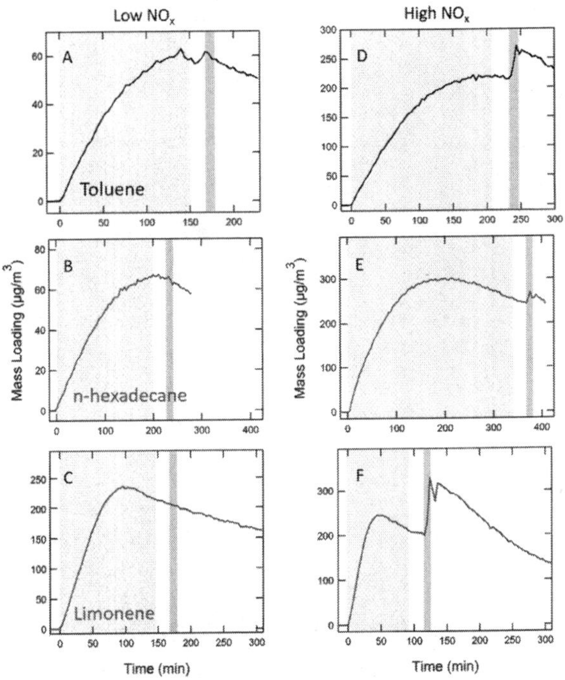

Figure 5. Mass concentration of particles during low-NO_x (left panels) and high-NO_x (right panels) photooxidation of toluene (a and d), hexadecane (b and e) and d-limonene (c and f). The yellow area indicates the time during which the chamber lamps were on. The purple area indicates the time during which a 50 ppb pulse of ammonia was introduced to the chamber. Reproduced with permission from the Ph.D. dissertation by Mallory Hinks (70). Copyright 2017 under creative commons attribution license.

Even though there is no apparent particle size change in low-NO_x experiments, the formation of NOC could be confirmed from the increase in the peaks attributable to $C_xH_yN_z^+$ ions in ToF-AMS mass spectra (Figure 6). The specific fragments that were observed to increase were CHN^+, $C_2H_3N^+$, $C_2H_4N^+$, and $C_3H_8N^+$. The lack of any effect on the NH family of fragments (NH^+, NH_2^+, NH_3^+, and NH_4^+) and the NO family of fragments (NO^+, NO_2^+, and NO_3^+) confirms that no ammonium nitrate impurity or ammonium carboxylates were produced during exposure of LIM/O_3 to ammonia. Since under low-NO_x conditions, the only possible source of nitrogen in the particles is the SOA + NH_3

reaction, this observation supports the assumption that ammonia is being taken up by the SOA particles and reacts with carbonyls in the particles forming NOC.

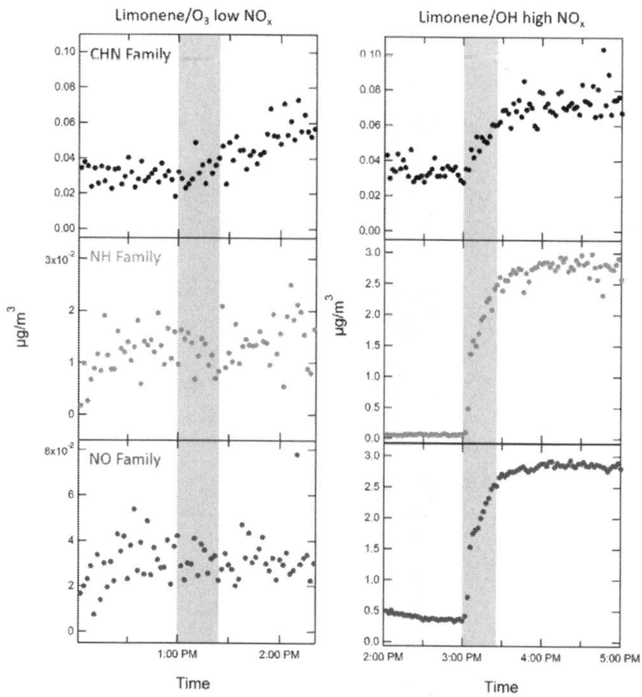

Figure 6. ToF-AMS data for LIM/O$_3$ SOA (left panels) and LIM/OH/NO$_x$ SOA (right panels) conditions. The yellow region indicates the time during which the light was on for photooxidation. The purple region indicates the time during which 200 ppb of ammonia was added to the chamber. The $C_xH_yN_z^+$ family of ToF-AMS fragments is plotted in black in the top panels. The NH family of fragments is plotted in green in the middle panels. The NO family of fragments is plotted in red in the bottom panels. Reproduced with permission from the Ph.D. dissertation by Mallory Hinks (70). Copyright 2017 under creative commons attribution license.

The $C_xH_yN_z^+$ ions also increased in the case of LIM/OH/NO$_x$ SOA (Figure 6), but some of these ions could conceivably be produced in the ToF-AMS electron impact ionizer because high-NO$_x$ SOA particles contained a large amount of ammonium nitrate (Figure 5). A similar increase in the $C_xH_yN_z^+$ ions in TOL/OH SOA and HEX/OH SOA exposed to ammonia could not be detected with sufficiently high signal-to-noise ratio, suggesting that different types of SOA have different reactivities towards ammonia.

The formation of NOC was additionally confirmed with off-line analysis of SOA particles collected on a filter by direct analysis in real time mass spectrometry (DART-MS) (*70*). The initial SOA compounds, before they were exposed to ammonia, consisted of only C, H, and O atoms, which typically appear at odd nominal *m/z* in DART-MS (*71*). However, NOC with one nitrogen atom appeared at even *m/z* in DART-MS. From the observed increase in the relative intensity of the even *m/z* peaks compared to the odd *m/z* peaks, we could estimate the NOC molar fraction in the exposed SOA. The estimated fraction was the highest (up to 20%) in the LIM/O_3 SOA and dropped to 5% for TOL/OH SOA and HEX/OH SOA.

Modeling Reactive Uptake of Ammonia by SOA

The equilibria between ammonia and ammonium ion driven by reactions (R1-R3) are very important for predicting regional air quality and global climate effects of aerosols. These equilibria have been incorporated in all major air quality models. In contrast, only two studies (both done by our group) have examined the effect of the ammonia + SOA chemistry on air quality (*72, 73*). While the ammonia + SOA → NOC reaction does not strongly affect the mass concentration of organic particulate compounds (e.g., see Figure 5), it can convert more basic ammonia ($pK_b \sim 5$) into less basic NOC such as imines ($pK_b \sim 13$) and pyrroles ($pK_b \sim 9$). This may lead to a reduction in the amount of ammonium nitrate produced by reaction (R3), and therefore a reduction in the overall concentration of $PM_{2.5}$.

Furthermore, the particle pH can also be affected by shifting the relative amount of sulfate and bisulfate anions in the particle in reactions (R1-R2). Sulfuric acid has a low volatility such that it readily finds its way into particles, even in the absence of ammonia. This results in highly acidic particles when the ammonia concentration is low. As the concentration of ammonia increases, the sulfuric acid becomes more neutralized and the pH is increased. However, because of the buffering between gaseous ammonia and particle phase ammonium ion, the dependence of the particle acidity on the ammonia concentration is highly non-linear. As mentioned above, unrealistically high concentrations of ammonia are needed to achieve pH-neutral conditions in ambient particles (*10, 11*).

Our modeling approach is illustrated in Figure 7. Since experimental information about the ammonia uptake by SOA particles is currently quite limited, we modeled this process in the simplest possible way as an irreversible conversion of ammonia into NOC on surfaces of SOA particles. The same reactive uptake coefficient, γ, for ammonia was assumed regardless of the SOA type, and no dependence of the uptake on particle acidity (*66*) or RH (*15*) noted in previous experiments was considered in our exploratory simulations. Furthermore, based on the experiments described above we assumed that no more than ~10% of SOA compounds can be converted into NOC. Even though this is arguably an oversimplified approach, it explores an important question: "How large must the reactive uptake coefficient for ammonia be for this chemistry to measurably affect air quality?"

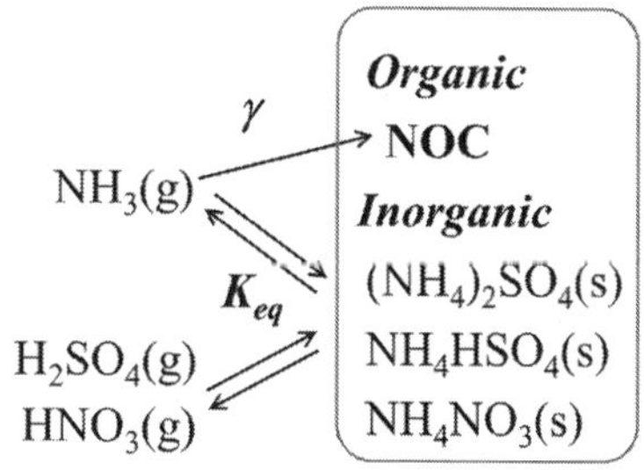

Figure 7. The simplest approach to modeling reactive uptake of ammonia by SOA. In addition to the equilibrium between ammonia and its inorganic salts in $PM_{2.5}$, an irreversible conversion of ammonia into NOC is added to the model, with its rate controlled by reactive uptake coefficient, γ. Since NOC is less basic than ammonia, formation of NOC reduces the amount of inorganic salts as well as affects the particle pH.

In our first study (*73*), we relied on the University of California, Irvine - California Institute of Technology (UCI-CIT) regional airshed model to investigate the potential impacts of this chemistry on the air quality in the SoCAB. The chemical mechanism used by the UCI-CIT model relies on the Caltech Atmospheric Chemical Mechanism (CACM) (*74–76*). The UCI-CIT model was previously used to simulate the effect of newly-discovered atmospheric chemistry on air quality in the SoCAB, for example in refs. (*77–79*). We added a surface reaction of ammonia with SOA particles to the base-case model and used uptake coefficients ranging from $\gamma = 10^{-5}$ to 10^{-2} to test the sensitivity of the model to this parameter. The largest effect on $PM_{2.5}$ was observed with $\gamma = 10^{-2}$, which resulted in a 15-40% reduction (depending on the time of day) in the domain-averaged ammonia concentration, and 2-12% reduction in the domain-averaged $PM_{2.5}$ concentration. In some locations, the reduction was considerably larger, as shown in Figure 8, particularly downwind from the large agricultural sources of ammonia in Chino, California. The predicted reduction in $PM_{2.5}$ concentrations was in excess of 10 μg/m^3 in the areas downwind from the major VOC and ammonia emissions, which is significant considering that the current U.S. National Ambient Air Quality Standards (NAAQS) for $PM_{2.5}$ concentrations is 35 μg/m^3 (24-hour average). The effect was still clearly observed in the model runs with $\gamma = 10^{-3}$, with the domain-averaged concentrations of ammonia and $PM_{2.5}$ dropping by 3-14% and 0.5-3%, respectively. For smaller values of the assumed γ values, the effect on ammonia and $PM_{2.5}$ concentrations was smaller, reducing to the level of the model noise when $\gamma = 10^{-5}$ was assumed. These model runs strongly suggested that if the uptake coefficient is indeed as large as reported by Liu et al. (2015), who reported the initial uptake coefficient of $\sim 4\times10^{-3}$ for α-pinene SOA and $\sim 1\times10^{-2}$ for xylene SOA (*66*), this could have a large effect on the distribution of both gaseous and particulate air pollutants in the SoCAB.

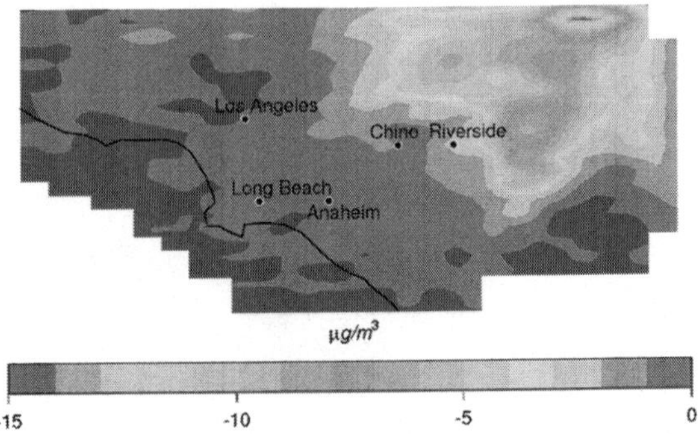

Figure 8. Difference from base case in 24-hour average of $PM_{2.5}$ when including ammonia uptake onto SOA using $\gamma = 10^{-2}$. The color bar is in the units of $\mu g/m^3$, with negative values indicating decreases in concentration with respect to the base case.

Inspired by the results of the regional airshed modeling with the UCI-CIT model, we extended the analysis to the continental United States (*72*). We added a first-order loss rate for ammonia onto SOA particles into the Community Multiscale Air Quality (CMAQ, version 5.2) model. The simulations covered the continental US in summer and winter periods. The initial CMAQ model simulations showed that predicted ammonia uptake with $\gamma = 10^{-2}$ was unrealistically large, exceeding the assumed 10% limit for the NOC conversion from SOA compounds. Therefore, simulations performed with the CMAQ model used uptake coefficients ranging from 10^{-5} to 10^{-3} to provide more realistic results.

Figure 9 shows a sample result of the simulation (the maps for this figure were prepared by the NCAR Command Language (*80*)). Similar to the SoCAB study, the simulation with the largest $\gamma = 10^{-3}$ predicted a reduction in the domain-averaged ammonia concentrations by about 30% in winter and 70% in summer. This reduction had two major effects. During the winter period, the model-predicted concentration of $PM_{2.5}$ dropped because of the smaller amount of ammonium nitrate particles formed, similar to the situation in the SoCAB simulations. However, in the summer period an increase in the amount SOA was predicted over the southeastern part of the US. This unanticipated result was driven by the reduction in the predicted particle pH (by up to two pH units), which led to an increase in the production of SOA by acid-catalyzed uptake of isoprene-derived epoxides (IEPOX). The IEPOX pathway was previously shown to improve SOA predictions in the southeastern US, and it is sensitive to the particle pH (*81*). We should note that the predicted drop in pH should be treated with caution and verified in future simulations as more accurate predictions of pH in aerosol particles become available. For example, it has been shown that particle pH is not especially sensitive to ammonia and particles in the southeastern United States remain acidic even when sulfate concentration drops and ammonia concentration increases (*11*).

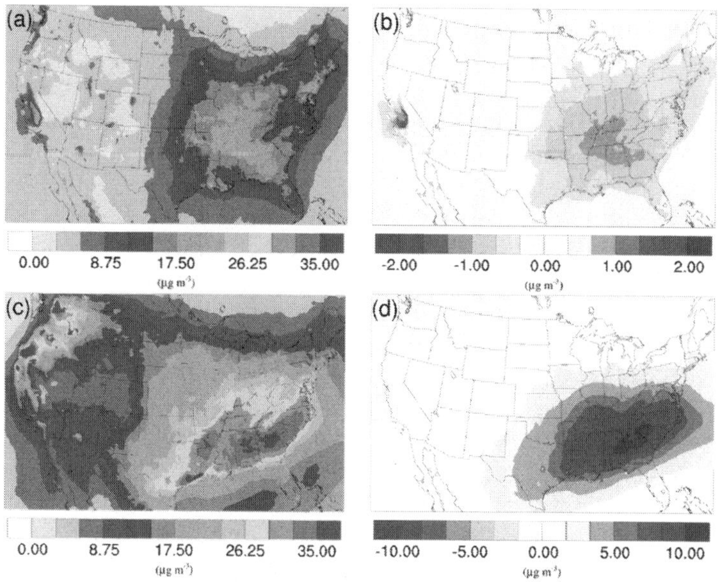

Figure 9. Spatial distribution of time-averaged $PM_{2.5}$ concentrations in the base case for (a) winter, and (c) summer. Spatial distribution of the difference in time-averaged $PM_{2.5}$ concentrations between the $\gamma = 10^{-3}$ case and the base case for (b) winter, and (d) summer. Positive values represent increases in concentration with respect to the base case, and negative values represent decreases in concentration with respect to the base case. Concentrations are expressed in units of $\mu g/m^3$. The figure is reproduced with permission from ref. (72). Copyright 2018 under creative commons attribution license.

Summary and Future Directions

The combination of experimental and modeling results described in this chapter show that it is critical to better understand the chemical processes involved in reactive uptake of ammonia by SOA particles. The predicted effects of these processes on air quality are substantial, and they can further increase in the future as POA concentrations decline due to control measures. The planet's population is expected to grow, with a steadily increasing demand for crop and cattle production. The concentration of ammonia will increase in parallel to the population growth and expanding agriculture. United Nations' Food and Agriculture Organization predicts that the annual ammonia emissions from livestock will increase by 60% by 2030 relative to the 1999 levels, with the largest increase occurring in the developing countries (*82*). The increased emissions of ammonia from livestock operations will be aided by warmer temperatures resulting from the changing climate because volatilization of ammonia is exponentially sensitive to temperature (*83–85*). For example, the anticipated increase in surface temperature predicted by an ensemble of seven different climate models could increase ammonia emissions in Europe by up to 40% by the end of the 21st century (*86*). The emissions of volatile organic compounds, which act as SOA precursors, also increase steeply with temperature (*87*). The simultaneous rise in both ammonia and SOA concentrations will increase the relative importance of the reactive uptake of ammonia by SOA particles.

We have just scratched the surface with respect to understanding the mechanism of the ammonia + SOA chemistry. Considerable experimental and modeling research is needed before these reactions can be incorporated into all air quality models. There is a need to develop more accurate ammonia emission inventories and to increase the frequency and accuracy of ammonia field measurements so that future models can be tested against observations. The effects of the particle phase state and acidity on the rate of ammonia uptake needs to be explored in more detail. The reversibility of the reactive uptake of ammonia by SOA particles, and the chemical stability of the resulting NOC with respect to hydrolysis, photolysis, and oxidation need to be better understood. The basicity of NOC produced in the ammonia + SOA reactions needs to be better constrained. In our modeling studies we made a first assumption that these NOC cannot form salts with inorganic acids, but some of these NOC were in fact detected as nitrogen-containing organic salts (*56*). Finally, the effect of ammonia on $PM_{2.5}$ needs to be directly tested in cleverly designed field campaigns. In summary, we can anticipate a number of new targeted laboratory experiments, field measurements, and air quality modeling studies on this rapidly developing topic in the coming years. This is an urgent area of research because ammonia emissions are already on the rise while our understanding of ammonia chemistry in the atmosphere remains woefully inadequate.

Acknowledgments

This publication was developed under Assistance Agreement No. EPA 83588101 awarded by the U.S. Environmental Protection Agency to the Regents of the University of California. It has not been formally reviewed by EPA. The views expressed in this document are solely those of the authors and do not necessarily reflect those of the Agency. EPA does not endorse any products or commercial services mentioned in this publication.

References

1. Schlesinger, W. H.; Hartley, A. E. A global budget for atmospheric NH_3. *Biogeochemistry* **1992**, *15*, 191–211.
2. Dentener, F. J.; Crutzen, P. J. A 3-dimensional model of the global ammonia cycle. *J. Atmos. Chem.* **1994**, *19*, 331–369.
3. Bouwman, A. F.; Lee, D. S.; Asman, W. A. H.; Dentener, F. J.; VanderHoek, K. W.; Olivier, J. G. J. A global high-resolution emission inventory for ammonia. *Global Biogeochem. Cycles* **1997**, *11*, 561–587.
4. Clarisse, L.; Clerbaux, C.; Dentener, F.; Hurtmans, D.; Coheur, P. F. Global ammonia distribution derived from infrared satellite observations. *Nat. Geosci.* **2009**, *2*, 479–483.

5. Nowak, J. B.; Neuman, J. A.; Bahreini, R.; Middlebrook, A. M.; Holloway, J. S.; McKeen, S. A.; Parrish, D. D.; Ryerson, T. B.; Trainer, M. Ammonia sources in the California South Coast Air Basin and their impact on ammonium nitrate formation. *Geophys. Res. Lett.* **2012**, *39*, L07804.
6. Dentener, F.; Drevet, J.; Lamarque, J. F.; Bey, I.; Eickhout, B.; Fiore, A. M.; Hauglustaine, D.; Horowitz, L. W.; Krol, M.; Kulshrestha, U. C.; Lawrence, M.; Galy-Lacaux, C.; Rast, S.; Shindell, D.; Stevenson, D.; Van Noije, T.; Atherton, C.; Bell, N.; Bergman, D.; Butler, T.; Cofala, J.; Collins, B.; Doherty, R.; Ellingsen, K.; Galloway, J.; Gauss, M.; Montanaro, V.; Muller, J. F.; Pitari, G.; Rodriguez, J.; Sanderson, M.; Solmon, F.; Strahan, S.; Schultz, M.; Sudo, K.; Szopa, S.; Wild, O. Nitrogen and sulfur deposition on regional and global scales: A multimodel evaluation. *Global Biogeochem. Cycles* **2006**, *20*, GB4003.
7. Galloway, J. N.; Dentener, F. J.; Capone, D. G.; Boyer, E. W.; Howarth, R. W.; Seitzinger, S. P.; Asner, G. P.; Cleveland, C. C.; Green, P. A.; Holland, E. A.; Karl, D. M.; Michaels, A. F.; Porter, J. H.; Townsend, A. R.; Vorosmarty, C. J. Nitrogen cycles: past, present, and future. *Biogeochemistry* **2004**, *70*, 153–226.
8. Kim, E.; Turkiewicz, K.; Zulawnick, S. A.; Magliano, K. L. Sources of fine particles in the South Coast area, California. *Atmos. Environ.* **2010**, *44*, 3095–3100.
9. Chow, J. C.; Watson, J. G.; Fujita, E. M.; Lu, Z. Q.; Lawson, D. R.; Ashbaugh, L. L. emporal and spatial variations of $PM_{2.5}$ and PM_{10} aerosol in the Southern California air quality study. *Atmos. Environ.* **1994**, *28*, 2061–2080.
10. Guo, H. Y.; Weber, R. J.; Nenes, A. High levels of ammonia do not raise fine particle pH sufficiently to yield nitrogen oxide-dominated sulfate production. *Sci. Rep.* **2017**, *7*, 12109.
11. Weber, R. J.; Guo, H. Y.; Russell, A. G.; Nenes, A. High aerosol acidity despite declining atmospheric sulfate concentrations over the past 15 years. *Nat. Geosci.* **2016**, *9*, 282–286.
12. Laskin, A.; Laskin, J.; Nizkorodov, S. A. Chemistry of Atmospheric Brown Carbon. *Chem. Rev.* **2015**, *115*, 4335–4382.
13. Li, Y. J.; Liu, P. F.; Gong, Z. H.; Wang, Y.; Bateman, A. P.; Bergoend, C.; Bertram, A. K.; Martin, S. T. Chemical Reactivity and Liquid/Nonliquid States of Secondary Organic Material. *Environ. Sci. Technol.* **2015**, *49*, 13264–13274.
14. Bell, D. M.; Imre, D.; Martin, S. T.; Zelenyuk, A. The properties and behavior of alpha-pinene secondary organic aerosol particles exposed to ammonia under dry conditions. *Phys. Chem. Chem. Phys.* **2017**, *19*, 6497–6507.
15. Liu, P.; Li, Y. J.; Wang, Y.; Bateman, A. P.; Zhang, Y.; Gong, Z.; Bertram, A. K.; Martin, S. T. Highly Viscous States Affect the Browning of Atmospheric Organic Particulate Matter. *ACS Cent. Sci.* **2018**, *4*, 207–215.
16. Mang, S. A.; Henricksen, D. K.; Bateman, A. P.; Andersen, M. P. S.; Blake, D. R.; Nizkorodov, S. A. Contribution of carbonyl photochemistry to

aging of atmospheric secondary organic aerosol. *J. Phys. Chem. A* **2008**, *112*, 8337–8344.

17. Bones, D. L.; Henricksen, D. K.; Mang, S. A.; Gonsior, M.; Bateman, A. P.; Nguyen, T. B.; Cooper, W. J.; Nizkorodov, S. A. Appearance of strong absorbers and fluorophores in limonene-O_3 secondary organic aerosol due to NH_4^+-mediated chemical aging over long time scales. *J. Geophys. Res. D* **2010**, *115*, D05203.
18. Updyke, K. M.; Nguyen, T. B.; Nizkorodov, S. A. Formation of brown carbon via reactions of ammonia with secondary organic aerosols from biogenic and anthropogenic precursors. *Atmos. Environ.* **2012**, *63*, 22–31.
19. Aiona, P. K.; Lee, H. J.; Lin, P.; Heller, F.; Laskin, A.; Laskin, J.; Nizkorodov, S. A. A Role for 2-Methyl Pyrrole in the Browning of 4-Oxopentanal and Limonene Secondary Organic Aerosol. *Environ. Sci. Technol.* **2017**, *51*, 11048–11056.
20. Laskin, J.; Laskin, A.; Nizkorodov, S. A.; Roach, P.; Eckert, P.; Gilles, M. K.; Wang, B.; Lee, H. J.; Hu, Q. Molecular Selectivity of Brown Carbon Chromophores. *Environ. Sci. Technol.* **2014**, *48*, 12047–12055.
21. Nguyen, T. B.; Laskin, A.; Laskin, J.; Nizkorodov, S. A. Brown carbon formation from ketoaldehydes of biogenic monoterpenes. *Faraday Discuss.* **2013**, *165*, 473–494.
22. Nguyen, T. B.; Lee, P. B.; Updyke, K. M.; Bones, D. L.; Laskin, J.; Laskin, A.; Nizkorodov, S. A. Formation of nitrogen- and sulfur-containing light-absorbing compounds accelerated by evaporation of water from secondary organic aerosols. *J. Geophys. Res. D* **2012**, *117*, D01207.
23. Laskin, J.; Laskin, A.; Roach, P. J.; Slysz, G. W.; Anderson, G. A.; Nizkorodov, S. A.; Bones, D. L.; Nguyen, L. Q. High-Resolution Desorption Electrospray Ionization Mass Spectrometry for Chemical Characterization of Organic Aerosols. *Anal. Chem.* **2010**, *82*, 2048–2058.
24. De Haan, D. O.; Tolbert, M. A.; Jimenez, J. L. Atmospheric condensed-phase reactions of glyoxal with methylamine. *Geophys. Res. Lett.* **2009**, *36*, L11819.
25. Galloway, M. M.; Chhabra, P. S.; Chan, A. W. H.; Surratt, J. D.; Flagan, R. C.; Seinfeld, J. H.; Keutsch, F. N. Glyoxal uptake on ammonium sulphate seed aerosol: reaction products and reversibility of uptake under dark and irradiated conditions. *Atmos. Chem. Phys.* **2009**, *9*, 3331–3345.
26. Shapiro, E. L.; Szprengiel, J.; Sareen, N.; Jen, C. N.; Giordano, M. R.; McNeill, V. F. Light-absorbing secondary organic material formed by glyoxal in aqueous aerosol mimics. *Atmos. Chem. Phys.* **2009**, *9*, 2289–2300.
27. Noziere, B.; Dziedzic, P.; Cordova, A. Inorganic ammonium salts and carbonate salts are efficient catalysts for aldol condensation in atmospheric aerosols. *Phys. Chem. Chem. Phys.* **2010**, *12*, 3864–3872.
28. Sareen, N.; Schwier, A. N.; Shapiro, E. L.; Mitroo, D.; McNeill, V. F. Secondary organic material formed by methylglyoxal in aqueous aerosol mimics. *Atmos. Chem. Phys.* **2010**, *10*, 997–1016.
29. Schwier, A. N.; Sareen, N.; Mitroo, D.; Shapiro, E. L.; McNeill, V. F. Glyoxal-Methylglyoxal Cross-Reactions in Secondary Organic Aerosol Formation. *Environ. Sci. Technol.* **2010**, *44*, 6174–6182.

30. Trainic, M.; Abo Riziq, A.; Lavi, A.; Flores, J. M.; Rudich, Y. The optical, physical and chemical properties of the products of glyoxal uptake on ammonium sulfate seed aerosols. *Atmos. Chem. Phys.* **2011**, *11*, 9697–9707.
31. Yu, G.; Bayer, A. R.; Galloway, M. M.; Korshavn, K. J.; Fry, C. G.; Keutsch, F. N. Glyoxal in aqueous ammonium sulfate solutions: products, kinetics and hydration effects. *Environ. Sci. Technol.* **2011**, *45*, 6336–6342.
32. Galloway, M. M.; Loza, C. L.; Chhabra, P. S.; Chan, A. W. H.; Yee, L. D.; Seinfeld, J. H.; Keutsch, F. N. Analysis of photochemical and dark glyoxal uptake: Implications for SOA formation. *Geophys. Res. Lett.* **2011**, *38*, L17811.
33. De Haan, D. O.; Hawkins, L. N.; Kononenko, J. A.; Turley, J. J.; Corrigan, A. L.; Tolbert, M. A.; Jimenez, J. L. Formation of Nitrogen-Containing Oligomers by Methylglyoxal and Amines in Simulated Evaporating Cloud Droplets. *Environ. Sci. Technol.* **2011**, *45*, 984–991.
34. Zarzana, K. J.; De Haan, D. O.; Freedman, M. A.; Hasenkopf, C. A.; Tolbert, M. A. Optical properties of the products of α-dicarbonyl and amine reactions in simulated cloud droplets. *Environ. Sci. Technol.* **2012**, *46*, 4845–4851.
35. Kampf, C. J.; Jakob, R.; Hoffmann, T. Identification and characterization of aging products in the glyoxal/ammonium sulfate system – implications for light-absorbing material in atmospheric aerosols. *Atmos. Chem. Phys.* **2012**, *12*, 6323–6333.
36. Sareen, N.; Moussa, S. G.; McNeill, V. F. Photochemical aging of light-absorbing secondary organic aerosol material. *J. Phys. Chem. A* **2013**, *117*, 2987–2996.
37. Hamilton, J. F.; Baeza-Romero, M. T.; Finessi, E.; Rickard, A. R.; Healy, R. M.; Peppe, S.; Adams, T. J.; Daniels, M. J. S.; Ball, S. M.; Goodall, I. C. A.; Monks, P. S.; Borras, E.; Munoz, A. Online and offline mass spectrometric study of the impact of oxidation and ageing on glyoxal chemistry and uptake onto ammonium sulfate aerosols. *Faraday Discuss.* **2013**, *165*, 447–472.
38. Sedehi, N.; Takano, H.; Blasic, V. A.; Sullivan, K. A.; De Haan, D. O. Temperature- and pH-dependent aqueous-phase kinetics of the reactions of glyoxal and methylglyoxal with atmospheric amines and ammonium sulfate. *Atmos. Environ.* **2013**, *77*, 656–663.
39. Lee, A. K. Y.; Zhao, R.; Li, R.; Liggio, J.; Li, S.-M.; Abbatt, J. P. D. Formation of Light Absorbing Organo-Nitrogen Species from Evaporation of Droplets Containing Glyoxal and Ammonium Sulfate. *Environ. Sci. Technol.* **2013**, *47*, 12819–12826.
40. Rossignol, S.; Aregahegn, K. Z.; Tinel, L.; Fine, L.; Noziere, B.; George, C. Glyoxal Induced Atmospheric Photosensitized Chemistry Leading to Organic Aerosol Growth. *Environ. Sci. Technol.* **2014**, *48*, 3218–3227.
41. Drozd, G. T.; McNeill, V. F. Organic matrix effects on the formation of light-absorbing compounds from α-dicarbonyls in aqueous salt solution. *Environ. Sci.: Processes Impacts* **2014**, *16*, 741–747.
42. Powelson, M. H.; Espelien, B. M.; Hawkins, L. N.; Galloway, M. M.; De Haan, D. O. Brown Carbon Formation by Aqueous-Phase Carbonyl

Compound Reactions with Amines and Ammonium Sulfate. *Environ. Sci. Technol.* **2014**, *48*, 985–993.
43. Hawkins, L. N.; Baril, M. J.; Sedehi, N.; Galloway, M. M.; De Haan, D. O.; Schill, G. P.; Tolbert, M. A. Formation of Semisolid, Oligomerized Aqueous SOA: Lab Simulations of Cloud Processing. *Environ. Sci. Technol.* **2014**, *48*, 2273–2280.
44. Lin, P.; Laskin, J.; Nizkorodov, S. A.; Laskin, A. Revealing Brown Carbon Chromophores Produced in Reactions of Methylglyoxal with Ammonium Sulfate. *Environ. Sci. Technol.* **2015**, *49*, 14257–14266.
45. Zhao, R.; Lee, A. K. Y.; Huang, L.; Li, X.; Yang, F.; Abbatt, J. P. D. Photochemical processing of aqueous atmospheric brown carbon. *Atmos. Chem. Phys.* **2015**, *15*, 6087–6100.
46. Maxut, A.; Noziere, B.; Fenet, B.; Mechakra, H. Formation mechanisms and yields of small imidazoles from reactions of glyoxal with NH_4^+ in water at neutral pH. *Phys. Chem. Chem. Phys.* **2015**, *17*, 20416–20424.
47. Tang, M.; Alexander, J. M.; Kwon, D.; Estillore, A. D.; Laskina, O.; Young, M. A.; Kleiber, P. D.; Grassian, V. H. Optical and Physicochemical Properties of Brown Carbon Aerosol: Light Scattering, FTIR Extinction Spectroscopy, and Hygroscopic Growth. *J. Phys. Chem. A* **2016**, *120*, 4155–4166.
48. Li, W. Y.; Li, X.; Jockusch, S.; Wang, H.; Xu, B. L.; Wu, Y. J.; Tsui, W. G.; Dai, H. L.; McNeill, V. F.; Rao, Y. Photoactivated Production of Secondary Organic Species from Isoprene in Aqueous Systems. *J. Phys. Chem. A* **2016**, *120*, 9042–9048.
49. Hawkins, L. N.; Lemire, A. N.; Galloway, M. M.; Corrigan, A. L.; Turley, J. J.; Espelien, B. M.; De Haan, D. O. Maillard Chemistry in Clouds and Aqueous Aerosol As a Source of Atmospheric Humic-Like Substances. *Environ. Sci. Technol.* **2016**, *50*, 7443–7452.
50. Wong, J. P. S.; Nenes, A.; Weber, R. J. Changes in Light Absorptivity of Molecular Weight Separated Brown Carbon Due to Photolytic Aging. *Environ. Sci. Technol.* **2017**, *51*, 8414–8421.
51. Aiona, P. K.; Lee, H. J.; Leslie, R.; Lin, P.; Laskin, A.; Laskin, J.; Nizkorodov, S. A. Photochemistry of Products of the Aqueous Reaction of Methylglyoxal with Ammonium Sulfate. *ACS Earth Space Chem.* **2017**, *1*, 522–532.
52. De Haan, D. O.; Hawkins, L. N.; Welsh, H. G.; Pednekar, R.; Casar, J. R.; Pennington, E. A.; de Loera, A.; Jimenez, N. G.; Symons, M. A.; Zauscher, M.; Pajunoja, A.; Caponi, L.; Cazaunau, M.; Formenti, P.; Gratien, A.; Pangui, E.; Doussin, J. F. Brown Carbon Production in Ammonium-or Amine-Containing Aerosol Particles by Reactive Uptake of Methylglyoxal and Photolytic Cloud Cycling. *Environ. Sci. Technol.* **2017**, *51*, 7458–7466.
53. Stangl, C. M.; Johnston, M. V. Aqueous Reaction of Dicarbonyls with Ammonia as a Potential Source of Organic Nitrogen in Airborne Nanoparticles. *J. Phys. Chem. A* **2017**, *121*, 3720–3727.
54. Ackendorf, J. M.; Ippolito, M. G.; Galloway, M. M. pH Dependence of the Imidazole-2-carboxaldehyde Hydration Equilibrium: Implications for

Atmospheric Light Absorbance. *Environ. Sci. Technol. Lett.* **2017**, *4*, 551–555.

55. Debus, H. Ueber die Einwirkung des Ammoniaks auf Glyoxal. *Justus Liebigs Ann. Chem.* **1858**, *107*, 199–208.
56. Wang, X.; Gao, S.; Yang, X.; Chen, H.; Chen, J.; Zhuang, G., Surratt, J. D.; Chan, M. N.; Seinfeld, J. H. Evidence for high molecular weight nitrogen-containing organic salts in urban aerosols. *Environ. Sci. Technol.* **2010**, *44*, 4441–4446.
57. Teich, M.; van Pinxteren, D.; Kecorius, S.; Wang, Z.; Herrmann, H. First Quantification of Imidazoles in Ambient Aerosol Particles: Potential Photosensitizers, Brown Carbon Constituents, and Hazardous Components. *Environ. Sci. Technol.* **2016**, *50*, 1166–1173.
58. Aregahegn, K. Z.; Noziere, B.; George, C. Organic aerosol formation photo-enhanced by the formation of secondary photosensitizers in aerosols. *Faraday Discuss.* **2013**, *165*, 123–134.
59. Palacios, L. G.; Arroyo, P. C.; Aregahegn, K. Z.; Steimer, S. S.; Bartels-Rausch, T.; Noziere, B.; George, C.; Ammann, M.; Volkamer, R. Heterogeneous photochemistry of imidazole-2-carboxaldehyde: HO2 radical formation and aerosol growth. *Atmos. Chem. Phys.* **2016**, *16*, 11823–11836.
60. Tsui, W. G.; Rao, Y.; Dai, H. L.; McNeill, V. F. Modeling Photosensitized Secondary Organic Aerosol Formation in Laboratory and Ambient Aerosols. *Environ. Sci. Technol.* **2017**, *51*, 7496–7501.
61. Velisek, J.; Davidek, T.; Davidek, J.; Trska, P.; Kvasnicka, F.; Velcova, K. New imidazoles formed in nonenzymic browning reactions. *J. Food Sci.* **1989**, *54*, 1544–6.
62. Rizzi, G. P. Chemical structure of colored Maillard reaction products. *Food Rev. Int.* **1997**, *13*, 1–28.
63. Meade Susie, J.; Miller Antonia, G.; Gerrard Juliet, A. The role of dicarbonyl compounds in non-enzymatic crosslinking: a structure-activity study. *Bioorg. Med. Chem.* **2003**, *11*, 853–62.
64. Wang, Y.; Ho, C. T. Flavour chemistry of methylglyoxal and glyoxal. *Chem. Soc. Rev.* **2012**, *41*, 4140–4149.
65. Kampf, C. J.; Filippi, A.; Zuth, C.; Hoffmann, T.; Opatz, T. Secondary brown carbon formation via the dicarbonyl imine pathway: nitrogen heterocycle formation and synergistic effects. *Phys. Chem. Chem. Phys.* **2016**, *18*, 18353–18364.
66. Liu, Y.; Liggio, J.; Staebler, R.; Li, S. M. Reactive uptake of ammonia to secondary organic aerosols: kinetics of organonitrogen formation. *Atmos. Chem. Phys.* **2015**, *15*, 13569–13584.
67. Song, M. J.; Liu, P. F. F.; Hanna, S. J.; Zaveri, R. A.; Potter, K.; You, Y.; Martin, S. T.; Bertram, A. K. Relative humidity-dependent viscosity of secondary organic material from toluene photo-oxidation and possible implications for organic particulate matter over megacities. *Atmos. Chem. Phys.* **2016**, *16*, 8817–8830.
68. Koop, T.; Bookhold, J.; Shiraiwa, M.; Poschl, U. Glass transition and phase state of organic compounds: dependency on molecular properties and

implications for secondary organic aerosols in the atmosphere. *Phys. Chem. Chem. Phys.* **2011**, *13*, 19238–19255.

69. Ye, Q.; Robinson, E. S.; Ding, X.; Ye, P.; Sullivan, R. C.; Donahue, N. M. Mixing of secondary organic aerosols versus relative humidity. *Proc. Natl. Acad. Sci. U. S. A.* **2016**, *113*, 12649–12654.
70. Hinks, M. L. Effect of Environmental Conditions on Composition and Photochemistry of Secondary Organic Aerosols. Ph.D. Thesis, University of California, Irvine, Irvine, CA, May 2017.
71. Nah, T.; Chan, M.; Leone, S. R.; Wilson, K. R. Real Time in Situ Chemical Characterization of Submicrometer Organic Particles Using Direct Analysis in Real Time-Mass Spectrometry. *Anal. Chem.* **2013**, *85*, 2087–2095.
72. Zhu, S. P.; Horne, J. R.; Montoya-Aguilera, J.; Hinks, M. L.; Nizkorodov, S. A.; Dabdub, D. Modeling reactive ammonia uptake by secondary organic aerosol in CMAQ: application to the continental US. *Atmos. Chem. Phys.* **2018**, *18*, 3641–3657.
73. Horne, J. R.; Zhu, S. P.; Montoya-Aguilera, J.; Hinks, M. L.; Wingen, L. M.; Nizkorodov, S. A.; Dabdub, D. Reactive Uptake of Ammonia by Secondary Organic Aerosols: Implications for Air Quality. *Atmos. Environ.* **2018**submitted.
74. Griffin, R. J.; Dabdub, D.; Seinfeld, J. H. Secondary organic aerosol 1. Atmospheric chemical mechanism for production of molecular constituents. *J. Geophys. Res. D* **2002**, *107*, 4332.
75. Griffin, R. J.; Dabdub, D.; Kleeman, M. J.; Fraser, M. P.; Cass, G. R.; Seinfeld, J. H. Secondary organic aerosol 3. Urban/regional scale model of size- and composition-resolved aerosols. *J. Geophys. Res. D* **2002**, *107*, 4334.
76. Dawson, M. L.; Xu, J. L.; Griffin, R. J.; Dabdub, D. Development of aroCACM/MPMPO 1.0: a model to simulate secondary organic aerosol from aromatic precursors in regional models. *Geosci. Model Dev.* **2016**, *9*, 2143–2151.
77. Montoya-Aguilera, J.; Horne, J. R.; Hinks, M. L.; Fleming, L. T.; Perraud, V.; Lin, P.; Laskin, A.; Laskin, J.; Dabdub, D.; Nizkorodov, S. A. Secondary organic aerosol from atmospheric photooxidation of indole. *Atmos. Chem. Phys.* **2017**, *17*, 11605–11621.
78. Perraud, V.; Horne, J. R.; Martinez, A. S.; Kalinowski, J.; Meinardi, S.; Dawson, M. L.; Wingen, L. M.; Dabdub, D.; Blake, D. R.; Gerber, R. B.; Finlayson-Pitts, B. J. The future of airborne sulfur-containing particles in the absence of fossil fuel sulfur dioxide emissions. *Proc. Natl. Acad. Sci. U. S. A.* **2015**, *112*, 13514–13519.
79. Carreras-Sospedra, M.; Lunden, M. M.; Brouwer, J.; Singer, B. C.; Dabdub, D. Air quality impacts of liquefied natural gas in the South Coast Air Basin of California. *J. Natural Gas Sci. Eng.* **2014**, *21*, 680–690.
80. *The NCAR Command Language*, Version 6.4.0 [Software]; UCAR/NCAR/CISL/TDD: Boulder, CO, 2017.
81. Pye, H. O. T.; Pinder, R. W.; Piletic, I. R.; Xie, Y.; Capps, S. L.; Lin, Y.-H.; Surratt, J. D.; Zhang, Z.; Gold, A.; Luecken, D. J.; Hutzell, W. T.; Jaoui, M.; Offenberg, J. H.; Kleindienst, T. E.; Lewandowski, M.; Edney, E. O. Epoxide

Pathways Improve Model Predictions of Isoprene Markers and Reveal Key Role of Acidity in Aerosol Formation. *Environ. Sci. Technol.* **2013**, *47*, 11056–11064.
82. Bruinsma, J. *World Agriculture: Towards 2015/2030 : an FAO Perspective*; Food and Agriculture Organization (FAO): London, UK, 2003; pp 1–432.
83. Zhang, G.; Strom, J. S.; Li, B.; Rom, H. B.; Morsing, S.; Dahl, P.; Wang, C. Emission of ammonia and other contaminant gases from naturally ventilated dairy cattle buildings. *Biosyst. Eng.* **2005**, *92*, 355–364.
84. Bleizgys, R.; Bagdoniene, I.; Balezentiene, L. Reduction of the Livestock Ammonia Emission under the Changing Temperature during the Initial Manure Nitrogen Biomineralization. *Sci. World J.* **2013**, *2013*, 825437.
85. Pereira, J.; Misselbrook, T. H.; Chadwick, D. R.; Coutinho, J.; Trindade, H. Effects of temperature and dairy cattle excreta characteristics on potential ammonia and greenhouse gas emissions from housing: A laboratory study. *Biosyst. Eng.* **2012**, *112*, 138–150.
86. Skjøth, C. A.; Geels, C. The effect of climate and climate change on ammonia emissions in Europe. *Atmos. Chem. Phys.* **2013**, *13*, 117–128.
87. Constable, J. V. H.; Guenther, A. B.; Schimel, D. S.; Monson, R. K. Modelling changes in VOC emission in response to climate change in the continental United States. *Glob. Change Biol.* **1999**, *5*, 791–806.

Chapter 8

Aqueous Aerosol Processing of Glyoxal and Methylglyoxal: Recent Measurements of Uptake Coefficients, SOA Production, and Brown Carbon Formation

David O. De Haan*

Department of Chemistry and Biochemistry, University of San Diego, 5998 Alcala Park, San Diego, California 92110, United States
***Phone: 011-1-619-260-6882; fax: 011-1-619-260-2211; e-mail: ddehaan@sandiego.edu**

Methylglyoxal uptake coefficients (γ) are needed in order to predict the amount of secondary organic aerosol (SOA) and brown carbon formed by heterogeneous processing of methylglyoxal. We compare recent laboratory and theoretical estimates of uptake coefficients for glyoxal and methylglyoxal on various aqueous aerosol surfaces. Measured methylglyoxal uptake coefficients on pre-reacted glycine aerosol particles increased from $\gamma = 0.0004$ to 0.0057 between 72 and 99% RH. At high ($\geq$ 95%) RH, the measured uptake coefficients for methylglyoxal are greater than those measured for glyoxal, and four orders of magnitude higher than theoretical estimates for methylglyoxal, likely due to irreversible reactions between methylglyoxal and ammonia or glycine. Both laboratory and theoretical studies of methylglyoxal uptake found similar dependence on RH, however, where γ increases with RH due to a "salting out" effect. Methylglyoxal uptake measured on cloud droplets was too rapid to allow uptake coefficients to be extracted from the data, but was largely reversible. Recent results have also demonstrated that in evaporating aqueous aerosol particles, brown carbon is formed much more rapidly and is much more resistant to photolytic bleaching than in bulk-phase simulations. Together, these results suggest that SOA and brown carbon formation by methylglyoxal in aqueous aerosol is larger than models currently predict, but SOA formation by methylglyoxal in cloud droplets may be smaller than current predictions.

Introduction

Particles entering clouds serve as sites for condensing water vapor, such that every cloud droplet must form on a pre-existing aerosol particle. However, many aerosol particles that enter a cloud are unable nucleate a stable cloud droplet. Only the larger and/or more hygroscopic aerosol particles will attract condensation from supersaturated water vapor, while the rest will be out-competed, remaining as "interstitial" aerosol inside the cloud (*1*). The aerosol particles that initiate cloud droplet formation contain organic and inorganic matter that is dissolved in (or at least surrounded by) water during the lifetime of the droplet, which has been estimated at ~10 min (*2*). During its lifetime, the cloud droplet may scavenge water-soluble gases at rates governed by uptake kinetics and Henry's law equilibrium. During the day, the droplet's dilute chemical contents will also be subjected to photolysis and to oxidation by dissolved OH radicals (*2, 3*) as described in more detail in the chapters by Herrmann & Tilgner and Nizkorodov *et al.* in this volume.

Cloud droplets may grow large enough to precipitate, or they may leave a cloud via updrafts or other air motion. In either case, once a droplet leaves a cloud, the relative humidity drops below 100% and some of the water rapidly evaporates to stay in equilibrium with subsaturated air. As water evaporates, other dissolved volatile species may evaporate, too, while dissolved *non-volatile* species increase greatly in concentration, affecting particle acidity and viscosity. It has been estimated that aqueous aerosol particles may contain salts such as ammonium sulfate at concentrations as high as 3 M (*4, 5*). Aqueous aerosol particles may also take up species from the gas phase, but this uptake may deviate significantly from estimates based on Henry's law coefficients due to interactions with high concentrations of solute species. For example, the effective Henry's law coefficient H^* of glyoxal in seawater is 3×10^5 (*6, 7*), but in aqueous ammonium sulfate aerosol H^* has been measured to be 4 orders of magnitude larger (*8–10*), due to methylglyoxal interactions with both ammonium and sulfate ions (*4, 5, 9, 10*).

Since at least the year 2000, the dicarbonyl compounds glyoxal and methylglyoxal have been suspected of increasing the mass of atmospheric aerosol particles through aqueous phase reactions (*3, 11*) with water (*12, 13*), oxidants (*2, 3, 14–16*), other aldehydes (*17, 18*), ammonium salts (*4, 5, 8, 19–21*), and amines (*22–26*). These dicarbonyl compounds are common, stable intermediates produced by the atmospheric oxidation of many different precursor species, including aromatic compounds (*27–31*) and isoprene (*32, 33*). Even though dicarbonyl yields from isoprene oxidation are much smaller than yields from aromatic precursors (*31, 33*), the large global emissions of isoprene (*34*) make it the dominant atmospheric source of dicarbonyls in all but the most heavily polluted locations. As a result, glyoxal and methylglyoxal can be detected in the gas, cloudwater, and aerosol phases just about anywhere in the troposphere.

The production of small aldehydes and dicarbonyls from the oxidation of isoprene by OH radical reactions is shown in Scheme 1 for high-NOx (>30 ppt) conditions. While formaldehyde is the dominant aldehyde product, methylglyoxal and glyoxal appear as 2nd and 3rd generation products via multiple pathways. Once gas-phase dicarbonyls are produced by such mechanisms, they can be further oxidized, photolyzed, or taken up into the aqueous phase where they may form hydrates and/or oligomers.

Uptake and reaction of glyoxal can take place in an aqueous phase medium as small as an adsorbed liquid (mono)layer on a solid particle (*35*, *36*). Several studies have estimated global SOA formed by aqueous-phase reactions of glyoxal and methylglyoxal at 3 – 13 TgC/year for glyoxal (*37–39*) and 1.5 – 8 TgC/year for methylglyoxal (*38–40*). These estimates represent ~6% to 28% of global SOA production (*41*), as well as significant fractions of total SOA in some urban areas (*42*). As a result, interest has grown in understanding the chemical processes involved in aqueous SOA (aqSOA) formation, and in detecting the hydrate, oligomer, and acidic products generated from dicarbonyl uptake into atmospheric aqSOA particles.

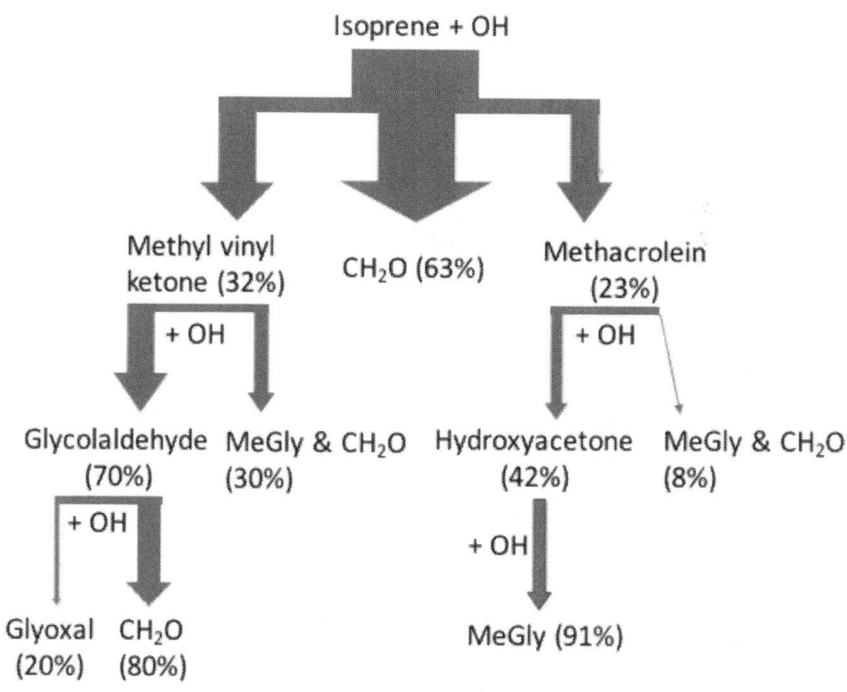

Scheme 1. Oxidation products of isoprene under high-NOx conditions (> 30 ppt), with arrow widths proportional to yields. Data taken from ref (33).

But are these estimates of aqSOA production correct? Because global clouds have much larger surface area and volume than that of aqueous aerosol particles, models predict that clouds take up more water-soluble gases and are the main site of aqueous SOA (aqSOA) production involving α-dicarbonyls (*39*). However, aqSOA production estimates for aldehydes have large uncertainties because of persistent questions about the reversibility and the mechanism of uptake (dependent on surface area or ion catalysis) (*43*), and the magnitude of the uptake coefficient (γ) itself. For example, studies of glyoxal uptake to aqueous aerosol have come to opposite conclusions about reversibility (*35, 44, 45*). Our studies of glyoxal-containing droplets generated in the laboratory indicated that glyoxal uptake to clouds must be at least partially reversible: at initial concentrations between 4 and 1000 μM, 50 – 65% of the glyoxal evaporated along with the water, while 35 – 50% of the glyoxal either self-reacted (*46*) or reacted with ammonium salts to form oligomer species with low volatility, Figure 1 (*40*). Similarly, lab studies of glyoxal uptake have measured $\gamma = 0.0004$ on solid glycine aerosol at 50% RH (*22*), $\gamma = 0.001$ on aqueous droplets (*47*), and as large as $\gamma = 0.016$ on non-hygroscopic ammonium sulfate / fulvic acid aerosol under photolytic conditions (*43, 48*). Glyoxal uptake coefficients depend not only on the type of aqueous aerosol, but also on relative humidity (RH) (*35, 45*) and the acidity (*45*) and ionic strength of the aerosol, which themselves depend on RH (*44*). While the glyoxal uptake coefficient measured by Liggio *et al.* (*45*) ($\gamma = 0.0029$ on non-acidified aerosol) has been used in several modeling studies (*39, 49, 50*), much higher values ($\gamma = 0.016$) were required to successfully model Mexico City $PM_{2.5}$ levels (*42*).

Even less is known about methylglyoxal uptake. Methylglyoxal uptake coefficients have been measured only on 55-85% H_2SO_4 solutions (*51*). Such highly acidic aerosol may be reasonable proxies for stratospheric aerosol, but are dissimilar to aged tropospheric particles, which contain organic species and are less acidic in comparison. Modelers have therefore used glyoxal uptake coefficients to estimate SOA formation by methylglyoxal (*39, 42, 49, 50*). While this may at first seem reasonable, it is important to recognize that a single methyl group gives methylglyoxal a much higher surface activity (*4, 5, 52*), lower hydration equilibrium constants (especially at the ketone functional group) (*53, 54*), a much lower Henry's law coefficient (*6, 7, 10, 53, 54*), and more diverse oligomerization pathways (aldol condensation in addition to acetal formation) (*55, 56*). Thus, in order to characterize the atmospheric significance of SOA formation by methylglyoxal, there is a clear need for better estimates and/or measurements of methylglyoxal uptake coefficients onto atmospherically-relevant aerosol and droplet surfaces.

Curry *et al.* recently estimated methylglyoxal uptake coefficients from first principles based on known aqueous reaction rates with OH radicals (*57*). Using in-cloud OH radical concentrations ranging from 5×10^{-15} to 5×10^{-12} M (*58*), they estimated methylglyoxal uptake coefficients of $< 1 \times 10^{-5}$ onto cloud droplets, and $< 1 \times 10^{-6}$ onto humidified sulfate/nitrate/ ammonium aerosol particles (*57*).

Finally, even when methylglyoxal is taken up into a cloud, Figure 1 shows that it is more likely to be returned to the gas phase upon droplet evaporation than is glyoxal. This data emphasizes the need for multiphase laboratory studies of methylglyoxal chemistry.

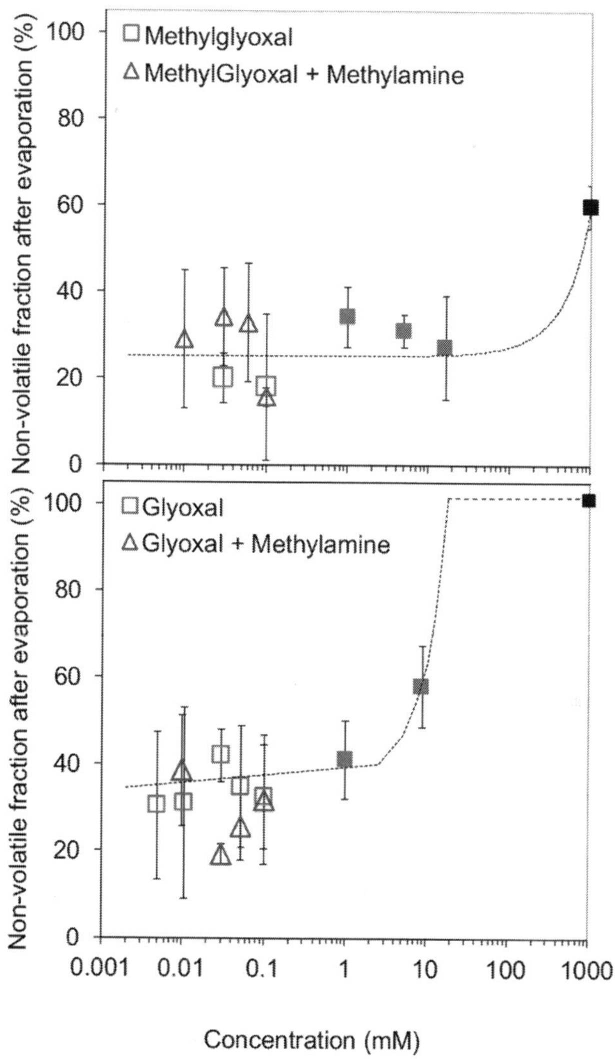

Figure 1. Measurements of non-volatile fraction produced by evaporation of droplets containing methylglyoxal (top) or glyoxal (bottom) as a function of concentration, with (triangles) or without (squares) methylamine at the same concentration. Open symbols: monodisperse droplet experiments. Solid symbols: polydisperse (1-20 mM) and large droplet (1 M) experiments. (Reprinted with permission from ref (40). Copyright 2014 American Chemical Society.)

Methylglyoxal Uptake on Aqueous Aerosol

In a recent study at the Chamber for Experimental Multiphase Atmospheric Simulation (CESAM) at Université Paris Est - Créteil, we measured methylglyoxal uptake coefficients on deliquesced aqueous aerosol containing ammonium sulfate or glycine (*59*). The methods used in these methylglyoxal experiments at the CESAM chamber were described in recent publications (*59*, *60*). CESAM is a 4 m^3 chamber with automated temperature and pressure control (*61*, *62*) that uses input flows of humidified or dry purified air to offset total sampling flows. Constant pressure is maintained at 5 to 100 mbar above ambient levels so that, even if a leak occurs, no outside air enters the chamber. The stirred chamber has a mixing time of 1 min, uncoated 304L stainless steel walls, and is pumped down to a few mTorr between each experiment. While uncoated steel walls may be more reactive towards gas-phase species than other surfaces (such as Teflon or halocarbon wax) used in atmospheric chambers, aerosol lifetimes in the chamber are lengthened to several hours because the conductive walls do not hold a static charge. The chamber walls were cleaned with pure ethanol (VWR, 99%) and ultrapure water (18.2 MΩ, ELGA Maxima) to remove contaminants after a set of experiments, before changing to new reactant gases. This is particularly important for aldehydes and amines, which can desorb from the walls, especially at high RH. During a set of experiments, however, deposited aerosol particles would build up on the chamber walls until they were removed by cleaning.

Ammonium sulfate (AS) and glycine seed aerosol particles were produced by atomizing 1 – 2 mM or 10 mM aqueous solutions, respectively, followed by addition of methylglyoxal gas. Methylglyoxal concentrations were monitored continuously by high-resolution PTR-MS (KORE Tech. Series II). RH-dependent PTR-MS methylglyoxal signals were corrected by dividing by the sum of the ^{18}O water and water cluster signals at 21 and 39, and were also corrected for dilution in the chamber. These traces of methylglyoxal concentration vs. time were then corrected for RH-dependent wall losses (Figure 2) and fit with 1st-order loss rate constants (Figure 3). The observed uptake coefficient γ, defined as the fraction of gas molecule collisions with a surface that lead to reactive uptake, is then calculated from the observed rate constant using the equation:

$$\gamma = \frac{4k}{\bar{c}\,SA}$$

where k is the 1st-order loss rate constant (s^{-1}) extracted from gas-phase methylglyoxal signals, SA is the aerosol surface area (m^2 surface / m^3 air), and c-bar is the mean speed of methylglyoxal molecules in m/s. Aerosol surface areas were derived from both scanning mobility particle sizing (SMPS, TSI 3080/3772, 20-900 nm, sampling via Nafion drying tube) and, at high RH, optical droplet scattering spectrometry (Welas, Palas Particle Tech., 0.25 to 15 μm, corrected for inlet losses (*63*)). In order to measure methylglyoxal uptake rates at high RH, the water vapor in the chamber was increased stepwise up to supersaturation by additions of high purity water vapor from a steel boiler.

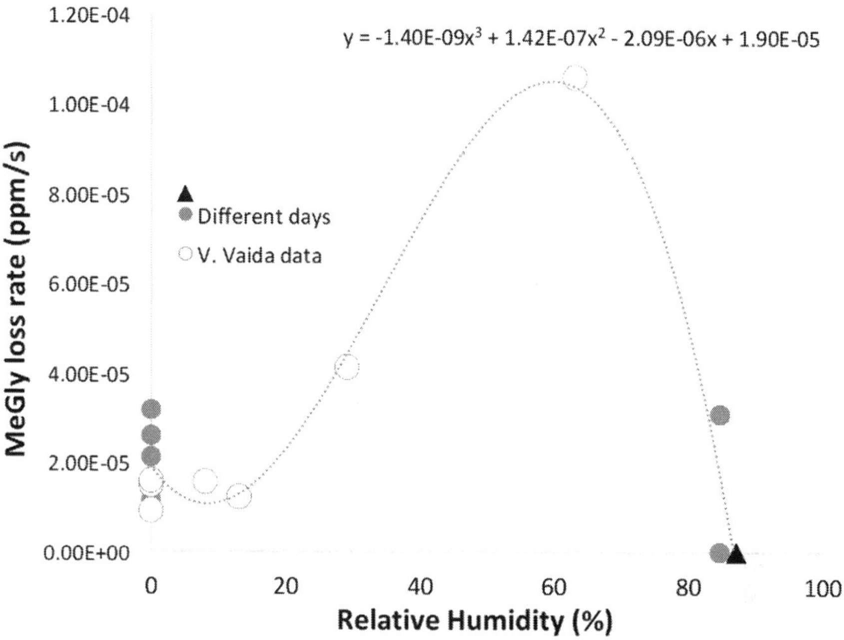

Figure 2. Methylglyoxal losses measured on steel walls, expressed as zero-order loss rates (ppm/s) as a function of relative humidity. Different symbols denote data measured on different days. Open circles are from experiments performed at CESAM by V. Vaida of the University of Colorado, Boulder. Polynomial fit (dotted line and equation) was used to correct for wall losses for aerosol experiments at RH between 15 and 87%. Aerosol experiments conducted below 15% RH were corrected by the average wall loss measured in this range (0.018 ±0.007 ppb s^{-1}). Above 87% RH, no wall losses were observed, so no wall loss correction was required. (Adapted with permission from ref (59). Copyright 2018 American Chemical Society.)

It is interesting to note from Figure 2 that losses of methylglyoxal to steel chamber walls were observed at RH up to 87%, but above this level an equilibrium with the walls was established within 2 minutes of the RH increase. In the absence of aerosol particles, this humid wall equilibrium could hold methylglyoxal concentrations at nearly constant levels for hours. A similar humid wall equilibrium has been previously noted in experiments with glyoxal (*44*, *45*).

In the presence of AS or glycine aerosol particles, methylglyoxal concentrations would decline, however, even at RH > 87%. Sample data for methylglyoxal in the presence of glycine aerosol at 72% RH is shown in Figure 3. It can be seen that methylglyoxal loss in the presence of aerosol particles is a long-term process (continuing for over 40 min.), and is much more rapid than wall losses (estimated from Figure 2) even when these wall losses were occurring at near-maximum rates.

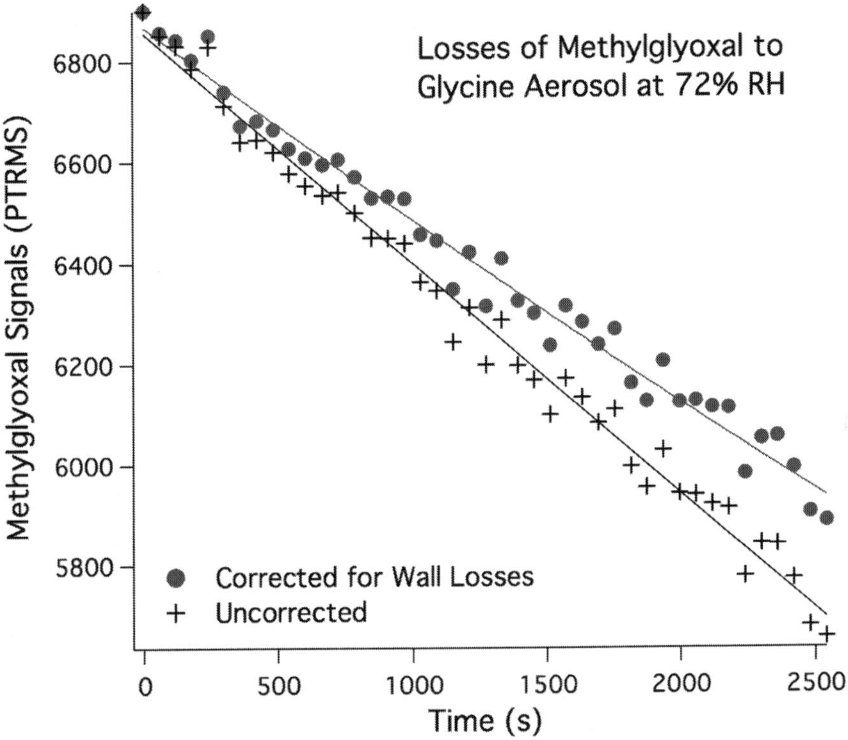

Figure 3. Methylglyoxal losses measured in the presence of glycine aerosol at 72% RH, before (gray +) and after correction for losses to steel chamber walls (filled circles). Fit lines are 1st-order exponential functions.

First-order rate constants for methylglyoxal uptake by aerosol extracted from datasets like Figure 3 were converted to uptake coefficients as described above, and are compared with uptake coefficients measured for glyoxal by Liggio et al. (*45*) in Figure 4. It is apparent that methylglyoxal uptake coefficients rapidly increase with increasing RH, as predicted by Curry *et al.* (*57*), who attributed this trend to "salting out" effects (*10*). However, our measured uptake coefficients are larger than those predicted by Curry *et al.* (*57*) by more than four orders of magnitude.

In making this comparison, it is important to recognize that Curry *et al.* (*57*) considered only the aqueous-phase reaction with OH radicals in their estimates of methylglyoxal uptake coefficients, and explicitly did not consider other irreversible chemical processes. Methylglyoxal is known to participate in reactions with ammonium salts, amines, and amino acids such as glycine to irreversibly form imidazoles and other oligomers (*5, 23, 26, 64–66*). It is evidently these processes that cause the unexpectedly high uptake rates observed in our study.

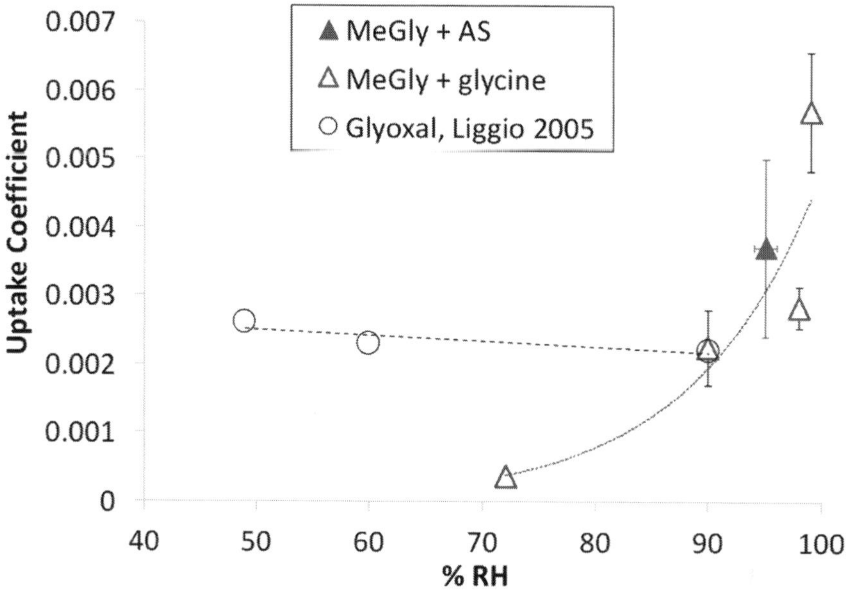

Figure 4. Comparison of uptake coefficients measured for methylglyoxal (filled triangle, AS; open triangles, pre-reacted glycine aerosol) from ref (59), and glyoxal (open circles, AS and sodium nitrate aerosol) from ref (45). (Reprinted with permission from ref (59). Copyright 2018 American Chemical Society)

We can assume that glyoxal uptake coefficients do not increase above 90% RH, since glyoxal is less soluble at low salt concentrations (high RH) than at high salt concentrations (low RH) (*9*, *10*, *57*). Our new measurements suggest that above 90% RH, methylglyoxal uptake by aqueous aerosol surfaces is more efficient than glyoxal uptake. This is unexpected, since glyoxal has an effective Henry's law coefficient that is 80× larger than that of methylglyoxal even in pure water (*7*). However, the uptake coefficients we measured over rather long exposure times (7 to 42 min.) were evidently limited by aqueous phase reactions, rather than Henry's law equilibria. The measured methylglyoxal uptake coefficient on cloud-processed, aqueous glycine aerosol at 99% RH is also 2× larger than the $\gamma_{glyoxal}$ = 0.0029 value used for methylglyoxal uptake to cloud droplets in some recent modeling studies (*39*, *50*). Incorporation of our RH-dependent values in future modeling efforts should improve the estimation of SOA formation in aqueous aerosol, but an uptake coefficient for methylglyoxal on cloud droplets is still needed.

Methylglyoxal / Cloud Interactions

Since it would be beneficial to know the uptake coefficient of methylglyoxal onto cloud droplets, we attempted to extract this information from data collected during multiple cloud events in the CESAM chamber. Each cloud event caused gas-phase methylglyoxal signals to decline initially, and then to recover after a few minutes as the cloud dissipated. The size of each temporary methylglyoxal decline was proportional to the peak droplet counts for each cloud event. Furthermore, because cloud size distributions were similar in each event, the methylglyoxal declines were also proportional to cloud surface area and total cloud water content. This means we cannot distinguish between surface and bulk uptake processes in cloud droplets. We also concluded that the 1-min time resolution in the large CESAM chamber is too slow to follow this fast and reversible uptake process in detail. It appears that gas / cloud droplet equilibrium is reached in < 1 min. for methylglyoxal, and uptake coefficients onto cloud surfaces could therefore not be determined. Furthermore, the Fuchs-Sutugin equation suggests that even if we were able to experimentally follow faster uptake processes in the chamber, methylglyoxal uptake onto cloud droplets would have a Knudsen number $K_n < 0.05$. This means the uptake measurement would be in the continuum regime, where uptake is limited by gas-phase diffusion.

The fact that methylglyoxal uptake was temporary, however, is useful, in that it demonstrates that the uptake of methylglyoxal by cloud droplets is largely (~80%) reversible. Furthermore, we noted that the methylglyoxal concentrations in the gas phase after a cloud event were always greater than what would be predicted by extrapolation of the pre-cloud first-order decline in methylglyoxal signals. We quantified this increase relative to the extrapolated trend for each cloud event, and found that each cloud event returned 3 to 8% of aerosol-phase methylglyoxal back to the gas phase (*59*). We attributed this process to the hydrolysis of some methylglyoxal oligomers during cloud events, which adds to the monomer pool, some of which can dehydrate and return to the gas phase in order to reestablish equilibrium between the gas and aqueous phases.

Light Absorption by Dicarbonyl Reaction Products

A number of studies have reported browning of concentrated mixtures of glyoxal and methylglyoxal with ammonium salts (*4*, *5*, *19*, *66*, *67*) and amines (*18*, *25*, *26*, *60*, *68*). Amines and high pH tend to accelerate brown carbon formation (*69*), while the presence of excess sugars (*70*), acids (*69*), and small aldehydes slow it down (*18*). Dicarbonyl / ammonium sulfate / amine brown mixtures produce a wide variety of nitrogen-containing oligomer molecules (*21*, *26*, *66*). Many similarities to oligomerized, humic-like substances (HULIS) extracted from atmospheric particles have been observed (*71*). (The N / C ratio in HULIS is noticeably lower, however.) Brown dicarbonyl / ammonium salt bulk aqueous mixtures are sometimes used in laboratory studies as proxies of atmospheric HULIS or brown carbon.

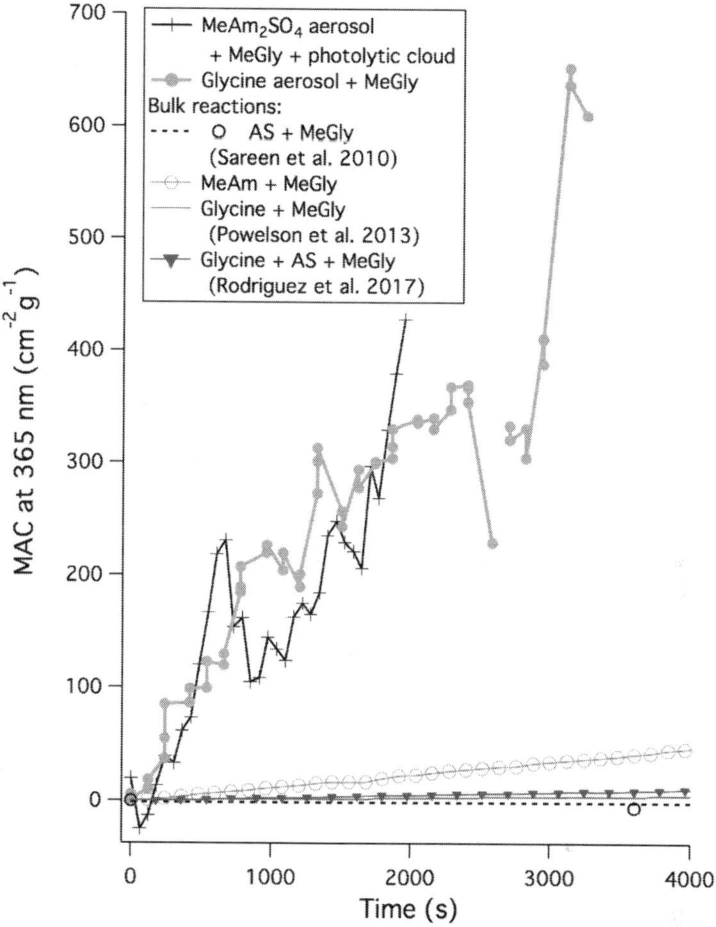

Figure 5. Faster brown carbon formation in aerosol than in bulk simulations. Comparison of time-dependent mass absorption coefficients measured at 365 nm during browning events in deliquesced glycine aerosol (thick line with filled circles) and deliquesced methylammonium sulfate (line with crosses), and in bulk aqueous phase experiments: 16 mM MeGly + 3.1 M AS (dotted line, ref (5)); 0.25 M MeGly + 0.25 M methylamine (line with open circles, ref (68)); 0.125 M MeGly + 0.25 M glycine (thin solid line, ref (68)); 0.25 M MeGly + 0.25 M AS + 0.05 glycine (line with triangles, ref (18)). (Reprinted with permission from ref (60). Copyright 2017 American Chemical Society.)

Until recently, almost all studies of brown carbon formation by glyoxal and methylglyoxal were performed in bulk solutions meant to simulate aqueous aerosol particles (*4*, *5*, *18*, *19*, *25*, *68*). By comparing bulk phase measurements with aerosol mass spectrometer measurements of aerosol in small chambers, we observed that reactant losses and first generation product formation were accelerated by over three orders of magnitude in suspended aerosol particles relative to bulk-phase measurements (*22*). This increase suggested that important steps in the chemical mechanism were taking place at aerosol surfaces, such as the dehydration of hydrated aldehyde functional groups, and therefore could not be accessed in bulk solution experiments. Lee *et al.* (*67*) then showed that glyoxal and ammonium sulfate mixtures produced substantially greater absorbance when aerosolized and dried than when left in solution, demonstrating that the entire reaction sequence forming brown carbon, not just the first few steps, could be greatly accelerated in evaporating aerosol particles (*72*).

In response, we have recently studied brown carbon formation in a variety of systems while comparing aerosol and bulk phase reaction rates. Measuring brown carbon formation in the aerosol phase can be done by simply filtering the aerosol and extracting the brown carbon, as done by Lee et al. (*67*) The disadvantage of this method is that a few hours are typically required to amass enough material for absorbance measurements. Some of the browning observed, therefore, may occur during the hours-long aerosol filtration process, rather than during the actual lifetime of the suspended aerosol particle, which is typically only a few minutes long. In order to measure aerosol browning rates in a near *in situ* fashion, we have been recently collaborating with Lelia Hawkins at Harvey Mudd College in the use of particle-into-liquid sampling (PILS) followed by absorption measurements in a waveguide spectrometer and online total organic carbon (TOC) measurements. Together, this data can be combined to determine the wavelength-dependent absorption of water-soluble brown carbon as it forms in a chamber, expressed as mass absorption coefficients (MAC) in cm^{-2} g^{-1}. An example of browning rates measured by PILS / waveguide spectroscopy in two CESAM chamber experiments involving methylglyoxal is shown in Figure 5, and compared with four published browning rates measured for methylglyoxal in bulk (*60*).

The aerosol-based browning measurements shown are 13 to 500× faster than the bulk phase measurements. Furthermore, the aerosol-based measurement shown as a black line (with + markers) in Figure 5 was recorded with the chamber solar simulator lights on. This is in striking contrast to bulk-phase methylglyoxal experiments, where light-absorbing brown carbon products are rapidly destroyed (*73*).

Summary

New measurements indicate that uptake of methylglyoxal to aqueous glycine and AS aerosol particles at high RH is larger than that of glyoxal, due mainly to irreversible reactions with amine and ammonia functional groups. This suggests that SOA formation via methylglyoxal processing will be larger than expected in aqueous aerosol. On the other hand, uptake to cloud droplets is fast but largely reversible, and can even return some aerosol-phase methylglyoxal back to the gas phase. Thus, in cloud droplets – thought to be the more important location for aqSOA production due to their higher water content – the new measurements indicate that aqSOA formation by methylglyoxal will be smaller than expected. In both locations, cloud droplets and aqueous aerosol particles, brown carbon produced by methylglyoxal uptake is likely to be larger than anticipated, because we find that bulk-phase simulation experiments tend to underproduce brown carbon relative to aerosol-phase experiments. Furthermore, when brown carbon produced by methylglyoxal reactions is photolyzed in bulk-phase experiments, it is rapidly bleached, while photolysis of aqueous, cloud-processed brown carbon aerosol can actually cause further browning by triggering further oligomer formation via radical-based reactions.

Acknowledgments

This work was supported by NSF grant AGS-1523178. The author acknowledges France's CNRS-INSU for supporting CESAM as an open facility through the National Instrument label. The CESAM chamber has received funding from the European Union's Horizon 2020 research and innovation programme through the EUROCHAMP-2020 Infrastructure Activity under grant 730997.

References

1. Twomey, S. Cloud Nucleation in the Atmosphere and Influence of Nucleus Concentration Levels in Atmospheric Physics. *J. Phys. Chem.* **1980**, *84*, 1459–1463.
2. Ervens, B.; Turpin, B. J.; Weber, R. J. Secondary Organic Aerosol Formation in Cloud Droplets and Aqueous Particles (Aqsoa): A Review of Laboratory, Field and Model Studies. *Atmos. Chem. Phys.* **2011**, *11*, 11069–11102.
3. Blando, J. D.; Turpin, B. J. Secondary Organic Aerosol Formation in Cloud and Fog Droplets: A Literature Evaluation of Plausibility. *Atmos. Environ.* **2000**, *34*, 1623–1632.
4. Shapiro, E. L.; Szprengiel, J.; Sareen, N.; Jen, C. N.; Giordano, M. R.; McNeill, V. F. Light-Absorbing Secondary Organic Material Formed by Glyoxal in Aqueous Aerosol Mimics. *Atmos. Chem. Phys.* **2009**, *9*, 2289–2300.
5. Sareen, N.; Schwier, A. N.; Shapiro, E. L.; Mitroo, D.; McNeill, V. F. Secondary Organic Material Formed by Methylglyoxal in Aqueous Aerosol Mimics. *Atmos. Chem. Phys.* **2010**, *10*, 997–1016.
6. Betterton, E. A.; Hoffmann, M. R. Henry's Law Constants of Some Environmentally Important Aldehydes. *Environ. Sci. Technol.* **1988**, *22*, 1415–1418.
7. Ip, H. S. S.; Huang, X. H. H.; Yu, J. Z. Effective Henry's Law Constants of Glyoxal, Glyoxylic Acid, and Glycolic Acid. *Geophys. Res. Lett.* **2009**, *36*, L01802.
8. Yu, G.; Bayer, A. R.; Galloway, M. M.; Korshavn, K. J.; Fry, C. G.; Keutsch, F. N. Glyoxal in Aqueous Ammonium Sulfate Solutions: Products, Kinetics, and Hydration Effects. *Environ. Sci. Technol.* **2011**, *45*, 6336–6342.
9. Kampf, C. J.; Waxman, E. M.; Slowik, J. G.; Dommen, J.; Pfaffenberger, L.; Praplan, A. P.; Prévôt, A. S. H.; Baltensperger, U.; Hoffmann, T.;

Volkamer, R. Effective Henry's Law Partitioning and the Salting Constant of Glyoxal in Aerosols Containing Sulfate. *Environ. Sci. Technol.* **2013**, *47*, 4236–4244.

10. Waxman, E. M.; Elm, J.; Kurtén, T.; Mikkelsen, K. V.; Ziemann, P. J.; Volkamer, R. Glyoxal and Methylglyoxal Setschenow Salting Constants in Sulfate, Nitrate, and Chloride Solutions: Measurements and Gibbs Energies. *Environ. Sci. Technol.* **2015**, *49*, 11500–11508.
11. Tolocka, M. P.; Jang, M.; Ginter, J. M.; Cox, F. J.; Kamens, R. M.; Johnston, M. V. Formation of Oligomers in Secondary Organic Aerosol. *Environ. Sci. Technol.* **2004**, *38*, 1428–1434.
12. Chastrette, F.; Bracoud, C.; Chastrette, M.; Mattioda, G.; Christidis, Y. Study of Aqueous Glyoxal Solutions by Carbon-13 Nmr. *Bull. Soc. Chim. Fr.* **1983**, *1-2* (Pt. 2), 33–40.
13. Chastrette, F.; Chastrette, M.; Bracoud, C. Acetalization of Aqueous Glyoxal Solutions: Structure Determination by Tandem Gas Chromatography, Mass Spectrometry, and Factorial Analysis. *Bull. Soc. Chim. Fr.* **1986**, *1986*, 822–836.
14. Altieri, K. E.; Seitzinger, S. P.; Carlton, A. G.; Turpin, B. J.; Klein, G. C.; Marshall, A. G. Oligomers Formed through in-Cloud Methylglyoxal Reactions: Chemical Composition, Properties, and Mechanisms Investigated by Ultra-High Resolution Ft-Icr Mass Spectrometry. *Atmos. Environ.* **2008**, *42*, 1476–1490.
15. Tan, Y.; Lim, Y. B.; Altieri, K. E.; Seitzinger, S. P.; Turpin, B. J. Mechanisms Leading to Oligomers and SOA through Aqueous Photooxidation: Insights from Oh Radical Oxidation of Acetic Acid and Methylglyoxal. *Atmos. Chem. Phys.* **2012**, *12*, 801–813.
16. Tan, Y.; Perri, M. J.; Seitzinger, S. P.; Turpin, B. J. Effects of Precursor Concentration and Acidic Sulfate in Aqueous Glyoxal - Oh Radical Oxidation and Implications for Secondary Organic Aerosol. *Environ. Sci. Technol.* **2009**, *43*, 8105–8112.
17. Li, Z.; Schwier, A. N.; Sareen, N.; McNeill, V. F. Reactive Processing of Formaldehyde and Acetaldehyde in Aqueous Aerosol Mimics: Surface Tension Depression and Secondary Organic Products. *Atmos. Chem. Phys.* **2011**, *11*, 11617–11629.
18. Rodriguez, A. A.; de Loera, A.; Powelson, M. H.; Galloway, M. M.; De Haan, D. O. Formaldehyde and Acetaldehyde Increase Aqueous-Phase Production of Imidazoles in Methylglyoxal - Amine Mixtures: Quantifying a Secondary Organic Aerosol Formation Mechanism. *Environ. Sci. Technol. Lett.* **2017**, *4*, 234–239.
19. Noziere, B.; Dziedzic, P.; Cordova, A. Products and Kinetics of the Liquid-Phase Reaction of Glyoxal Catalyzed by Ammonium Ions (Nh_4^+). *J. Phys. Chem. A* **2009**, *113*, 231–237.
20. Galloway, M. M.; Chhabra, P. S.; Chan, A. W. H.; Surratt, J. D.; Flagan, R. C.; Seinfeld, J. H.; Keutsch, F. N. Glyoxal Uptake on Ammonium Sulphate Seed Aerosol: Reaction Products and Reversibility of Uptake under Dark and Irradiated Conditions. *Atmos. Chem. Phys.* **2009**, *9*, 3331–3345.

21. Kampf, C. J.; Jakob, R.; Hoffmann, T. Identification and Characterization of Aging Products in the Glyoxal/Ammonium Sulfate System -- Implications for Light-Absorbing Material in Atmospheric Aerosols. *Atmos. Chem. Phys.* **2012**, *12*, 6323–6333.
22. De Haan, D. O.; Corrigan, A. L.; Smith, K. W.; Stroik, D. R., Turley, J. T.; Lee, F. E.; Tolbert, M. A.; Jimenez, J. L.; Cordova, K. E.; Ferrell, G. R. Secondary Organic Aerosol-Forming Reactions of Glyoxal with Amino Acids. *Environ. Sci. Technol.* **2009**, *43*, 2818–2824.
23. De Haan, D. O.; Hawkins, L. N.; Kononenko, J. A.; Turley, J. J.; Corrigan, A. L.; Tolbert, M. A.; Jimenez, J. L. Formation of Nitrogen-Containing Oligomers by Methylglyoxal and Amines in Simulated Evaporating Cloud Droplets. *Environ. Sci. Technol.* **2011**, *45*, 984–991.
24. De Haan, D. O.; Tolbert, M. A.; Jimenez, J. L. Atmospheric Condensed-Phase Reactions of Glyoxal with Methylamine. *Geophys. Res. Lett.* **2009**, *36*, L11819.
25. Trainic, M.; Abo Riziq, A.; Lavi, A.; Rudich, Y. Role of Interfacial Water in the Heterogeneous Uptake of Glyoxal by Mixed Glycine and Ammonium Sulfate Aerosols. *J. Phys. Chem.* **2012**, *116*, 5948–5957.
26. De Haan, D. O.; Tapavicza, E.; Riva, M.; Cui, T.; Surratt, J.; Smith, A. C.; Jordan, M.-C.; Nilakantan, S.; Almodovar, M.; Stewart, T. N.; de Loera, A.; De Haan, A. C.; Cazaunau, M.; Gratien, A.; Pangui, E.; Doussin, J. F. Nitrogen-Containing, Light-Absorbing Oligomers Produced in Aerosol Particles Exposed to Methylglyoxal, Photolysis, and Cloud Cycling. *Environ. Sci. Technol.* **2018**, *52*, 4061–4071.
27. Volkamer, R.; Platt, U.; Wirtz, K. Primary and Secondary Glyoxal Formation from Aromatics: Experimental Evidence for the Bicycloalkyl-Radical Pathway from Benzene, Toluene, and *P*-Xylene. *J. Phys. Chem. A* **2001**, *105*, 7865–7874.
28. Motta, F.; Ghigo, G.; Tonachini, G. Oxidative Degradation of Benzene in the Troposphere. Theoretical Mechanistic Study of the Formation of Unsaturated Dialdehydes and Dialdehyde Epoxides. *J. Phys. Chem. A* **2002**, *106*, 4411–4422.
29. Dechapanya, W.; Eusebi, A.; Kimura, Y.; Allen, D. T. Secondary Organic Aerosol Formation from Aromatic Precursors. 1. Mechanisms for Individual Hydrocarbons. *Environ. Sci. Technol.* **2003**, *37*, 3662–3670.
30. Paulsen, D.; Dommen, J.; Kalberer, M.; Prevot, A. S. H.; Richter, R.; Sax, M.; Steinbacher, M.; Weingartner, E.; Baltensperger, U. Secondary Organic Aerosol Formation by Irradiation of 1,3,5-Trimethylbenzene-Nox-H2o in a New Reaction Chamber for Atmospheric Chemistry and Physics. *Environ. Sci. Technol.* **2005**, *39*, 2668–2678.
31. Martin-Reviejo, M.; Wirtz, K. Is Benzene a Precursor for Secondary Organic Aerosol? *Environ. Sci. Technol.* **2005**, *39*, 1045–1054.
32. Yu, J.; Jeffries, H. E.; Le Lacheur, R. M. Identifying Airborne Carbonyl Compounds in Isoprene Atmospheric Photooxidation Products by Their Pfbha Oximes Using Gas Chromatography / Ion Trap Mass Spectrometry. *Environ. Sci. Technol.* **1995**, *29*, 1923–32.

33. Spaulding, R. S.; Schade, G. W.; Goldstein, A. H.; Charles, M. J. Characterization of Secondary Atmospheric Photooxidation Products: Evidence for Biogenic and Anthropogenic Sources. *J. Geophys. Res. Atmos.* **2003**, *108*, 4247.
34. Guenther, A.; Hewitt, C. N.; Erickson, D.; Fall, R.; Geron, C.; Graedel, T.; Harley, P.; Klinger, L.; Lerdau, M.; McKay, W. A.; Pierce, T.; Scholes, B.; Steinbrecher, R.; Tallamraju, R.; Taylor, J.; Zimmerman, P. A Global Model of Natural Volatile Organic Compound Emissions. *J. Geophys. Res. Atmos.* **1995**, *100*, 8873–8892.
35. Corrigan, A. L.; Hanley, S. W.; De Haan, D. O. Uptake of Glyoxal by Organic and Inorganic Aerosol. *Environ. Sci. Technol.* **2008**, *42*, 4428–4433.
36. Connelly, B. M.; De Haan, D. O.; Tolbert, M. A. Heterogeneous Glyoxal Oxidation: A Potential Source of Secondary Organic Aerosol. *J. Phys. Chem. A* **2012**, *116*, 6180–6187.
37. Stavrakou, T.; Müller, J.-F.; De Smedt, I.; Van Roozendael, M.; Kanakidou, M.; Vrekoussis, M.; Wittrock, F.; Richter, A.; Burrows, J. P. The Continental Source of Glyoxal Estimated by the Synergistic Use of Spaceborne Measurements and Inverse Modelling. *Atmos. Chem. Phys.* **2009**, *9*, 8431–8446.
38. Myriokefalitakis, S.; Tsigaridis, K.; Mihalopoulos, N.; Sciare, J.; Nenes, A.; Kawamura, K.; Segers, A.; Kanakidou, M. In-Cloud Oxalate Formation in the Global Troposphere: A 3-D Modeling Study. *Atmos. Chem. Phys.* **2011**, *11*, 5761–5782.
39. Fu, T.-M.; Jacob, D. J.; Wittrock, F.; Burrows, J. P.; Vrekoussis, M.; Henze, D. K. Global Budgets of Atmospheric Glyoxal and Methylglyoxal, and Implications for Formation of Secondary Organic Aerosols. *J. Geophys. Res. Atmos.* **2008**, *113*, D15303.
40. Galloway, M. M.; Powelson, M. H.; Sedehi, N.; Wood, S. E.; Millage, K. D.; Kononenko, J. A.; Rynaski, A. D.; De Haan, D. O. Secondary Organic Aerosol Formation During Evaporation of Droplets Containing Atmospheric Aldehydes, Amines, and Ammonium Sulfate. *Environ. Sci. Technol.* **2014**, *48*, 14417–14425.
41. Hodzic, A.; Kasibhatla, P. S.; Jo, D. S.; Cappa, C. D.; Jimenez, J. L.; Madronich, S.; Park, R. J. Rethinking the Global Secondary Organic Aerosol (SOA) Budget: Stronger Production, Faster Removal, Shorter Lifetime. *Atmos. Chem. Phys.* **2016**, *16*, 7917–7941.
42. Ying, Q.; Cureño, I. V.; Chen, G.; Ali, S.; Zhang, H.; Malloy, M.; Bravo, H. A.; Sosa, R. Impacts of Stabilized Criegee Intermediates, Surface Uptake Processes and Higher Aromatic Secondary Organic Aerosol Yields on Predicted Pm2.5 Concentrations in the Mexico City Metropolitan Zone. *Atmos. Environ.* **2014**, *94*, 438–447.
43. Ervens, B.; Volkamer, R. Glyoxal Processing by Aerosol Multiphase Chemistry: Towards a Kinetic Modeling Framework of Secondary Organic Aerosol Formation in Aqueous Particles. *Atmos. Chem. Phys.* **2010**, *10*, 8219–8244.
44. Kroll, J. H.; Ng, N. L.; Murphy, S. M.; Varutbangkul, V.; Flagan, R. C.; Seinfeld, J. H. Chamber Studies of Secondary Organic Aerosol Growth by

Reactive Uptake of Simple Carbonyl Compounds. *J. Geophys. Res. Atmos.* **2005**, *110*, D23207.

45. Liggio, J.; Li, S.-M.; McLaren, R. Reactive Uptake of Glyoxal by Particulate Matter. *J. Geophys. Res. Atmos.* **2005**, *110*, D10304.
46. De Haan, D. O.; Corrigan, A. L.; Tolbert, M. A.; Jimenez, J. L.; Wood, S. E.; Turley, J. J. Secondary Organic Aerosol Formation by Self-Reactions of Methylglyoxal and Glyoxal in Evaporating Droplets. *Environ. Sci. Technol.* **2009**, *43*, 8184–8190.
47. Schweitzer, F.; Magi, L.; Mirabel, P.; George, C. Uptake Rate Measurements of Methanesulfonic Acid and Glyoxal by Aqueous Droplets. *J. Phys. Chem. A* **1998**, *102*, 593–600.
48. Volkamer, R.; Ziemann, P. J.; Molina, M. J. Secondary Organic Aerosol Formation from Acetylene (C_2h_2): Seed Effect on SOA Yields Due to Organic Photochemistry in the Aerosol Aqueous Phase. *Atmos. Chem. Phys.* **2009**, *9*, 1907–1928.
49. Parikh, H. M.; Carlton, A. G.; Zhou, Y.; Zhang, H.; Kamens, R. M.; Vizuete, W. Modeling Secondary Organic Aerosol Formation from Xylene and Aromatic Mixtures Using a Dynamic Partitioning Approach Incorporating Particle Aqueous-Phase Chemistry (Ii). *Atmos. Environ.* **2012**, *56*, 250–260.
50. Li, J.; Cleveland, M.; Ziemba, L. D.; Griffin, R. J.; Barsanti, K. C.; Pankow, J. F.; Ying, Q. Modeling Regional Secondary Organic Aerosol Using the Master Chemical Mechanism. *Atmos. Environ.* **2015**, *102*, 52–61.
51. Zhao, J.; Levitt, N. P.; Zhang, R.; Chen, J. Heterogeneous Reactions of Methylglyoxal in Acidic Media: Implications for Secondary Organic Aerosol Formation. *Environ. Sci. Technol.* **2006**, *40*, 7682–7687.
52. Sareen, N.; Schwier, A. N.; Lathem, T. L.; Nenes, A.; McNeill, V. F. Surfactants from the Gas Phase May Promote Cloud Droplet Formation. *Proc. Natl. Acad. Sci. U.S.A.* **2013**, *110*, 2723–2728.
53. Doussin, J. F.; Monod, A. Structure–Activity Relationship for the Estimation of Oh-Oxidation Rate Constants of Carbonyl Compounds in the Aqueous Phase. *Atmos. Chem. Phys.* **2013**, *13*, 11625–11641.
54. Raventos-Duran, T.; Camredon, M.; Valorso, R.; Mouchel-Vallon, C.; Aumont, B. Structure-Activity Relationships to Estimate the Effective Henry's Law Constants of Organics of Atmospheric Interest. *Atmos. Chem. Phys.* **2010**, *10*, 7643–7654.
55. Kua, J.; Hanley, S. W.; De Haan, D. O. Thermodynamics and Kinetics of Glyoxal Dimer Formation: A Computational Study. *J. Phys. Chem. A* **2008**, *112*, 66–72.
56. Krizner, H. E.; De Haan, D. O.; Kua, J. Thermodynamics and Kinetics of Methylglyoxal Dimer Formation: A Computational Study. *J. Phys. Chem. A* **2009**, *113*, 6994–7001.
57. Curry, L. A.; Tsui, W. G.; McNeill, V. F. Technical Note: Updated Parameterization of the Reactive Uptake of Glyoxal and Methylglyoxal by Atmospheric Aerosols and Cloud Droplets. *Atmos. Chem. Phys. Discuss.* **2018**, *2018*, 1–13.

58. Herrmann, H.; Hoffmann, D.; Schaefer, T.; Bräuer, P.; Tilgner, A. Tropospheric Aqueous-Phase Free-Radical Chemistry: Radical Sources, Spectra, Reaction Kinetics and Prediction Tools. *ChemPhysChem* **2010**, *11*, 3796–3822.
59. De Haan, D. O.; Jimenez, N. G.; de Loera, A.; Cazaunau, M.; Gratien, A.; Pangui, E.; Doussin, J. F. Methylglyoxal Uptake Coefficients on Aqueous Aerosol Surfaces. *J. Phys. Chem. A* **2018**, *122*, 4854–4860.
60. De Haan, D. O.; Hawkins, L. N.; Welsh, H. G.; Pednekar, R.; Casar, J. R.; Pennington, E. A.; de Loera, A.; Jimenez, N. G.; Symons, M. A.; Zauscher, M.; Pajunoja, A.; Caponi, L.; Cazaunau, M.; Formenti, P.; Gratien, A.; Pangui, E.; Doussin, J. F. Brown Carbon Production in Ammonium- or Amine-Containing Aerosol Particles by Reactive Uptake of Methylglyoxal and Photolytic Cloud Cycling. *Environ. Sci. Technol.* **2017**, *51*, 7458–7466.
61. Wang, J.; Doussin, J. F.; Perrier, S.; Perraudin, E.; Katrib, Y.; Pangui, E.; Picquet-Varrault, B. Design of a New Multi-Phase Experimental Simulation Chamber for Atmospheric Photosmog, Aerosol and Cloud Chemistry Research. *Atmos. Meas. Tech.* **2011**, *4*, 2465–2494.
62. Denjean, C.; Formenti, P.; Picquet-Varrault, B.; Katrib, Y.; Pangui, E.; Zapf, P.; Doussin, J. F. A New Experimental Approach to Study the Hygroscopic and Optical Properties of Aerosols: Application to Ammonium Sulfate Particles. *Atmos. Meas. Tech.* **2014**, *7*, 183–197.
63. von der Weiden, S.-L.; Drewnick, F.; Borrmann, S. Particle Loss Calculator — a New Software Tool for the Assessment of the Performance of Aerosol Inlet Systems. *Atmos. Meas. Tech.* **2009**, *2*, 479–494.
64. Bonsignore, A.; Leoncini, G.; Ricci, D.; Siri, A.; Bignardi, G. Properties of the Polymer Formed from Methylglyoxal in the Presence of Lysine. *Ital. J. Biochem.* **1973**, *22*, 105–116.
65. Wang, Y.; Ho, C.-T. Formation of 2,5-Dimethyl-4-Hydroxy-3(2h)-Furanone through Methylglyoxal: A Maillard Reaction Intermediate. *J. Agric. Food Chem.* **2008**, *56*, 7405–7409.
66. Lin, P.; Laskin, J.; Nizkorodov, S. A.; Laskin, A. Revealing Brown Carbon Chromophores Produced in Reactions of Methylglyoxal with Ammonium Sulfate. *Environ. Sci. Technol.* **2015**, *49*, 14257–14266.
67. Lee, A. K. Y.; Zhao, R.; Li, R.; Liggio, J.; Li, S.-M.; Abbatt, J. P. D. Formation of Light Absorbing Organo-Nitrogen Species from Evaporation of Droplets Containing Glyoxal and Ammonium Sulfate. *Environ. Sci. Technol.* **2013**, *47*, 12819–12826.
68. Powelson, M. H.; Espelien, B. M.; Hawkins, L. N.; Galloway, M. M.; De Haan, D. O. Brown Carbon Formation by Aqueous-Phase Aldehyde Reactions with Amines and Ammonium Sulfate. *Environ. Sci. Technol.* **2014**, *48*, 985–993.
69. Sedehi, N.; Takano, H.; Blasic, V. A.; Sullivan, K. A.; De Haan, D. O. Temperature- and pH-Dependent Aqueous-Phase Kinetics of the Reactions of Glyoxal and Methylglyoxal with Atmospheric Amines and Ammonium Sulfate. *Atmos. Environ.* **2013**, *77*, 656–663.

70. Drozd, G. T.; McNeill, V. F. Organic Matrix Effects on the Formation of Light-Absorbing Compounds from α-Dicarbonyls in Aqueous Salt Solution. *Environ. Sci.: Processes Impacts* **2014**, *16*, 741–747.
71. Hawkins, L. N.; Lemire, A. N.; Galloway, M. M.; Corrigan, A. L.; Turley, J. J.; Espelien, B. M.; De Haan, D. O. Maillard Chemistry in Clouds and Aqueous Aerosol as a Source of Atmospheric Humic-Like Substances. *Environ. Sci. Technol.* **2016**, *50*, 7443–7452.
72. Nguyen, T. B.; Lee, P. B.; Updyke, K. M.; Bones, D. L.; Laskin, J.; Laskin, A.; Nizkorodov, S. A. Formation of Nitrogen- and Sulfur-Containing Light-Absorbing Compounds Accelerated by Evaporation of Water from Secondary Organic Aerosols. *J. Geophys. Res. Atmos.* **2012**, *117*, D01207.
73. Lee, H. J.; Aiona, P. K.; Laskin, A.; Laskin, J.; Nizkorodov, S. A. Effect of Solar Radiation on the Optical Properties and Molecular Composition of Laboratory Proxies of Atmospheric Brown Carbon. *Environ. Sci. Technol.* **2014**, *48*, 10217–10226.

Physical Properties Impacting Multiphase Chemistry

Chapter 9

Aerosol Acidity: Direct Measurement from a Spectroscopic Method

R. L. Craig[1] and A. P. Ault[1,2,*]

[1]Department of Chemistry, University of Michigan, Ann Arbor, Michigan 48109, United States
[2]Department of Environmental Health Science, University of Michigan, Ann Arbor, Michigan 48109, United States
[*]E-mail: aulta@umich.edu

Acidity is a critical chemical property of atmospheric aerosols that governs key atmospheric processes, such as secondary organic aerosol formation. However, aerosol acidity is challenging to measure due to the non-conservative nature of H^+ ion concentration, which changes with relative humidity and aerosol water content. Current methods used to determine acidity include ion balance, phase partitioning, and thermodynamic modeling, which are all indirect and have limitations. Herein, we discuss the recent development of a direct method to measure aerosol acidity using vibrational spectroscopy. For this spectroscopic method, the vibrational modes of several organic and inorganic acids and their conjugate bases are distinct (often ~ 50 cm^{-1} apart) and measurable, despite structurally differing by only a proton. Using Raman microspectroscopy, the pH of individual particles (> 1 μm) has been determined for a range of chemical equilibrium systems at pH values near the pK_a, including bisulfate-sulfate, bicarbonate-carbonate, nitric acid-nitrate, bioxalate-oxalate, and acetic acid-acetate, as well as a mixture of bisulfate-sulfate and bioxalate-oxalate. The integrated peak areas for the acid and conjugate base vibrational modes are calibrated for concentration and combined with thermodynamic calculations of activity coefficients to determine H^+ activity, and, ultimately, pH. The development of direct methods for measuring aerosol acidity is needed to mechanistically understand multiphase chemistry in a range of important atmospheric systems.

Introduction

Acidity impacts many multiphase chemical processes of atmospheric aerosols; the most well-known and studied being secondary organic aerosol (SOA) formation. Overall, increased acidity levels are associated with enhanced SOA yields due to acid-catalyzed heterogeneous reactions, such as epoxide ring-opening reactions (*1*, *2*) or reactions of carbonyls (*3*). This phenomenon has been observed for a wide range of SOA precursors, including isoprene (*4–8*), specifically its photooxidation products methacroelin (*9*), methyl vinyl ketone (*10*), and isoprene epoxydiols (IEPOX) (*11*), monoterpenes (such as α-pinene) (*12–16*), sesquiterpenes (such as β-caryophyllene) (*12*, *17*), 1,3-butadiene (*18*), 2-methyl-3-buten-2-ol (MBO) (*19*, *20*), aromatics (such as toluene) (*21*), and aldehydes (*22*). In addition to increased kinetics of acid-catalyzed reactions, SOA yield can be enhanced in terms of mass due to the formation of larger oligomers (*6*, *10*, *23*).

One particular class of chemical species that is formed via pH-dependent reactions in SOA is organosulfates. Organosulfate abundance is strongly correlated with acidity, as formation often occurs only under acidic conditions (*24–26*). Although acidic conditions dictate organosulfate formation in SOA, these compounds can form via photooxidation and/or ozonolysis of many precursors, including α-pinene (*15*, *25–28*), β-pinene (*24–26*), limonene (*26*, *29*) and other monoterpenes, as well as isoprene (*6*, *26*, *27*, *30*, *31*), β-caryophyllene (*17*), MBO (*2*, *19*), alkanes (*32*), alkylamines (*33*), and polycyclic aromatic hydrocarbons (PAHs), such as naphthalene (NAP) and 2-methylnaphthalene (2-MeNAP) (*34*). While organosulfate formation has been observed and studied to a greater extent, formation and hydrolysis of organic nitrate compounds also occurs under acidic conditions (*2*, *16*, *35*). Though acid-catalyzed SOA formation has primarily been studied through laboratory and chamber work, it has been observed in ambient data globally (*8*, *36–40*), including the formation of oligomers (*41*), organosulfates (*26*), and organic nitrates (*42*).

In addition to SOA formation, pH also plays an important role in many other multiphase chemical and physical atmospheric processes. Examples include changes to gas-particle partitioning equilibria leading to increased reactive uptake of organic compounds (*43–48*), heterogeneous reactions of non-SOA particle types, such as chloride depletion in sea spray aerosol (SSA) (*49–52*), acid-catalyzed hydrolysis reactions (*35*, *53–56*), and increased metal ion dissolution and solubility under acidic conditions (*57–59*). Other multiphase processes affected by particle acidity include water uptake (*60–62*), liquid-liquid phase separation (LLPS) (*63–66*), light absorption (*67*, *68*), as well as photolysis and OH radical reaction chemistry (*69–71*). Despite the evidence for enhanced SOA formation and other multiphase atmospheric processes that occur under acidic conditions, more specific correlation to acidity is not well characterized, in part due to the challenges associated with measuring aerosol pH.

Aerosol pH is difficult to determine due to the small size of particles (attoliter or smaller volumes) and the non-conservative nature of H^+ with respect to other chemical species present, as it is dependent on aerosol liquid water content and relative humidity (RH). As a result of these challenges, indirect measurements, proxy methods, and thermodynamic equilibrium models have often been used to predict aerosol acidity (*72*). Indirect measurements from filter-based extraction and extrapolation are associated with high uncertainty, often due to changes in ion distribution during extraction or sampling artifacts (*72–75*). Proxy methods, such as ion balance and molar ratio, which infer H^+ concentration, and subsequently pH, by balancing measured concentrations of inorganic anions and cations (*76–79*), are also associated with high uncertainty and often cannot predict aerosol pH with more precision than "acidic" or "basic" classification (*72*). Additionally, the ion balance and molar ratio methods typically disagree with thermodynamic equilibrium model predictions of pH, and the discrepancy between the two methods can be attributed to lack of considerations for aerosol liquid water content, ion activity coefficients and the inability to differentiate between free and bound H^+ (e.g. protons associated with bisulfate, HSO_4^-, or other inorganic ions) (*72*, *76*, *77*, *80*, *81*).

In contrast, a third proxy method, phase partitioning, which uses measurements of semivolatile compounds in the gas and aerosol phase, such as NH_3/NH_4^+, to indirectly predict pH, has yielded much better agreement with models (*72*, *81*, *82*). Thermodynamic models, such as E-AIM (*83*, *84*) and ISORROPIA-II (*85*, *86*), use measurements of both gas and aerosol phase chemical species, temperature, and RH to predict aerosol pH. Although the phase partitioning method and thermodynamic models show the best agreement and are widely considered to be the most accurate of the current methods for indirect prediction of aerosol pH, they are not without limitations (*72*). Both methods are most accurate when constrained by measurements of gas and aerosol phase chemical components (*72*, *87*), but are sensitive to the respective measurement input values and their associated uncertainties. Additionally, both methods assume gas-particle phase equilibrium, which is not always accurate under conditions of low liquid water content or high ionic strength (*72*, *81*, *88–90*), and neither method accounts for the potential influence of organic components (*72*, *91*), such as organic acids, which are ubiquitous in ambient aerosol (*92*, *93*).

The phase partitioning method and thermodynamic models have been applied to evaluate ambient aerosol acidity and variability globally (*94–98*). Overall, aerosol particles are often acidic, but there is variability due to differing source contributions (*99–102*), regional location, such as urban versus rural areas (*99*, *103*), and seasonality (*104*, *105*). These predictions indicate that acidity-dependent chemistry can occur readily in most ambient aerosol, as pH levels are low enough, despite decreasing atmospheric gaseous SO_2 emissions and subsequent condensed-phase sulfate concentrations in some regions, such as the southeast United States, where SOA formation is prevalent (*106*). Direct measurement of aerosol pH is needed to constrain model and proxy method predictions of acidity, particularly when there is disagreement, and will help further understanding of the impact of pH on multiphase atmospheric processes. Additionally, measurement of individual particle pH is also key, as atmospheric processes occur on the single particle level and measurements of bulk aerosol pH are not necessarily representative of all particles within an aerosol population, as illustrated by Figure 1.

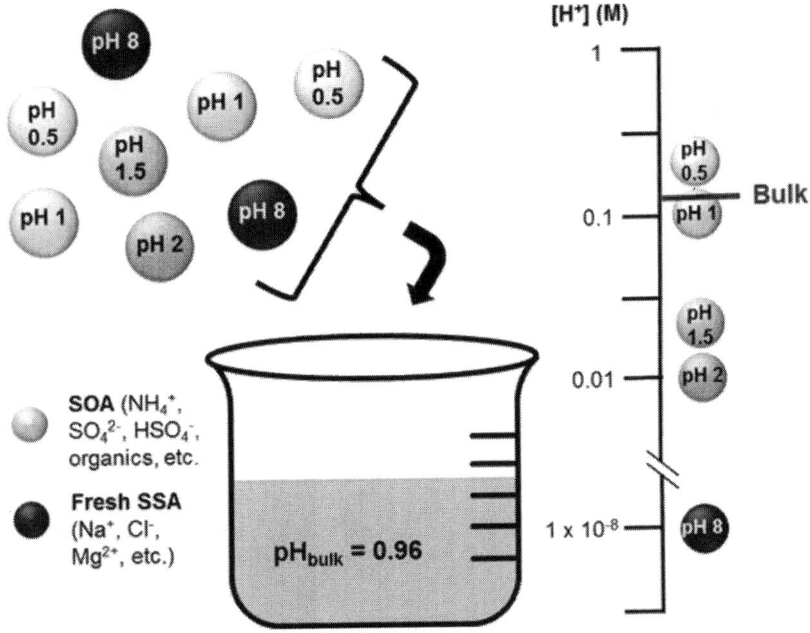

Figure 1. Schematic illustrating the difference between single particle and bulk aerosol pH measurements for a simplified system of a mixture of SOA and SSA particles of equal volume, but varying pH. Although individual particle pH ranges from 0.5 to 8, bulk pH is 0.96 and does not capture the variability in acidity levels within the particles. The scale on the right highlights the difference in [H+] for each of the particles compared to the bulk measurement.

Currently, there are few methods for direct measurement of aerosol acidity. One method measures proton concentration in particle samples collected on dyed filters through a colorimetric analysis integrated with a reflectance UV-Visible spectrometer (*107*, *108*). A second method monitors acidity via fluorescence spectroscopy with a pH-sensitive dye (*65*). However, application of these methods has been limited. Herein, a recently developed Raman microspectroscopic method for direct measurement of individual aerosol particle pH and its applications are discussed in further detail.

Raman Microspectroscopic Method for Measuring Single Particle Aerosol pH

The pH of individual aerosol particles can be determined using the Raman microspectroscopic method described in Rindelaub et al. and Craig et al. (*109*,

110) This method builds on the expansion in the use of Raman microspectroscopy for aerosol analysis in recent years (*111–119*). Briefly, Raman spectra are collected for individual aerosol particles and the peak areas of the vibrational modes corresponding to an acid (HA) and conjugate base (A^-) are integrated. The integrated peak areas for HA and A^- are related to concentration using calibration curves. For other ions not directly involved in the acid-base equilibrium, but still present in the particle, concentration is determined from the ratio of [ion]:[acid + conjugate base]. Ionic strength (I) can be calculated via Equation 1 once the concentration of all ions present is determined. C_i and z_i represent the concentration of each ion and its corresponding charge, respectively.

Equation 1 $$I = \frac{1}{2}\sum C_i z_i^2$$

Molality units are used for concentration of each species, determined by conversion from molarity using the density of the solution mixture. The solution densities are found by using the Laliberté model (*120*), and are iteratively solved during molality conversions. The density calculations require concentrations of each solute in the solution mixture, so equivalent concentrations of each cation and anion are found by Clegg's equivalent fraction method (*121*), which assumes that all possible combinations of cation and anion are present as solute components.

The extended Debye-Hückel relationship (Equation 2) can then be applied to calculate the activity coefficient for each species in the acid-base equilibrium. A and B represent constants characteristic of the solvent (water) and $å_i$ is the effective diameter of the ion in solution (*122*, *123*).

Equation 2 $$-log\gamma_i = \frac{Az_i^2\sqrt{I}}{1+å_i B\sqrt{I}}$$

The measured concentrations of acid and conjugate base and their respective calculated activity coefficients can then be related to the acid dissociation constant, K_a, to calculate $[H^+]$ (Equation 3), and subsequently, pH (Equation 4).

Equation 3 $$K_a = \frac{a_{H^+} * a_{A^-}}{a_{HA}} = \frac{([H^+]*\gamma_{H^+})([A^-]*\gamma_{A^-})}{[HA]*\gamma_{HA}}$$

Equation 4 $$pH = -log(a_{H^+}) = -log\ (\gamma_{H^+} * [H^+])$$

When solving Equations 1 – 3, the value of $[H^+]$ is not known and is needed in the calculation of ionic strength and activity coefficients and so an iterative method is required. Clegg's equivalent fraction method (*121*) is used to find the initial value for $[H^+]$, in a similar manner to the density calculations. The $[H^+]$ value can then be used to solve for ionic strength (Equation 1) and the activity coefficients (Equation 2). Using the activity coefficients, Equation 3 is applied to calculate a new value for $[H^+]$. The initial value for $[H^+]$ is then iteratively changed until it equals that from the Equation 3 calculations.

Thus far, this spectroscopic method has been applied to laboratory-generated aerosol particles (d_{pa} 1 – 30 μm) composed of the following equilibrium systems: nitric acid/nitrate (HNO_3/NO_3^-, pK_a -1.3), bisulfate/sulfate (HSO_4^-/SO_4^{2-}, pK_a 1.99), bioxalate/oxalate ($HC_2O_4^-/C_2O_4^{2-}$, pK_a 3.81), acetic acid/acetate (CH_3COOH/CH_3COO^-, pK_a 4.76), and bicarbonate/carbonate (HCO_3^-/CO_3^{2-}, pK_a 10.30), as well as a two-component mixture of HSO_4^-/SO_4^{2-} and $HC_2O_4^-/C_2O_4^{2-}$. These equilibrium systems cover a wide pH range, including the typical range for atmospheric aerosols (pH -0.5 – 5), as predicted by thermodynamic equilibrium models for several field campaigns, as shown in Figure 2 (*72*, *94–96*, *98*).

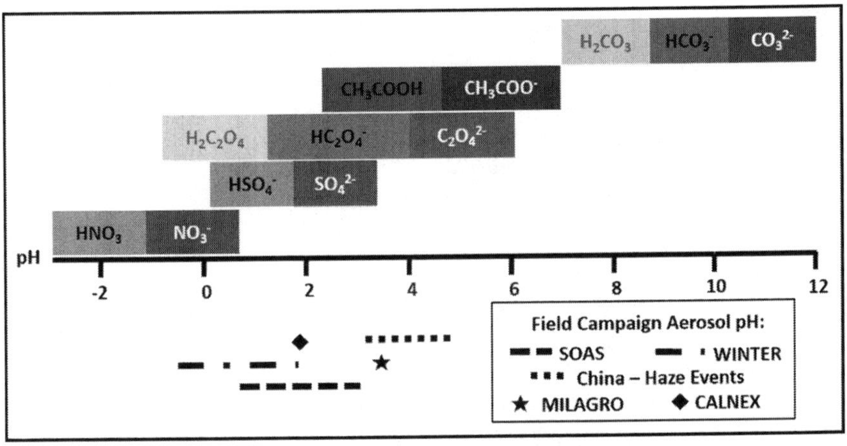

Figure 2. Schematic showing dominant species present as a function of pH for each acid-base system studied, as well as a comparison to the aerosol pH predicted by thermodynamic models for several field campaigns. $H_2C_2O_4$ and H_2CO_3 are included, but cannot be quantified with this method. (Reprinted with permission from Ref. (110), Copyright 2017 American Chemical Society.)

Through pH measurements with this method, the effect of RH on particle acidity, gas-particle partitioning of acidic chemical species, and the relationship between ionic strength and H^+ activity could also be studied. For RH impacts on aerosol acidity, after initial pH measurements, particles were exposed to varying RH conditions and changes in pH were monitored. With increasing RH, water uptake readily occurs, leading to a general dilution in concentration as well as dissociation of HSO_4^- and an increase in SO_4^{2-} concentration. Evidence for this increase in [SO_4^{2-}] is found in the observed changes in HSO_4^- and SO_4^{2-} peak intensity, as well as an increase in average particle pH, as shown in Figure 3. For all particles across a range of acidities, pH increases with increasing RH (up to ~1 pH unit). The slopes of linear regressions representing the relationship between average aerosol particle pH and RH across all acidity levels are comparable, ~0.01 pH unit/% RH, indicating a similar degree of change in aerosol pH with changing RH, regardless of initial pH.

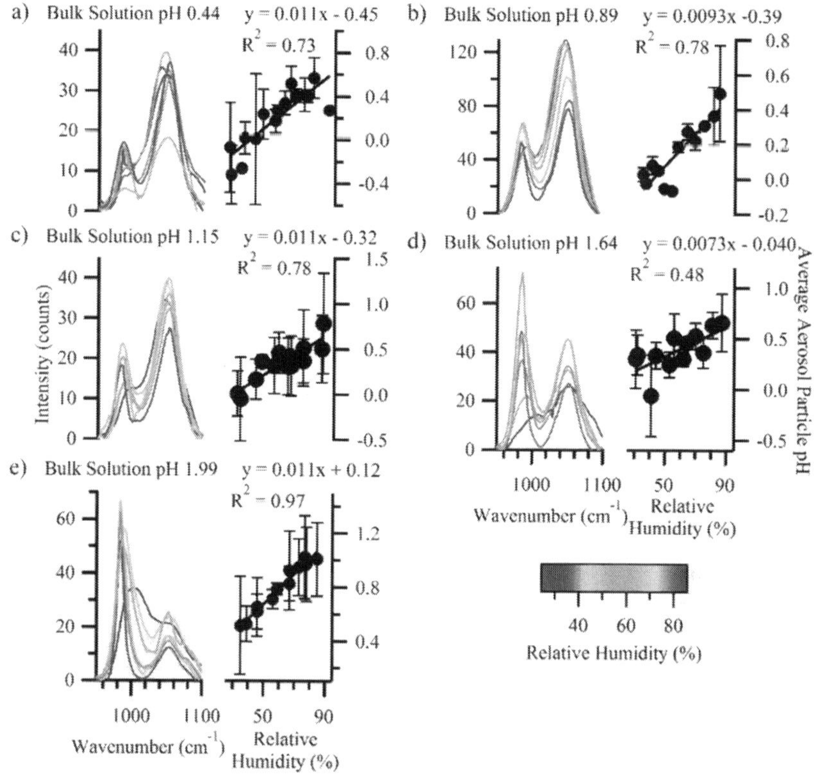

Figure 3. Raman spectra (left) of $v_s(SO_4^{2-})$ and $v_s(HSO_4^-)$ at varying RH for aerosol particles generated from bulk solution pH (a) 0.44, (b) 0.89, (c) 1.15, (d)1.64, and (e) 1.99. Average aerosol pH as a function of RH (right) for each bulk solution. Error bars are based on the standard deviation for multiple trials. (Reprinted with permission from Ref. (109). Copyright 2016 American Chemical Society.)

In terms of gas-particle partitioning, for the HNO_3/NO_3^- and CH_3COOH/CH_3COO^- equilibrium systems, HNO_3 and CH_3COOH are not conserved in the generated aerosol particles relative to their respective conjugate bases. Figure 4 shows a comparison of normalized Raman spectra of HNO_3/NO_3^- and CH_3COOH/CH_3COO^- particles to normalized Raman spectra of the bulk solutions from which the particles were generated. The vibrational modes corresponding to the acid for both systems are significantly more intense in the bulk solution than in the particle, indicating the [acid]:[base] ratio in the bulk solution is larger than that of the particle. This observation is attributed to partitioning of nitric acid and acetic acid from the particle to gas phase during the atomization process. Probable causes include quicker loss of vapor because of increased surface to volume ratio for the aerosol particles versus the solution droplets and/or heating from the laser. Nitric acid and acetic acid exhibit this behavior more than the other compounds because of their higher volatility. The partitioning of higher volatility acid species from particles could have implications for particle phase reactions, such as acid-displacement in mixed sea salt and weak organic acid particles (*52*), as well as the impact of acidity as a function of particle size.

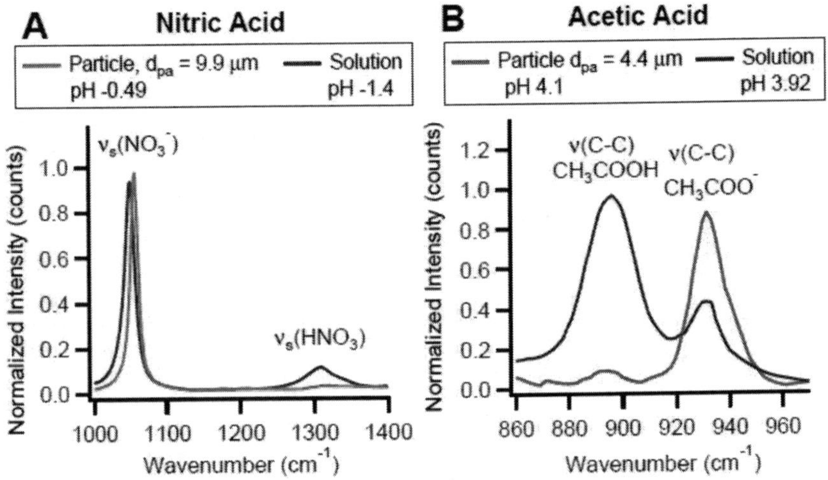

Figure 4. Raman spectra of a representative particle and the bulk solution for the A) HNO_3/NO_3^- and B) CH_3COOH/CH_3COO^- equilibrium systems. The spectra for the HNO_3/NO_3^- system were normalized to the $v_s(NO_3^-)$ mode and the spectra for the CH_3COOH/CH_3COO^- system were normalized to the v(C-C) mode of CH_3COO^-. (Reprinted with permission from Ref. (110). Copyright 2017 American Chemical Society.)

In addition to measuring the pH of aerosol particles for each of these systems, trends were observed for H^+ activity coefficient, $\gamma(H^+)$, in relation to ionic strength through measurements with this spectroscopic method. For all of the acid-base systems, there was a negative relationship for $\gamma(H^+)$ as a function of ionic strength (Figure 5), with increased $\gamma(H^+)$ sensitivity at lower ionic strength values. More specifically, the inorganic systems have larger, more varied ionic strength with lower, less variable $\gamma(H^+)$, while the inverse holds for the organic systems, which show larger, more varied $\gamma(H^+)$ with smaller and less variability in ionic strength. For the $HC_2O_4^-/C_2O_4^{2-}$ and HSO_4^-/SO_4^{2-} mixed system, particles had higher average pH than those composed of only $HC_2O_4^-/C_2O_4^{2-}$, even though the particles were generated from bulk solutions of similar pH and the $HC_2O_4^-/C_2O_4^{2-}$ system was used to determine pH. Because of the inorganic component in the $HC_2O_4^-/C_2O_4^{2-}$ and HSO_4^-/SO_4^{2-} mixed particles, $\gamma(H^+)$ was lower, leading to decreased activity of H^+ and less acidic particles overall. Conversely, the presence of the organic component in the $HC_2O_4^-/C_2O_4^{2-}$ and HSO_4^-/SO_4^{2-} mixed particles led to higher $\gamma(H^+)$ and lower ionic strength values than the particles composed of only HSO_4^-/SO_4^{2-}. It should be noted that for the sulfate-containing particles, inorganic sulfate is more soluble under higher ionic strength conditions, which could influence the concentration and activity of the chemical species present, including H^+. These observations of $\gamma(H^+)$ and ionic strength support that all chemical species present dictate ion behavior, and thus particle pH, in mixed organic and inorganic particles.

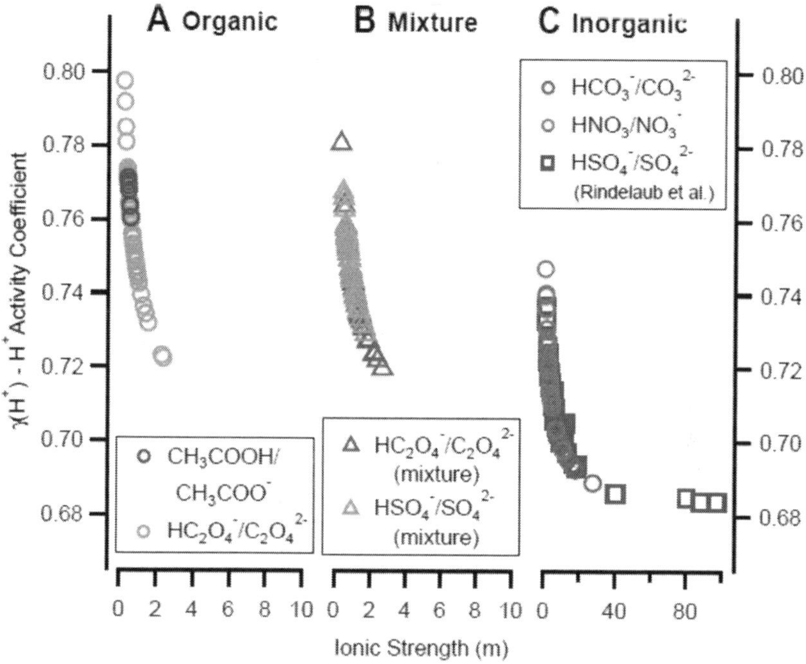

Figure 5. H^+ activity coefficient ($\gamma(H^+)$) as a function of ionic strength for each organic (A), mixture (B), and inorganic (C) acid-base system. Data for the HSO_4^-/SO_4^{2-} equilibrium from Rindelaub et al. is also included for comparison. Note the differing scales for ionic strength. (Reprinted with permission from Ref. (110). Copyright 2017 American Chemical Society.)

Conclusions

The development of direct measurement methods for aerosol pH are important to constrain indirect measurements via proxy methods and thermodynamic model predictions in order to improve understanding of the many pH-dependent multiphase atmospheric processes, as these processes have varying dependence on pH. To provide a hypothetical example, SOA formation is known to be acid-catalyzed, but differences between SOA formation at pH 2 versus pH 5 are not well quantified. Less SOA formation occurring at higher pH could have implications for haze formation, and ultimately radiative forcing effects. A recently developed method for direct measurement of single particle aerosol pH has been discussed. While further work is necessary for this method to be fully applicable to ambient aerosol pH measurements, studies with laboratory-generated aerosol particles have already provided insight into gas-particle partitioning of volatile acid species, and ion behavior in relation to ionic strength and H^+ activity.

Given the complex nature of aerosol particles in terms of chemical composition and atmospheric conditions, it is likely that there are many factors that influence aerosol acidity and pH-dependent atmospheric processes. Beyond its application here, Raman microspectroscopy can be a powerful tool for probing the protonation state of important species, such as carboxylic acid versus carboxylate groups in organosulfates, above and below the pK_a (*114*). Simpler methods suitable for field deployment, such as the use of pH indicator paper for aqueous aerosol pH can also be benchmarked by using the Raman method described above (*124*). Further studies of aerosol acidity and other important properties, including chemical morphology (*125*), phase state (*90, 126, 127*) and phase effects on kinetics (*128–130*) and surface tension (*131*), are needed to elucidate the potential mechanisms driving multiphase aerosol chemistry.

References

1. Eddingsaas, N. C.; VanderVelde, D. G.; Wennberg, P. O. Kinetics and Products of the Acid-Catalyzed Ring-Opening of Atmospherically Relevant Butyl Epoxy Alcohols. *J. Phys. Chem. A* **2010**, *114*, 8106–8113.
2. Mael, L. E.; Jacobs, M. I.; Elrod, M. J. Organosulfate and Nitrate Formation and Reactivity from Epoxides Derived from 2-Methyl-3-Buten-2-Ol. *J. Phys. Chem. A* **2015**, *119*, 4464–4472.
3. Jang, M.; Czoschke, N. M.; Lee, S.; Kamens, R. M. Heterogeneous Atmospheric Aerosol Production by Acid-Catalyzed Particle-Phase Reactions. *Science* **2002**, *298*, 814–817.
4. Surratt, J. D.; Lewandowski, M.; Offenberg, J. H.; Jaoui, M.; Kleindienst, T. E.; Edney, E. O.; Seinfeld, J. H. Effect of Acidity on Secondary Organic Aerosol Formation from Isoprene. *Environ. Sci. Technol.* **2007**, *41*, 5363–5369.
5. Jaoui, M.; Corse, E. W.; Lewandowski, M.; Offenberg, J. H.; Kleindienst, T. E.; Edney, E. O. Formation of Organic Tracers for Isoprene SOA under Acidic Conditions. *Atmos. Environ.* **2010**, *44*, 1798–1805.
6. Kuwata, M.; Liu, Y.; McKinney, K.; Martin, S. T. Physical State and Acidity of Inorganic Sulfate Can Regulate the Production of Secondary Organic Material from Isoprene Photooxidation Products. *Phys. Chem. Chem. Phys.* **2015**, *17*, 5670–5678.
7. Xu, L.; Middlebrook, A. M.; Liao, J.; de Gouw, J. A.; Guo, H.; Weber, R. J.; Nenes, A.; Lopez-Hilfiker, F. D.; Lee, B. H.; Thornton, J. A.; Brock, C. A.; Neuman, J. A.; Nowak, J. B.; Pollack, I. B.; Welti, A.; Graus, M.; Warneke, C.; Ng, N. L. Enhanced Formation of Isoprene-Derived Organic Aerosol in Sulfur-Rich Power Plant Plumes during Southeast Nexus. *J. Geophys. Res. Atmos.* **2016**, *121*, 11137–11153.
8. Edney, E. O.; Kleindienst, T. E.; Jaoui, M.; Lewandowski, M.; Offenberg, J. H.; Wang, W.; Claeys, M. Formation of 2-Methyl Tetrols and 2-Methylglyceric Acid in Secondary Organic Aerosol from Laboratory Irradiated Isoprene/NO_X/SO_2/Air Mixtures and Their Detection in Ambient

PM2.5 Samples Collected in the Eastern United States. *Atmos. Environ.* **2005**, *39*, 5281–5289.

9. Zhang, H.; Lin, Y.-H.; Zhang, Z.; Zhang, X.; Shaw, S. L.; Knipping, E. M.; Weber, R. J.; Gold, A.; Kamens, R. M.; Surratt, J. D. Secondary Organic Aerosol Formation from Methacrolein Photooxidation: Roles of NO_x Level, Relative Humidity and Aerosol Acidity. *Environ. Chem.* **2012**, *9*, 247–262.
10. Chan, K. M.; Huang, D. D.; Li, Y. J.; Chan, M. N.; Seinfeld, J. H.; Chan, C. K. Oligomeric Products and Formation Mechanisms from Acid-Catalyzed Reactions of Methyl Vinyl Ketone on Acidic Sulfate Particles. *J. Atmos. Chem.* **2013**, *70*, 1–18.
11. Riva, M.; Bell, D. M.; Hansen, A.-M. K.; Drozd, G. T.; Zhang, Z.; Gold, A.; Imre, D.; Surratt, J. D.; Glasius, M.; Zelenyuk, A. Effect of Organic Coatings, Humidity and Aerosol Acidity on Multiphase Chemistry of Isoprene Epoxydiols. *Environ. Sci. Technol.* **2016**, *50*, 5580–5588.
12. Offenberg, J. H.; Lewandowski, M.; Edney, E. O.; Kleindienst, T. E.; Jaoui, M. Influence of Aerosol Acidity on the Formation of Secondary Organic Aerosol from Biogenic Precursor Hydrocarbons. *Environ. Sci. Technol.* **2009**, *43*, 7742–7747.
13. Czoschke, N. M.; Jang, M. Acidity Effects on the Formation of α-Pinene Ozone SOA in the Presence of Inorganic Seed. *Atmos. Environ.* **2006**, *40*, 4370–4380.
14. Northcross, A. L.; Jang, M. Heterogeneous SOA Yield from Ozonolysis of Monoterpenes in the Presence of Inorganic Acid. *Atmos. Environ.* **2007**, *41*, 1483–1493.
15. Lal, V.; Khalizov, A. F.; Lin, Y.; Galvan, M. D.; Connell, B. T.; Zhang, R. Heterogeneous Reactions of Epoxides in Acidic Media. *J. Phys. Chem. A* **2012**, *116*, 6078–6090.
16. Han, Y.; Stroud, C. A.; Liggio, J.; Li, S. M. The Effect of Particle Acidity on Secondary Organic Aerosol Formation from α-Pinene Photooxidation under Atmospherically Relevant Conditions. *Atmos. Chem. Phys.* **2016**, *16*, 13929–13944.
17. Chan, M. N.; Surratt, J. D.; Chan, A. W. H.; Schilling, K.; Offenberg, J. H.; Lewandowski, M.; Edney, E. O.; Kleindienst, T. E.; Jaoui, M.; Edgerton, E. S.; Tanner, R. L.; Shaw, S. L.; Zheng, M.; Knipping, E. M.; Seinfeld, J. H. Influence of Aerosol Acidity on the Chemical Composition of Secondary Organic Aerosol from Beta-Caryophyllene. *Atmos. Chem. Phys.* **2011**, *11*, 1735–1751.
18. Lewandowski, M.; Jaoui, M.; Offenberg, J. H.; Krug, J. D.; Kleindienst, T. E. Atmospheric Oxidation of Isoprene and 1,3-Butadiene: Influence of Aerosol Acidity and Relative Humidity on Secondary Organic Aerosol. *Atmos. Chem. Phys.* **2015**, *15*, 3773–3783.
19. Zhang, H.; Worton, D. R.; Lewandowski, M.; Ortega, J.; Rubitschun, C. L.; Park, J.-H.; Kristensen, K.; Campuzano-Jost, P.; Day, D. A.; Jimenez, J. L.; Jaoui, M.; Offenberg, J. H.; Kleindienst, T. E.; Gilman, J.; Kuster, W. C.; de Gouw, J.; Park, C.; Schade, G. W.; Frossard, A. A.; Russell, L.; Kaser, L.; Jud, W.; Hansel, A.; Cappellin, L.; Karl, T.; Glasius, M.; Guenther, A.; Goldstein, A. H.; Seinfeld, J. H.; Gold, A.; Kamens, R. M.; Surratt, J. D.

Organosulfates as Tracers for Secondary Organic Aerosol (SOA) Formation from 2-Methyl-3-Buten-2-Ol (MBO) in the Atmosphere. *Environ. Sci. Technol.* **2012**, *46*, 9437–9446.

20. Zhang, H.; Zhang, Z.; Cui, T.; Lin, Y.-H.; Bhathela, N. A.; Ortega, J.; Worton, D. R.; Goldstein, A. H.; Guenther, A.; Jimenez, J. L.; Gold, A.; Surratt, J. D. Secondary Organic Aerosol Formation via 2-Methyl-3-Buten-2-Ol Photooxidation: Evidence of Acid-Catalyzed Reactive Uptake of Epoxides. *Environ. Sci. Technol. Lett.* **2014**, *1*, 242–247.
21. Cao, G.; Jang, M. Effects of Particle Acidity and UV Light on Secondary Organic Aerosol Formation from Oxidation of Aromatics in the Absence of NOx. *Atmos. Environ.* **2007**, *41*, 7603–7613.
22. Jang, M.; Lee, S.; Kamens, R. M. Organic Aerosol Growth by Acid-Catalyzed Heterogeneous Reactions of Octanal in a Flow Reactor. *Atmos. Environ.* **2003**, *37*, 2125–2138.
23. Gao, S.; Ng, N. L.; Keywood, M.; Varutbangkul, V.; Bahreini, R.; Nenes, A.; He, J.; Yoo, K. Y.; Beauchamp, J. L.; Hodyss, R. P.; Flagan, R. C.; Seinfeld, J. H. Particle Phase Acidity and Oligomer Formation in Secondary Organic Aerosol. *Environ. Sci. Technol.* **2004**, *38*, 6582–6589.
24. Iinuma, Y.; Müller, C.; Berndt, T.; Böge, O.; Claeys, M.; Herrmann, H. Evidence for the Existence of Organosulfates from β-Pinene Ozonolysis in Ambient Secondary Organic Aerosol. *Environ. Sci. Technol.* **2007**, *41*, 6678–6683.
25. Iinuma, Y.; Boge, O.; Kahnt, A.; Herrmann, H. Laboratory Chamber Studies on the Formation of Organosulfates from Reactive Uptake of Monoterpene Oxides. *Phys. Chem. Chem. Phys.* **2009**, *11*, 7985–7997.
26. Surratt, J. D.; Gomez-Gonzalez, Y.; Chan, A. W. H.; Vermeylen, R.; Shahgholi, M.; Kleindienst, T. E.; Edney, E. O.; Offenberg, J. H.; Lewandowski, M.; Jaoui, M.; Maenhaut, W.; Claeys, M.; Flagan, R. C.; Seinfeld, J. H. Organosulfate Formation in Biogenic Secondary Organic Aerosol. *J. Phys. Chem. A* **2008**, *112*, 8345–8378.
27. Surratt, J. D.; Kroll, J. H.; Kleindienst, T. E.; Edney, E. O.; Claeys, M.; Sorooshian, A.; Ng, N. L.; Offenberg, J. H.; Lewandowski, M.; Jaoui, M.; Flagan, R. C.; Seinfeld, J. H. Evidence for Organosulfates in Secondary Organic Aerosol. *Environ. Sci. Technol.* **2007**, *41*, 517–527.
28. Duporté, G.; Flaud, P.-M.; Geneste, E.; Augagneur, S.; Pangui, E.; Lamkaddam, H.; Gratien, A.; Doussin, J.-F.; Budzinski, H.; Villenave, E.; Perraudin, E. Experimental Study of the Formation of Organosulfates from α-Pinene Oxidation. Part I: Product Identification, Formation Mechanisms and Effect of Relative Humidity. *J. Phys. Chem. A* **2016**, *120*, 7909–7923.
29. Iinuma, Y.; Müller, C.; Böge, O.; Gnauk, T.; Herrmann, H. The Formation of Organic Sulfate Esters in the Limonene Ozonolysis Secondary Organic Aerosol (SOA) under Acidic Conditions. *Atmos. Environ.* **2007**, *41*, 5571–5583.
30. Darer, A. I.; Cole-Filipiak, N. C.; O'Connor, A. E.; Elrod, M. J. Formation and Stability of Atmospherically Relevant Isoprene-Derived Organosulfates and Organonitrates. *Environ. Sci. Technol.* **2011**, *45*, 1895–1902.

31. Li, J.; Jang, M.; Beardsley, R. L. Dialkylsulfate Formation in Sulfuric Acid-Seeded Secondary Organic Aerosol Produced Using an Outdoor Chamber under Natural Sunlight. *Environ. Chem.* **2016**, *13*, 590–601.
32. Riva, M.; Da Silva Barbosa, T.; Lin, Y. H.; Stone, E. A.; Gold, A.; Surratt, J. D. Chemical Characterization of Organosulfates in Secondary Organic Aerosol Derived from the Photooxidation of Alkanes. *Atmos. Chem. Phys.* **2016**, *16*, 11001–11018.
33. Wang, L.; Lal, V.; Khalizov, A. F.; Zhang, R. Heterogeneous Chemistry of Alkylamines with Sulfuric Acid: Implications for Atmospheric Formation of Alkylaminium Sulfates. *Environ. Sci. Technol.* **2010**, *44*, 2461–2465.
34. Riva, M.; Tomaz, S.; Cui, T.; Lin, Y. H.; Perraudin, E.; Gold, A.; Stone, E. A.; Villenave, E.; Surratt, J. D. Evidence for an Unrecognized Secondary Anthropogenic Source of Organosulfates and Sulfonates: Gas-Phase Oxidation of Polycyclic Aromatic Hydrocarbons in the Presence of Sulfate Aerosol. *Environ. Sci. Technol.* **2015**, *49*, 6654–6664.
35. Rindelaub, J. D.; McAvey, K. M.; Shepson, P. B. The Photochemical Production of Organic Nitrates from α-Pinene and Loss via Acid-Dependent Particle Phase Hydrolysis. *Atmos. Environ.* **2015**, *100*, 193–201.
36. Zhang, Q.; Jimenez, J. L.; Worsnop, D. R.; Canagaratna, M. A Case Study of Urban Particle Acidity and Its Influence on Secondary Organic Aerosol. *Environ. Sci. Technol.* **2007**, *41*, 3213–3219.
37. Budisulistiorini, S. H.; Li, X.; Bairai, S. T.; Renfro, J.; Liu, Y.; Liu, Y. J.; McKinney, K. A.; Martin, S. T.; McNeill, V. F.; Pye, H. O. T.; Nenes, A.; Neff, M. E.; Stone, E. A.; Mueller, S.; Knote, C.; Shaw, S. L.; Zhang, Z.; Gold, A.; Surratt, J. D. Examining the Effects of Anthropogenic Emissions on Isoprene-Derived Secondary Organic Aerosol Formation during the 2013 Southern Oxidant and Aerosol Study (SOAS) at the Look Rock, Tennessee Ground Site. *Atmos. Chem. Phys.* **2015**, *15*, 8871–8888.
38. Pathak, R. K.; Wang, T.; Ho, K. F.; Lee, S. C. Characteristics of Summertime PM2.5 Organic and Elemental Carbon in Four Major Chinese Cities: Implications of High Acidity for Water-Soluble Organic Carbon (WSOC). *Atmos. Environ.* **2011**, *45*, 318–325.
39. Ding, X.; Wang, X. M.; Zheng, M. The Influence of Temperature and Aerosol Acidity on Biogenic Secondary Organic Aerosol Tracers: Observations at a Rural Site in the Central Pearl River Delta Region, South China. *Atmos. Environ.* **2011**, *45*, 1303–1311.
40. Tanner, R. L.; Olszyna, K. J.; Edgerton, E. S.; Knipping, E.; Shaw, S. L. Searching for Evidence of Acid-Catalyzed Enhancement of Secondary Organic Aerosol Formation Using Ambient Aerosol Data. *Atmos. Environ.* **2009**, *43*, 3440–3444.
41. Liggio, J.; Li, S.-M. A New Source of Oxygenated Organic Aerosol and Oligomers. *Atmos. Chem. Phys.* **2013**, *13*, 2989–3002.
42. Kiendler-Scharr, A.; Mensah, A. A.; Friese, E.; Topping, D.; Nemitz, E.; Prevot, A. S. H.; Äijälä, M.; Allan, J.; Canonaco, F.; Canagaratna, M.; Carbone, S.; Crippa, M.; Dall Osto, M.; Day, D. A.; De Carlo, P.; Di Marco, C. F.; Elbern, H.; Eriksson, A.; Freney, E.; Hao, L.; Herrmann, H.; Hildebrandt, L.; Hillamo, R.; Jimenez, J. L.; Laaksonen, A.; McFiggans, G.;

Mohr, C.; O'Dowd, C.; Otjes, R.; Ovadnevaite, J.; Pandis, S. N.; Poulain, L.; Schlag, P.; Sellegri, K.; Swietlicki, E.; Tiitta, P.; Vermeulen, A.; Wahner, A.; Worsnop, D.; Wu, H. C. Ubiquity of Organic Nitrates from Nighttime Chemistry in the European Submicron Aerosol. *Geophys. Res. Lett.* **2016**, *43*, 7735–7744.

43. Gaston, C. J.; Riedel, T. P.; Zhang, Z.; Gold, A.; Surratt, J. D.; Thornton, J. A. Reactive Uptake of an Isoprene-Derived Epoxydiol to Submicron Aerosol Particles. *Environ. Sci. Technol.* **2014**, *48*, 11178–11186.
44. Chen, H.; Wang, M.; Yao, L.; Chen, J.; Wang, L. Uptake of Gaseous Alkylamides by Suspended Sulfuric Acid Particles: Formation of Ammonium/Aminium Salts. *Environ. Sci. Technol.* **2017**, *51*, 11710–11717.
45. Liggio, J.; Li, S.-M. Reactive Uptake of Pinonaldehyde on Acidic Aerosols. *J. Geophys. Res. Atmos.* **2006**, *111*, D24303.
46. Liu, Y.; Liggio, J.; Staebler, R.; Li, S.-M. Reactive Uptake of Ammonia to Secondary Organic Aerosols: Kinetics of Organonitrogen Formation. *Atmos. Chem. Phys.* **2015**, *15*, 13569–13584.
47. Wang, T.; Liu, Z.; Wang, W.; Ge, M. Uptake Kinetics of Three Epoxides into Sulfuric Acid Solution. *Atmos. Environ.* **2012**, *56*, 58–64.
48. Marais, E. A.; Jacob, D. J.; Jimenez, J. L.; Campuzano-Jost, P.; Day, D. A.; Hu, W.; Krechmer, J.; Zhu, L.; Kim, P. S.; Miller, C. C.; Fisher, J. A.; Travis, K.; Yu, K.; Hanisco, T. F.; Wolfe, G. M.; Arkinson, H. L.; Pye, H. O. T.; Froyd, K. D.; Liao, J.; McNeill, V. F. Aqueous-Phase Mechanism for Secondary Organic Aerosol Formation from Isoprene: Application to the Southeast United States and Co-Benefit of SO2 Emission Controls. *Atmos. Chem. Phys.* **2016**, *16*, 1603–1618.
49. Gard, E. E.; Kleeman, M. J.; Gross, D. S.; Hughes, L. S.; Allen, J. O.; Morrical, B. D.; Fergenson, D. P.; Dienes, T.; E. Gälli, M.; Johnson, R. J.; Cass, G. R.; Prather, K. A. Direct Observation of Heterogeneous Chemistry in the Atmosphere. *Science* **1998**, *279*, 1184–1187.
50. Ault, A. P.; Guasco, T. L.; Baltrusaitis, J.; Ryder, O. S.; Trueblood, J. V; Collins, D. B.; Ruppel, M. J.; Cuadra-Rodriguez, L. A.; Prather, K. A.; Grassian, V. H. Heterogeneous Reactivity of Nitric Acid with Nascent Sea Spray Aerosol: Large Differences Observed between and within Individual Particles. *J. Phys. Chem. Lett.* **2014**, *5*, 2493–2500.
51. Bondy, A. L.; Wang, B.; Laskin, A.; Craig, R. L.; Nhliziyo, M. V; Bertman, S. B.; Pratt, K. A.; Shepson, P. B.; Ault, A. P. Inland Sea Spray Aerosol Transport and Incomplete Chloride Depletion: Varying Degrees of Reactive Processing Observed during SOAS. *Environ. Sci. Technol.* **2017**, *51*, 9533–9542.
52. Laskin, A.; Moffet, R. C.; Gilles, M. K.; Fast, J. D.; Zaveri, R. A.; Wang, B.; Nigge, P.; Shutthanandan, J. Tropospheric Chemistry of Internally Mixed Sea Salt and Organic Particles: Surprising Reactivity of NaCl with Weak Organic Acids. *J. Geophys. Res. - Atmos.* **2012**, *117*, D15302.
53. Rindelaub, J. D.; Borca, C. H.; Hostetler, M. A.; Slade, J. H.; Lipton, M. A.; Slipchenko, L. V; Shepson, P. B. The Acid-Catalyzed Hydrolysis of an Alpha-Pinene-Derived Organic Nitrate: Kinetics, Products, Reaction

Mechanisms, and Atmospheric Impact. *Atmos. Chem. Phys.* **2016**, *16*, 15425–15432.

54. Hu, K. S.; Darer, A. I.; Elrod, M. J. Thermodynamics and Kinetics of the Hydrolysis of Atmospherically Relevant Organonitrates and Organosulfates. *Atmos. Chem. Phys.* **2011**, *11*, 8307–8320.
55. Jacobs, M. I.; Burke, W. J.; Elrod, M. J. Kinetics of the Reactions of Isoprene-Derived Hydroxynitrates: Gas Phase Epoxide Formation and Solution Phase Hydrolysis. *Atmos. Chem. Phys.* **2014**, *14*, 8933–8946.
56. Cortés, D. A.; Elrod, M. J. Kinetics of the Aqueous Phase Reactions of Atmospherically Relevant Monoterpene Epoxides. *J. Phys. Chem. A* **2017**, *121*, 9297–9305.
57. Fang, T.; Guo, H.; Zeng, L.; Verma, V.; Nenes, A.; Weber, R. J. Highly Acidic Ambient Particles, Soluble Metals, and Oxidative Potential: A Link between Sulfate and Aerosol Toxicity. *Environ. Sci. Technol.* **2017**, *51*, 2611–2620.
58. Longo, A. F.; Feng, Y.; Lai, B.; Landing, W. M.; Shelley, R. U.; Nenes, A.; Mihalopoulos, N.; Violaki, K.; Ingall, E. D. Influence of Atmospheric Processes on the Solubility and Composition of Iron in Saharan Dust. *Environ. Sci. Technol.* **2016**, *50*, 6912–6920.
59. Chen, H.; Laskin, A.; Baltrusaitis, J.; Gorski, C. A.; Scherer, M. M.; Grassian, V. H. Coal Fly Ash as a Source of Iron in Atmospheric Dust. *Environ. Sci. Technol.* **2012**, *46*, 2112–2120.
60. Prenni, A. J.; DeMott, P. J.; Kreidenweis, S. M. Water Uptake of Internally Mixed Particles Containing Ammonium Sulfate and Dicarboxylic Acids. *Atmos. Environ.* **2003**, *37*, 4243–4251.
61. Ghorai, S.; Laskin, A.; Tivanski, A. V. Spectroscopic Evidence of Keto−enol Tautomerism in Deliquesced Malonic Acid Particles. *J. Phys. Chem. A* **2011**, *115*, 4373–4380.
62. Ghorai, S.; Wang, B.; Tivanski, A.; Laskin, A. Hygroscopic Properties of Internally Mixed Particles Composed of NaCl and Water-Soluble Organic Acids. *Environ. Sci. Technol.* **2014**, *48*, 2234–2241.
63. Losey, D. J.; Parker, R. G.; Freedman, M. A. PH Dependence of Liquid–Liquid Phase Separation in Organic Aerosol. *J. Phys. Chem. Lett.* **2016**, *7*, 3861–3865.
64. You, Y.; Smith, M. L.; Song, M.; Martin, S. T.; Bertram, A. K. Liquid–liquid Phase Separation in Atmospherically Relevant Particles Consisting of Organic Species and Inorganic Salts. *Int. Rev. Phys. Chem.* **2014**, *33*, 43–77.
65. Dallemagne, M. A.; Huang, X. Y.; Eddingsaas, N. C. Variation in PH of Model Secondary Organic Aerosol during Liquid–Liquid Phase Separation. *J. Phys. Chem. A* **2016**, *120*, 2868–2876.
66. Ault, A. P.; Guasco, T. L.; Ryder, O. S.; Baltrusaitis, J.; Cuadra-Rodriguez, L. A.; Collins, D. B.; Ruppel, M. J.; Bertram, T. H.; Prather, K. A.; Grassian, V. H. Inside versus Outside: Ion Redistribution in Nitric Acid Reacted Sea Spray Aerosol Particles as Determined by Single Particle Analysis. *J. Am. Chem. Soc.* **2013**, *135*, 14528–14531.
67. Song, C.; Gyawali, M.; Zaveri, R. A.; Shilling, J. E.; Arnott, W. P. Light Absorption by Secondary Organic Aerosol from α-Pinene: Effects of

Oxidants, Seed Aerosol Acidity, and Relative Humidity. *J. Geophys. Res. Atmos.* **2013**, *118*, 11741–11749.

68. Lin, Y.-H.; Budisulistiorini, S. H.; Chu, K.; Siejack, R. A.; Zhang, H.; Riva, M.; Zhang, Z.; Gold, A.; Kautzman, K. E.; Surratt, J. D. Light-Absorbing Oligomer Formation in Secondary Organic Aerosol from Reactive Uptake of Isoprene Epoxydiols. *Environ. Sci. Technol.* **2014**, *48*, 12012–12021.
69. Rapf, R. J.; Dooley, M. R.; Kappes, K.; Perkins, R. J.; Vaida, V. PH Dependence of the Aqueous Photochemistry of α-Keto Acids. *J. Phys. Chem. A* **2017**, *121*, 8368–8379.
70. Witkowski, B.; Gierczak, T. Cis-Pinonic Acid Oxidation by Hydroxyl Radicals in the Aqueous Phase under Acidic and Basic Conditions: Kinetics and Mechanism. *Environ. Sci. Technol.* **2017**, *51*, 9765–9773.
71. Liu, M. J.; Wiegel, A. A.; Wilson, K. R.; Houle, F. A. Aerosol Fragmentation Driven by Coupling of Acid–Base and Free-Radical Chemistry in the Heterogeneous Oxidation of Aqueous Citric Acid by OH Radicals. *J. Phys. Chem. A* **2017**, *121*, 5856–5870.
72. Hennigan, C. J.; Izumi, J.; Sullivan, A. P.; Weber, R. J.; Nenes, A. A Critical Evaluation of Proxy Methods Used to Estimate the Acidity of Atmospheric Particles. *Atmos. Chem. Phys.* **2015**, *15*, 2775–2790.
73. Keene, W. C.; Pszenny, A. A. P.; Maben, J. R.; Sander, R. Variation of Marine Aerosol Acidity with Particle Size. *Geophys. Res. Lett.* **2002**, *29*, 4–5.
74. Koutrakis, P.; Wolfson, J. M.; Spengler, J. D. An Improved Method for Measuring Aerosol Strong Acidity: Results from a Nine-Month Study in St Louis, Missouri and Kingston, Tennessee. *Atmos. Environ.* **1988**, *22*, 157–162.
75. Pathak, R. K.; Yao, X. H.; Chan, C. K. Sampling Artifacts of Acidity and Ionic Species in $PM_{2.5}$. *Environ. Sci. Technol.* **2004**, *38*, 254–259.
76. Metzger, S.; Mihalopoulos, N.; Lelieveld, J. Importance of Mineral Cations and Organics in Gas-Aerosol Partitioning of Reactive Nitrogen Compounds: Case Study Based on MINOS Results. *Atmos. Chem. Phys.* **2006**, *6*, 2549–2567.
77. Trebs, I.; Metzger, S.; Meixner, F. X.; Helas, G.; Hoffer, A.; Rudich, Y.; Falkovich, A. H.; Moura, M. A. L.; da Silva, R. S.; Artaxo, P.; Slanina, J.; Andreae, M. O. The NH_4^+-NO_3^--Cl^--SO_4^{2-}-H_2O Aerosol System and Its Gas Phase Precursors at a Pasture Site in the Amazon Basin: How Relevant Are Mineral Cations and Soluble Organic Acids? *J. Geophys. Res. - Atmos.* **2005**, *110*, D07303.
78. Kerminen, V.-M.; Hillamo, R.; Teinilä, K.; Pakkanen, T.; Allegrini, I.; Sparapani, R. Ion Balances of Size-Resolved Tropospheric Aerosol Samples: Implications for the Acidity and Atmospheric Processing of Aerosols. *Atmos. Environ.* **2001**, *35*, 5255–5265.
79. Feng, J. L.; Guo, Z. G.; Zhang, T. R.; Yao, X. H.; Chan, C. K.; Fang, M. Source and Formation of Secondary Particulate Matter in PM2.5 in Asian Continental Outflow. *J. Geophys. Res. - Atmos.* **2012**, *117*, D03302.

80. Pathak, R. K.; Wu, W. S.; Wang, T. Summertime PM2.5 Ionic Species in Four Major Cities of China: Nitrate Formation in an Ammonia-Deficient Atmosphere. *Atmos. Chem. Phys.* **2009**, *9*, 1711–1722.
81. Keene, W. C.; Pszenny, A. A. P.; Maben, J. R.; Stevenson, E.; Wall, A. Closure Evaluation of Size Resolved Aerosol pH in the New England Coastal Atmosphere during Summer. *J. Geophys. Res. - Atmos.* **2004**, *109*, D23307.
82. Young, A. H.; Keene, W. C.; Pszenny, A. A. P.; Sander, R.; Thornton, J. A.; Riedel, T. P.; Maben, J. R. Phase Partitioning of Soluble Trace Gases with Size-Resolved Aerosols in near-Surface Continental Air over Northern Colorado, USA, during Winter. *J. Geophys. Res. - Atmos.* **2013**, *118*, 9414–9427.
83. Clegg, S. L.; Brimblecombe, P.; Wexler, A. S. Thermodynamic Model of the System $H^+–NH_4^+–SO_4^{2-}–NO_3^-–H_2O$ at Tropospheric Temperatures. *J. Phys. Chem. A* **1998**, *102*, 2137–2154.
84. Wexler, A. S.; Clegg, S. L. Atmospheric Aerosol Models for Systems Including the Ions H^+, NH_4^+, Na^+, SO_4^{2-}, NO_3^-, Cl^-, Br^-, and H_2O. *J. Geophys. Res. - Atmos.* **2002**, *107*, ACH 14-1–ACH 14-14.
85. Fountoukis, C.; Nenes, A. ISORROPIA II: A Computationally Efficient Thermodynamic Equilibrium Model for K^+-Ca^{2+}-Mg^{2+}-NH_4^+-Na^+-SO_4^{2-}-NO_3^--Cl^--H_2O Aerosols. *Atmos. Chem. Phys.* **2007**, *7*, 4639–4659.
86. Nenes, A.; Pandis, S. N.; Pilinis, C. ISORROPIA: A New Thermodynamic Equilibrium Model for Multiphase Multicomponent Inorganic Aerosols. *Aquat. Geochem.* **1998**, *4*, 123–152.
87. Murphy, J. G.; Gregoire, P. K.; Tevlin, A. G.; Wentworth, G. R.; Ellis, R. A.; Markovic, M. Z.; VandenBoer, T. C. Observational Constraints on Particle Acidity Using Measurements and Modelling of Particles and Gases. *Faraday Discuss.* **2017**, *200*, 379–395.
88. Perraud, V.; Bruns, E. A.; Ezell, M. J.; Johnson, S. N.; Yu, Y.; Alexander, M. L.; Zelenyuk, A.; Imre, D.; Chang, W. L.; Dabdub, D.; Pankow, J. F.; Finlayson-Pitts, B. J. Nonequilibrium Atmospheric Secondary Organic Aerosol Formation and Growth. *Proc. Natl. Acad. Sci. U. S. A.* **2012**, *109*, 2836–2841.
89. Virtanen, A.; Joutsensaari, J.; Koop, T.; Kannosto, J.; Yli-Pirila, P.; Leskinen, J.; Makela, J. M.; Holopainen, J. K.; Poeschl, U.; Kulmala, M.; Worsnop, D. R.; Laaksonen, A. An Amorphous Solid State of Biogenic Secondary Organic Aerosol Particles. *Nature* **2010**, *467*, 824–827.
90. Shiraiwa, M.; Li, Y.; Tsimpidi, A. P.; Karydis, V. A.; Berkemeier, T.; Pandis, S. N.; Lelieveld, J.; Koop, T.; Pöschl, U. Global Distribution of Particle Phase State in Atmospheric Secondary Organic Aerosols. *Nat. Commun.* **2017**, *8*, 15002.
91. Silvern, R. F.; Jacob, D. J.; Kim, P. S.; Marais, E. A.; Turner, J. R.; Campuzano-Jost, P.; Jimenez, J. L. Inconsistency of Ammonium-Sulfate Aerosol Ratios with Thermodynamic Models in the Eastern US: A Possible Role of Organic Aerosol. *Atmos. Chem. Phys.* **2017**, *17*, 5107–5118.
92. Zhang, Q.; Jimenez, J. L.; Canagaratna, M. R.; Allan, J. D.; Coe, H.; Ulbrich, I.; Alfarra, M. R.; Takami, A.; Middlebrook, A. M.; Sun, Y. L.; Dzepina, K.; Dunlea, E.; Docherty, k.; DeCarlo, P. F.; Salcedo, D.;

Onasch, T.; Jayne, J. T.; Miyoshi, T.; Shimono, A.; Hatakeyama, S.; Takegawa, N.; kondo, Y.; Schneider, J.; Drewnick, F.; Borrmann, S.; Weimer, S.; Demerjian, K.; Williams, P.; Bower, K.; Bahreini, R.; Cottrell, L.; Griffin, R. J.; Rautiainen, J.; Sun, J. Y.; Zhang, Y. M.; Worsnop, D. R. Ubiquity and Dominance of Oxygenated Species in Organic Aerosols in Anthropogenically-Influence Norther Hemisphere Midlatitudes. *Geophys. Res. Lett.* **2007**, *34*, LI3801.

93. Kolb, C. E.; Worsnop, D. R. Chemistry and Composition of Atmospheric Aerosol Particles. *Annu. Rev. Phys. Chem.* **2012**, *63*, 471–491.
94. Guo, H.; Xu, L.; Bougiatioti, A.; Cerully, K. M.; Capps, S. L.; Hite, J. R., Jr.; Carlton, A. G.; Lee, S. H.; Bergin, M. H.; Ng, N. L.; Nenes, A.; Weber, R. J. Fine-Particle Water and PH in the Southeastern United States. *Atmos. Chem. Phys.* **2015**, *15*, 5211–5228.
95. Guo, H.; Sullivan, A. P.; Campuzano-Jost, P.; Schroder, J. C.; Lopez-Hilfiker, F. D.; Dibb, J. E.; Jimenez, J. L.; Thornton, J. A.; Brown, S. S.; Nenes, A.; Weber, R. J. Fine Particle PH and the Partitioning of Nitric Acid during Winter in the Northeastern United States. *J. Geophys. Res. - Atmos.* **2016**, *121*, 10355–10376.
96. Liu, M.; Song, Y.; Zhou, T.; Xu, Z.; Yan, C.; Zheng, M.; Wu, Z.; Hu, M.; Wu, Y.; Zhu, T. Fine Particle PH during Severe Haze Episodes in Northern China. *Geophys. Res. Lett.* **2017**, *44*, 5213–5221.
97. Vieira-Filho, M.; Pedrotti, J. J.; Fornaro, A. Water-Soluble Ions Species of Size-Resolved Aerosols: Implications for the Atmospheric Acidity in São Paulo Megacity, Brazil. *Atmos. Res.* **2016**, *181*, 281–287.
98. Guo, H.; Liu, J.; Froyd, K. D.; Roberts, J. M.; Veres, P. R.; Hayes, P. L.; Jimenez, J. L.; Nenes, A.; Weber, R. J. Fine Particle pH and Gas–particle Phase Partitioning of Inorganic Species in Pasadena, California, during the 2010 CalNex Campaign. *Atmos. Chem. Phys.* **2017**, *17*, 5703–5719.
99. Bougiatioti, A.; Nikolaou, P.; Stavroulas, I.; Kouvarakis, G.; Weber, R.; Nenes, A.; Kanakidou, M.; Mihalopoulos, N. Particle Water and PH in the Eastern Mediterranean: Source Variability and Implications for Nutrient Availability. *Atmos. Chem. Phys.* **2016**, *16*, 4579–4591.
100. Bougiatioti, A.; Stavroulas, I.; Fourtziou, L.; Nikolaou, P.; Paraskevopoulou, D.; Kouvarakis, G.; Nenes, A.; Mihalopoulos, N. Fine Particle Water and PH in an Urban and Remote Location and the Role of Biomass Burning. In *Perspectives on Atmospheric Science*; Karacostas, T. S., Bais, A. F., Nastos, P., Ed.; Springer Atmospheric Sciences: Dordrecht, Netherlands, 2017; pp 837−843.
101. Shi, G.; Xu, J.; Peng, X.; Xiao, Z.; Chen, K.; Tian, Y.; Guan, X.; Feng, Y.; Yu, H.; Nenes, A.; Russell, A. G. pH of Aerosols in a Polluted Atmosphere: Source Contributions to Highly Acidic Aerosol. *Environ. Sci. Technol.* **2017**, *51*, 4289–4296.
102. Pozzer, A.; Tsimpidi, A. P.; Karydis, V. A.; de Meij, A.; Lelieveld, J. Impact of Agricultural Emission Reductions on Fine-Particulate Matter and Public Health. *Atmos. Chem. Phys.* **2017**, *17*, 12813–12826.
103. Battaglia, M. A.; Douglas, S.; Hennigan, C. J. Effect of the Urban Heat Island on Aerosol PH. *Environ. Sci. Technol.* **2017**, *51*, 13095–13103.

104. Wu, X.; Deng, J.; Chen, J.; Hong, Y.; Xu, L.; Yin, L.; Du, W.; Hong, Z.; Dai, N.; Yuan, C.-S. Characteristics of Water-Soluble Inorganic Components and Acidity of PM2.5 in a Coastal City of China. *Aerosol Air Qual. Res.* **2017**, *17*, 2152–2164.

105. Kumar, S.; Raman, R. S. Inorganic Ions in Ambient Fine Particles over a National Park in Central India: Seasonality, Dependencies between SO_4^{2-}, NO_3^-, and NH_4^+, and Neutralization of Aerosol Acidity. *Atmos. Environ.* **2016**, *143*, 152–163.

106. Weber, R. J.; Guo, H.; Russell, A. G.; Nenes, A. High Aerosol Acidity despite Declining Atmospheric Sulfate Concentrations over the Past 15 Years. *Nat. Geosci.* **2016**, *9*, 282–285.

107. Jang, M.; Cao, G.; Paul, J. Colorimetric Particle Acidity Analysis of Secondary Organic Aerosol Coating on Submicron Acidic Aerosols. *Aerosol Sci. Technol.* **2008**, *42*, 409–420.

108. Li, J.; Jang, M. Aerosol Acidity Measurement Using Colorimetry Coupled With a Reflectance UV-Visible Spectrometer. *Aerosol Sci. Technol.* **2012**, *46*, 833–842.

109. Rindelaub, J. D.; Craig, R. L.; Nandy, L.; Bondy, A. L.; Dutcher, C. S.; Shepson, P. B.; Ault, A. P. Direct Measurement of PH in Individual Particles via Raman Microspectroscopy and Variation in Acidity with Relative Humidity. *J. Phys. Chem. A* **2016**, *120*, 911–917.

110. Craig, R. L.; Nandy, L.; Axson, J. L.; Dutcher, C. S.; Ault, A. P. Spectroscopic Determination of Aerosol PH from Acid–Base Equilibria in Inorganic, Organic, and Mixed Systems. *J. Phys. Chem. A* **2017**, *121*, 5690–5699.

111. Ault, A. P.; Axson, J. L. Atmospheric Aerosol Chemistry: Spectroscopic and Microscopic Advances. *Anal. Chem.* **2017**, *89*, 430–452.

112. Ebben, C. J.; Ault, A. P.; Ruppel, M. J.; Ryder, O. S.; Bertram, T. H.; Grassian, V. H.; Prather, K. A.; Geiger, F. M. Size-Resolved Sea Spray Aerosol Particles Studied by Vibrational Sum Frequency Generation. *J. Phys. Chem. A* **2013**, *117*, 6589.

113. Craig, R. L.; Bondy, A. L.; Ault, A. P. Computer-Controlled Raman Microspectroscopy (CC-Raman): A Method for the Rapid Characterization of Individual Atmospheric Aerosol Particles. *Aerosol Sci. Technol.* **2017**, *51*, 1099–1112.

114. Bondy, A. L.; Craig, R. L.; Zhang, Z.; Gold, A.; Surratt, J. D.; Ault, A. P. Isoprene-Derived Organosulfates: Vibrational Mode Analysis by Raman Spectroscopy, Acidity-Dependent Spectral Modes, and Observation in Individual Atmospheric Particles. *J. Phys. Chem. A* **2018**, *122*, 303–315.

115. Craig, R. L.; Bondy, A. L.; Ault, A. P. Surface Enhanced Raman Spectroscopy Enables Observations of Previously Undetectable Secondary Organic Aerosol Components at the Individual Particle Level. *Anal. Chem.* **2015**, *87*, 7510–7514.

116. Creamean, J. M.; Axson, J. L.; Bondy, A. L.; Craig, R. L.; May, N. W.; Shen, H.; Weber, M. H.; Pratt, K. A.; Ault, A. P. Changes in Precipitating Snow Chemistry with Location and Elevation in the California Sierra Nevada. *J. Geophys. Res. Atmos.* **2016**, *121*, 2015JD024700.

117. Baustian, K. J.; Cziczo, D. J.; Wise, M. E.; Pratt, K. A.; Kulkarni, G.; Hallar, A. G.; Tolbert, M. A. Importance of Aerosol Composition, Mixing State, and Morphology for Heterogeneous Ice Nucleation: A Combined Field and Laboratory Approach. *J. Geophys. Res.* **2012**, *117*, D06217.
118. Laskina, O.; Young, M. A.; Kleiber, P. D.; Grassian, V. H. Infrared Extinction Spectroscopy and Micro-Raman Spectroscopy of Select Components of Mineral Dust Mixed with Organic Compounds. *J. Geophys. Res.* **2013**, *118*, 6593–6606.
119. Ault, A. P.; Zhao, D.; Ebben, C. J.; Tauber, M. J.; Geiger, F. M.; Prather, K. A.; Grassian, V. H. Raman Microspectroscopy and Vibrational Sum Frequency Generation Spectroscopy as Probes of the Bulk and Surface Compositions of Size-Resolved Sea Spray Aerosol Particles. *Phys. Chem. Chem. Phys.* **2013**, *15*, 6206–6214.
120. Laliberte, M.; Cooper, W. E. Model for Calculating the Density of Aqueous Electrolyte Solutions. *J. Chem. Eng. Data* **2004**, *49*, 1141–1151.
121. Clegg, S. L.; Simonson, J. M. A BET Model of the Thermodynamics of Aqueous Multicomponent Solutions at Extreme Concentration. *J. Chem. Thermodyn.* **2001**, *33*, 1457–1472.
122. Garrels, R. M.; Christ, C. L. *Solutions, Minerals, and Equilibria*; Harper & Row Publishers, Inc.: New York City, NY, 1965.
123. Kielland, J. Individual Activity Coefficients of Ions in Aqueous Solutions. *J. Am. Chem. Soc.* **1937**, *59*, 1675–1678.
124. Craig, R. L.; Peterson, P. K.; Nandy, L.; Lei, Z.; Hossain, M.; Camarena, S.; Dodson, R. A.; Cook, R. D.; Dutcher, C. S.; Ault, A. P. Direct Determination of Aerosol PH: Size-Resolved Measurements of Submicron and Supermicron Aqueous Particles. *Anal. Chem.* **2018**, submitted.
125. Morenz, K. J.; Donaldson, D. J. Chemical Morphology of Frozen Mixed Nitrate–Salt Solutions. *J. Phys. Chem. A* **2017**, *121*, 2166–2171.
126. Lee, H. D.; Ray, K. K.; Tivanski, A. V. Solid, Semisolid, and Liquid Phase States of Individual Submicrometer Particles Directly Probed Using Atomic Force Microscopy. *Anal. Chem.* **2017**, *89*, 12720–12726.
127. Bondy, A. L.; Kirpes, R. M.; Merzel, R. L.; Pratt, K. A.; Banaszak Holl, M. M.; Ault, A. P. Atomic Force Microscopy-Infrared Spectroscopy of Individual Atmospheric Aerosol Particles: Subdiffraction Limit Vibrational Spectroscopy and Morphological Analysis. *Anal. Chem.* **2017**, *89*, 8594–8598.
128. Grossman, J. N.; Stern, A. P.; Kirich, M. L.; Kahan, T. F. Anthracene and Pyrene Photolysis Kinetics in Aqueous, Organic, and Mixed Aqueous-Organic Phases. *Atmos. Environ.* **2016**, *128*, 158–164.
129. Malley, P. P. A.; Grossman, J. N.; Kahan, T. F. Effects of Chromophoric Dissolved Organic Matter on Anthracene Photolysis Kinetics in Aqueous Solution and Ice. *J. Phys. Chem. A* **2017**, *121*, 7619–7626.
130. Zhang, Y.; Chen, Y.; Lambe, A. T.; Olson, N. E.; Lei, Z.; Craig, R. L.; Zhang, Z.; Gold, A.; Onasch, T. B.; Jayne, J. T.; Worsnop, D. R.; Gaston, C. J.; Thornton, J. A.; Vizuete, W.; Ault, A. P.; Surratt, J. D. Effect of the Aerosol-Phase State on Secondary Organic Aerosol Formation from the

Reactive Uptake of Isoprene-Derived Epoxydiols (IEPOX). *Environ. Sci. Technol. Lett.* **2018**, *5*, 167–174.

131. Lee, H. D.; Estillore, A. D.; Morris, H. S.; Ray, K. K.; Alejandro, A.; Grassian, V. H.; Tivanski, A. V. Direct Surface Tension Measurements of Individual Sub-Micrometer Particles Using Atomic Force Microscopy. *J. Phys. Chem. A* **2017**, *121*, 8296–8305.

Chapter 10

Chemical Morphology and Reactivity at Environmental Interfaces

D. James Donaldson,[1,2,*] Jessica T. Clouthier,[1] Karen J. Morenz,[1] and Adam Marr[1]

[1]Department of Chemistry, University of Toronto, Toronto, Ontario, Canada M5S 3H6
[2]Department of Physical and Environmental Sciences, University of Toronto, Toronto, Ontario, Canada M1C 1A4
[*]E-mail: jdonalds@chem.utoronto.ca

The ability of reactions to take place at atmospheric interfaces depends upon the presence of reagents at (or near) those interfaces; that is, upon their location with respect to the overlying atmosphere. In particular, the release of gas phase products from the photolysis of nitrate anions, present in snow and ice, as well as in urban grime, depends upon the location and the local chemical environment of the nitrate. We use Raman microscopy to both locate the region where nitrate and sulfate (as a comparison species) are present, as well as whether these anions are primarily in a crystalline or solution local environment. Nitrate shows a strong propensity for the air-ice surface, whereas sulfate does not; this may relate to the photoreactivity of nitrate in frozen media. Raman microscopy further reveals that several of the components of urban grime are very effective at maintaining nitrate in a "solution-like" state, even under "dry" ambient conditions. This may go a long way to explaining the observation that nitrate in urban grime is significantly more photoactive at higher relative humidities.

Introduction

The specific chemical environment experienced by a reacting species is a key element in determining its reactivity. The degree of solvation, the physical phase, as well as local dielectric constant, viscosity and solvent polarizability all factor into the potential for chemical reaction, either unimolecular (typically photochemical) or bimolecular. In addition, the physical location of reagents within a matrix will certainly affect their availability to react with atmospheric gas-phase species, and the efficacy with which reaction products may be released to the atmosphere.

In the lower atmosphere, the formation of ozone - a key precursor for OH, as well as being itself a powerful oxidant and atmospheric pollutant – is tied directly to the photolysis of gas phase NO_2. Near Earth's surface, removal of atmospheric NO_2 is primarily via its reaction with OH, to form nitric acid, or NO_3, forming dinitrogen pentoxide, followed by the hydrolysis of these product compounds at atmospheric interfaces. The nitrate anion thus formed undergoes a very slow photolysis in aqueous solution, reforming NO_2 and HONO which may be recycled back to the atmosphere. However, the low photolysis rate constant has meant that this recycling process is not considered to be important to local atmospheric oxidation chemistry.

Even in pristine snowpacks, but certainly in laboratory experiments, nitrate present in ice and snow samples produces gas phase nitrogen oxides when illuminated with solar radiation. In frozen aqueous matrices, anions such as nitrate are believed to be excluded into liquid regions, whose chemical composition should somehow reflect the phase diagram for the system (*1*). A concentrated nitrate solution thus formed at the air interface is believed to be responsible for observations of gas phase nitrogen oxide emission from illuminated snowpacks (*2*). The assumption that the brine concentration at the interface is that predicted by the bulk phase diagram has not been convincingly demonstrated experimentally (*3*); indeed in the case of magnesium nitrate, the brine is less concentrated than expected according to its phase diagram (*4*).

The nitrate observed as one constituent of the "urban grime" present on urban impervious surfaces also undergoes photolysis and thus recycles gas phase nitrogen oxide oxidants to the urban atmosphere. In this case, the photochemistry is observed to be orders of magnitude faster than that in solution (*5*) and shows a strong dependence on ambient relative humidity (*6*). The reasons for the enhanced photochemistry and its humidity dependence are not known at this time. However, given the measured hygroscopiscity of real urban grime (*6*), one possibility for enhanced reactivity may be whether the nitrate is present in grime in a solution or solid state.

In the following, we report on Raman microscopic studies of nitrate and sulfate anions that track their physical state (solution phase or crystalline) as a function of their location in frozen solutions and urban grime, and attempt to relate this to their observed photoreactivity.

Experimental Methods

(a) Frozen Samples

Precursor solutions were made to be 250 mM in nitrate or sulfate and were prepared from stock solutions of salts dissolved in ultrapure deionized water. Sodium nitrate, sodium sulfate and sodium chloride were of reagent grade and used without purification. Instant Ocean (Spectrum Brands) has stated composition: chloride (47.53 wt %), sodium (26.45 wt %), sulfate (6.41 wt %), magnesium (3.19 wt %), calcium (1.00 wt %), potassium (0.952 wt %), bicarbonate (0.356 wt %), and bromide (0.16 wt %) (*7*). The NaCl-containing solutions were 500 mM in chloride; the Instant Ocean solutions (29.22 g/L) were made to be the same mass per unit volume as the NaCl. Liquid samples were pipetted into round Teflon or Cu sample holders with a volume of 3 microlitres (1 mm interior diameter, 2 mm interior depth). Samples for freezing were pipetted into the sample holders and either placed in a freezer at 261 K or dipped into liquid nitrogen prior to Raman measurements. No significant difference in results between the two freezing methods was observed in these studies.

Raman spectra were measured using a Bruker Raman Microscope, with a 532 nm laser at 20 mW, with a 50-1000 μm aperture. All frozen samples were immersed briefly in liquid nitrogen directly prior to measurement, and some were kept chilled during measurement using a salt-ice bath, which maintained the temperature at approximately 255 K. In all cases, the OH stretching region of the water Raman spectrum indicated that the sample was frozen. Spectra for each sample were measured at 4 different points across the surface at least 200 μm apart; at each such point in the horizontal plane, spectra from 5 different depths were collected, spaced 250 μm apart in depth. The "surface" point was determined visually, by observing the camera output of the microscope. At least 4 separate trials were performed for all frozen samples, and at least 2 for all liquid samples, resulting in 16 and 8 spectra respectively for each type of solution. OPUS software package was used for baseline correction and normalization of spectra. Igor pro or Matlab routines were used for calculation of nitrate peak position, area and height.

(b) Grime and Grime Proxy Samples

Grime proxies were made by spreading a very thin layer of hydrocarbon vacuum grease over a Pyrex slide, then depositing small volumes of solutions of sodium nitrate, ammonium or sodium sulfate or calcium chloride using a Pasteur pipette. Samples were dried thoroughly using the output from a hot air gun; some samples were re-humidified by exposing the grease-ionic salt mix to stream from a kettle. Real grime samples were collected onto the same glass slides in downtown Toronto over the period of 2-4 weeks. The slides were shielded from both rainfall and sunlight during collection. Raman spectra were collected using the same instrument as that described above.

Chemical Morphology and Reactivity of Frozen Nitrate and Sulfate Solutions

The exclusion of nitrate anions to the air-ice interface was observed by Wren and Donaldson (*4*) using glancing-angle Raman to probe the near-interface region during freezing of a nitrate solution. Those authors noted that although exclusion to the interface was clearly seen, the inferred nitrate concentration there was not in accord with that expected from the phase diagram; that is, there must exist significant amounts of nitrate within the ice matrix, either "frozen in", or dissolved in liquid pockets. Morenz and Donaldson (*8*) explored this idea using Raman microscopy to follow the nitrate concentration as a function of depth below the interface. The spectral resolution of this tool allowed assignment of the nitrate to either its solution or crystalline phase, based upon the different nitrate symmetric stretching frequencies of nitrate in these phases (*9*).

The results, displayed in Figure 1a, indicate that there is a distinct concentration enhancement of nitrate at the ice-air interface, and that nitrate does exist throughout the frozen sample. Moreover, the spectra clearly show that the nitrate present within the ice matrix is exclusively in solution phase, consistent with its exclusion to liquid pockets within the matrix during the freezing process. Interestingly, Figure 1b illustrates that at the air interface solution-phase and crystalline nitrate co-exist, at least under the conditions of this measurement (~-15 C and 250 mM sodium nitrate). When the same concentration of nitrate is present in a frozen solution of "Instant Ocean", made up to seawater salt concentration to simulate fresh sea ice, the same strong preference of nitrate for the interface is shown as in the simple nitrate solution, as displayed in Figure 1c. Using NaCl in place of Instant Ocean gives essentially the same result. However, nitrate is now observed exclusively in its solution phase throughout the sample, as shown in Figure 1d. This observation is consistent with the formation of a brine layer, governed by the temperature and amount of added salts present, being formed at the interface, in which the nitrate is dissolved. The formation of such a salty brine at the air-ice interface is well documented (*1*) and has been shown to have strong effects on the ice surface chemistry (*10*).

Morenz et al (*11*) investigated whether the presence of other salts within the frozen nitrate system could affect the release of gas phase photolysis products (NO, NO_2 and HONO). The points in Figure 2a illustrate the dependence of the steady-state NO_2 concentration released as a function of the pre-freezing nitrate concentration during constant illumination of frozen nitrate-containing snow from a filtered Xe lamp. The solid line represents the expected concentration of nitrate in an excluded brine. There is a clear dependence of the amount of gas phase NO_2 released on the predicted nitrate concentration excluded to liquid regions. Although not probed in these experiments, it seems very reasonable to assume that most of the observed gas phase products are formed from nitrate that is present at the air-snow interface. The results of Figure 1c imply that this nitrate is present in both solution and crystalline phases; again, it seems reasonable to assume that it is the solution phase nitrate that is most photolabile, and so primarily responsible for the observed gas phase products.

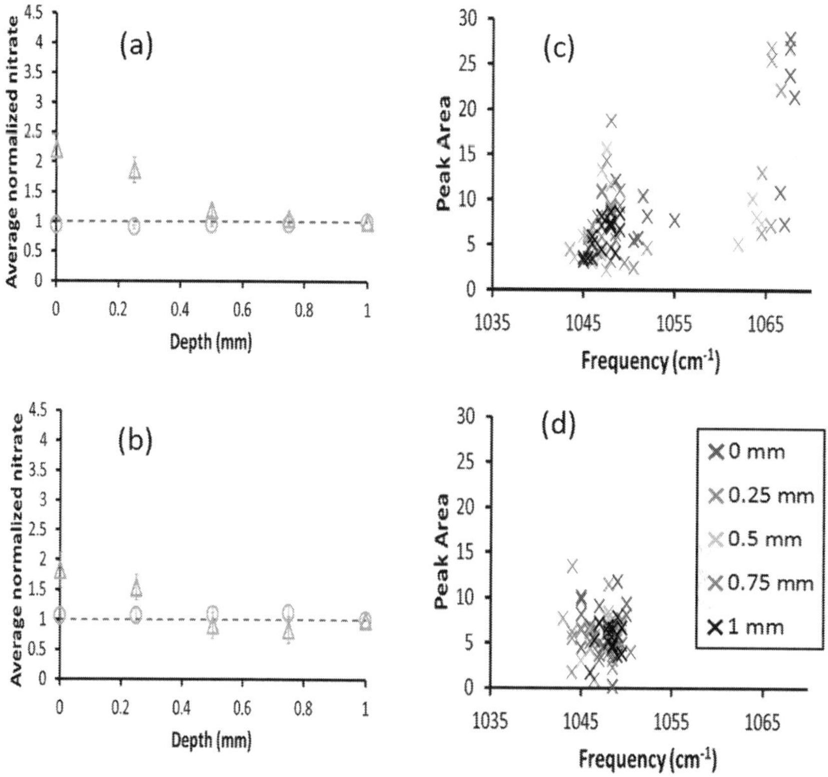

Figure 1. Location and phase of nitrate in frozen solutions. (a) and (c) show the relative nitrate concentration as a function of depth from the air interface for 250 mM $NaNO_3$ and 250 mM ($NaNO_3$ + 500 mM Instant Ocean) solutions, respectively, each normalized to its value at 1 mm deep (shown by the dashed line). The circles show liquid phase nitrate and the triangles represent the frozen data. (b) and (d) show the nitrate symmetric stretch intensity as a function of wavenumber and depth (see key) from the interface. The region around 1045 cm^{-1} is assigned to solution-phase nitrate; that around 1065 cm^{-1} is associated with crystalline nitrate. See original work for details. Reprinted with permission from Reference (8). Copyright 2017, American Chemical Society. (see color insert)

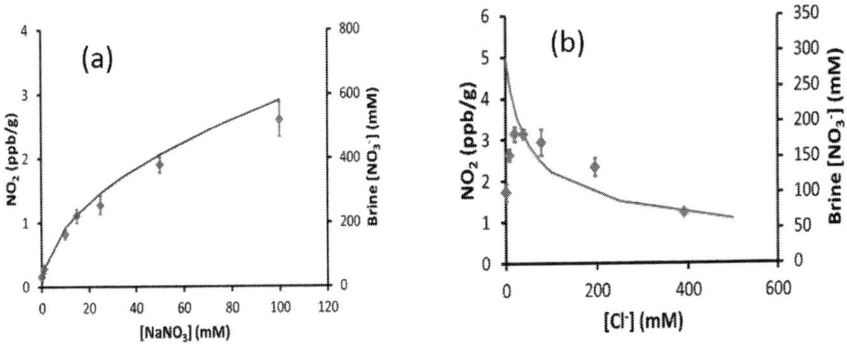

Figure 2. Gas phase NO_2 released from frozen solutions of (a) $NaNO_3$ and (b) (25 mM $NaNO_3$+Instant Ocean) as a function of salt concentration. The solid lines show the dependence expected based on simple exclusion models. See text and original reference for details. Reprinted with permission from Reference (11). Copyright 2016, American Chemical Society. (see color insert)

Figure 1d shows that the addition of Instant Ocean to the nitrate solution forms a solution brine throughout the frozen system. The points in Figure 2b display the amount of gas phase NO_2 released from snow made by freezing 25 mM nitrate with various concentration of Instant Ocean. Similar results are observed using NaCl. Here, the solid line shows the expected dilution effect on the nitrate by forming a larger and larger quantity of brine via the addition of more salt. At lower Instant Ocean concentrations, there is a dramatically higher amount of product observed than can be explained by the simple dilution effect. Some of this is well modelled by solution phase reaction taking place within the liquid brine (*11*), but there may be an additional effect of increased mobility and photolability of the nitrate at the air interface, due to the brine there.

Frozen sulfate solutions display a different chemical morphology from that observed with nitrate. Although sulfate is observed throughout the frozen sample, Figure 3a illustrates quite different behavior in the sulfate concentration as a function of distance from the air interface. The contrast with Figure 1a is clear: sulfate concentrations are not higher near the air-ice interface than within the bulk. Furthermore, Figure 3b shows that, unlike the nitrate case, sulfate exists primarily or exclusively in its crystalline form, with symmetric stretch frequencies between 990 and 995 cm^{-1}, *throughout* the frozen sample. Liquid solutions (not shown in the Figure) uniformly display this feature below about 985 cm^{-1}. The latter observation may reflect the significantly lower aqueous solubility of sodium sulfate than sodium nitrate; concentration of the salt into liquid regions during freezing may well give rise to supersaturation of the sulfate. The reason for the difference in exclusion to the air interface is not known, but may be related to the different propensities for the liquid water surface of sulfate vs. nitrate (*12*, *13*). Preliminary results suggest that the very different behavior from the nitrate case is maintained upon the addition of NaCl or Instant Ocean; work is continuing to explore this further.

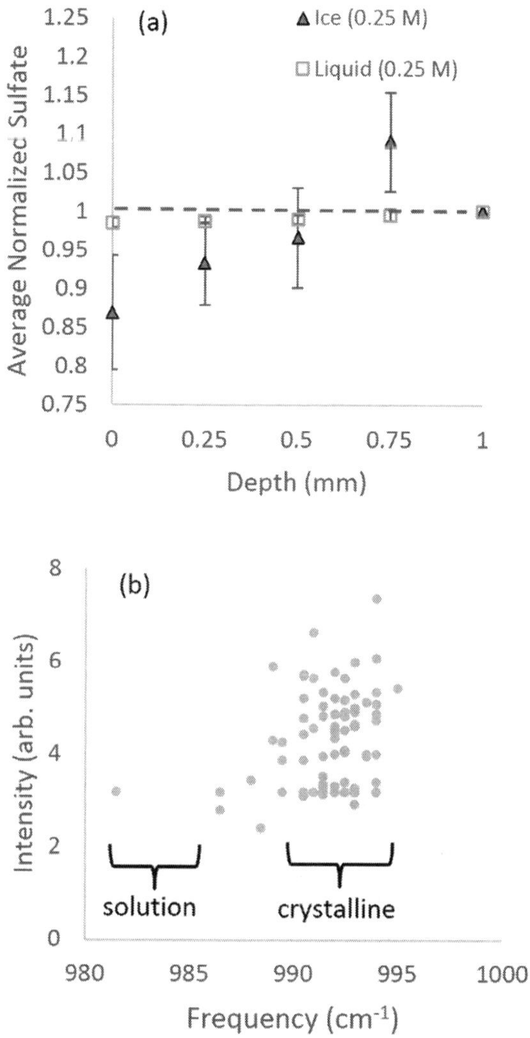

Figure 3. Location and phase of sulfate in frozen solutions of 250 mM sodium sulfate. (a) shows the relative sulfate concentration as a function of depth from the air interface normalized to its value at 1 mm deep (shown by the dashed line). The squares show liquid phase sulfate and the triangles represent the frozen data. (b) shows the sulfate symmetric stretch intensity in frozen samples as a function of wavenumber, collected together for all depths. (see color insert)

Chemical Morphology and Reactivity of Urban Grime and Urban Grime Proxies

The hypothesis that the physical state of nitrate (whether in solution or crystalline phase) strongly affects its photoreactivity is suggested as well by our work on urban grime photochemistry. The film deposited onto impervious urban surfaces ("urban grime") is a complex mixture of organic and inorganic compounds, but contains significant levels of inorganic nitrate and sulfate (*14*, *15*). The grime-deposited nitrate undergoes very rapid photochemical recycling, releasing gas phase nitrogen oxides (NO_2 and HONO) back into the atmosphere (*5*, *6*). The release shows a strong dependence on relative humidity (*6*), as illustrated in Figure 4. This dependence could be a consequence of the mobility or phase of species held in the grime changing with different amounts of water uptake. The upper trace in Figure 5 displays the water uptake of urban grime as a function of ambient relative humidity, as measured by a quartz crystal microbalance. The lower trace shows the corresponding result using hydrocarbon vacuum grease as a grime proxy. It is clear that real urban grime is more hygroscopic than the grease, probably as a consequence of its rich chemical composition. This property of the grime may allow nitrate to be present in the solution phase over a wide range of ambient humidities, perhaps enhancing its photoreactivity.

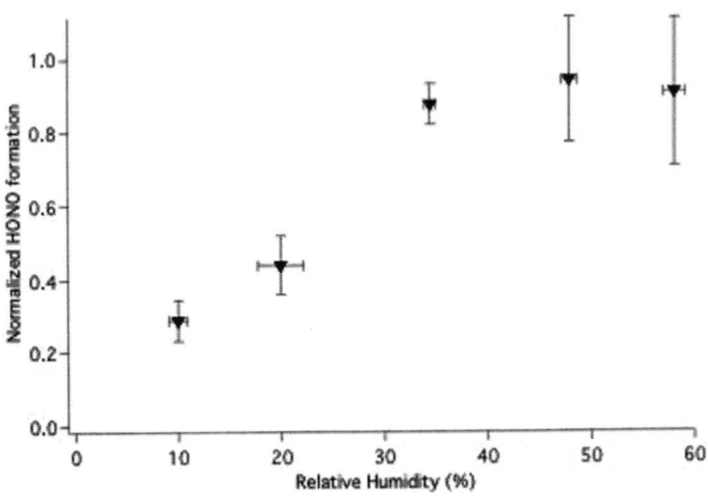

Figure 4. Normalized production of gas phase HONO as a function of relative humidity, from illumination of real urban grime using a filtered Xe lamp. Copied with permission from Reference (6), Copyright 2016, Copernicus Publications. Consult the original work for experimental details.

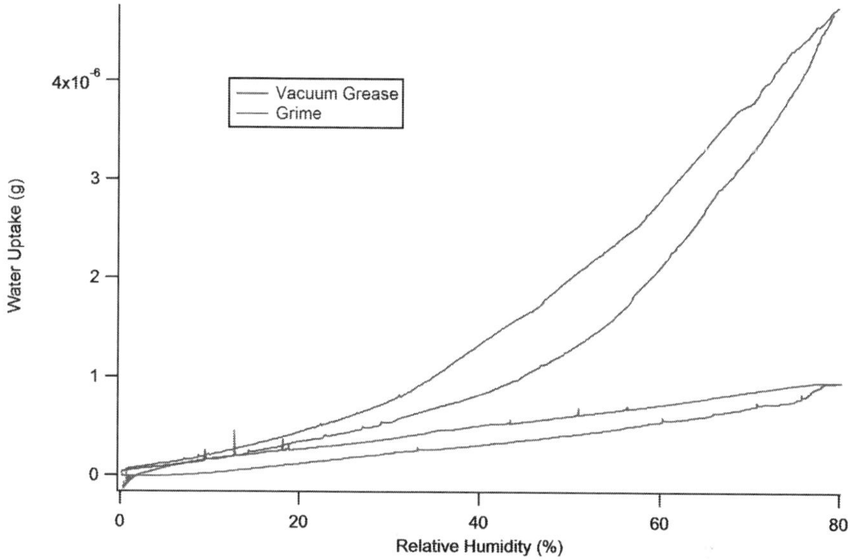

Figure 5. Water uptake onto samples of collected urban grime (blue trace) and hydrocarbon vacuum grease (red trace) as a function of relative humidity, as measured using a quartz crystal microbalance. In each trace, the upper leg is the wetting cycle and the lower indicates the drying cycle. See Reference (19) for details of how these measurements are made. (see color insert)

We explore this idea using the same Raman spectroscopic method as described above for the frozen nitrate solutions. Pyrex slides were left outdoors for one week to collect grime, then the morphology and spectrum were recorded using the Raman microscope. Figure 6a displays a photograph of the grime sample, taken through the microscope. The sample is clearly non-homogeneous with particles clearly evident over the surface. The lowermost (blue) traces in Figure 7a and 7b show the Raman spectra of (respectively) a particle region and a "clear" region of the grime. Nitrate is evident in both, showing a broad peak in the 1050 cm^{-1} region; it is not possible to resolve solution and crystalline nitrate from these peaks. Exposure of the samples to vapor taken from the headspace above a concentrated nitric acid solution yields the morphology shown in Figure 6b. Clearly the particles are quite hygroscopic; examination of the Raman spectra in Figure 7a demonstrates that they have also taken up nitrate, and that this nitrate is primarily in the solution phase. The spectra shown in Figure 7b suggest that not so much nitrate is taken up in the grime region away from the particles; the broad nature of the nitrate peak persists, suggesting a mix of solution and crystalline nitrate. Illumination with a filtered Xe lamp produces a decrease in particle-associated, solution phase nitrate, as displayed in Figure 7a, but a smaller decrease, and no change in the peak shape for the non-particulate nitrate, as shown in Figure 7b. These observations are consistent with the idea that it is the solution-phase nitrate that is most photolabile, in the urban grime matrix as well.

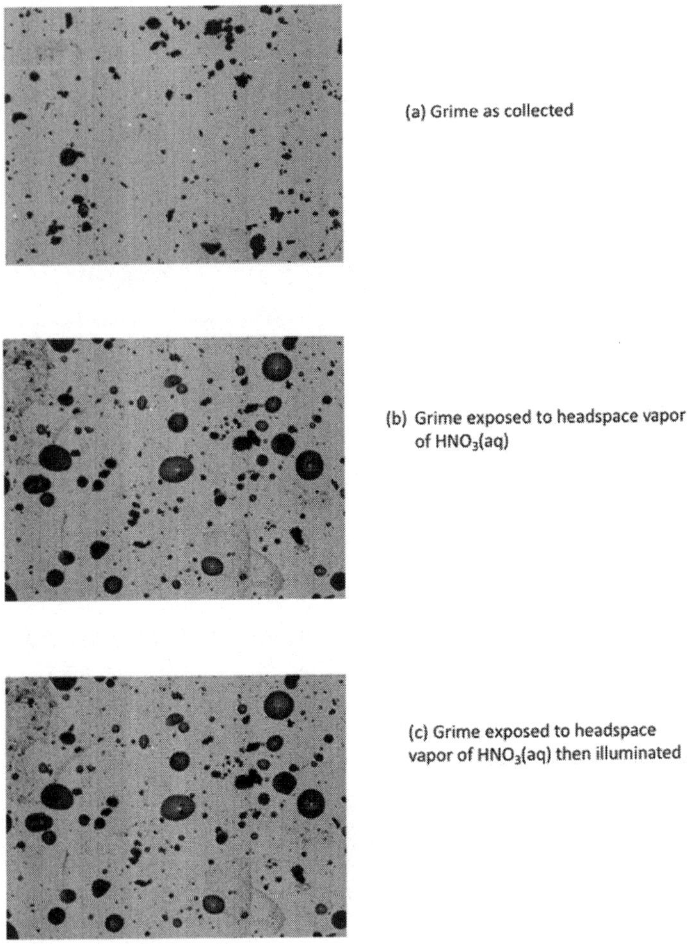

Figure 6. Photographs of real grime samples, taken using a Raman microscope. (a) The non-homogeneous nature of the grime is evident from the distribution of particulate material seen. (b) This shows hygroscopic behavior when exposed to the headspace vapor above a concentrated nitric acid solution, and (c) exposure to illumination from a filtered Xe lamp does not cause complete evaporation of the sample. (see color insert)

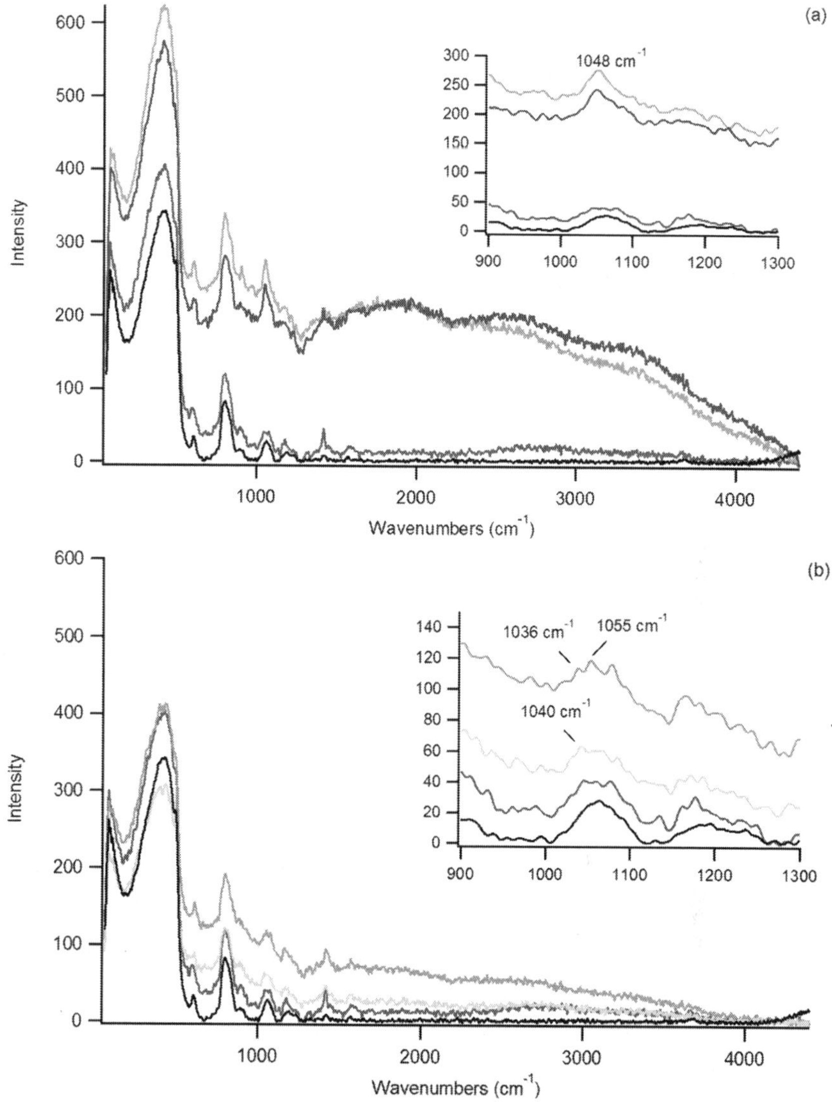

Figure 7. Raman spectra of the samples shown in Figure 6. In both frames, the collected grime spectrum is shown as the lowermost, blue trace. In all traces, the nitrate symmetric stretch is evident near 1050 cm^{-1}. (a) Spectra of one of the particulates, as collected (lower blue trace; after exposure to nitric acid vapor (upper gold trace); following illumination (purple trace). The nitrate peak sharpens and grows in the solution region upon exposure to the acid vapor. A loss of approx. 6 % in intensity is observed in the nitrate after illumination. (b) Spectra collected in the region between particles. The nitrate peak stays broad upon exposure to gas phase acid (grey trace) and only a modest decrease in its intensity is seen after illumination (light blue trace). (see color insert)

Finally, we use the nitrate probe to interrogate whether different constituents within real urban grime may play a role in maintaining a solution-like environment around nitrate, even at somewhat lower ambient relative humidities. For these experiments, we use hydrocarbon vacuum grease as a grime proxy, and dope it with nitrate and some other components of urban grime (discussed below), then perform humidification / drying cycles while following the phase behavior of the nitrate via its Raman spectrum.

The data in Figure 5 indicate that the grease alone is not effective at attracting or holding water at lower humidities. Indeed, as illustrated in Figure 8, upon drying, there is no liquid water present and the nitrate that was deposited to the film in solution phase is completely transformed to the crystalline phase. Upon repeated re-humidification and drying, the same result is obtained. The same complete desiccation upon drying is observed using 1:1 mole ratio mixtures of sodium sulfate, to rest whether sulfate-nitrate mixes behave differently. Interestingly, this is not seen using ammonium sulfate; this observation will be reported elsewhere. Clearly, the two atmospherically-relevant anions most abundant in urban grime (*14, 15*) do not give rise in mixtures to an increase in the local hygroscopicity.

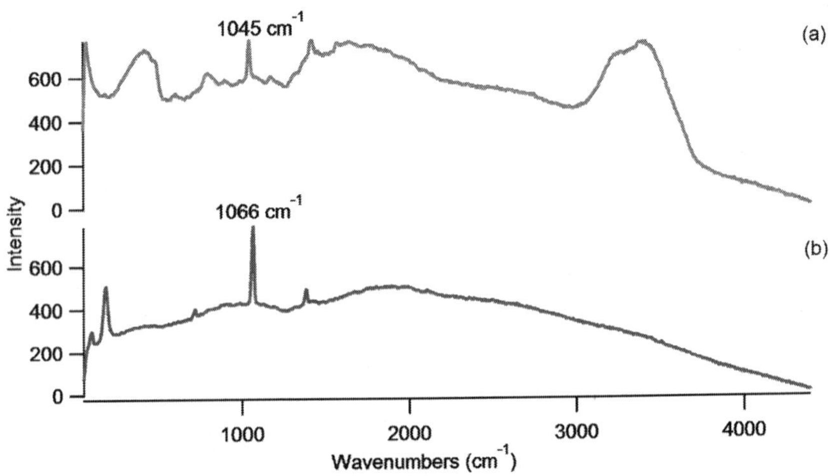

Figure 8. Raman spectra of nitrate-doped vacuum grease. The trace in (a) shows the spectrum measured immediately following deposition of solution nitrate onto the grease; note the intense liquid water band near 3400 cm^{-1}. The nitrate symmetric stretch peak is present at 1045 cm^{-1}. The lower trace, (b), shows the spectrum obtained following drying of the sample above; the liquid water band is no longer present and the nitrate peak now appears at 1066 cm^{-1}. (see color insert)

The situation is markedly different, however, when mixtures of $CaCl_2$ with $NaNO_3$ are used. Calcium and chloride are both measured to some extent in urban grimes from both Toronto, Canada (*16*) and Leipzig, Germany (*14*). As shown in Figure 9, even after three humidification / drying cycles, the nitrate peak appears at

~1050 cm^{-1}, indicative of a concentrated solution environment (*9*), and perhaps a tiny liquid water band persists. Calcium chloride (and calcium nitrate) both exhibit very low efflorescence relative humidities (*17*, *18*), meaning that their solutions will persist without crystallization into very dry conditions. It appears that the strong hygroscopicity of the $CaCl_2$ component of the mixtures in the proxy also governs the local environment of the nitrate present there. It then seems reasonable to hypothesize that calcium (and similarly hygroscopic grime components) are responsible for the enhanced water uptake observed with real urban grime over that of hydrocarbon grease, displayed in Figure 4, and may establish local "solution-like" regions for nitrate, as suggested in Figure 7.

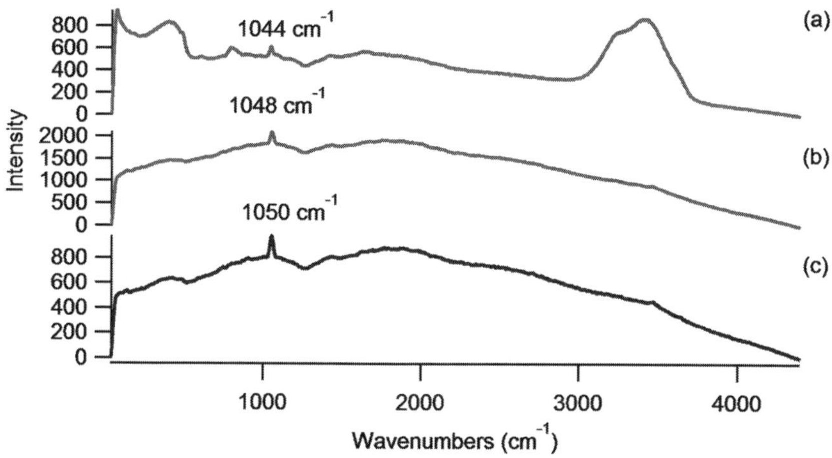

Figure 9. Raman spectra of vacuum grease doped with a mixed solution of sodium nitrate and calcium chloride. (a) The spectrum measured immediately following deposition of solution. (b) Spectrum measured after drying the sample above. (c) The spectrum obtained following three drying-rehumidification cycles of the sample. Note that the liquid water band has almost entirely vanished, but the nitrate peak remains in the solution region near 1050 cm^{-1}. (see color insert)

Overall Conclusions

The studies outlined above have provided some hints about how reactivity is related to location and physical state at some environmentally important atmospheric interfaces, although many mysteries remain. Although nitrate is clearly excluded preferentially to the frozen air interface upon freezing, sulfate is not, and may indeed show a depressed concentration there. The addition of other salts to the solution before freezing gives rise to a liquid environment experienced by nitrate, but not by sulfate. Clearly more work is necessary to understand how and where salt components are excluded during freezing – not to mention organic solutes! It appears that the rich chemical composition of urban grime gives rise to its hygroscopic nature – and may thus allow nitrate to remain in a solution-like environment even at lower ambient humidities, enhancing its photolysis rate. More work is needed to track and understand the seasonality and temperature dependence of the photochemical recycling of nitrogen oxides on real urban grime.

Acknowledgments

This work was supported financially by NSERC, to whom we are grateful. We thank Prof. Jennifer Murphy for the loan of some equipment.

References

1. Bartels-Rausch, T.; Jacobi, H.-W.; Kahan, T. F.; Thomas, J. L.; Thomson, E. S.; Abbatt, J. P. D.; Ammann, M.; Blackford, J. R.; Bluhm, H.; Boxe, C.; Dominé, F.; Frey, M. M.; Gladich, I.; Guzmán, I.; Heger, D.; Huthwelkder, Th.; Klán, P.; Kuhs, W. F.; Kuo, M. H.; Maus, S.; Moussa, S. G.; McNeill, V. F.; Newberg, J. T.; Petterson, J. B. C.; Roeselová, M.; Sodeau, J. R. A Review of Air-Ice Chemical and Physical Interactions (AICI): Liquids, Quasi-liquids, and Solids in Snow. *Atmos. Chem. Phys.* **2014**, *14*, 1587–1633.
2. Boxe, C. S.; Saiz-Lopez, A. Multiphase Modeling of Nitrate Photochemistry in the Quasi-Liquid Layer (QLL): Implications for NO_x Release from the Arctic and Coastal Antarctic Snowpack. *Atmos. Chem. Phys.* **2008**, *8*, 4855–4864.
3. Dominé, F.; Bock, J.; Voisin, D.; Donaldson, D. J. Can We Model Snow Photochemistry? Problems with the Current Approaches. *J. Phys. Chem. A* **2013**, *117*, 4733–474.
4. Wren, S. N.; Donaldson, D. J. The Exclusion of Nitrate to the Air-Ice Interface During Freezing. *J. Phys. Chem. Lett.* **2011**, *2*, 1967–1971.
5. Baergen, A. M.; Donaldson, D. J. Photochemical Renoxification of Nitric Acid on Real Urban Grime. *Environ. Sci. Technol.* **2013**, *47*, 815–820.
6. Baergen, A. M.; Donaldson, D. J. Formation of Reactive Nitrogen Oxides from Urban Grime Photochemistry. *Atmos. Chem. Phys.* **2016**, *16*, 6355–6363.
7. Langer, S.; Pemberton, R. S.; Finlayson-Pitts, B. J. Diffuse Reflectance Infrared Studies of the Reaction of Synthetic Sea Salt Mixtures with NO_2: A Key Role for Hydrates in the Kinetics and Mechanism. *J. Phys. Chem. A* **1997**, *101*, 1277–1286.
8. Morenz, K. J.; Donaldson, D. J. Chemical Morphology of Frozen Mixed Nitrate-Salt Solutions. *J. Phys. Chem. A* **1989**, *121*, 2166–2171.
9. Rusli, I. T.; Schrader, G. L.; Larson, M. A. Raman Spectroscopic Study of $NaNO_3$ Solution System - Solute Clustering in Supersaturated Solutions. *J. Crystal Growth* **1989**, *97*, 345–351.
10. Kahan, T. F.; Kwamena, N.-O. A.; Donaldson, D. J. Different Photolysis Kinetics at the Surface of Frozen Freshwater vs. Frozen Salt Solutions. *Atmos. Chem. Phys.* **2010**, *10*, 10917–10922.
11. Morenz, K. J.; Shi, Q.; Murphy, J. G.; Donaldson, D. J. Nitrate Photolysis in Salty Snow. *J. Phys. Chem. A* **2016**, *120*, 7902–7908.

12. Salvador, P.; Curtis, J. E.; Tobias, D. J.; Jungwirth, P. Polarizability of the Nitrate Anion and its Solvation at the Air/Water Interface. *Phys. Chem. Chem. Phys.* **2003**, *5*, 3752–3757.
13. Tian, C.; Byrnes, S. J.; Han, H.-L.; Shen, Y. R. Surface Propensities of Atmospherically Relevant Ions in Salt Solutions Revealed by Phase-Sensitive Sum Frequency Vibrational Spectroscopy. *J. Phys. Chem. Lett.* **2011**, *2*, 1946–1949.
14. Baergen, A. M.; Styler, S. A.; van Pinxteren, D.; Müller, K.; Herrmann, H.; Donaldson, D. J. Chemistry of Urban Grime: Inorganic Ion Composition of Grime vs. Particles in Leipzig, Germany. *Environ. Sci. Technol.* **2015**, *49*, 12688–12696.
15. Lam, B.; Diamond, M. L.; Simpson, A. J.; Makar, P. A.; Truong, J.; Hernandez-Martinez, N. A. Chemical Composition of Surface Films on Glass Windows and Implications for Atmospheric Chemistry. *Atmos. Environ.* **2005**, *39*, 6578–6586.
16. Baergen, A. M.; Donaldson, D. J. Unpublished data.
17. Martin, S. T. Phase Transitions of Aqueous Atmospheric Particles. *Chem. Rev.* **2000**, *100* (9), 3503–3453.
18. Liu, Y. J.; Zhu, T.; Zhao, D. F.; Zhang, Z. F. Investigation of the Hygroscopic Properties of $Ca(NO_3)_2$ and Internally Mixed $Ca(NO_3)_2/CaCO_3$ Particles by Micro-Raman Spectrometry. *Atmos. Chem. Phys.* **2008**, *8*, 7205–7215.
19. Demou, E.; Visram, H.; Donaldson, D. J.; Makar, P. A. Uptake of Water by Organic Films: the Dependence on the Film Oxidation State. *Atmos. Environ.* **2003**, *37*, 3527–3535.

Chapter 11

Molecular Corridors, Volatility and Particle Phase State in Secondary Organic Aerosols

Ying Li and Manabu Shiraiwa*

Department of Chemistry, University of California Irvine, Irvine, California 92697-2025, United States
***E-mail: m.shiraiwa@uci.edu**

Secondary organic aerosols (SOA) derived from the multigenerational oxidation of gaseous precursors are major components in atmospheric aerosols. Chemical evolution of SOA from a variety of volatile organic compounds (VOC) adheres to characteristic "molecular corridors" with an inverse correlation between volatility and molar mass. Parameterizations were developed to predict the saturation mass concentration of organic compounds containing oxygen, nitrogen, and sulfur from the elemental composition that can be measured by soft-ionization high-resolution mass spectrometry. The chemical nature of organic compounds observed in field measurements and laboratory experiments was characterized by mapping them into molecular corridors. SOA can occur in amorphous solid or semi-solid phase states depending on chemical composition, relative humidity (RH), and temperature. The phase transition between amorphous solid and semi-solid states occurs at the glass transition temperature (*T*g). A method was developed to estimate *T*g of pure compounds containing carbon, hydrogen, and oxygen atoms based on their elemental composition. Viscosity can be predicted using the *T*g-scaled Arrhenius plot of fragility by accounting for hygroscopic growth of SOA and applying the Gordon-Taylor mixing rule. Applying this method in an air quality model, recent global simulations have shown that SOA particles should be mostly liquid or semi-solid in the planetary boundary layer and glassy solid in the middle and upper troposphere.

Introduction

Organic aerosols are major components of atmospheric fine particulate matter (*1*, *2*). They are introduced into the atmosphere either by being directly emitted by primary sources such as fossil fuel combustion and biomass burning, or being formed through secondary processes such as multigenerational oxidation of gaseous precursors. The evolution of secondary organic aerosols (SOA) is a complex process involving both chemical reactions and mass transport in the gas and particle phases; this complexity makes interpretation of field measurements and laboratory experiments as well as accurate representation of OA evolution in chemical transport models challenging (*3–5*). Here we introduce a model framework, referred as "molecular corridors", which can help to constrain and describe the properties of SOA compounds and their fomration pathways and kinetics. This framework can also be applied in air quality and climate models (*6*).

Current two-dimensional (2D) frameworks proposed for efficient representation of SOA properties and the associated multiphase processes were built based on various SOA properties, including volatility, number of carbon and oxygen atoms in a molecule, mean carbon oxidation state, atomic O:C or H:C ratios, polarity, solubility, or the equilibrium partitioning coefficient (*2*, *7–16*). Among current 2-D frameworks, the volatility basis set (VBS) approach has been extensively applied in chemical transport models and significantly improves predictions of SOA concentrations (*17–20*).

In the VBS method, the total organic mass is classified into volatility bins and their distribution between gas and particle phases is calculated according to absorptive equilibrium partitioning, which implicitly assumes gas-phase formation of semi-volatile organic compounds as the limiting step of SOA formation (*7*, *10*, *21*). However, other processes can also be the dominant rate-limiting steps in SOA formation. Shiraiwa et al. (2014) found that SOA from a variety of biogenic and anthropogenic precursors can be represented well by the concept of molecular corridors with a tight inverse correlation between molar mass (M) and volatility of SOA oxidation products. Molecular corridors can help to constrain chemical and physical properties as well as reaction rates and pathways with characteristic kinetic regimes of reaction-, diffusion-, or accommodation-limited multiphase chemical kinetics involved in SOA evolution (*6*).

The saturation vapor pressure or pure compound saturation mass concentration (C_0) is a critical thermodynamic property describing the equilibrium gas-particle partitioning of organic compounds (*21–25*). Multiplying C_0 with an activity coefficient (γ) gives the effective saturation mass concentration (C^*) that includes the effect of non-ideal thermodynamic mixing (*10*, *26*). The term volatility often refers to C^* and can also be used for C_0 under the assumption of ideal thermodynamic mixing. According to their volatility, organic compounds can be classified as volatile organic compounds (VOC), intermediate volatility OC (IVOC), semi-volatile OC (SVOC), low-volatile OC (LVOC), and extremely low-volatile OC (ELVOC) (*27*, *28*). Quantitative estimates of the volatility of ambient OA are uncertain, especially for compounds with low volatility. Large discrepancies often exist in volatility distributions measured by different techniques (*29–33*). Current computational methods to estimate vapor pressure utilizes a functional group contribution approach that needs the information of molecular structure (*34–40*). In terms of ambient SOA, chemical composition is complex and molecular specificity is often unknown (*16*). An alternative way to assess vapor pressures of individual SOA constituents is based on their elemental composition, which can be measured by soft-ionization high-resolution mass spectrometry (HRMS) (*10, 41*).

In addition to volatility, viscosity is another important physical property of SOA. SOA phase state can be liquid (low dynamic viscosity η; $\eta < 10^2$ Pa s), semi-solid (highly viscous 'liquid'; $10^2 \leq \eta \leq 10^{12}$ Pa s), or solid (crystalline or glass; $\eta > 10^{12}$ Pa s), depending on particle composition and ambient conditions (*42–47*). Particle-phase water can act as a plasticizer to decrease viscosity (*48–50*). SOA particles containing high molar mass compounds tend to have higher viscosity (*43, 51, 52*). The relationship between viscosity and O:C is more complex (*47*). Increasing the oxidation degree, on one hand can generally lead to the increase of glass transition temperature (T_g) and thus increase viscosity (*43*); on the other hand aerosol hygroscopicity can be enhanced at the same time (*53*) and thus result in liquefying particles (*43, 50, 54*). SOA formation conditions can also affect the particle phase state (*55, 56*). The viscous behavior may also arise from highly ordered three-dimensional self-assembly of surface-active species (*57*). A mayonnaise effect on viscosity shows that structuring caused by the mixing of liquids or the addition of solutes to a solvent could increase the viscosity dramatically (*58*). A number of experimental approaches have been developed to infer particle viscosity, i.e., particle rebound (*44, 47, 59*), poke flow (*60*), bead mobility (*61*), aerosol optical tweezers (*62, 63*), fluorescence lifetime imaging (*64*), light scattering (*65*), dimer relaxation (*66, 67*) and atomic force microscopy (*68*). These methods are complementary in their capabilities accessing different viscosity ranges, particle sizes and compositions, and requiring different sample volumes (*69*).

Understanding the phase state of organic particles is important in many aspects related to SOA formation mechanisms and associated air quality, climate and health effects. The semisolid or solid phase states can limit the diffusion of condensable gas-phase molecules from the surface into the particle bulk and disturb the equilibrium in gas–particle partitioning (*54, 70–76*). Current aerosol models generally use either a thermodynamic partitioning approach (assuming instant equilibrium between semi-volatile oxidation products and the particle phase) (*7, 17–22*) or a kinetic approach (accounting for the size dependence of condensation) (*77–84*). Global aerosol microphysics models show that applying a kinetic approach rather than a thermodynamic approach in biogenic SOA formation could enhance the global mean first aerosol indirect effect by 24 % (*82*). The direct radiative effect due to biogenic SOA is less sensitive to the way gas–particle partitioning is treated (*82*). The particle phase state also affects ice nucleation (IN) processes: liquid particles can freeze homogeneously, whereas (semi-)solid particles can form ice crystals heterogeneously via deposition nucleation or via immersion freezing on partially liquefied particles that undergo a kinetic transition from glassy to liquid (*69, 85–91*). The rates of heterogeneous reactions and photochemistry are also dependent on ambient particle viscosity

(*69, 92–95*), as molecular motion can be hindered in a highly viscous matrix (*96–98*). Phase state also plays an important role in particle size distribution (*71, 80, 99, 100*). Low bulk diffusivity inside viscous semi-solid SOA can affect the growth of the smallest particles, i.e. those in the nucleation mode (*71, 101*). Low-temperature oxidation in a glassy solid state contributes to an increase of cloud droplet number concentrations (*102*). Phase state also affects the life-time of reactive pesticides (*103*) and estimations of aerosol acidity (*104*). Most of the current regional and global aerosol models treat particles as liquid droplets considering no particle phase diffusion limitations (*54, 105, 106*). Chamber experiments probing mixing timescales of SOA particles derived by oxidation of various precursors such as isoprene, terpene, and toluene have observed strong kinetic limitations at low RH, but not at moderate and high RH (*107, 108*). Gorkowski et al. (2017) did not observe significant diffusion limitations for glycerol and squalene in α-pinene SOA (*109*). Quasi-equilibrium versus kinetically-limited or non-equilibrium SOA growth is an open issue which is subject to further investigations.

Here we outline the molecular corridor approach to constrain chemical and physical properties of SOA, parameterizations to predict the saturation mass concentration and the glass transition temperature, and their application in a chemistry climate model to present a global picture of atmospheric SOA phase state.

Molecular Corridor

Figure 1 shows two-dimensional maps of molar mass (M) plotted against saturation mass concentration for pure organic compounds (C_0) for identified SOA compounds formed from oxidation of dodecane, α-pinene and isoprene (*6*). These compounds, detected in laboratory experiments and the related references, were summarized in Shiraiwa et al. (2014) (*6*). Values of C_0 were estimated using the EVAPORATION model (*36*). The markers in Figure 1 are color-coded with atomic O:C ratio. Generally, volatility decreases and molar mass increases with chemical aging of SOA both in the gas and particle phases. Molar mass of oxidation products tightly correlates with volatility with a high coefficient of determination (R^2). The 95% prediction intervals (dashed lines in Figure 1) can be regarded as molecular corridors within which additional unidentified oxidation products are likely to fall. The negative slope of the fit lines corresponds to the increase in molar mass required to decrease volatility by one order of magnitude, $-dM/d\log C_0$. It increases from ~15 g mol^{-1} for isoprene to ~25 g mol^{-1} for dodecane, depending on the molecular size of the SOA precursor and the O:C ratio of the reaction products (*6*). Figure 1(d) shows the ensemble of molecular corridors with a total of 909 identified oxidation products from seven different SOA precursors (dodecane, cyclododecane, hexylcyclohexane, α-pinene, limonene, isoprene, and glyoxal) (*6*). They are constrained by two boundary lines corresponding to the volatility of linear alkanes C_nH_{2n+2} and sugar alcohols $C_nH_{2n+2}O_n$. These lines illustrate the regular dependence of volatility on the molar mass of organic compounds; the different slopes of 30 g mol^{-1} for C_nH_{2n+2} and 12 g mol^{-1} for $C_nH_{2n+2}O_n$ reflect that the decrease of volatility with increasing molar mass is stronger for polar compounds.

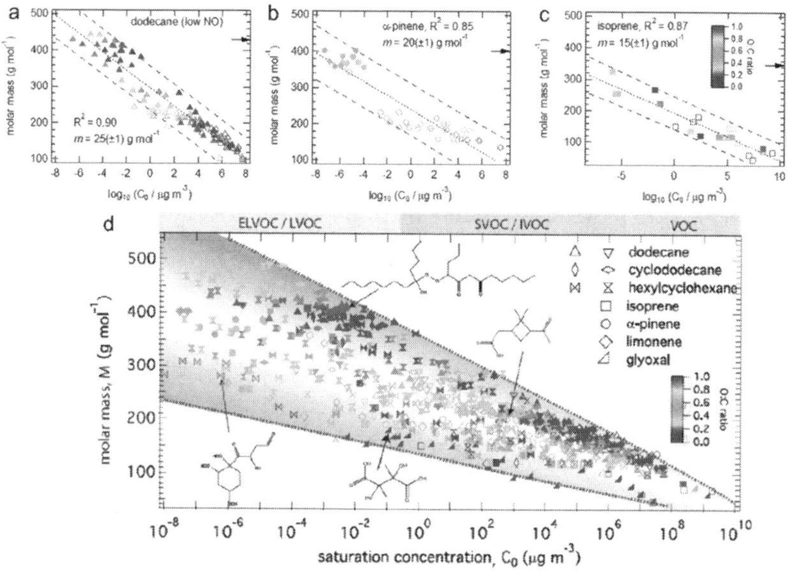

Figure 1. Molar mass vs. volatility (saturation mass concentration of pure compounds, C_0) at 298 K for oxidation products of (a) dodecane at low NO condition, (b) α-pinene and (c) isoprene. The open and solid markers correspond to the gas- and particle-phase products, respectively, color-coded by atomic O:C ratio. With a linear regression analysis, the coefficient of determination (R^2), fitted lines (dotted lines) and their slopes (m), and prediction intervals with 95% confidence (dashed lines) are shown. (d) Ensemble of molecular corridors. The dotted lines represent linear alkanes C_nH_{2n+2} (purple with O:C = 0) and sugar alcohols $C_nH_{2n+2}O_n$ (red with O:C = 1). Chemical structures of some representative products are shown. Adapted from reference (6). Copyright 2014, Copernicus Publications.

Many early generation gas-phase oxidation products of alkanes as well as dimers or oligomers with low O:C ratio (LOC) fall close to the C_nH_{2n+2} line, which we designate as the LOC corridor ($-dM/d\log C_0 \geq \sim 25$ g mol^{-1}, blue shaded area). Aqueous-phase reaction and autoxidation products with high O:C ratio (HOC), e.g. isoprene and glyoxal oxidation products, tend to fall into a corridor near the $C_nH_{2n+2}O_n$ line, designated as the HOC corridor ($-dM/d\log C_0$ of $\leq \sim 15$ g mol^{-1}, red shaded area). The area in between is characterized by intermediate O:C ratios and accordingly designated as IOC corridor ($-dM/d\log C_0 \approx \sim 20$ g mol^{-1}), which, for example, can be occupied by oxidation products of terpene such as α-pinene and limonene.

The three main reaction types of SOA evolution are functionalization, oligomerization, and fragmentation, which can also be represented in Figure 1(d). Single-step functionalization usually leads to a small increase in molar mass, corresponding to one order of decrease in volatility (*6*). Particle-phase dimerization and oligomerization involving two or more molecules usually leads to the formation of compounds with low volatility and high molar mass lying in the upper left area in the 2D space. The formation of such particle-phase products is likely limited by reaction or diffusion in the particle bulk or by gas-to-particle mass transfer when the reactions are sufficiently fast, such as those catalyzed by acid (*4*, *70*, *110–116*). Fragmentation, on the other hand, can lead to a substantial decrease of molar mass and increase in volatility (*117–121*). As a result, simple gas-phase oxidation products are confined to the lower right area in the 2D space.

Because molar mass and O:C ratio also correlate with the glass transition temperature of organic compounds (*43*), presenting identified SOA products in a molecular corridor encapsulates fundamental aspects of SOA formation and aging: volatility, molar mass, O:C ratio, and phase state. The clustering of identified reaction products in molecular corridors may facilitate estimation of the relative importance of gas- vs. particle-phase routes to SOA formation. For example, Vogel et al. (2016) studied the new particle formation and the particle growth during a field campaign conducted at a rural mountain top station in central Germany. They found that the compounds of the smaller *m/z* (mass-to-charge ratio) mode were less present in the particle phase during nucleation while they became important during particle growth. Vogel et al. (2016) explained the above phenomenon applying the relationship between molar mass and volatility indicated by the molecular corridor (*122*). Pye et al. (2017) used the molecular corridor to constrain SOA species in the CMAQ (Community Multiscale Air Quality) model (Figure 2). Historically, in CMAQ v5.1 and prior, the number of carbons, saturation mass concentration, and the organic matter to organic carbon ratios (OM/OC) were set independently, which could lead to some apparent contradictions (*123*, *124*). CMAQ v5.2 relates C_0, molar mass, and OM/OC, updating molar mass as a function of pure species C_0 and OM/OC (*123*). The updates can lead to 5–8% decrease in OC concentrations across the southeast US (*123*). Based on data from over 30 000 organic compounds, Li et al. (2016) showed that almost all of those 30 000 organics covering a wide variety of functional groups were located into the molecular corridor bounded by the two lines representing linear alkanes and sugar alcohols, except some compounds with branched structures or with O:C ratios higher than 1 (*41*). In the following sections the molecular corridor refers to the 2-D space bounded by molar mass and volatility of linear alkanes and alcohols.

Volatility Estimation

Parameterization

Following Donahue et al. (2011) in which C_0 is estimated as a function of numbers of carbon and oxygen for compounds containing oxygen (*10*), we broadened their parameterization to $\log_{10}C_0 = f(n_C, n_O, n_N, n_S,)$ to be applicable to the N and S-containing compounds:

$$\log_{10}C_0 = (n_C^0 - n_C)b_C - n_O b_O - 2\frac{n_C n_O}{n_C + n_O} b_{CO} - n_N b_N - n_S b_S \quad (1)$$

where n_C^0 is the reference carbon number; n_C, n_O, n_N, and n_S denote the numbers of carbon, oxygen, nitrogen, and sulfur atoms, respectively; b_C, b_O, b_N, and b_S denote the contribution of each atom to $\log_{10}C_0$, respectively, and b_{CO} is the carbon-oxygen nonideality (*41*). Table 1 showed the values of n_C^0 and b for each chemical composition class (CH, CHO, CHN, CHON, CHOS, and CHONS) at 298 K (*41*).

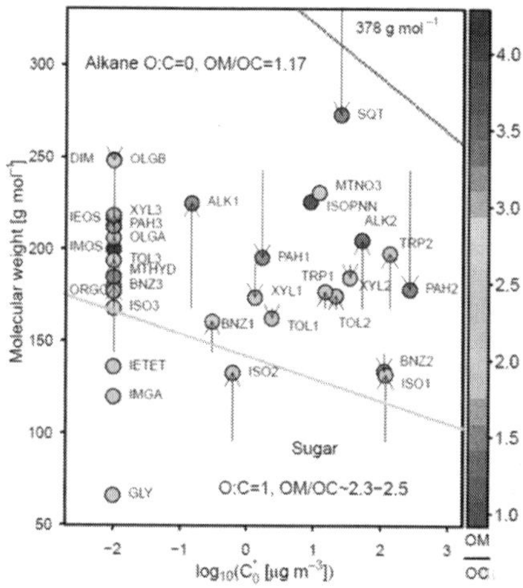

Figure 2. Molecular corridor representation of SOA species in CMAQ (Community Multiscale Air Quality) Model. The arrows start at the old molecular weights assumed in CMAQ v5.1 and end at the new molecular weights in CMAQ v5.2. Reprinted from reference (123). Copyright 2017, Copernicus Publications.

Table 1. Composition Classes and the n_C^0 and b Values for Saturation Mass Concentration Parameterizations (*41*).

Classes	n_C^0	b_C	b_O	b_{CO}	b_N	b_S
CH	23.80	0.4861				
CHO	22.66	0.4481	1.656	-0.7790		
CHN	24.59	0.4066			0.9619	
CHON	24.13	0.3667	0.7732	-0.07790	1.114	
CHOS	24.06	0.3637	1.327	-0.3988		0.7579
CHONS	28.50	0.3848	1.011	0.2921	1.053	1.316

The dataset used for deriving the above best-fit parameters is the National Cancer Institute (NCI) open database (http://cactus.nci.nih.gov/download/nci/), which contains 31,066 organic compounds. These compounds cover a molar mass from 41 to 1779 g mol^{-1}. The Estimation Programs Interface (EPI) Suite software (version 4.1) is applied to estimate the saturation vapor pressure of each compound. The EPI Suite applies three separate methods, which use the normal boiling point to estimate vapor pressure. The estimation error of EPI Suite increases as the vapor pressure decreases, especially when the vapor pressure decreases below 10^{-4} Pa (http://www.epa.gov/sab/pdf/sab-07-011.pdf). Comparing to the EPI Suite, the EVAPORATION model calculating the vapor pressure from molecular structure may give better estimations but this model does not cover all the compounds included in the NCI database, particularly some compounds containing heteroatoms (*36*).

For CHO compounds, when comparing to the EVAPORATION predictions, as shown in Figure 3a, both our parameterization and Donahue et al. (2011) give similar performance for compounds falling in IVOC and SVOC. For LVOC and ELVOC compounds, our estimates are generally higher than EVAPORATION predictions, while estimates of Donahue et al. (2011) tend to be lower. This may be partly because most estimation methods, including the EPI Suite, are constrained by databases heavily biased toward mono-functional compounds with vapor pressures in the range of ~10^3—10^5 Pa (*10, 24, 125*). Relatively large errors are also found for compounds with low volatility when comparing C_0 estimated by our parameterization with the experimental data (Figure 3b). An accurate prediction of low vapor pressure is difficult due, in large part, to the limitation caused by measurement challenges. Measurements of vapor pressures could span orders of magnitude across different techniques, especially for low volatility compounds (*16, 40, 126, 127*). The effects of functionalization, phase states, and molecular structure need to be considered in future experimental studies (*24, 25, 126, 128*).

Despite some of the limitations as described above, our new estimation parameterization derived from a large dataset is sufficiently good to predict C_0 for various structural organic classes (see Table S2 in Li et al. 2016). Since aerosol chemical composition is complex and often does not provide molecular structure specificity (*16*), estimations of C_0 based on the elemental composition are well suited for the high-resolution mass spectra data as they provide an experimental distribution of molecular formulas in the SOA sample (*129*).

Application in Laboratory, Ambient, and Modeled Organic Aerosol

Li et al. (2016) compiled data sets with 9053 organic compounds detected in chamber experiments, and field and indoor measurements of OA formed from different sources. These large data sets were mapped into the molecular corridor applying Eq. (1). Figure 4 is an example of the molecular corridor constraining nitrogen or sulfur containing compounds found in ambient air. Among these species, nitroxy-organosulfates have the highest O:C ratio (> 0.9) and the lowest volatility falling into the ELVOC group with molar mass up to 400 g mol^{-1}.

Organosulfates and organonitrates have an O:C ratio generally higher than 0.7, covering the range of IVOC to ELVOC with a broad molar mass range (100 – 600 g mol^{-1}) to occupy the high O:C corridor. Reduced sulfur compounds have a low O:C ratio (< 0.4) and are located close to the alkane line. Amine and N-heterocyclic alkaloid compounds found during new particle formation and biomass burning have the lowest O:C ratio and molar mass and the highest volatility (in VOC and IVOC groups), following the low O:C corridor.

Hatch et al. (2017) applied Eq. (1) estimating the volatility distribution of the non-methane organic gases (NMOGs) identified in laboratory biomass burning (BB) measurements, showing 6–11% of NMOGs were associated with intermediate-volatility organic compounds (Figure 5a), which historically had been unresolved in BB smoke measurements and should be considered in future updates of emission inventories (*130*). Applying Eq. (1), Koss et al. (2018) showed the volatility distribution of NMOGs can change considerably over the course of a fire. Species emitted from lower-temperature processes during the fire had a higher fraction of compounds with low volatility compared to the high-temperature processes (Figure 5b) (*131*). Considering freshly emitted BB organic aerosol (BBOA), Lin et al. (2016) investigated the molecular composition achieved during test burns of sawgrass, peat, ponderosa pine, and black spruce. The compounds in the fresh BBOA spanned a wide range of molecular weights, structures, and light absorption properties, and could be well constrained in the space characterized by molecular corridor (Figure 5c). It was estimated by Eq. (1) that many brown carbon (BrC) chromophores had low saturation mass concentration (<1 μg m^{-3}) and were expected to be retained in the particle phase under atmospherically relevant conditions. Some BrC chromophores with higher saturation mass concentration were expected to evaporate from the particles (*132*).

Aqueous processing is an important pathway of SOA formation and traditionally considered to lower the volatility of organic compounds (*133*, *134*). For example, Nie et al. (2017) used Eq.(1) and found that LVOC and ELVOC accounted for a large fraction of compounds found in water-soluble humic-like substances (HULIS) collected in a polluted megacity in eastern China (*135*). Romonosky et al. (2017) showed a complex effect of photooxidation on the volatility of dissolved SOA compounds. They prepared SOA particles formed from α-pinene and α-humulene oxidized with O_3, OH or NO_3 radicals in a smog chamber. The water-soluble compounds were extracted in water, then the resulting aqueous solutions were photolyzed. Eq. (1) was used to estimate the effect of photolysis on dissolved particle volatility (Figure 5d). It was found that the SOA compounds became more volatile on average during the photolysis, which could lead to a reduction of SOA loading (*129*). As atmospheric water could act as a reactant, product, or solvent in several ways, recent attention has focused on the role of relative humidity on chemical composition and mass yields of SOA as well as its associated climate effects (*136–138*). Hinks et al. (2018) used Eq. (1) to estimate the effect of RH on the volatility distributions of SOA formed from low-NO_x toluene oxidation in the absence of seed particles. It was found that the compounds falling in the lower volatility bins were suppressed at high RH due to a significant reduction in the fraction of oligomers which however, did not occur under dry conditions (*139*). Budisulistiorini et al (2017) applied Eq.

(1) in the CMAQ model to estimate the volatility of organosulfates during the simulation of aqueous-phase isoprene-epoxydiol (IEPOX) SOA formation (*140*).

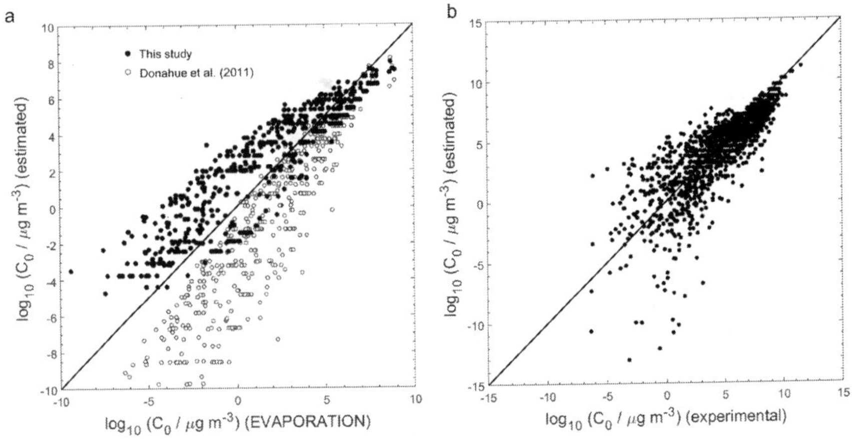

Figure 3. (a) Saturation mass concentration (C_0) of CHO compounds predicted by Eq. (1) with the coefficients from Li et al. (2016) and Donahue et al. (2011) plotted against C_0 computed by the EVAPORATION model (36). The data comprise 704 SOA oxidation products from biogenic and anthropogenic precursors (6). (b) Comparison of C_0 predicted by Eq. (1) and experimental values taken from PHYSPROP Database in the EPI Suite. Adapted from reference (41). Copyright 2016, Copernicus Publications.

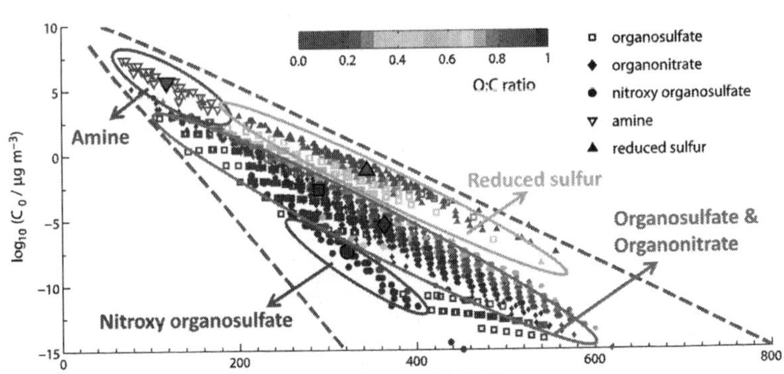

Figure 4. Molecular corridors of saturation mass concentration of pure compounds (C_0) vs. molar mass (M) for nitrogen or sulfur containing compounds collected from outdoor observations. The upper dashed line indicates the low O:C bound of the molecular corridor (linear alkanes C_nH_{2n+2} with O:C=0), the lower dotted line indicates the high O:C bound (sugar alcohols $C_nH_{2n+2}O_n$ with O:C=1).The small markers represent individual compounds color-coded by O:C ratio. The larger symbols indicate surrogate compounds with the mean values of M, C_0, and O:C ratio. Reprinted from reference (41). Copyright 2016, Copernicus Publications.

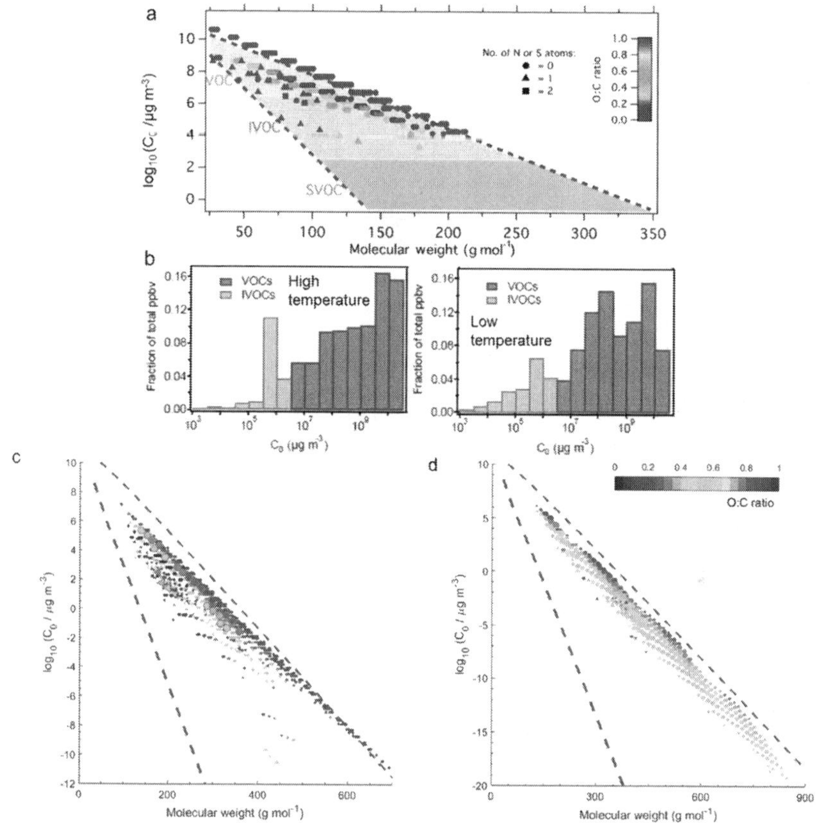

Figure 5. Application of the molecular corridors of volatility (based on the parameterization of Li et al., 2016) vs. molar mass for various measurements: (a) non-methane organic gases (NMOGs) in laboratory biomass burning measurements (reprinted and adapted from reference (130). Copyright 2017, Copernicus Publications); (b) NMOGs from high- and low-temperature processes during burning of ponderosa pine (reprinted from reference (131). Copyright 2018, Copernicus Publications); (c) BBOA samples collected from burning of sawgrass ponderosa pine (adapted from reference (132). Copyright 2016, American Chemical Society); and (d) SOA compounds formed from α-pinene oxidized with O_3. (adapted from reference (129). Copyright 2017, American Chemical Society).

Glass Transition Temperature and Viscosity Estimation

Glass Transition Temperature of Organic Compounds

Shiraiwa et al. (2017) developed a parameterization for T_g estimation based on the molar mass and atomic O:C ratio of CH and CHO compounds with $M < 450$ g mol^{-1} (*90*):

$$T_g = A + BM + CM^2 + D\,(\text{O:C}) + E\,M\,(\text{O:C}) \qquad (2.1)$$

where $A = -21.57$ [K], $B = 1.51$ [K mol g^{-1}], $C = -1.7\times10^{-3}$ [K mol^2 g^{-2}], $D = 131.4$ [K] and $E = -0.25$ [K mol g^{-1}], respectively. These values were obtained by fitting the experimentally measured T_g of 179 CH and CHO compounds with $M < 450$ g mol^{-1} (*43, 141*).

The estimation of T_g as a function of the molar mass and atomic O:C ratio is based on a strong dependence of T_g on the molar mass and a relatively weak dependence on the atomic O:C ratio (Figure 6). However, application of Eq. (2.1) may provide unreasonable T_g values for compounds with $M > 500$ g mol^{-1} due to a lack of information on experimentally measured T_g for high molar mass compounds. Recently, Rothfuss and Petters (2017) compiled an experimental dataset of T_g for compounds with M up to 1153 g mol^{-1}, providing an avenue to improve the T_g parameterization (*52*). As shown in Figure 6, when M increases above ~500 g mol^{-1}, the slope of T_g decreases, making it challenging to extrapolate the formulation of Eq. (2.1) to higher M values. When M increases to ~1000 g mol^{-1}, the corresponding T_g appears to level at around 420 K.

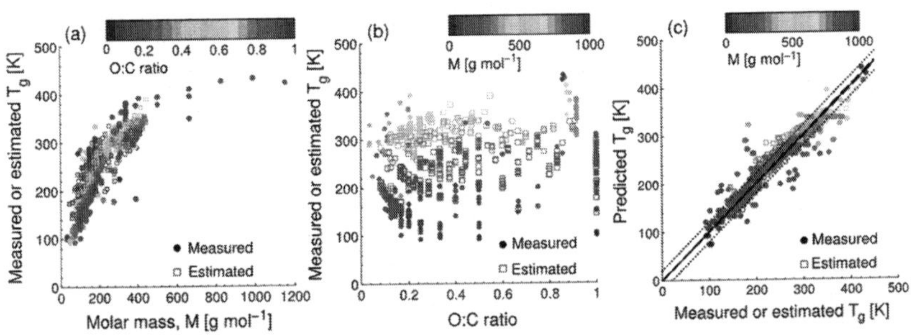

Figure 6. Characteristic relationships between molecular properties and the glass transition temperature (T_g) of organic compounds. (a) T_g of organic compounds as measured (circles) and estimated with the Boyer-Kauzmann rule (squares) plotted against molar mass. The markers are color-coded by atomic O:C ratio. (b) Measured (circles) and estimated (squares) T_g of organic compounds plotted against O:C ratio. The markers are color-coded by molar mass. (c) Predicted T_g for CHO compounds using a parameterization (Eq. 2.2) developed in DeRieux et al. (2018) compared to measured (circles) and estimated T_g by the Boyer-Kauzmann rule (squares). The solid line shows 1:1 line and the dashed and dotted lines show 68% (one standard deviation) confidence and prediction bands, respectively. Reprinted from reference (142). Copyright 2018, Copernicus Publications.

Motivated by a good correlation between T_g and volatility, another parameterization (Eq. 2.2) to estimate T_g of CH and CHO compounds was developed using the number of carbon (n_C), hydrogen (n_H), and oxygen (n_O) atoms (*142*), similar to the formulation used to estimate the saturation mass concentration (Eq. 1) (Figure 6a). Eq. (2.2) can be applied to compounds with $M > 450$ g mol^{-1}.

$$T_g = (n_C^0 + \ln(n_C))\, b_C + \ln(n_H)\, b_H + \ln(n_C)\ln(n_H)\, b_{CH} + \ln(n_O)\, b_O + \ln(n_C)\ln(n_O)\, b_{CO} \qquad (2.2)$$

where n_C^0 is the reference carbon number, b_C, b_H and b_O denote the contribution of each atom to T_g, and b_{CH} and b_{CO} are coefficients that reflect contributions from carbon-hydrogen and carbon-oxygen bonds, respectively. These values were obtained by fitting the measured T_g of 42 CH compounds and 258 CHO compounds included in Koop et al. (2011), Dette et al. (2014) and Rothfuss and Petters (2017). The best-fit parameters are summarized in Table 2. Because the evaluation dataset used to derive Eq. (2.2) contains CH compounds with $M < 260$ g mol^{-1}, the application of Eq. (2.2) to higher molar mass CH compounds requires further investigations.

Table 2. Composition Classes and the n_C^0 and b Values (K) for Glass Transition Temperature Parameterizations (*142*).

Classes	n_C^0	b_C	b_H	b_{CH}	b_O	b_{CO}
CH	1.96	61.99	−113.33	28.74		
	(±1.81)	(±53.65)	(±44.47)	(±20.86)		
CHO	12.13	10.95	−41.82	21.61	118.96	−24.38
	(±2.66)	(±13.60)	(±14.78)	(±5.30)	(±9.72)	(±4.21)

Figure 6c shows that the T_g values predicted using Eq. (2.2) are in good agreement with the T_g values measured in experiments or estimated by the Boyer-Kauzmann rule as indicated by the high correlation coefficient of 0.95. The Boyer-Kauzmann rule applied $T_g = g \cdot T_m$ where T_m is the melting temperature and g is adopted as 0.7 (*43*). T_g of individual compounds can be predicted by Eq. (2.2) to within ±21 K as indicated by the prediction band (dotted lines in Figure 6c); however, this uncertainty may be much smaller for multicomponent SOA mixtures under ideal mixing conditions as indicated in the confidence band (dashed lines, almost overlapping with the 1:1 line).

It is shown that Eq. (2.1) and Eq. (2.2) give similar performance for compounds with $M < 450$ g mol^{-1} (*142*). Comparing with Eq. (2.1), Eq. (2.2) is more flexible to be potentially expanded to include compounds containing heteroatoms (e.g., nitrogen or sulfur), once substantial sets of experimental values of T_g for such compounds become available. Regarding the application in air quality and climate models, Eq. (2.1) can be applied in the volatility basis set (VBS)[7] to predict the T_g of SOA particles (*90*). Eq. (2.2) is suitable for coupling with the statistical oxidation model (SOM) which characterizes the SOA evolution as a function of n_C and n_O (*12, 143*).

Glass Transition Temperature of SOA-Water Mixtures

SOA particles contain a number of organic compounds as well as a variable amount of liquid water. Estimations of T_g for SOA-water mixtures can be made using the Gordon-Taylor equation, which has been validated for a wide range of mixtures of organics, polymer, and water (*42*, *141*, *144*, *145*). In Shiraiwa et al. (2017) (*90*), T_g of mixtures of SOA compounds under dry conditions ($T_{g,org}$) were calculated assuming the Gordon-Taylor constant (k_{GT}) of 1 (*141*). Under humid conditions, the mass concentrations of water (m_{H2O}) absorbed by SOA particles can be estimated using the effective hygroscopicity parameter (κ) as (*146*):

$$m_{H2O} = \frac{\kappa \rho_w m_{SOA}}{\rho_{SOA}\left(\frac{1}{a_w}-1\right)} \tag{3}$$

where ρ_w and ρ_{SOA} are the density of water and SOA particles, respectively. m_{SOA} is the total mass concentrations of SOA, and a_w is the water activity calculated as a_w = RH/100.

T_g of organic - water mixtures can be simulated using the Gordon-Taylor equation (*43*):

$$T_g(w_{org}) = \frac{(1-w_{org})T_{g,w}+\frac{1}{k_{GT}}w_{org}T_{g,org}}{(1-w_{org})+\frac{1}{k_{GT}}w_{org}} \tag{4}$$

where w_{org} is the mass fraction of organics in particles, equal to m_{SOA} / (m_{SOA} + m_{H2O}). $T_{g,w}$ is the glass transition temperature of pure water (136 K), and the Gordon-Taylor constant k_{GT} is assumed as 2.5 (±1.0) (*43*, *147*).

It has been pointed out previously that the Gordon-Taylor approach may fail if T_g of each of the compounds are in very similar range or in the case of adduct or complex formation (*43*, *67*). Complex formation may occur in binary or ternary mixtures, where the two strongly interacting compounds occur at high mole fractions. However, adduct or complex formation is unlikely in multicomponent mixtures such as SOA with hundreds of compounds, which would favor a mean-field type Gordon-Taylor approach (*90*).

Viscosity Estimation

When the glass transition temperature of SOA particles could be estimated, either the Vogel-Tammann-Fulcher (VTF) equation (*148*) or the Williams-Landel-Ferry (WLF) equation (*149*) can be used to predict the viscosity.

Angell (1991) shows a modified VTF equation as: $\eta = \eta_{\infty} e^{\frac{T_0 D}{T - T_0}}$, where η_{∞} is viscosity at infinite temperature, e.g., 10^{-5} Pa s as suggested (*150*). *D* is called the fragility parameter, characterizing how rapidly the dynamics of a material slows down as *T* approaches T_g. Smaller *D* values indicate that viscosity is sensitive to temperature change (fragile behavior); while larger *D* values indicate that viscosity is less sensitive to temperature change (strong or Arrhenius behavior). Typical *D* values for organic compounds are in the range of ~5 – 20 (*151*) and the value of 10 is used in the base simulations of Shiraiwa et al. (2017) (*90*). T_0 is the Vogel temperature, which is related to T_g through $T_0 = \frac{39.17\,T_g}{D + 39.17}$, which was derived assuming $\eta = 10^{12}$ Pa s at $T = T_g$ (*150, 152*).

For viscosity estimations, both VTF and WLF equations are applied in the atmospheric community to predict the viscosity of SOA mixtures. Wang et al. (2015) applied the WLF equation to predict the viscosity of α-pinene SOA assuming the T_g of α-pinene SOA was 310 ± 15 K ranging from 295 to 328 K (*153*). Schill et al. (2013) applied the WLF equation to predict the viscosity of the mixture of 1,2,6-hexanetriol and 2,2,6,6-tetrakis (hydroxymethyl)cyclohexanol in their ice nucleation study (*154*). Maclean et al. (2017) used the WLF equation and the Gordon–Taylor equation to predict the viscosity of α-pinene SOA (*155*). Rothfuss and Petters (2017) used an adapted VTF equation and the Gordon–Taylor mixing rule to model the viscosity of sucrose under a wide range of *T* and RH (*67*). Pratap et al. (2018) applied Eq. (2.1) and the VTF equation in a hybrid model accounting for the effects of temperature and relative humidity on the lifetime of biomass burning molecular markers (*156*).

Global Phase State Distribution

Shiraiwa et al. (2017) presented a global distribution of atmospheric SOA phase state using a global model EMAC (*157*) coupled with the organic aerosol module ORACLE (*158*). ORACLE uses the volatility basis set framework (*7*) for distributing SOA oxidation products into logarithmically spaced volatility bins (*158*). Once the values of molar mass and O:C ratio of oxidation products in different volatility bins are assigned based either on the molecular corridor approach (*6, 41*) or previous studies (*159*), T_g of dry SOA products in each volatility bin can be predicted using Eq. (2.1). T_g of SOA mixed with water due to hygroscopic growth at given RH is estimated using the Gordon-Taylor equation.

The SOA phase state can be inferred using the ratio of T_g and ambient temperature *T*: $T_g/T \geq 1$ indicates an amorphous solid phase, and the threshold between semi-solid and liquid states is estimated at T_g/T of around 0.8 (*90*). Figure 7 shows that the annually average T_g/T ratio within the planetary boundary layer, indicates that SOA is mostly liquid in tropical and polar air with high relative humidity, semi-solid in the mid-latitudes, and solid over dry lands.

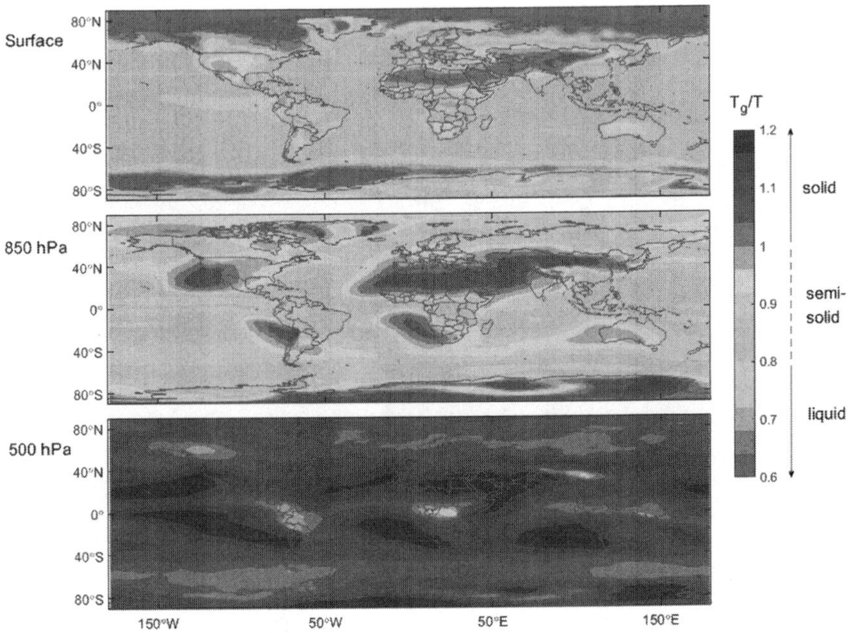

Figure 7. SOA Phase state in the global atmosphere. Modeled annual averages of the inverse ambient temperature (1/T) scaled by the glass transition temperature (T_g) of SOA (T_g/T) at the surface, 850 hPa and 500 hPa, respectively, for the years 2005-2009. T_g/T is an indicator of the particle phase state: $T_g/T \geq 1$, solid; $\sim 0.8 < T_g/T < 1$, semi-solid; $T_g/T \leq \sim 0.8$, liquid. Adapted from reference (90). Copyright 2017, Springer Nature.

Ambient measurements of particle phase state are still scarce. For background conditions of the Amazonian tropical forest, a region dominated by isoprene-derived SOA and high RH, particles are mostly liquid while with the anthropogenic influence, they occur as a nonliquid phase (*45*, *46*). Submicrometer particles in the highly polluted atmosphere of Beijing, China are reported to be in the liquid state during heavy haze episodes with RH often higher than 60% (*160*). At typical ambient RH and temperature, organic-dominated particles stay mostly liquid in the atmospheric conditions in the southeastern US, but they often turn semisolid when dried below ~50% RH in the sampling inlets (*54*). Organic particles collected in California, Mexico City and Chile may have higher viscosities and could occur in glassy states (*161*). Particles over the boreal forest in Finland have been shown to be amorphous solids at low RH (< ~30%) (*44*). Simulated results of SOA phase state in Shiraiwa et al. (2017) are consistent with those ambient measurements.

Figure 7 also shows transport of SOA particles to higher altitudes leads to more frequent occurrence of solid or semi-solid phases. At 850 hPa solid particles prevail over most of the continents at low and mid latitudes, while particles remain semi-solid or liquid over the continents in the tropics and at high latitudes as well as over the oceans. After further uplifting to 500 hPa, almost all SOA particles are expected to undergo phase transition into a glassy solid state. The occurrence of viscous states at high altitudes with low temperature is consistent with recent chamber experiments, showing that α-pinene-derived SOA particles exist in a viscous state at low temperatures corresponding to the cirrus cloud region of the free troposphere (*65*).

For the estimation of bulk diffusivity of water through a (semi-)solid matrix, the Stokes–Einstein equation is not applicable (*63, 70, 72, 86, 162*), but diffusivity can be estimated using a semi-empirical method of Berkemeier et al. (*86*) similar to the Vogel-Tammann-Fulcher (VTF) approach. The characteristic diffusion timescales of water molecules in SOA particles are predicted on the order of microseconds at the Earth's surface and seconds at 850 hPa, and can range from several minutes to days at 500 hPa (*90*). Thus, the mixing timescales (τ_{mix}) can be longer than typical cloud activation time periods, potentially inhibiting full deliquescence and allowing the OA to serve as a substrate for ice nucleation (*86, 89, 91*).

The mixing timescales of organic molecules within the SOA matrix, which can be estimated using the Stokes-Einstein equation, are much longer than mixing timescales of water molecules. Recent research has shown that mixing times of organic molecules within SOA particles can be of the order of hours at room temperature and at low RH (*55, 107, 112, 163–166*). Our global simulations predict that τ_{mix} are shorter than minutes over oceans, tropics and high latitudes at the surface and 850 hPa, indicating that particles are homogeneously mixed and are likely to be in equilibrium between the gas and particulate phases. However, τ_{mix} are more than a day over dry regions at the surface, over most continental regions at 850 hPa, and over the entire globe at 500 hPa (*90*). Maclean et al. (2017) calculate that τ_{mix} are mostly < 1 h within the planetary boundary layer during both January and July when their parameterization uses experimental viscosity data of α-pinene SOA generated in the laboratory at mass concentrations of ~1000 $\mu g\ m^{-3}$. However, the occurrences of τ_{mix} less than 1 h are significantly decreased when calculations use the experimental viscosity data of α-pinene SOA generated at a lower mass concentration of ~70 $\mu g\ m^{-3}$ (*155*). These results suggest that viscosity measurements of SOA generated at atmospheric-relevant mass concentrations are needed (*55*). In addition, the Stokes–Einstein equation may underpredict diffusion coefficients in highly viscous SOA (*95, 167–169*).

The predicted global mixing timescales of organic molecules within the SOA matrix also have important implications for long-range transport of persistent organic pollutants and polycyclic aromatic hydrocarbons (*90, 170, 171*). Toxic compounds can be embedded within glassy SOA matrices with low bulk diffusivities and long mixing timescales (*169*), which can effectively shield them from chemical degradation by atmospheric photo-oxidants, facilitating efficient long-range transport in the atmosphere (*113, 172*). Recent experiments show that the sufficiently slow diffusion of reactant molecules in the semi-solid or solid physical states of organic aerosols can inhibit browning reactions in the atmosphere, thus may influence the atmospheric energy balance (*173*).

Conclusions and Outlook

A full characterization of the physical and chemical properties of SOA is desirable but challenging, given the complexity of SOA formation and the limitation of current analytical techniques. Based on molecular identification of SOA oxidation products, it was shown that the chemical evolution of SOA from a variety of VOC precursors adheres to characteristic "molecular corridors" with a tight inverse correlation between volatility and molar mass. The model framework of molecular corridors can provide insights into physical properties of SOA compounds, formation pathways, and kinetic regimes. Parameterizations predicting the saturation mass concentration and the glass transition temperature of SOA were developed based on molar mass, atomic O:C ratio or elemental composition. These parameterizations have been applied to laboratory and ambient organic aerosols to predict SOA volatility and viscosity and the predicted values are consistent with the observations. The utility of this approach for SOA components measured by soft-ionization high-resolution mass spectrometry to predict the bulk volatility and its evolution with aging needs further investigation (*174*, *175*).

Several important aspects should be further explored in dedicated studies. A large fraction of inorganic species present in atmospheric particles can affect the partitioning of organic substances via non-ideal mixing and salt effects, which must be considered when simulating the phase state of ambient aerosols (*72*, *123*, *176–187*). Calculations indicate that comparing to a one-phase assumption, liquid–liquid phase separation can lead to either enhanced or reduced partitioning of organics to the particulate phase, depending on the overall composition of a system and the particle water content (*72*, *177*, *188*). When the phase separation occurs, the inorganic-rich and organic-rich phases may undergo glass transition at different temperatures (*189*). At ideal mixing conditions with one phase, the presence of inorganic salts (which often have lower T_g compared to SOA compounds) would lead to lower viscosity (*189*). Changes in the activity coefficient, due to the interactions of organic compounds and inorganic salts in aerosols, result in changes of the particle volatility (*10*). In addition, phase separation would also affect the saturation mass concentration by including water and other inorganics in the absorbing phase (*26*, *188*).

Other topics to be further explored include relating functional group analysis to volatility and viscosity parameterizations (*51*, *52*, *62*, *190*), diel and seasonal variations of the phase state, the RH history of particles (*191*), dependence of phase state on particle size (*192*) and various anthropogenic and biogenic precursors, the impact of temperature-modulated particle phase on cloud condensation nuclei activity (*102*), the effects of extremely low volatile organic compounds (*193*, *194*) and highly oxidized multifunctional organic compounds (HOM compounds) (*195–198*) as well as particle- and aqueous-phase chemistry (*4*, *199*, *200*). Considering the lifetime of atmospheric aerosols, laboratory measurements indicate that atmospheric chemical aging processes can increase the viscosity of particles, due to the formation of high molar mass compounds via oligomerization (*64*, *121*). Experiments and simulations to track viscosity changes during the atmospheric processing of SOA are also subject to future studies. Further development of advanced and detailed formalisms for the SOA lifecycle is required for better understanding and quantification of SOA effects on climate, air quality, and public health (*90*, *201*).

Acknowledgments

This work was funded by the National Science Foundation (AGS-1654104) and the Department of Energy (DE-SC0018349). We thank a number of our collaborators who contributed to the development of the molecular corridor approach including Ulrich Pöschl, John Seinfeld, Thomas Koop, Sergey Nizkorodov, Alex Laskin, Julia Laskin, Peng Lin, Allan Bertram, Thomas Berkemeier, Wing-Sy DeRieux, Spyros Pandis, Jos Lelieveld, Alexandra Tsimpidi, Vlassis Karydis, and Katherine Schilling-Fahnestock.

References

1. Hallquist, M.; Wenger, J.; Baltensperger, U.; Rudich, Y.; Simpson, D.; Claeys, M.; Dommen, J.; Donahue, N.; George, C.; Goldstein, A. The formation, properties and impact of secondary organic aerosol: current and emerging issues. *Atmos. Chem. Phys.* **2009**, *9*, 5155–5236.
2. Jimenez, J. L.; Canagaratna, M. R.; Donahue, N. M.; Prevot, A. S. H.; Zhang, Q.; Kroll, J. H.; DeCarlo, P. F.; Allan, J. D.; Coe, H.; Ng, N. L.; Aiken, A. C.; Docherty, K. S.; Ulbrich, I. M.; Grieshop, A. P.; Robinson, A. L.; Duplissy, J.; Smith, J. D.; Wilson, K. R.; Lanz, V. A.; Hueglin, C.; Sun, Y. L.; Tian, J.; Laaksonen, A.; Raatikainen, T.; Rautiainen, J.; Vaattovaara, P.; Ehn, M.; Kulmala, M.; Tomlinson, J. M.; Collins, D. R.; Cubison, M. J.; Dunlea, E. J.; Huffman, J. A.; Onasch, T. B.; Alfarra, M. R.; Williams, P. I.; Bower, K.; Kondo, Y.; Schneider, J.; Drewnick, F.; Borrmann, S.; Weimer, S.; Demerjian, K.; Salcedo, D.; Cottrell, L.; Griffin, R.; Takami, A.; Miyoshi, T.; Hatakeyama, S.; Shimono, A.; Sun, J. Y.; Zhang, Y. M.; Dzepina, K.; Kimmel, J. R.; Sueper, D.; Jayne, J. T.; Herndon, S. C.; Trimborn, A. M.; Williams, L. R.; Wood, E. C.; Middlebrook, A. M.; Kolb, C. E.; Baltensperger, U.; Worsnop, D. R. Evolution of Organic Aerosols in the Atmosphere. *Science* **2009**, *326*, 1525–1529.
3. Kroll, J. H.; Seinfeld, J. H. Chemistry of secondary organic aerosol: Formation and evolution of low-volatility organics in the atmosphere. *Atmos. Environ.* **2008**, *42*, 3593–3624.
4. Ziemann, P. J.; Atkinson, R. Kinetics, products, and mechanisms of secondary organic aerosol formation. *Chem. Soc. Rev.* **2012**, *41*, 6582–6605.
5. Shrivastava, M.; Cappa, C. D.; Fan, J.; Goldstein, A. H.; Guenther, A. B.; Jimenez, J. L.; Kuang, C.; Laskin, A.; Martin, S. T.; Ng, N. L.; Petaja, T.; Pierce, J. R.; Rasch, P. J.; Roldin, P.; Seinfeld, J. H.; Shilling, J.; Smith, J. N.; Thornton, J. A.; Volkamer, R.; Wang, J.; Worsnop, D. R.; Zaveri, R. A.; Zelenyuk, A.; Zhang, Q. Recent advances in understanding secondary

organic aerosol: Implications for global climate forcing. *Rev. Geophys.* **2017**, *55*, 509–559.

6. Shiraiwa, M.; Berkemeier, T.; Schilling-Fahnestock, K. A.; Seinfeld, J. H.; Pöschl, U. Molecular corridors and kinetic regimes in the multiphase chemical evolution of secondary organic aerosol. *Atmos. Chem. Phys.* **2014**, *14*, 8323–8341.
7. Donahue, N.; Robinson, A.; Stanier, C.; Pandis, S. Coupled partitioning, dilution, and chemical aging of semivolatile organics. *Environ. Sci. Technol.* **2006**, *40*, 2635–2643.
8. Pankow, J. F.; Barsanti, K. C. The carbon number-polarity grid: A means to manage the complexity of the mix of organic compounds when modeling atmospheric organic particulate matter. *Atmos. Environ.* **2009**, *43*, 2829–2835.
9. Heald, C. L.; Kroll, J. H.; Jimenez, J. L.; Docherty, K. S.; DeCarlo, P. F.; Aiken, A. C.; Chen, Q.; Martin, S. T.; Farmer, D. K.; Artaxo, P. A simplified description of the evolution of organic aerosol composition in the atmosphere. *Geophys. Res. Lett.* **2010**, *37*, L08803.
10. Donahue, N.; Epstein, S.; Pandis, S.; Robinson, A. A two-dimensional volatility basis set: 1. organic-aerosol mixing thermodynamics. *Atmos. Chem. Phys.* **2011**, *11*, 3303–3318.
11. Kroll, J. H.; Donahue, N. M.; Jimenez, J. L.; Kessler, S. H.; Canagaratna, M. R.; Wilson, K. R.; Altieri, K. E.; Mazzoleni, L. R.; Wozniak, A. S.; Bluhm, H.; Mysak, E. R.; Smith, J. D.; Kolb, C. E.; Worsnop, D. R. Carbon oxidation state as a metric for describing the chemistry of atmospheric organic aerosol. *Nat. Chem.* **2011**, *3*, 133–139.
12. Cappa, C. D.; Wilson, K. R. Multi-generation gas-phase oxidation, equilibrium partitioning, and the formation and evolution of secondary organic aerosol. *Atmos. Chem. Phys.* **2012**, *12*, 9505–9528.
13. Zhang, X.; Seinfeld, J. H. A functional group oxidation model (FGOM) for SOA formation and aging. *Atmos. Chem. Phys.* **2013**, *13*, 5907–5926.
14. Riipinen, I.; Rastak, N.; Pandis, S. N. Connecting the solubility and CCN activation of complex organic aerosols: a theoretical study using solubility distributions. *Atmos. Chem. Phys.* **2015**, *15*, 6305–6322.
15. Wania, F.; Lei, Y. D.; Wang, C.; Abbatt, J. P. D.; Goss, K. U. Using the chemical equilibrium partitioning space to explore factors influencing the phase distribution of compounds involved in secondary organic aerosol formation. *Atmos. Chem. Phys.* **2015**, *15*, 3395–3412.
16. Bzdek, B. R.; Reid, J. P. Perspective: Aerosol microphysics: From molecules to the chemical physics of aerosols. *J. Chem. Phys.* **2017**, *147*, 220901.
17. Lane, T. E.; Donahue, N. M.; Pandis, S. N. Simulating secondary organic aerosol formation using the volatility basis-set approach in a chemical transport model. *Atmos. Environ.* **2008**, *42*, 7439–7451.
18. Tsimpidi, A. P.; Karydis, V. A.; Pandis, S. N.; Lelieveld, J. Global-scale combustion sources of organic aerosols: sensitivity to formation and removal mechanisms. *Atmos. Chem. Phys.* **2017**, *17*, 7345–7364.

19. Murphy, B. N.; Pandis, S. N. Simulating the formation of semivolatile primary and secondary organic aerosol in a regional chemical transport model. *Environ. Sci. Technol.* **2009**, *43*, 4722–4728.
20. Ahmadov, R.; McKeen, S.; Robinson, A.; Bahreini, R.; Middlebrook, A.; Gouw, J.; Meagher, J.; Hsie, E.; Edgerton, E.; Shaw, S. A volatility basis set model for summertime secondary organic aerosols over the eastern United States in 2006. *J. Geophys. Res.* **2012**, *117*, D06301.
21. Pankow, J. F. An absorption-model of the gas aerosol partitioning involved in the formation of secondary organic aerosol. *Atmos. Environ.* **1994**, *28*, 189–193.
22. Odum, J. R.; Hoffmann, T.; Bowman, F.; Collins, D.; Flagan, R. C.; Seinfeld, J. H. Gas/particle partitioning and secondary organic aerosol yields. *Environ. Sci. Technol.* **1996**, *30*, 2580–2585.
23. Donahue, N. M.; Robinson, A. L.; Trump, E. R.; Riipinen, I.; Kroll, J. H. Volatility and Aging of Atmospheric Organic Aerosol. In *Atmospheric and Aerosol Chemistry. Topics in Current Chemistry*; McNeill, V. F., Ariya, P. A., Eds.; Springer: Berlin, Heidelberg, 2012; Vol. 339, pp 97−143.
24. Krieger, U. K.; Marcolli, C.; Reid, J. P. Exploring the complexity of aerosol particle properties and processes using single particle techniques. *Chem. Soc. Rev.* **2012**, *41*, 6631–6662.
25. Bilde, M.; Barsanti, K.; Booth, M.; Cappa, C. D.; Donahue, N. M.; Emanuelsson, E. U.; McFiggans, G.; Krieger, U. K.; Marcolli, C.; Topping, D.; Ziemann, P.; Barley, M.; Clegg, S.; Dennis-Smither, B.; Hallquist, M.; Hallquist, Å. M.; Khlystov, A.; Kulmala, M.; Mogensen, D.; Percival, C. J.; Pope, F.; Reid, J. P.; Ribeiro da Silva, M. A. V.; Rosenoern, T.; Salo, K.; Soonsin, V. P.; Yli-Juuti, T.; Prisle, N. L.; Pagels, J.; Rarey, J.; Zardini, A. A.; Riipinen, I. Saturation vapor pressures and transition enthalpies of low-volatility organic molecules of atmospheric relevance: from dicarboxylic acids to complex mixtures. *Chem. Rev.* **2015**, *115*, 4115–4156.
26. Zuend, A.; Seinfeld, J. H. Modeling the gas-particle partitioning of secondary organic aerosol: the importance of liquid-liquid phase separation. *Atmos. Chem. Phys.* **2012**, *12*, 3857–3882.
27. Donahue, N. M.; Kroll, J. H.; Pandis, S. N.; Robinson, A. L. A two-dimensional volatility basis set – Part 2: Diagnostics of organic-aerosol evolution. *Atmos. Chem. Phys.* **2012**, *12*, 615–634.
28. Murphy, B. N.; Donahue, N. M.; Robinson, A. L.; Pandis, S. N. A naming convention for atmospheric organic aerosol. *Atmos. Chem. Phys.* **2014**, *14*, 5825–5839.
29. Lopez-Hilfiker, F. D.; Mohr, C.; D'Ambro, E. L.; Lutz, A.; Riedel, T. P.; Gaston, C. J.; Iyer, S.; Zhang, Z.; Gold, A.; Surratt, J. D.; Lee, B. H.; Kurten, T.; Hu, W. W.; Jimenez, J.; Hallquist, M.; Thornton, J. A. Molecular Composition and Volatility of Organic Aerosol in the Southeastern U.S.: Implications for IEPOX Derived SOA. *Environ. Sci. Technol.* **2016**, *50*, 2200–2209.

30. Saha, P. K.; Khlystov, A.; Yahya, K.; Zhang, Y.; Xu, L.; Ng, N. L.; Grieshop, A. P. Quantifying the volatility of organic aerosol in the southeastern US. *Atmos. Chem. Phys.* **2017**, *17*, 501–520.
31. Stark, H.; Yatavelli, R. L.; Thompson, S. L.; Kang, H.; Krechmer, J. E.; Kimmel, J. R.; Palm, B. B.; Hu, W.; Hayes, P. L.; Day, D. A. Impact of thermal decomposition on thermal desorption instruments: advantage of thermogram analysis for quantifying volatility distributions of organic species. *Environ. Sci. Technol.* **2017**, *51*, 8491–8500.
32. Louvaris, E. E.; Florou, K.; Karnezi, E.; Papanastasiou, D. K.; Gkatzelis, G. I.; Pandis, S. N. Volatility of source apportioned wintertime organic aerosol in the city of Athens. *Atmos. Environ.* **2017**, *158*, 138–147.
33. Saha, P. K.; Khlystov, A.; Grieshop, A. P. Downwind evolution of the volatility and mixing state of near-road aerosols near a US interstate highway. *Atmos. Chem. Phys.* **2018**, *18*, 2139–2154.
34. Myrdal, P. B.; Yalkowsky, S. H. Estimating pure component vapor pressures of complex organic molecules. *Ind. Eng. Chem. Res.* **1997**, *36*, 2494–2499.
35. Nannoolal, Y.; Rarey, J.; Ramjugernath, D. Estimation of pure component properties: Part 3. Estimation of the vapor pressure of non-electrolyte organic compounds via group contributions and group interactions. *Fluid Phase Equilib.* **2008**, *269*, 117–133.
36. Compernolle, S.; Ceulemans, K.; Müller, J. F. EVAPORATION: a new vapour pressure estimation methodfor organic molecules including non-additivity and intramolecular interactions. *Atmos. Chem. Phys.* **2011**, *11*, 9431–9450.
37. Moller, B.; Rarey, J.; Ramjugernath, D. Estimation of the vapour pressure of non-electrolyte organic compounds via group contributions and group interactions. *J. Mol. Liq.* **2008**, *143*, 52–63.
38. Pankow, J. F.; Asher, W. E. SIMPOL.1: a simple group contribution method for predicting vapor pressures and enthalpies of vaporization of multifunctional organic compounds. *Atmos. Chem. Phys.* **2008**, *8*, 2773–2796.
39. Capouet, M.; Müller, J.-F. A group contribution method for estimating the vapour pressures of α-pinene oxidation products. *Atmos. Chem. Phys.* **2006**, *6*, 1455–1467.
40. O'Meara, S.; Booth, A. M.; Barley, M. H.; Topping, D.; McFiggans, G. An assessment of vapour pressure estimation methods. *Phys. Chem. Chem. Phys.* **2014**, *16*, 19453–19469.
41. Li, Y.; Pöschl, U.; Shiraiwa, M. Molecular corridors and parameterizations of volatility in the chemical evolution of organic aerosols. *Atmos. Chem. Phys.* **2016**, *16*, 3327–3344.
42. Zobrist, B.; Marcolli, C.; Pedernera, D.; Koop, T. Do atmospheric aerosols form glasses? *Atmos. Chem. Phys.* **2008**, *8*, 5221–5244.
43. Koop, T.; Bookhold, J.; Shiraiwa, M.; Poschl, U. Glass transition and phase state of organic compounds: dependency on molecular properties and implications for secondary organic aerosols in the atmosphere. *Phys. Chem. Chem. Phys.* **2011**, *13*, 19238–19255.

44. Virtanen, A.; Joutsensaari, J.; Koop, T.; Kannosto, J.; Yli-Pirilä, P.; Leskinen, J.; Mäkelä, J. M.; Holopainen, J. K.; Pöschl, U.; Kulmala, M. An amorphous solid state of biogenic secondary organic aerosol particles. *Nature* **2010**, *467*, 824–827.
45. Bateman, A. P.; Gong, Z. H.; Liu, P. F.; Sato, B.; Cirino, G.; Zhang, Y.; Artaxo, P.; Bertram, A. K.; Manzi, A. O.; Rizzo, L. V.; Souza, R. A. F.; Zaveri, R. A.; Martin, S. T. Sub-micrometre particulate matter is primarily in liquid form over Amazon rainforest. *Nat. Geosci.* **2016**, *9*, 34–37.
46. Bateman, A. P.; Gong, Z.; Harder, T. H.; de Sá, S. S.; Wang, B.; Castillo, P.; China, S.; Liu, Y.; O'Brien, R. E.; Palm, B. B.; Shiu, H. W.; Cirino, G. G.; Thalman, R.; Adachi, K.; Alexander, M. L.; Artaxo, P.; Bertram, A. K.; Buseck, P. R.; Gilles, M. K.; Jimenez, J. L.; Laskin, A.; Manzi, A. O.; Sedlacek, A.; Souza, R. A. F.; Wang, J.; Zaveri, R.; Martin, S. T. Anthropogenic influences on the physical state of submicron particulate matter over a tropical forest. *Atmos. Chem. Phys.* **2017**, *17*, 1759–1773.
47. Saukko, E.; Lambe, A. T.; Massoli, P.; Koop, T.; Wright, J. P.; Croasdale, D. R.; Pedernera, D. A.; Onasch, T. B.; Laaksonen, A.; Davidovits, P.; Worsnop, D. R.; Virtanen, A. Humidity-dependent phase state of SOA particles from biogenic and anthropogenic precursors. *Atmos. Chem. Phys.* **2012**, *12*, 7517–7529.
48. Zobrist, B.; Soonsin, V.; Luo, B. P.; Krieger, U. K.; Marcolli, C.; Peter, T.; Koop, T. Ultra-slow water diffusion in aqueous sucrose glasses. *Phys. Chem. Chem. Phys.* **2011**, *13*, 3514–3526.
49. Mikhailov, E.; Vlasenko, S.; Martin, S. T.; Koop, T.; Pöschl, U. Amorphous and crystalline aerosol particles interacting with water vapor: conceptual framework and experimental evidence for restructuring, phase transitions and kinetic limitations. *Atmos. Chem. Phys.* **2009**, *9*, 9491–9522.
50. Pajunoja, A.; Lambe, A. T.; Hakala, J.; Rastak, N.; Cummings, M. J.; Brogan, J. F.; Hao, L.; Paramonov, M.; Hong, J.; Prisle, N. L.; Malila, J.; Romakkaniemi, S.; Lehtinen, K. E. J.; Laaksonen, A.; Kulmala, M.; Massoli, P.; Onasch, T. B.; Donahue, N. M.; Riipinen, I.; Davidovits, P.; Worsnop, D. R.; Petäjä, T.; Virtanen, A. Adsorptive uptake of water by semisolid secondary organic aerosols. *Geophys. Res. Lett.* **2015**, *42*, 3063–3068.
51. Grayson, J. W.; Evoy, E.; Song, M.; Chu, Y.; Maclean, A.; Nguyen, A.; Upshur, M. A.; Ebrahimi, M.; Chan, C. K.; Geiger, F. M.; Thomson, R. J.; Bertram, A. K. The effect of hydroxyl functional groups and molar mass on the viscosity of non-crystalline organic and organic–water particles. *Atmos. Chem. Phys.* **2017**, *17*, 8509–8524.
52. Rothfuss, N. E.; Petters, M. D. Influence of Functional Groups on the Viscosity of Organic Aerosol. *Environ. Sci. Technol.* **2017**, *51*, 271–279.
53. Duplissy, J.; DeCarlo, P. F.; Dommen, J.; Alfarra, M. R.; Metzger, A.; Barmpadimos, I.; Prevot, A. S. H.; Weingartner, E.; Tritscher, T.; Gysel, M.; Aiken, A. C.; Jimenez, J. L.; Canagaratna, M. R.; Worsnop, D. R.; Collins, D. R.; Tomlinson, J.; Baltensperger, U. Relating hygroscopicity and composition of organic aerosol particulate matter. *Atmos. Chem. Phys.* **2011**, *11*, 1155–1165.

54. Pajunoja, A.; Hu, W.; Leong, Y. J.; Taylor, N. F.; Miettinen, P.; Palm, B. B.; Mikkonen, S.; Collins, D. R.; Jimenez, J. L.; Virtanen, A. Phase state of ambient aerosol linked with water uptake and chemical aging in the southeastern US. *Atmos. Chem. Phys.* **2016**, *16*, 11163–11176.
55. Grayson, J. W.; Zhang, Y.; Mutzel, A.; Renbaum-Wolff, L.; Böge, O.; Kamal, S.; Herrmann, H.; Martin, S. T.; Bertram, A. K. Effect of varying experimental conditions on the viscosity of α-pinene derived secondary organic material. *Atmos. Chem. Phys.* **2016**, *16*, 6027–6040.
56. Li, Y. J.; Liu, P.; Gong, Z.; Wang, Y.; Bateman, A. P.; Bergoend, C.; Bertram, A. K.; Martin, S. T. Chemical reactivity and liquid/nonliquid states of secondary organic material. *Environ. Sci. Technol.* **2015**, *49*, 13264–13274.
57. Pfrang, C.; Rastogi, K.; Cabrera-Martinez, E. R.; Seddon, A. M.; Dicko, C.; Labrador, A.; Plivelic, T.; Cowieson, N.; Squires, A. M. Complex three-dimensional self-assembly in proxies for atmospheric aerosols. *Nat. Commun.* **2017**, *8*, 1724.
58. Wynne, K. The mayonnaise effect. *J. Phys. Chem. Lett.* **2017**, *8*, 6189–6192.
59. Bateman, A. P.; Belassein, H.; Martin, S. T. Impactor apparatus for the study of particle rebound: relative humidity and capillary forces. *Aerosol Sci. Technol.* **2014**, *48*, 42–52.
60. Renbaum-Wolff, L.; Grayson, J. W.; Bateman, A. P.; Kuwata, M.; Sellier, M.; Murray, B. J.; Shilling, J. E.; Martin, S. T.; Bertram, A. K. Viscosity of α-pinene secondary organic material and implications for particle growth and reactivity. *Proc. Natl. Acad. Sci. U.S.A.* **2013**, *110*, 8014–8019.
61. Renbaum-Wolff, L.; Grayson, J. W.; Bertram, A. K. Technical Note: New methodology for measuring viscosities in small volumes characteristic of environmental chamber particle samples. *Atmos. Chem. Phys.* **2013**, *13*, 791–802.
62. Song, Y. C.; Haddrell, A. E.; Bzdek, B. R.; Reid, J. P.; Bannan, T.; Topping, D. O.; Percival, C.; Cai, C. Measurements and predictions of binary component aerosol particle viscosity. *J. Chem. Phys. A* **2016**, *120*, 8123–8137.
63. Power, R. M.; Simpson, S. H.; Reid, J. P.; Hudson, A. J. The transition from liquid to solid-like behaviour in ultrahigh viscosity aerosol particles. *Chem. Sci.* **2013**, *4*, 2597–2604.
64. Hosny, N.; Fitzgerald, C.; Vyšniauskas, A.; Athanasiadis, A.; Berkemeier, T.; Uygur, N.; Pöschl, U.; Shiraiwa, M.; Kalberer, M.; Pope, F. Direct imaging of changes in aerosol particle viscosity upon hydration and chemical aging. *Chem. Sci.* **2016**, *7*, 1357–1367.
65. Järvinen, E.; Ignatius, K.; Nichman, L.; Kristensen, T. B.; Fuchs, C.; Hoyle, C. R.; Höppel, N.; Corbin, J. C.; Craven, J.; Duplissy, J.; Ehrhart, S.; El Haddad, I.; Frege, C.; Gordon, H.; Jokinen, T.; Kallinger, P.; Kirkby, J.; Kiselev, A.; Naumann, K. H.; Petäjä, T.; Pinterich, T.; Prevot, A. S. H.; Saathoff, H.; Schiebel, T.; Sengupta, K.; Simon, M.; Slowik, J. G.; Tröstl, J.; Virtanen, A.; Vochezer, P.; Vogt, S.; Wagner, A. C.; Wagner, R.; Williamson, C.; Winkler, P. M.; Yan, C.; Baltensperger, U.; Donahue, N. M.; Flagan, R. C.; Gallagher, M.; Hansel, A.; Kulmala, M.; Stratmann, F.;

Worsnop, D. R.; Möhler, O.; Leisner, T.; Schnaiter, M. Observation of viscosity transition in α-pinene secondary organic aerosol. *Atmos. Chem. Phys.* **2016**, *16*, 4423–4438.

66. Rothfuss, N. E.; Petters, M. D. Coalescence-based assessment of aerosol phase state using dimers prepared through a dual-differential mobility analyzer technique. *Aerosol Sci. Technol.* **2016**, *50*, 1294–1305.
67. Rothfuss, N. E.; Petters, M. D. Characterization of the temperature and humidity-dependent phase diagram of amorphous nanoscale organic aerosols. *Phys. Chem. Chem. Phys.* **2017**, *19*, 6532–6545.
68. Lee, H. D.; Ray, K. K.; Tivanski, A. V. Solid, Semisolid, and Liquid Phase States of Individual Submicrometer Particles Directly Probed Using Atomic Force Microscopy. *Anal. Chem.* **2017**, *89*, 12720–12726.
69. Reid, J. P.; Bertram, A. K.; Topping, D. O.; Laskin, A.; Martin, S. T.; Petters, M. D.; Pope, F. D.; Rovelli, G. The viscosity of atmospherically relevant organic particles. *Nat. Commun.* **2018**, *9*, 956.
70. Shiraiwa, M.; Ammann, M.; Koop, T.; Poschl, U. Gas uptake and chemical aging of semisolid organic aerosol particles. *Proc. Natl. Acad. Sci. U.S.A.* **2011**, *108*, 11003–11008.
71. Riipinen, I.; Yli-Juuti, T.; Pierce, J. R.; Petaja, T.; Worsnop, D. R.; Kulmala, M.; Donahue, N. M. The contribution of organics to atmospheric nanoparticle growth. *Nat. Geosci.* **2012**, *5*, 453–458.
72. Shiraiwa, M.; Zuend, A.; Bertram, A. K.; Seinfeld, J. H. Gas-particle partitioning of atmospheric aerosols: interplay of physical state, non-ideal mixing and morphology. *Phys. Chem. Chem. Phys.* **2013**, *15*, 11441–11453.
73. Lienhard, D. M.; Huisman, A. J.; Bones, D. L.; Te, Y.-F.; Luo, B. P.; Krieger, U. K.; Reid, J. P. Retrieving the translational diffusion coefficient of water from experiments on single levitated aerosol droplets. *Phys. Chem. Chem. Phys.* **2014**, *16*, 16677–16683.
74. Bell, D. M.; Imre, D.; T. Martin, S.; Zelenyuk, A. The properties and behavior of [small alpha]-pinene secondary organic aerosol particles exposed to ammonia under dry conditions. *Phys. Chem. Chem. Phys.* **2017**, *19*, 6497–6507.
75. Ingram, S.; Cai, C.; Song, Y.-C.; Glowacki, D. R.; Topping, D. O.; O'Meara, S.; Reid, J. P. Characterising the evaporation kinetics of water and semi-volatile organic compounds from viscous multicomponent organic aerosol particles. *Phys. Chem. Chem. Phys.* **2017**, *19*, 31634–31646.
76. Krechmer, J. E.; Day, D. A.; Ziemann, P. J.; Jimenez, J. L. Direct measurements of gas/particle partitioning and mass accommodation coefficients in environmental chambers. *Environ. Sci. Technol.* **2017**, *51*, 11867–11875.
77. Riipinen, I.; Pierce, J. R.; Yli-Juuti, T.; Nieminen, T.; Hakkinen, S.; Ehn, M.; Junninen, H.; Lehtipalo, K.; Petaja, T.; Slowik, J.; Chang, R.; Shantz, N. C.; Abbatt, J.; Leaitch, W. R.; Kerminen, V. M.; Worsnop, D. R.; Pandis, S. N.; Donahue, N. M.; Kulmala, M. Organic condensation: a vital link connecting aerosol formation to cloud condensation nuclei (CCN) concentrations. *Atmos. Chem. Phys.* **2011**, *11*, 3865–3878.

78. Yu, F. A secondary organic aerosol formation model considering successive oxidation aging and kinetic condensation of organic compounds: global scale implications. *Atmos. Chem. Phys.* **2011**, *11*, 1083–1099.
79. Shiraiwa, M.; Pfrang, C.; Koop, T.; Pöschl, U. Kinetic multi-layer model of gas-particle interactions in aerosols and clouds (KM-GAP): linking condensation, evaporation and chemical reactions of organics, oxidants and water. *Atmos. Chem. Phys.* **2012**, *12*, 2777–2794.
80. Zaveri, R. A.; Easter, R. C.; Shilling, J. E.; Seinfeld, J. H. Modeling kinetic partitioning of secondary organic aerosol and size distribution dynamics: representing effects of volatility, phase state, and particle-phase reaction. *Atmos. Chem. Phys.* **2014**, *14*, 5153–5181.
81. Roldin, P.; Eriksson, A. C.; Nordin, E. Z.; Hermansson, E.; Mogensen, D.; Rusanen, A.; Boy, M.; Swietlicki, E.; Svenningsson, B.; Zelenyuk, A.; Pagels, J. Modelling non-equilibrium secondary organic aerosol formation and evaporation with the aerosol dynamics, gas- and particle-phase chemistry kinetic multilayer model ADCHAM. *Atmos. Chem. Phys.* **2014**, *14*, 7953–7993.
82. Scott, C. E.; Spracklen, D. V.; Pierce, J. R.; Riipinen, I.; D'Andrea, S. D.; Rap, A.; Carslaw, K. S.; Forster, P. M.; Artaxo, P.; Kulmala, M.; Rizzo, L. V.; Swietlicki, E.; Mann, G. W.; Pringle, K. J. Impact of gas-to-particle partitioning approaches on the simulated radiative effects of biogenic secondary organic aerosol. *Atmos. Chem. Phys.* **2015**, *15*, 12989–13001.
83. Gatzsche, K.; Iinuma, Y.; Tilgner, A.; Mutzel, A.; Berndt, T.; Wolke, R. Kinetic modeling studies of SOA formation from α-pinene ozonolysis. *Atmos. Chem. Phys.* **2017**, *17*, 13187–13211.
84. Sunol, A. M.; Charan, S. M.; Seinfeld, J. H. Computational simulation of the dynamics of secondary organic aerosol formation in an environmental chamber. *Aerosol Sci. Technol.* **2018**, *52*, 470–482.
85. Wang, B.; Lambe, A. T.; Massoli, P.; Onasch, T. B.; Davidovits, P.; Worsnop, D. R.; Knopf, D. A. The deposition ice nucleation and immersion freezing potential of amorphous secondary organic aerosol: Pathways for ice and mixed-phase cloud formation. *J. Geophys. Res.* **2012**, *117*, D16209.
86. Berkemeier, T.; Shiraiwa, M.; Pöschl, U.; Koop, T. Competition between water uptake and ice nucleation by glassy organic aerosol particles. *Atmos. Chem. Phys.* **2014**, *14*, 12513–12531.
87. Lienhard, D. M.; Huisman, A. J.; Krieger, U. K.; Rudich, Y.; Marcolli, C.; Luo, B. P.; Bones, D. L.; Reid, J. P.; Lambe, A. T.; Canagaratna, M. R.; Davidovits, P.; Onasch, T. B.; Worsnop, D. R.; Steimer, S. S.; Koop, T.; Peter, T. Viscous organic aerosol particles in the upper troposphere: diffusivity-controlled water uptake and ice nucleation? *Atmos. Chem. Phys.* **2015**, *15*, 13599–13613.
88. Ignatius, K.; Kristensen, T. B.; Järvinen, E.; Nichman, L.; Fuchs, C.; Gordon, H.; Herenz, P.; Hoyle, C. R.; Duplissy, J.; Garimella, S.; Dias, A.; Frege, C.; Höppel, N.; Tröstl, J.; Wagner, R.; Yan, C.; Amorim, A.; Baltensperger, U.; Curtius, J.; Donahue, N. M.; Gallagher, M. W.; Kirkby, J.; Kulmala, M.; Möhler, O.; Saathoff, H.; Schnaiter, M.; Tomé, A.; Virtanen, A.; Worsnop, D.; Stratmann, F. Heterogeneous ice nucleation of

viscous secondary organic aerosol produced from ozonolysis of α-pinene. *Atmos. Chem. Phys.* **2016**, *16*, 6495–6509.

89. Knopf, D. A.; Alpert, P. A.; Wang, B. The role of organic aerosol in atmospheric ice nucleation: a review. *ACS Earth Space Chem.* **2018**, *2*, 168–202.
90. Shiraiwa, M.; Li, Y.; Tsimpidi, A. P.; Karydis, V. A.; Berkemeier, T.; Pandis, S. N.; Lelieveld, J.; Koop, T.; Pöschl, U. Global distribution of particle phase state in atmospheric secondary organic aerosols. *Nat. Commun.* **2017**, *8*, 15002.
91. Charnawskas, J. C.; Alpert, P. A.; Lambe, A. T.; Berkemeier, T.; O'Brien, R. E.; Massoli, P.; Onasch, T. B.; Shiraiwa, M.; Moffet, R. C.; Gilles, M. K.; Davidovits, P.; Worsnop, D. R.; Knopf, D. A. Condensed-phase biogenic-anthropogenic interactions with implications for cold cloud formation. *Faraday Discuss.* **2017**, *200*, 165–194.
92. Kuwata, M.; Martin, S. T. Phase of atmospheric secondary organic material affects its reactivity. *Proc. Natl. Acad. Sci. U.S.A.* **2012**, *109*, 17354–17359.
93. Steimer, S. S.; Berkemeier, T.; Gilgen, A.; Krieger, U. K.; Peter, T.; Shiraiwa, M.; Ammann, M. Shikimic acid ozonolysis kinetics of the transition from liquid aqueous solution to highly viscous glass. *Phys. Chem. Chem. Phys.* **2015**, *17*, 31101–31109.
94. Berkemeier, T.; Steimer, S. S.; Krieger, U. K.; Peter, T.; Poschl, U.; Ammann, M.; Shiraiwa, M. Ozone uptake on glassy, semi-solid and liquid organic matter and the role of reactive oxygen intermediates in atmospheric aerosol chemistry. *Phys. Chem. Chem. Phys.* **2016**, *18*, 12662–12674.
95. Marshall, F. H.; Miles, R. E.; Song, Y.-C.; Ohm, P. B.; Power, R. M.; Reid, J. P.; Dutcher, C. S. Diffusion and reactivity in ultraviscous aerosol and the correlation with particle viscosity. *Chem. Sci.* **2016**, *7*, 1298–1308.
96. Lignell, H.; Hinks, M. L.; Nizkorodov, S. A. Exploring matrix effects on photochemistry of organic aerosols. *Proc. Natl. Acad. Sci. U.S.A.* **2014**, *111*, 13780–13785.
97. Wong, J. P. S.; Zhou, S.; Abbatt, J. P. D. Changes in secondary organic aerosol composition and mass due to photolysis: relative humidity dependence. *J. Phys. Chem. A* **2015**, *119*, 4309–4316.
98. Hinks, M. L.; Brady, M. V.; Lignell, H.; Song, M.; Grayson, J. W.; Bertram, A. K.; Lin, P.; Laskin, A.; Laskin, J.; Nizkorodov, S. A. Effect of viscosity on photodegradation rates in complex secondary organic aerosol materials. *Phys. Chem. Chem. Phys.* **2016**, *18*, 8785–8793.
99. Shiraiwa, M.; Yee, L. D.; Schilling, K. A.; Loza, C. L.; Craven, J. S.; Zuend, A.; Ziemann, P. J.; Seinfeld, J. H. Size distribution dynamics reveal particle-phase chemistry in organic aerosol formation. *Proc. Natl. Acad. Sci. U.S.A.* **2013**, *110*, 11746–11750.
100. Apsokardu, M. J.; Johnston, M. V. Nanoparticle growth by particle-phase chemistry. *Atmos. Chem. Phys.* **2018**, *18*, 1895–1907.
101. Zaveri, R. A.; Shilling, J. E.; Zelenyuk, A.; Liu, J.; Bell, D. M.; D'Ambro, E. L.; Gaston, C. J.; Thornton, J. A.; Laskin, A.; Lin, P. Growth kinetics and size distribution dynamics of viscous secondary organic aerosol. *Environ. Sci. Technol.* **2018**, *52*, 1191–1199.

102. Slade, J. H.; Shiraiwa, M.; Arangio, A.; Su, H.; Pöschl, U.; Wang, J.; Knopf, D. A. Cloud droplet activation through oxidation of organic aerosol influenced by temperature and particle phase state. *Geophys. Res. Lett.* **2017**, *44*, 1583–1591.

103. Socorro, J.; Lakey, P. S. J.; Han, L.; Berkemeier, T.; Lammel, G.; Zetzsch, C.; Pöschl, U.; Shiraiwa, M. Heterogeneous OH oxidation, shielding effects, and implications for the atmospheric fate of terbuthylazine and other pesticides. *Environ. Sci. Technol.* **2017**, *51*, 13749–13754.

104. Craig, R. L.; Nandy, L.; Axson, J. L.; Dutcher, C. S.; Ault, A. P. Spectroscopic determination of aerosol pH from acid–base equilibria in inorganic, organic, and mixed systems. *J. Phys. Chem. A* **2017**, *121*, 5690–5699.

105. O'Meara, S.; Topping, D. O.; Zaveri, R. A.; McFiggans, G. An efficient approach for treating composition-dependent diffusion within organic particles. *Atmos. Chem. Phys.* **2017**, *17*, 10477–10494.

106. Murphy, B. N.; Woody, M. C.; Jimenez, J. L.; Carlton, A. M. G.; Hayes, P. L.; Liu, S.; Ng, N. L.; Russell, L. M.; Setyan, A.; Xu, L.; Young, J.; Zaveri, R. A.; Zhang, Q.; Pye, H. O. T. Semivolatile POA and parameterized total combustion SOA in CMAQv5.2: impacts on source strength and partitioning. *Atmos. Chem. Phys.* **2017**, *17*, 11107–11133.

107. Ye, Q.; Robinson, E. S.; Ding, X.; Ye, P.; Sullivan, R. C.; Donahue, N. M. Mixing of secondary organic aerosols versus relative humidity. *Proc. Natl. Acad. Sci. U.S.A.* **2016**, *113*, 12649–12654.

108. Ye, Q.; Upshur, M. A.; Robinson, E. S.; Geiger, F. M.; Sullivan, R. C.; Thomson, R. J.; Donahue, N. M. Following Particle-Particle Mixing in Atmospheric Secondary Organic Aerosols by Using Isotopically Labeled Terpenes. *Chem* **2018**, *4*, 318–333.

109. Gorkowski, K.; Donahue, N. M.; Sullivan, R. C. Emulsified and Liquid–Liquid Phase-Separated States of α-Pinene Secondary Organic Aerosol Determined Using Aerosol Optical Tweezers. *Environ. Sci. Technol.* **2017**, *51*, 12154–12163.

110. Trump, E. R.; Donahue, N. M. Oligomer formation within secondary organic aerosols: equilibrium and dynamic considerations. *Atmos. Chem. Phys.* **2014**, *14*, 3691–3701.

111. Pfrang, C.; Shiraiwa, M.; Pöschl, U. Chemical ageing and transformation of diffusivity in semi-solid multi-component organic aerosol particles. *Atmos. Chem. Phys.* **2011**, *11*, 7343–7354.

112. Abramson, E.; Imre, D.; Beranek, J.; Wilson, J.; Zelenyuk, A. Experimental determination of chemical diffusion within secondary organic aerosol particles. *Phys. Chem. Chem. Phys.* **2013**, *15*, 2983–2991.

113. Zhou, S.; Shiraiwa, M.; McWhinney, R. D.; Pöschl, U.; Abbatt, J. P. Kinetic limitations in gas-particle reactions arising from slow diffusion in secondary organic aerosol. *Faraday Discuss.* **2013**, *165*, 391–406.

114. Surratt, J. D.; Chan, A. W. H.; Eddingsaas, N. C.; Chan, M.; Loza, C. L.; Kwan, A. J.; Hersey, S. P.; Flagan, R. C.; Wennberg, P. O.; Seinfeld, J. H. Reactive intermediates revealed in secondary organic aerosol formation from isoprene. *Proc. Natl. Acad. Sci. U.S.A.* **2010**, *107*, 6640–6645.

115. Offenberg, J. H.; Lewandowski, M.; Edney, E. O.; Kleindienst, T. E.; Jaoui, M. Influence of aerosol acidity on the formation of secondary organic aerosol from biogenic precursor hydrocarbons. *Environ. Sci. Technol.* **2009**, *43*, 7742–7747.
116. Tolocka, M. P.; Jang, M.; Ginter, J. M.; Cox, F. J.; Kamens, R. M.; Johnston, M. V. Formation of oligomers in secondary organic aerosol. *Environ. Sci. Technol.* **2004**, *38*, 1428–1434.
117. Bertram, A. K.; Ivanov, A. V.; Hunter, M.; Molina, L. T.; Molina, M. J. The reaction probability of OH on organic surfaces of tropospheric interest. *J. Phys. Chem. A* **2001**, *105*, 9415–9421.
118. Yee, L. D.; Craven, J. S.; Loza, C. L.; Schilling, K. A.; Ng, N. L.; Canagaratna, M. R.; Ziemann, P. J.; Flagan, R. C.; Seinfeld, J. H. Secondary organic aerosol formation from low-NO_x photooxidation of dodecane: evolution of multigeneration gas-phase chemistry and aerosol composition. *J. Phys. Chem. A* **2012**, *116*, 6211–6230.
119. Schilling Fahnestock, K. A.; Yee, L. D.; Loza, C. L.; Coggon, M. M.; Schwantes, R.; Zhang, X.; Dalleska, N. F.; Seinfeld, J. H. Secondary organic aerosol composition from C12 alkanes. *J. Phys. Chem. A* **2015**, *119*, 4281–4297.
120. Ruehl, C. R.; Nah, T.; Isaacman, G.; Worton, D. R.; Chan, A. W. H.; Kolesar, K. R.; Cappa, C. D.; Goldstein, A. H.; Wilson, K. R. The influence of molecular structure and aerosol phase on the heterogeneous oxidation of normal and branched alkanes by OH. *J. Phys. Chem. A* **2013**, *117*, 3990–4000.
121. Isaacman-VanWertz, G.; Massoli, P.; O'Brien, R.; Lim, C.; Franklin, J. P.; Moss, J. A.; Hunter, J. F.; Nowak, J. B.; Canagaratna, M. R.; Misztal, P. K.; Arata, C.; Roscioli, J. R.; Herndon, S. T.; Onasch, T. B.; Lambe, A. T.; Jayne, J. T.; Su, L.; Knopf, D. A.; Goldstein, A. H.; Worsnop, D. R.; Kroll, J. H. Chemical evolution of atmospheric organic carbon over multiple generations of oxidation. *Nat. Chem.* **2018**, *10*, 462–468.
122. Vogel, A. L.; Schneider, J.; Müller-Tautges, C.; Klimach, T.; Hoffmann, T. Aerosol chemistry resolved by mass spectrometry: insights into particle growth after ambient new particle formation. *Environ. Sci. Technol.* **2016**, *50*, 10814–10822.
123. Pye, H. O. T.; Murphy, B. N.; Xu, L.; Ng, N. L.; Carlton, A. G.; Guo, H.; Weber, R.; Vasilakos, P.; Appel, K. W.; Budisulistiorini, S. H.; Surratt, J. D.; Nenes, A.; Hu, W.; Jimenez, J. L.; Isaacman-VanWertz, G.; Misztal, P. K.; Goldstein, A. H. On the implications of aerosol liquid water and phase separation for organic aerosol mass. *Atmos. Chem. Phys.* **2017**, *17*, 343–369.
124. Baker, K. R.; Carlton, A. G.; Kleindienst, T. E.; Offenberg, J. H.; Beaver, M. R.; Gentner, D. R.; Goldstein, A. H.; Hayes, P. L.; Jimenez, J. L.; Gilman, J. B.; de Gouw, J. A.; Woody, M. C.; Pye, H. O. T.; Kelly, J. T.; Lewandowski, M.; Jaoui, M.; Stevens, P. S.; Brune, W. H.; Lin, Y. H.; Rubitschun, C. L.; Surratt, J. D. Gas and aerosol carbon in California: comparison of measurements and model predictions in Pasadena and Bakersfield. *Atmos. Chem. Phys.* **2015**, *15*, 5243–5258.

125. Barley, M. H.; McFiggans, G. The critical assessment of vapour pressure estimation methods for use in modelling the formation of atmospheric organic aerosol. *Atmos. Chem. Phys.* **2010**, *10*, 749–767.
126. Huisman, A. J.; Krieger, U. K.; Zuend, A.; Marcolli, C.; Peter, T. Vapor pressures of substituted polycarboxylic acids are much lower than previously reported. *Atmos. Chem. Phys.* **2013**, *13*, 6647–6662.
127. Krieger, U. K.; Siegrist, F.; Marcolli, C.; Emanuelsson, E. U.; Gøbel, F. M.; Bilde, M.; Marsh, A.; Reid, J. P.; Huisman, A. J.; Riipinen, I.; Hyttinen, N.; Myllys, N.; Kurtén, T.; Bannan, T.; Percival, C. J.; Topping, D. A reference data set for validating vapor pressure measurement techniques: homologous series of polyethylene glycols. *Atmos. Meas. Tech.* **2018**, *11*, 49–63.
128. Clegg, S. L.; Kleeman, M. J.; Griffin, R. J.; Seinfeld, J. H. Effects of uncertainties in the thermodynamic properties of aerosol components in an air quality model – Part 2: Predictions of the vapour pressures of organic compounds. *Atmos. Chem. Phys.* **2008**, *8*, 1087–1103.
129. Romonosky, D. E.; Li, Y.; Shiraiwa, M.; Laskin, A.; Laskin, J.; Nizkorodov, S. A. Aqueous photochemistry of secondary organic aerosol of α-pinene and α-humulene oxidized with ozone, hydroxyl radical, and nitrate radical. *J. Phys. Chem. A* **2017**, *121*, 1298–1309.
130. Hatch, L. E.; Yokelson, R. J.; Stockwell, C. E.; Veres, P. R.; Simpson, I. J.; Blake, D. R.; Orlando, J. J.; Barsanti, K. C. Multi-instrument comparison and compilation of non-methane organic gas emissions from biomass burning and implications for smoke-derived secondary organic aerosol precursors. *Atmos. Chem. Phys.* **2017**, *17*, 1471–1489.
131. Koss, A. R.; Sekimoto, K.; Gilman, J. B.; Selimovic, V.; Coggon, M. M.; Zarzana, K. J.; Yuan, B.; Lerner, B. M.; Brown, S. S.; Jimenez, J. L.; Krechmer, J.; Roberts, J. M.; Warneke, C.; Yokelson, R. J.; de Gouw, J. Non-methane organic gas emissions from biomass burning: identification, quantification, and emission factors from PTR-ToF during the FIREX 2016 laboratory experiment. *Atmos. Chem. Phys.* **2018**, *18*, 3299–3319.
132. Lin, P.; Aiona, P. K.; Li, Y.; Shiraiwa, M.; Laskin, J.; Nizkorodov, S. A.; Laskin, A. Molecular characterization of brown carbon in biomass burning aerosol particles. *Environ. Sci. Technol.* **2016**, *50*, 11815–11824.
133. Ervens, B.; Turpin, B.; Weber, R. Secondary organic aerosol formation in cloud droplets and aqueous particles(aqSOA): a review of laboratory, field and model studies. *Atmos. Chem. Phys.* **2011**, *11*, 11069–11102.
134. Herrmann, H.; Schaefer, T.; Tilgner, A.; Styler, S. A.; Weller, C.; Teich, M.; Otto, T. Tropospheric aqueous-phase chemistry: kinetics, mechanisms, and its coupling to a changing gas phase. *Chem. Rev.* **2015**, *115*, 4259–4334.
135. Nie, W.; Hong, J.; Häme, S. A. K.; Ding, A.; Li, Y.; Yan, C.; Hao, L.; Mikkilä, J.; Zheng, L.; Xie, Y.; Zhu, C.; Xu, Z.; Chi, X.; Huang, X.; Zhou, Y.; Lin, P.; Virtanen, A.; Worsnop, D. R.; Kulmala, M.; Ehn, M.; Yu, J.; Kerminen, V. M.; Petäjä, T. Volatility of mixed atmospheric humic-like substances and ammonium sulfate particles. *Atmos. Chem. Phys.* **2017**, *17*, 3659–3672.
136. Rastak, N.; Pajunoja, A.; Acosta Navarro, J. C.; Ma, J.; Song, M.; Partridge, D. G.; Kirkevåg, A.; Leong, Y.; Hu, W. W.; Taylor, N. F.;

Lambe, A.; Cerully, K.; Bougiatioti, A.; Liu, P.; Krejci, R.; Petäjä, T.; Percival, C.; Davidovits, P.; Worsnop, D. R.; Ekman, A. M. L.; Nenes, A.; Martin, S.; Jimenez, J. L.; Collins, D. R.; Topping, D. O.; Bertram, A. K.; Zuend, A.; Virtanen, A.; Riipinen, I. Microphysical explanation of the RH-dependent water affinity of biogenic organic aerosol and its importance for climate. *Geophys. Res. Lett.* **2017**, *44*, 5167–5177.

137. Stirnweis, L.; Marcolli, C.; Dommen, J.; Barmet, P.; Frege, C.; Platt, S. M.; Bruns, E. A.; Krapf, M.; Slowik, J. G.; Wolf, R.; Prévôt, A. S. H.; Baltensperger, U.; El-Haddad, I. Assessing the influence of NOx concentrations and relative humidity on secondary organic aerosol yields from α-pinene photo-oxidation through smog chamber experiments and modelling calculations. *Atmos. Chem. Phys.* **2017**, *17*, 5035–5061.
138. Faust, J. A.; Wong, J. P. S.; Lee, A. K. Y.; Abbatt, J. P. D. Role of aerosol liquid water in secondary organic aerosol formation from volatile organic compounds. *Environ. Sci. Technol.* **2017**, *51*, 1405–1413.
139. Hinks, M. L.; Montoya-Aguilera, J.; Ellison, L.; Lin, P.; Laskin, A.; Laskin, J.; Shiraiwa, M.; Dabdub, D.; Nizkorodov, S. A. Effect of relative humidity on the composition of secondary organic aerosol from the oxidation of toluene. *Atmos. Chem. Phys.* **2018**, *18*, 1643–1652.
140. Budisulistiorini, S. H.; Nenes, A.; Carlton, A. G.; Surratt, J. D.; McNeill, V. F.; Pye, H. O. Simulating aqueous-phase isoprene-epoxydiol (IEPOX) secondary organic aerosol production during the 2013 Southern Oxidant and Aerosol Study (SOAS). *Environ. Sci. Technol.* **2017**, *51*, 5026–5034.
141. Dette, H. P.; Qi, M.; Schröder, D. C.; Godt, A.; Koop, T. Glass-forming properties of 3-methylbutane-1,2,3-tricarboxylic acid and its mixtures with water and pinonic acid. *J. Phys. Chem. A* **2014**, *118*, 7024–7033.
142. DeRieux, W. S. W.; Li, Y.; Lin, P.; Laskin, J.; Laskin, A.; Bertram, A. K.; Nizkorodov, S. A.; Shiraiwa, M. Predicting the glass transition temperature and viscosity of secondary organic material using molecular composition. *Atmos. Chem. Phys.* **2018**, *18*, 6331–6351.
143. Jathar, S. H.; Cappa, C. D.; Wexler, A. S.; Seinfeld, J. H.; Kleeman, M. J. Multi-generational oxidation model to simulate secondary organic aerosol in a 3-D air quality model. *Geosci. Model Dev.* **2015**, *8*, 2553–2567.
144. Roos, Y. Melting and glass transitions of low molecular weight carbohydrates. *Carbohydr. Res.* **1993**, *238*, 39–48.
145. Hancock, B. C.; Zografi, G. The relationship between the glass transition temperature and the water content of amorphous pharmaceutical solids. *Pharm. Res.* **1994**, *11*, 471–477.
146. Petters, M. D.; Kreidenweis, S. M. A single parameter representation of hygroscopic growth and cloud condensation nucleus activity. *Atmos. Chem. Phys.* **2007**, *7*, 1961–1971.
147. Zobrist, B.; Marcolli, C.; Pedernera, D. A.; Koop, T. Do atmospheric aerosols form glasses? *Atmos. Chem. Phys.* **2008**, *8*, 5221–5244.
148. Fulcher, G. S. Analysis of recent measurements of the viscosity of glasses. *J. Am. Ceram. Soc.* **1925**, *8*, 339–355.

149. Williams, M. L.; Landel, R. F.; Ferry, J. D. The temperature dependence of relaxation mechanisms in amorphous polymers and other glass-forming liquids. *J. Am. Ceram. Soc.* **1955**, *77*, 3701–3707.
150. Angell, C. Relaxation in liquids, polymers and plastic crystals—strong/fragile patterns and problems. *J. Non-Cryst. Solids* **1991**, *131*, 13–31.
151. Angell, C. A. Entropy and fragility in supercooling liquids. *J. Res. Natl. Inst. Stand. Technol.* **1997**, *102*, 171–185.
152. Angell, C. A. Liquid fragility and the glass transition in water and aqueous solutions. *Chem. Rev.* **2002**, *102*, 2627–2650.
153. Wang, B.; O'Brien, R. E.; Kelly, S. T.; Shilling, J. E.; Moffet, R. C.; Gilles, M. K.; Laskin, A. Reactivity of liquid and semisolid secondary organic carbon with chloride and nitrate in atmospheric aerosols. *J. Phys. Chem. A* **2015**, *119*, 4498–4508.
154. Schill, G. P.; Tolbert, M. A. Heterogeneous ice nucleation on phase-separated organic-sulfate particles: effect of liquid vs. glassy coatings. *Atmos. Chem. Phys.* **2013**, *13*, 4681–4695.
155. Maclean, A. M.; Butenhoff, C. L.; Grayson, J. W.; Barsanti, K.; Jimenez, J. L.; Bertram, A. K. Mixing times of organic molecules within secondary organic aerosol particles: a global planetary boundary layer perspective. *Atmos. Chem. Phys.* **2017**, *17*, 13037–13048.
156. Pratap, V.; Chen, Y.; Yao, G.; Nakao, S. Temperature effects on multiphase reactions of organic molecular markers: A modeling study. *Atmos. Environ.* **2018**, *179*, 40–48.
157. Jöckel, P.; Tost, H.; Pozzer, A.; Brühl, C.; Buchholz, J.; Ganzeveld, L.; Hoor, P.; Kerkweg, A.; Lawrence, M.; nbsp; G; Sander, R.; Steil, B.; Stiller, G.; Tanarhte, M.; Taraborrelli, D.; van Aardenne, J.; Lelieveld, J. The atmospheric chemistry general circulation model ECHAM5/MESSy1: consistent simulation of ozone from the surface to the mesosphere. *Atmos. Chem. Phys.* **2006**, *6*, 5067–5104.
158. Tsimpidi, A. P.; Karydis, V. A.; Pozzer, A.; Pandis, S. N.; Lelieveld, J. ORACLE (v1.0): module to simulate the organic aerosol composition and evolution in the atmosphere. *Geosci. Model Dev.* **2014**, *7*, 3153–3172.
159. Murphy, B.; Donahue, N.; Fountoukis, C.; Pandis, S. Simulating the oxygen content of ambient organic aerosol with the 2D volatility basis set. *Atmos. Chem. Phys.* **2011**, *11*, 7859–7873.
160. Liu, Y.; Wu, Z.; Wang, Y.; Xiao, Y.; Gu, F.; Zheng, J.; Tan, T.; Shang, D.; Wu, Y.; Zeng, L.; Hu, M.; Bateman, A. P.; Martin, S. T. Submicrometer particles are in the liquid state during heavy haze episodes in the urban atmosphere of Beijing, China. *Environ. Sci. Technol. Lett.* **2017**, *4*, 427–432.
161. O'Brien, R. E.; Neu, A.; Epstein, S. A.; MacMillan, A. C.; Wang, B.; Kelly, S. T.; Nizkorodov, S. A.; Laskin, A.; Moffet, R. C.; Gilles, M. K. Physical properties of ambient and laboratory-generated secondary organic aerosol. *Geophys. Res. Lett.* **2014**, *41*, 4347–4353.
162. Price, H.; Murray, B.; Mattsson, J.; O'Sullivan, D.; Wilson, T.; Baustian, K.; Benning, L. Quantifying water diffusion in high-viscosity and glassy aqueous

solutions using a Raman isotope tracer method. *Atmos. Chem. Phys.* **2014**, *14*, 3817–3830.

163. Liu, P.; Li, Y. J.; Wang, Y.; Gilles, M. K.; Zaveri, R. A.; Bertram, A. K.; Martin, S. T. Lability of secondary organic particulate matter. *Proc. Natl. Acad. Sci. U.S.A.* **2016**, *113*, 12643–12648.
164. Zhang, Y.; Sanchez, M. S.; Douet, C.; Wang, Y.; Bateman, A. P.; Gong, Z.; Kuwata, M.; Renbaum-Wolff, L.; Sato, B. B.; Liu, P. F.; Bertram, A. K.; Geiger, F. M.; Martin, S. T. Changing shapes and implied viscosities of suspended submicron particles. *Atmos. Chem. Phys.* **2015**, *15*, 7819–7829.
165. Song, M.; Liu, P. F.; Hanna, S. J.; Zaveri, R. A.; Potter, K.; You, Y.; Martin, S. T.; Bertram, A. K. Relative humidity-dependent viscosity of secondary organic material from toluene photo-oxidation and possible implications for organic particulate matter over megacities. *Atmos. Chem. Phys.* **2016**, *16*, 8817–8830.
166. Perraud, V.; Bruns, E. A.; Ezell, M. J.; Johnson, S. N.; Yu, Y.; Alexander, M. L.; Zelenyuk, A.; Imre, D.; Chang, W. L.; Dabdub, D.; Pankow, J. F.; Finlayson-Pitts, B. J. Nonequilibrium atmospheric secondary organic aerosol formation and growth. *Proc. Natl. Acad. Sci. U.S.A.* **2012**, *109*, 2836–2841.
167. Chenyakin, Y.; Ullmann, D. A.; Evoy, E.; Renbaum-Wolff, L.; Kamal, S.; Bertram, A. K. Diffusion coefficients of organic molecules in sucrose–water solutions and comparison with Stokes–Einstein predictions. *Atmos. Chem. Phys.* **2017**, *17*, 2423–2435.
168. Price, H. C.; Mattsson, J.; Murray, B. J. Sucrose diffusion in aqueous solution. *Phys. Chem. Chem. Phys.* **2016**, *18*, 19207–19216.
169. Bastelberger, S.; Krieger, U. K.; Luo, B.; Peter, T. Diffusivity measurements of volatile organics in levitated viscous aerosol particles. *Atmos. Chem. Phys.* **2017**, *17*, 8453–8471.
170. Shrivastava, M.; Lou, S.; Zelenyuk, A.; Easter, R. C.; Corley, R. A.; Thrall, B. D.; Rasch, P. J.; Fast, J. D.; Simonich, S. L. M.; Shen, H. Global long-range transport and lung cancer risk from polycyclic aromatic hydrocarbons shielded by coatings of organic aerosol. *Proc. Natl. Acad. Sci. U.S.A.* **2017**, *114*, 1246–1251.
171. Mu, Q.; Shiraiwa, M.; Octaviani, M.; Ma, N.; Ding, A.; Su, H.; Lammel, G.; Pöschl, U.; Cheng, Y. Temperature effect on phase state and reactivity controls atmospheric multiphase chemistry and transport of PAHs. *Sci. Adv.* **2018**, *4*, eaap7314.
172. Zelenyuk, A.; Imre, D.; Beránek, J.; Abramson, E.; Wilson, J.; Shrivastava, M. Synergy between secondary organic aerosols and long-range transport of polycyclic aromatic hydrocarbons. *Environ. Sci. Technol.* **2012**, *46*, 12459–12466.
173. Liu, P.; Li, Y. J.; Wang, Y.; Bateman, A. P.; Zhang, Y.; Gong, Z.; Bertram, A. K.; Martin, S. T. Highly viscous states affect the browning of atmospheric organic particulate matter. *ACS Cent. Sci.* **2018**, *4*, 207–215.
174. Laskin, A.; Gilles, M. K.; Knopf, D. A.; Wang, B.; China, S. Progress in the analysis of complex atmospheric particles? *Annu. Rev. Anal. Chem.* **2016**, *9*, 117–143.

175. Laskin, J.; Laskin, A.; Nizkorodov, S. A. Mass spectrometry analysis in atmospheric chemistry? *Anal. Chem.* **2018**, *90*, 166–189.
176. Erdakos, G. B.; Pankow, J. F. Gas/particle partitioning of neutral and ionizing compounds to single- and multi-phase aerosol particles. 2. Phase separation in liquid particulate matter containing both polar and low-polarity organic compounds. *Atmos. Environ.* **2004**, *38*, 1005–1013.
177. Chang, E. I.; Pankow, J. F. Prediction of activity coefficients in liquid aerosol particles containing organic compounds, dissolved inorganic salts, and water—Part 2: Consideration of phase separation effects by an X-UNIFAC model. *Atmos. Environ.* **2006**, *40*, 6422–6436.
178. Marcolli, C.; Krieger, U. K. Phase changes during hygroscopic cycles of mixed organic/inorganic model systems of tropospheric aerosols. *J. Phys. Chem. A* **2006**, *110*, 1881–1893.
179. Ovadnevaite, J.; Zuend, A.; Laaksonen, A.; Sanchez, K. J.; Roberts, G.; Ceburnis, D.; Decesari, S.; Rinaldi, M.; Hodas, N.; Facchini, M. C.; Seinfeld, J. H.; O'Dowd, C. Surface tension prevails over solute effect in organic-influenced cloud droplet activation. *Nature* **2017**, *546*, 637–641.
180. Fard, M. M.; Krieger, U. K.; Peter, T. Kinetic limitation to inorganic ion diffusivity and to coalescence of inorganic inclusions in viscous liquid–liquid phase-separated particles. *J. Phys. Chem. A* **2017**, *121*, 9284–9296.
181. Freedman, M. A. Phase separation in organic aerosol. *Chem. Soc. Rev.* **2017**, *46*, 7694–7705.
182. You, Y.; Bertram, A. K. Effects of molecular weight and temperature on liquid-liquid phase separation in particles containing organic species and inorganic salts. *Atmos. Chem. Phys.* **2015**, *15*, 1351–1365.
183. Losey, D. J.; Parker, R. G.; Freedman, M. A. pH dependence of liquid–liquid phase separation in organic aerosol. *J. Phys. Chem. Lett.* **2016**, *7*, 3861–3865.
184. Renbaum-Wolff, L.; Song, M.; Marcolli, C.; Zhang, Y.; Liu, P. F.; Grayson, J. W.; Geiger, F. M.; Martin, S. T.; Bertram, A. K. Observations and implications of liquid–liquid phase separation at high relative humidities in secondary organic material produced by α-pinene ozonolysis without inorganic salts. *Atmos. Chem. Phys.* **2016**, *16*, 7969–7979.
185. Song, M.; Liu, P.; Martin, S. T.; Bertram, A. K. Liquid–liquid phase separation in particles containing secondary organic material free of inorganic salts. *Atmos. Chem. Phys.* **2017**, *17*, 11261–11271.
186. You, Y.; Renbaum-Wolff, L.; Bertram, A. Liquid–liquid phase separation in particles containing organics mixed with ammonium sulfate, ammonium bisulfate, ammonium nitrate or sodium chloride. *Atmos. Chem. Phys.* **2013**, *13*, 11723–11734.
187. Wang, C.; Lei, Y. D.; Endo, S.; Wania, F. Measuring and modeling the salting-out effect in ammonium sulfate solutions. *Environ. Sci. Technol.* **2014**, *48*, 13238–13245.
188. Zuend, A.; Marcolli, C.; Peter, T.; Seinfeld, J. H. Computation of liquid-liquid equilibria and phase stabilities: implications for RH-dependent gas/particle partitioning of organic-inorganic aerosols. *Atmos. Chem. Phys.* **2010**, *10*, 7795–7820.

189. Dette, H. P.; Koop, T. Glass formation processes in mixed inorganic/organic aerosol particles. *J. Phys. Chem. A* **2015**, *119*, 4552–4561.
190. Takahama, S.; Ruggeri, G. Relating functional group measurements to carbon types for improved model–measurement comparisons of organic aerosol composition. *Atmos. Chem. Phys.* **2017**, *17*, 4433–4450.
191. Altaf, M. B.; Freedman, M. A. Effect of drying rate on aerosol particle morphology. *J. Phys. Chem. Lett.* **2017**, *8*, 3613–3618.
192. Cheng, Y.; Su, H.; Koop, T.; Mikhailov, E.; Pöschl, U. Size dependence of phase transitions in aerosol nanoparticles. *Nat. Commun.* **2015**, *6*, 5923.
193. Ehn, M.; Thornton, J. A.; Kleist, E.; Sipila, M.; Junninen, H.; Pullinen, I.; Springer, M.; Rubach, F.; Tillmann, R.; Lee, B.; Lopez-Hilfiker, F.; Andres, S.; Acir, I.-H.; Rissanen, M.; Jokinen, T.; Schobesberger, S.; Kangasluoma, J.; Kontkanen, J.; Nieminen, T.; Kurten, T.; Nielsen, L. B.; Jorgensen, S.; Kjaergaard, H. G.; Canagaratna, M.; Dal Maso, M.; Berndt, T.; Petaja, T.; Wahner, A.; Kerminen, V.-M.; Kulmala, M.; Worsnop, D. R.; Wildt, J.; Mentel, T. F. A large source of low-volatility secondary organic aerosol. *Nature* **2014**, *506*, 476–479.
194. Barsanti, K. C.; Kroll, J. H.; Thornton, J. A. Formation of low-volatility organic compounds in the atmosphere: recent advancements and insights. *J. Phys. Chem. Lett.* **2017**, *8*, 1503–1511.
195. Tröstl, J.; Chuang, W. K.; Gordon, H.; Heinritzi, M.; Yan, C.; Molteni, U.; Ahlm, L.; Frege, C.; Bianchi, F.; Wagner, R.; Simon, M.; Lehtipalo, K.; Williamson, C.; Craven, J. S.; Duplissy, J.; Adamov, A.; Almeida, J.; Bernhammer, A.-K.; Breitenlechner, M.; Brilke, S.; Dias, A.; Ehrhart, S.; Flagan, R. C.; Franchin, A.; Fuchs, C.; Guida, R.; Gysel, M.; Hansel, A.; Hoyle, C. R.; Jokinen, T.; Junninen, H.; Kangasluoma, J.; Keskinen, H.; Kim, J.; Krapf, M.; Kürten, A.; Laaksonen, A.; Lawler, M.; Leiminger, M.; Mathot, S.; Möhler, O.; Nieminen, T.; Onnela, A.; Petäjä, T.; Piel, F. M.; Miettinen, P.; Rissanen, M. P.; Rondo, L.; Sarnela, N.; Schobesberger, S.; Sengupta, K.; Sipilä, M.; Smith, J. N.; Steiner, G.; Tomè, A.; Virtanen, A.; Wagner, A. C.; Weingartner, E.; Wimmer, D.; Winkler, P. M.; Ye, P.; Carslaw, K. S.; Curtius, J.; Dommen, J.; Kirkby, J.; Kulmala, M.; Riipinen, I.; Worsnop, D. R.; Donahue, N. M.; Baltensperger, U. The role of low-volatility organic compounds in initial particle growth in the atmosphere. *Nature* **2016**, *533*, 527–531.
196. Kirkby, J.; Duplissy, J.; Sengupta, K.; Frege, C.; Gordon, H.; Williamson, C.; Heinritzi, M.; Simon, M.; Yan, C.; Almeida, J.; Tröstl, J.; Nieminen, T.; Ortega, I. K.; Wagner, R.; Adamov, A.; Amorim, A.; Bernhammer, A.-K.; Bianchi, F.; Breitenlechner, M.; Brilke, S.; Chen, X.; Craven, J.; Dias, A.; Ehrhart, S.; Flagan, R. C.; Franchin, A.; Fuchs, C.; Guida, R.; Hakala, J.; Hoyle, C. R.; Jokinen, T.; Junninen, H.; Kangasluoma, J.; Kim, J.; Krapf, M.; Kürten, A.; Laaksonen, A.; Lehtipalo, K.; Makhmutov, V.; Mathot, S.; Molteni, U.; Onnela, A.; Peräkylä, O.; Piel, F.; Petäjä, T.; Praplan, A. P.; Pringle, K.; Rap, A.; Richards, N. A. D.; Riipinen, I.; Rissanen, M. P.; Rondo, L.; Sarnela, N.; Schobesberger, S.; Scott, C. E.; Seinfeld, J. H.; Sipilä, M.; Steiner, G.; Stozhkov, Y.; Stratmann, F.; Tomé, A.; Virtanen, A.; Vogel, A. L.; Wagner, A. C.; Wagner, P. E.; Weingartner, E.; Wimmer, D.;

Winkler, P. M.; Ye, P.; Zhang, X.; Hansel, A.; Dommen, J.; Donahue, N. M.; Worsnop, D. R.; Baltensperger, U.; Kulmala, M.; Carslaw, K. S.; Curtius, J. Ion-induced nucleation of pure biogenic particles. *Nature* **2016**, *533*, 521–526.

197. Cappa, C. Unexpected player in particle formation. *Nature* **2016**, *533*, 478–479.
198. Bianchi, F.; Tröstl, J.; Junninen, H.; Frege, C.; Henne, S.; Hoyle, C. R.; Molteni, U.; Herrmann, E.; Adamov, A.; Bukowiecki, N. New particle formation in the free troposphere: A question of chemistry and timing. *Science* **2016**, *352*, 1109–1112.
199. McNeill, V. F. Atmospheric aerosols: clouds, chemistry, and climate. *Annu. Rev. Chem. Biomol. Eng.* **2017**, *8*, 427–444.
200. Huang, W.; Saathoff, H.; Pajunoja, A.; Shen, X.; Naumann, K. H.; Wagner, R.; Virtanen, A.; Leisner, T.; Mohr, C. α-Pinene secondary organic aerosol at low temperature: chemical composition and implications for particle viscosity. *Atmos. Chem. Phys.* **2018**, *18*, 2883–2898.
201. Shiraiwa, M.; Ueda, K.; Pozzer, A.; Lammel, G.; Kampf, C. J.; Fushimi, A.; Enami, S.; Arangio, A. M.; Fröhlich-Nowoisky, J.; Fujitani, Y.; Furuyama, A.; Lakey, P. S. J.; Lelieveld, J.; Lucas, K.; Morino, Y.; Pöschl, U.; Takahama, S.; Takami, A.; Tong, H.; Weber, B.; Yoshino, A.; Sato, K. Aerosol health effects from molecular to global scales. *Environ. Sci. Technol.* **2017**, *51*, 13545–13567.

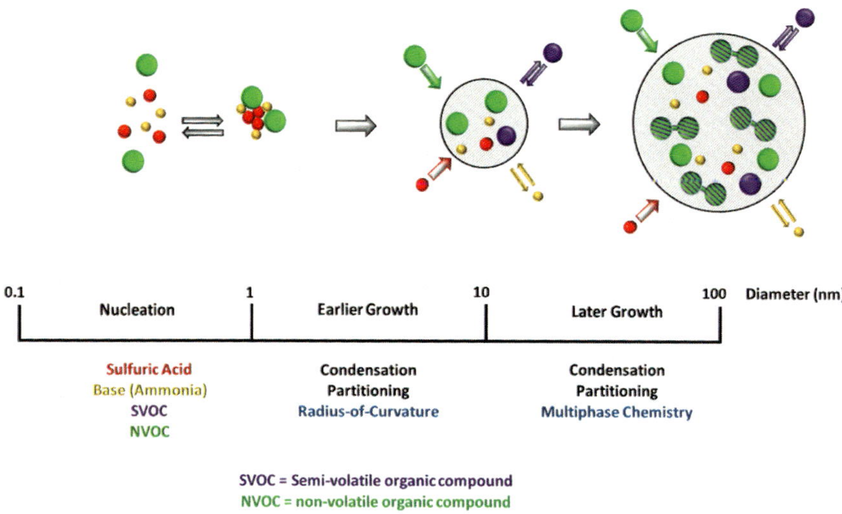

Figure 1. Summary of molecules and processes relevant to particle formation and growth in the atmosphere.

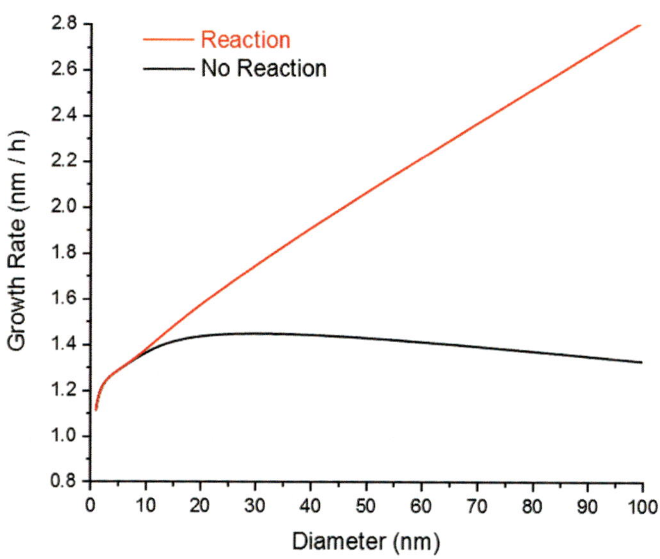

Figure 2. Growth rate vs. particle diameter for calculations with (red; $k_{DIMER} = 10^{-2}\ M^{-1}s^{-1}$) or without (black; $k_{DIMER} = 0\ M^{-1}s^{-1}$) accretion reaction in the particle phase.

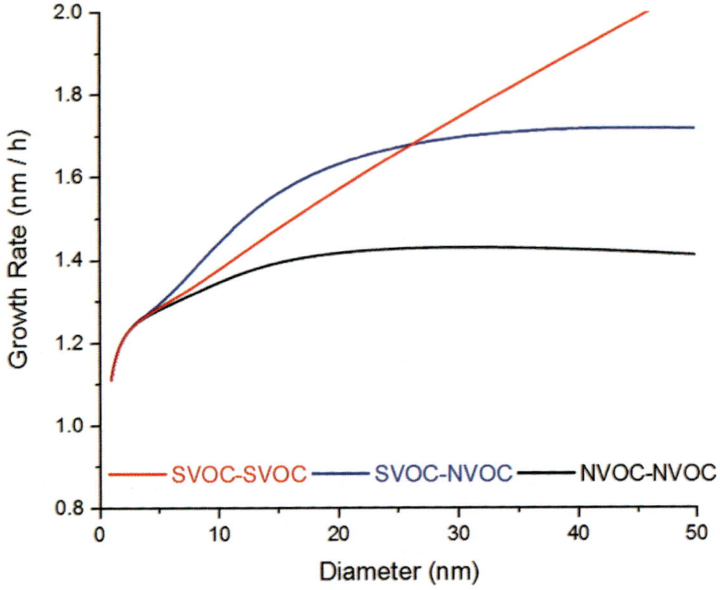

Figure 4. Growth rate calculations for different accretion reaction types: NVOC-NVOC (black), SVOC-NVOC (blue), and SVOC-SVOC (red). For all reactions, k = 10^{-2} $M^{-1}s^{-1}$. Reprinted from reference (59). Copyright 2018.

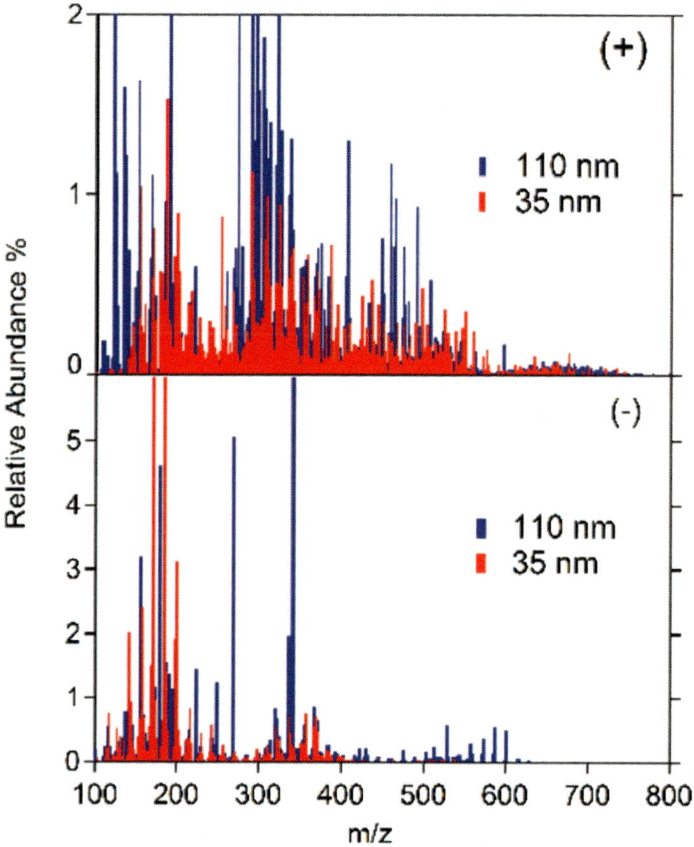

Figure 5. Positive (+) and negative (-) ion ESI mass spectra of 35 nm (red) and 110 nm (blue) monodisperse SOA samples. Reprinted from reference (64). Copyright 2017.

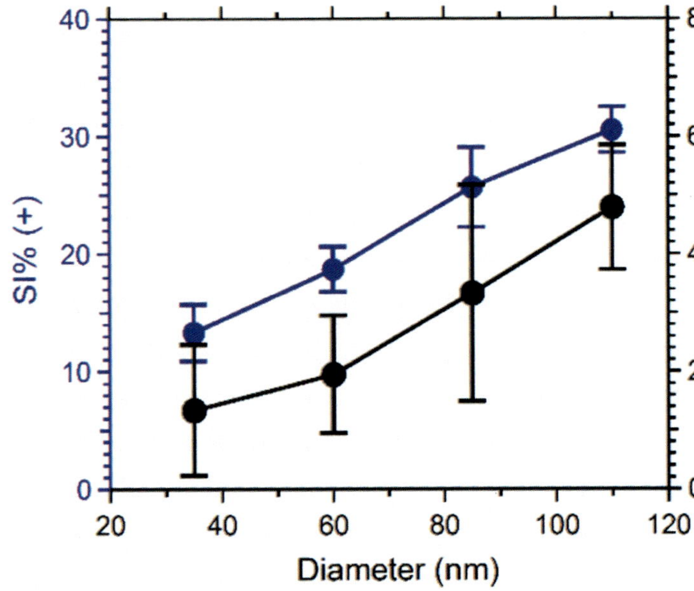

Figure 6. Percentage of total signal intensity (SI %) in positive (+) and negative (-) ion ESI mass spectra from higher order oligomers vs. particle diameter. Reprinted from reference (64). Copyright 2017.

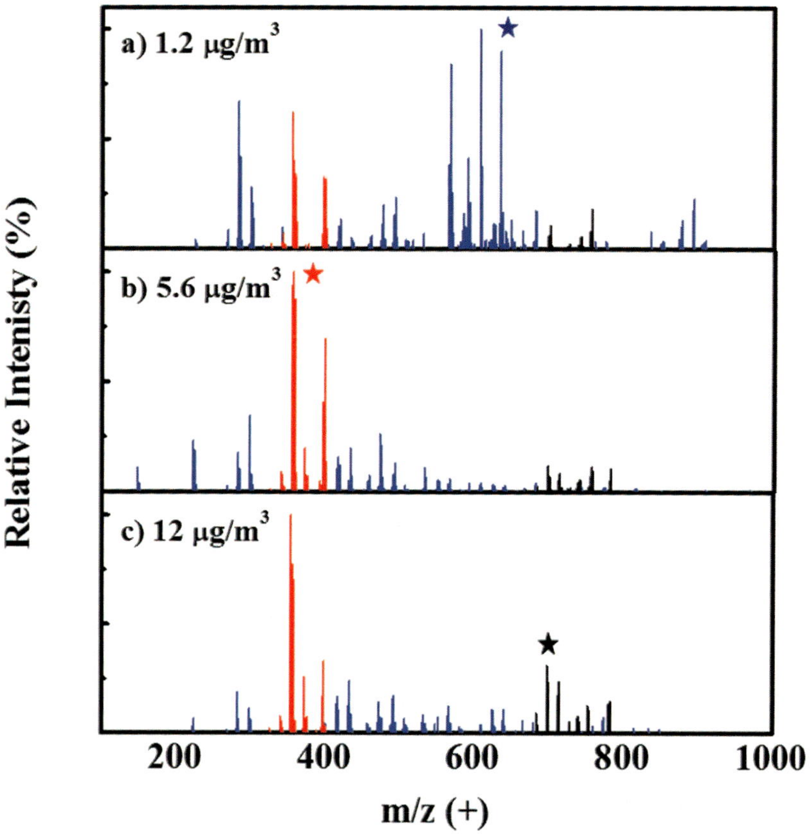

Figure 7. ESI mass spectra of secondary aerosol from D_5 oxidation under various aerosol mass loadings: a) 1.2 μg m^{-3}, b) 5.6 μg m^{-3}, c) 12 g m^{-3}. Peaks considered as ring-opened products are coded blue, dimers are coded black, and monomers are coded red. Stars highlight prominent ions of each type. Reprinted from reference (73). Copyright 2017, the American Chemical Society.

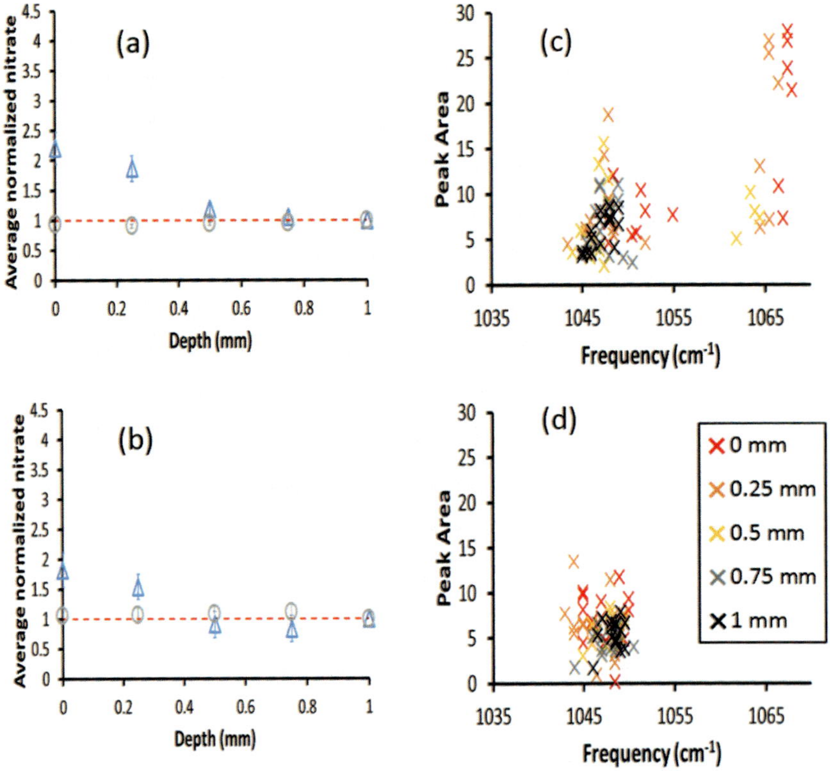

Figure 1. Location and phase of nitrate in frozen solutions. (a) and (c) show the relative nitrate concentration as a function of depth from the air interface for 250 mM $NaNO_3$ and 250 mM ($NaNO_3$ + 500 mM Instant Ocean) solutions, respectively, each normalized to its value at 1 mm deep (shown by the dashed line). The circles show liquid phase nitrate and the triangles represent the frozen data. (b) and (d) show the nitrate symmetric stretch intensity as a function of wavenumber and depth (see key) from the interface. The region around 1045 cm^{-1} is assigned to solution-phase nitrate; that around 1065 cm^{-1} is associated with crystalline nitrate. See original work for details. Reprinted with permission from Reference (8). Copyright 2017, American Chemical Society.

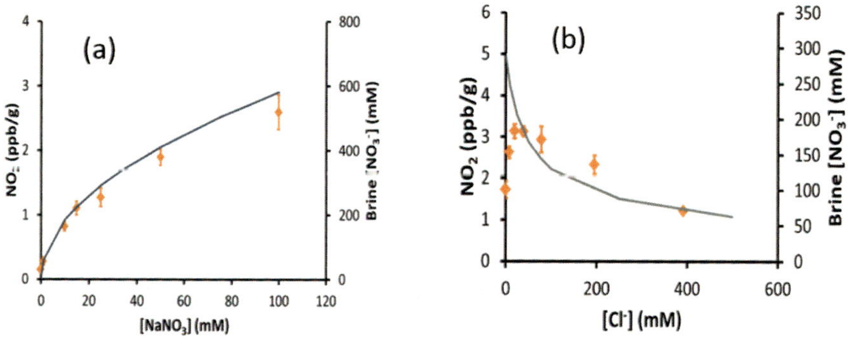

Figure 2. Gas phase NO_2 released from frozen solutions of (a) $NaNO_3$ and (b) (25 mM $NaNO_3$+Instant Ocean) as a function of salt concentration. The solid lines show the dependence expected based on simple exclusion models. See text and original reference for details. Reprinted with permission from Reference (11). Copyright 2016, American Chemical Society.

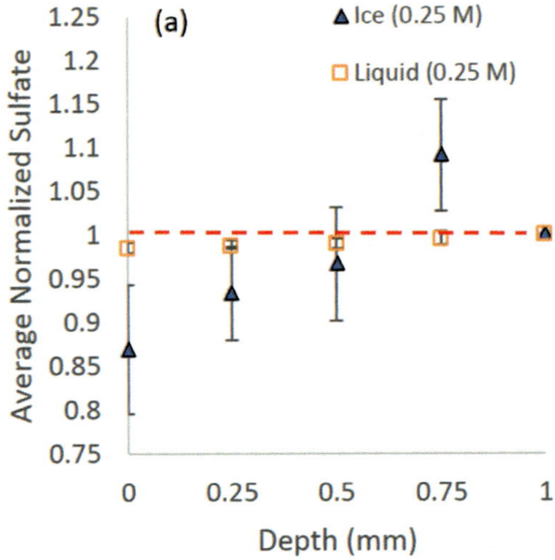

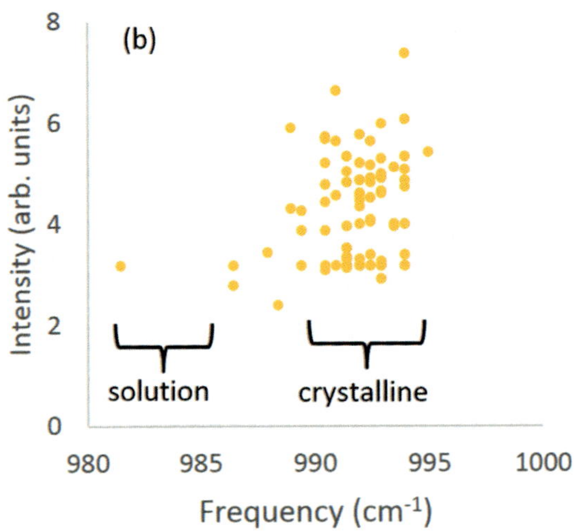

Figure 3. Location and phase of sulfate in frozen solutions of 250 mM sodium sulfate. (a) shows the relative sulfate concentration as a function of depth from the air interface normalized to its value at 1 mm deep (shown by the dashed line). The squares show liquid phase sulfate and the triangles represent the frozen data. (b) shows the sulfate symmetric stretch intensity in frozen samples as a function of wavenumber, collected together for all depths.

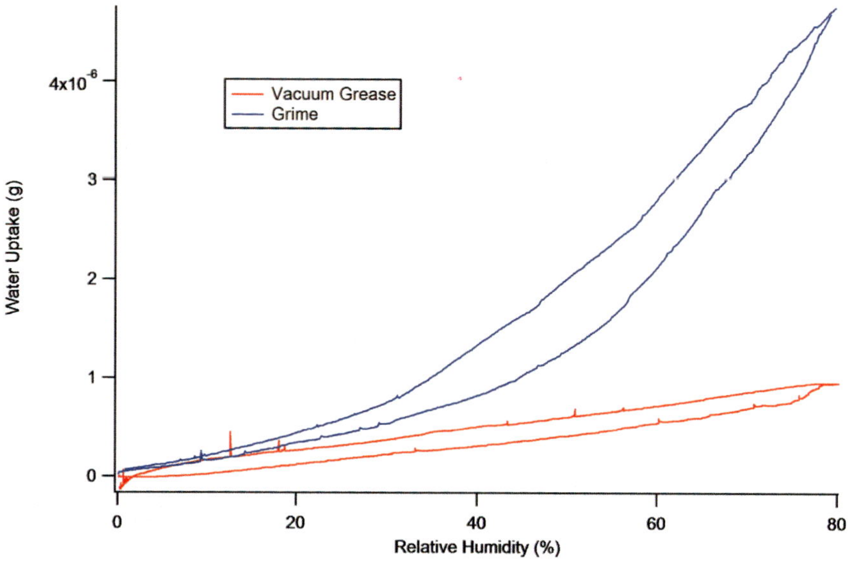

Figure 5. Water uptake onto samples of collected urban grime (blue trace) and hydrocarbon vacuum grease (red trace) as a function of relative humidity, as measured using a quartz crystal microbalance. In each trace, the upper leg is the wetting cycle and the lower indicates the drying cycle. See Reference (19) for details of how these measurements are made.

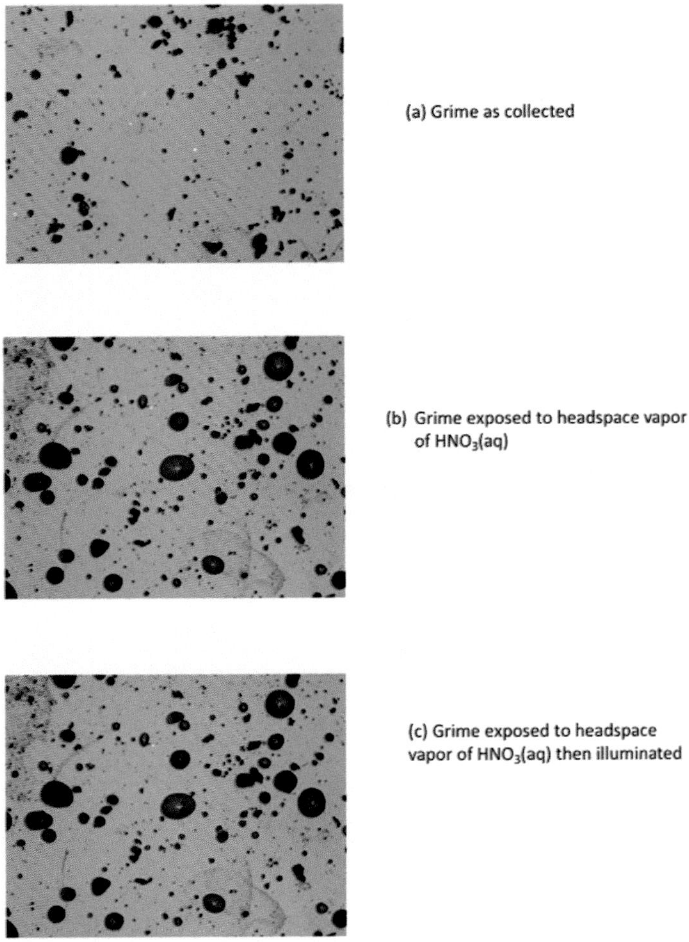

Figure 6. Photographs of real grime samples, taken using a Raman microscope. (a) The non-homogeneous nature of the grime is evident from the distribution of particulate material seen. (b) This shows hygroscopic behavior when exposed to the headspace vapor above a concentrated nitric acid solution, and (c) exposure to illumination from a filtered Xe lamp does not cause complete evaporation of the sample.

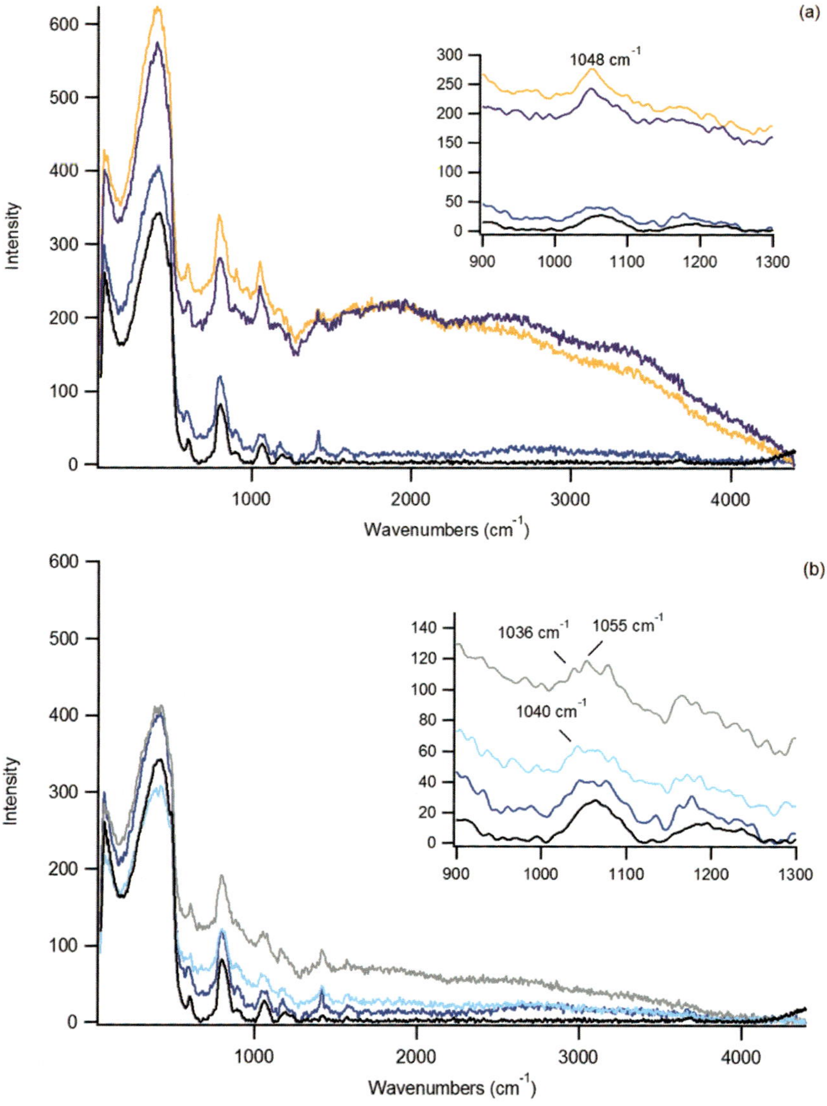

Figure 7. Raman spectra of the samples shown in Figure 6. In both frames, the collected grime spectrum is shown as the lowermost, blue trace. In all traces, the nitrate symmetric stretch is evident near 1050 cm^{-1}. (a) Spectra of one of the particulates, as collected (lower blue trace; after exposure to nitric acid vapor (upper gold trace); following illumination (purple trace). The nitrate peak sharpens and grows in the solution region upon exposure to the acid vapor. A loss of approx. 6 % in intensity is observed in the nitrate after illumination. (b) Spectra collected in the region between particles. The nitrate peak stays broad upon exposure to gas phase acid (grey trace) and only a modest decrease in its intensity is seen after illumination (light blue trace).

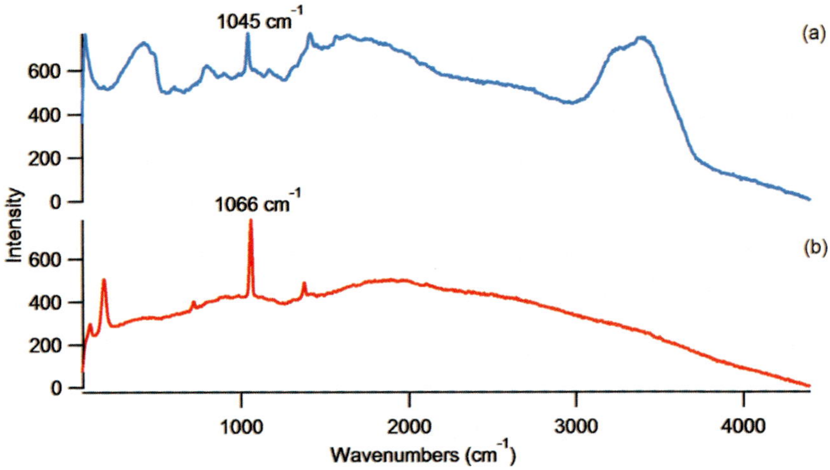

Figure 8. Raman spectra of nitrate-doped vacuum grease. The trace in (a) shows the spectrum measured immediately following deposition of solution nitrate onto the grease; note the intense liquid water band near 3400 cm^{-1}. The nitrate symmetric stretch peak is present at 1045 cm^{-1}. The lower trace, (b), shows the spectrum obtained following drying of the sample above; the liquid water band is no longer present and the nitrate peak now appears at 1066 cm^{-1}.

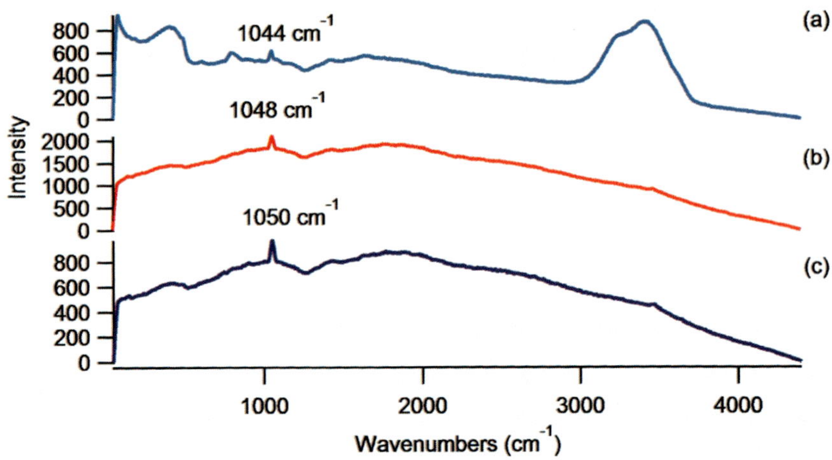

Figure 9. Raman spectra of vacuum grease doped with a mixed solution of sodium nitrate and calcium chloride. (a) The spectrum measured immediately following deposition of solution. (b) Spectrum measured after drying the sample above. (c) The spectrum obtained following three drying-rehumidification cycles of the sample. Note that the liquid water band has almost entirely vanished, but the nitrate peak remains in the solution region near 1050 cm^{-1}.

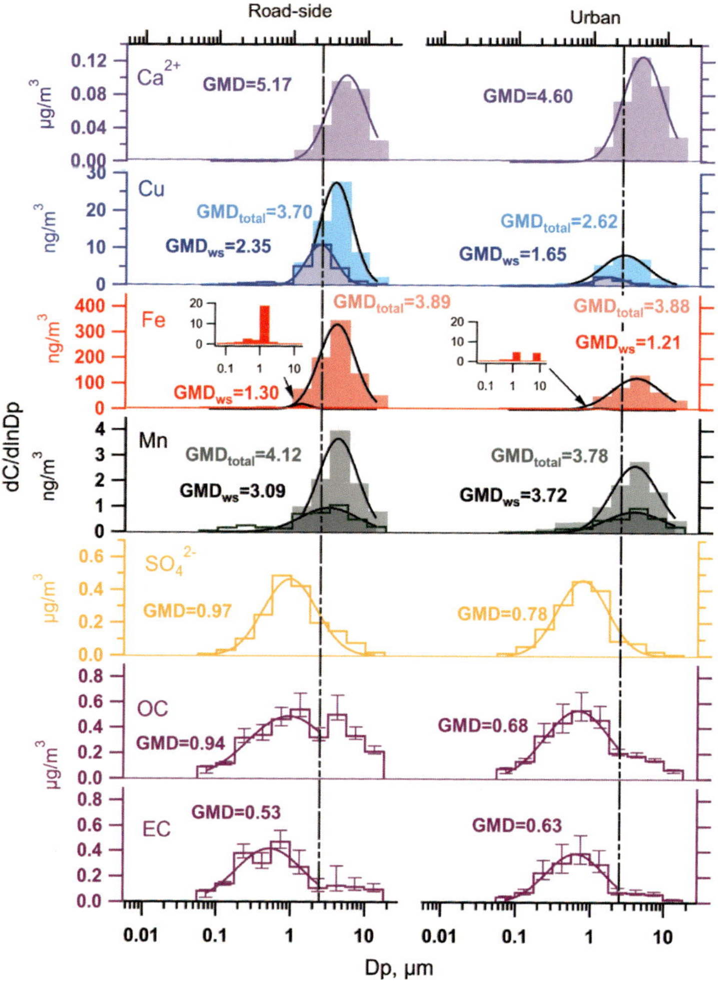

Figure 6. Size distributions of various aerosol species measured at the road-side (3/28 – 4/4/2016) and urban site (3/16 – 3/23/2016). The inserts in the Fe panel are enlarged plots of water-soluble Fe, which is a small fraction of total Fe. A particle diameter of 2.5 μm is shown as the vertical line. Lognormal fits to the data are also plotted with the corresponding geometric mean diameter (GMD). Since Ca^{2+} was measured by ion chromatography, it is the water-soluble fraction of calcium. Both total (elemental) and water-soluble (ws subscript) concentrations were measured for Cu and Fe.

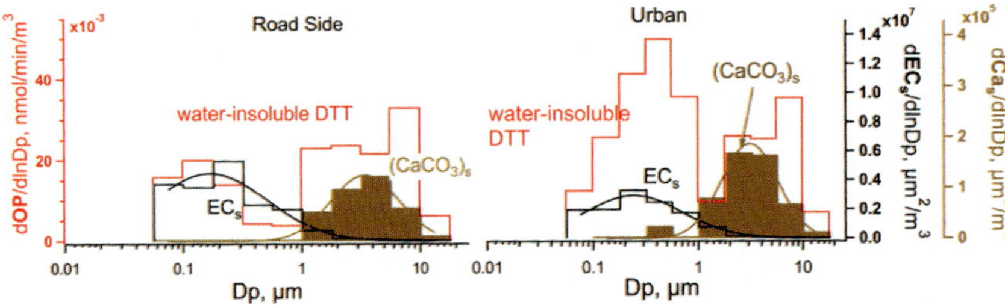

Figure 7. Measurements of water-insoluble DTT ($OP^{WI\text{-}DTT}$) size distributions and estimated particle surface area distributions, based on mass concentration measurements and assuming spherical particles, for elemental carbon (EC_s) and calcium carbonate ($CaCO_{3(s)}$, is inferred from Ca^{2+} measurements). The plots are to provide comparisons between the shapes of the distributions.

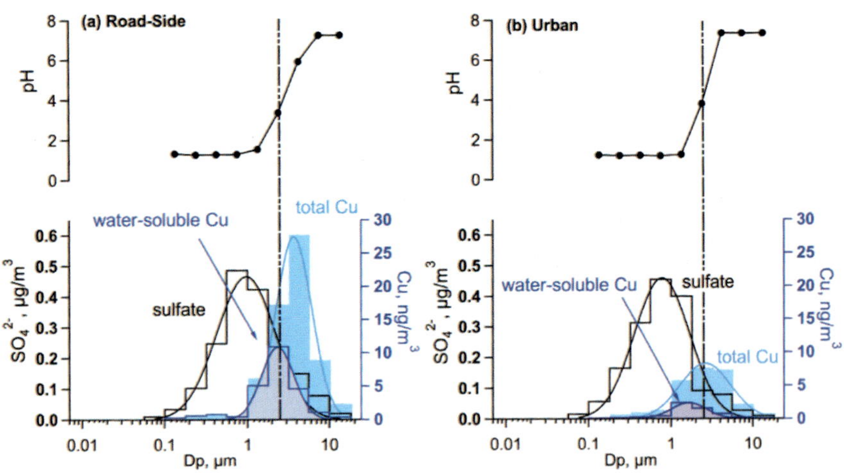

Figure 8. Evidence for the dissolution of copper (Cu) by sulfate under acidic conditions based on measured size distributions at a) the road-side and b) the urban sites. The vertical dotted line is aerodynamic diameter (Dp) at 2.5 µm, the upper limit of $PM_{2.5}$. Particle pH was estimated for each MOUDI stage from ISORROPIA-II based on measured ionic species from MOUDI samples collected on 3/28−4/4/2016 and 3/16−3/23/2016 at the road-side and urban site. Differences in geometric mean diameters (GMD) between water-soluble and total Cu can be found in Figure 4.

Chapter 12

Directly Probing the Phase States and Surface Tension of Individual Submicrometer Particles Using Atomic Force Microscopy

Hansol D. Lee, Kamal K. Ray, and Alexei V. Tivanski*

Department of Chemistry, University of Iowa, Iowa City, Iowa 52242, United States
***E-mail: alexei-tivanski@uiowa.edu**

Large uncertainties still remain in our ability to accurately predict the overall aerosol effect on the climate and atmosphere. One reason for this comes from the paucity of relevant experimental data for submicrometer-sized aerosol particles that can be in solid, semisolid, and liquid phases. Therefore, this becomes a multiphase chemistry problem that requires experimental techniques to overcome the inherent size limitations currently present in the scientific community and directly measure atmospherically relevant physical-chemical properties of aerosols. Here, we review the latest atomic force microscopy (AFM) based methodologies that provide the capability to identify the 3D morphology of aerosol particles and directly probe their important properties on a single particle basis. Herein the focus is on two physical-chemical properties of model aerosol systems: phase state and surface tension. Novel nanoindentation and nano-Wilhelmy AFM techniques are used to directly measure the two properties, while modulating the relative humidity and thus controlling the concentration and viscosity. Overall, the established methodologies discussed herein can be used to better understand the role of multiphase aerosols on the climate and atmosphere.

Introduction

Our world is full of aerosols that are more than a thousand times smaller than a single human hair strand. These small and perhaps even seemingly harmless aerosols, or particles, can have a paradoxically large impact on the climate. In the atmosphere, they can contribute to global cooling directly by reflecting the solar radiation and indirectly by seeding clouds (*1–4*). Examples of these primary particles can include mineral dust, volcanic ash, soot, and sea spray aerosols. Secondary aerosol particles also exist and form from volatile compounds that underwent chemical reactions that define their size, shape, and composition. A well-known example is secondary organic aerosols (*5–13*).

These aerosols of diverse origins in nature are chemically and physically complex. At first, the former may seem more intuitive than the latter; aerosols that come from trees have different chemical makeup in comparison to those that come from the ocean or a volcano. But what about physical complexity? The scientific community now understands that, sometimes due to the significant amount of oxygenated organic compounds in aerosols, the phase state (solid, semisolid, and liquid) can vary from particle-to-particle, depend on source, and vary as a function of relative humidity (RH) and temperature (*7*, *14–17*). This variability is in direct opposition to our simplified assumption, which we held in the past decades, that organic aerosols regardless of origin or atmospheric conditions were liquid in phase state (*18*).

The implication of this relatively new discovery is the following: from both experimental and theoretical perspective, the varying particle phase states bring a new dimension to an already difficult problem of accurately predicting the overall aerosol effect on the climate and atmosphere. Thus, our ability to accurately represent this complexity both experimentally and theoretically is absolutely paramount, and it must be done in a manner in which the small, yet unique, differences of individual particles are fully represented, instead of being simplified by being averaged out in an ensemble of a countless number of particles. An example of when this individual particle information is needed is atmospherically relevant particles, where only one out of a million particles actually form ice clouds and globally cool the climate (*19*, *20*). We therefore need to explore single particle techniques that can measure atmospherically relevant, physical-chemical properties of aerosols.

To better understand the implications that arise from the physical-chemical heterogeneity of individual aerosols, we now see a relatively new avenue of scientific research that specifically focuses on studying one particle at a time. For example, single particle experimental techniques such as atomic force microscopy (AFM) can uniquely study submicrometer-sized (less than 1 micrometer in size) particles, by imaging their 3D morphology and measuring the interaction forces with picoNewton force sensitivity that can be related to their physical-chemical properties. AFM uses a sharp probe or a tip, typically 2 – 10 nm at the apex, in order to raster scan over the surface (*21–25*). In 2016, AFM was used, for the very first time, to directly quantify the hygroscopic growth, or increase in the size of a substrate-deposited particle due to water uptake, on an individual particle basis at room temperature and varying RH (Figure 1A) (*26*). When comparing the individual particle AFM water uptake data to bulk ensemble average data from a conventional hygroscopic tandem differential mobility analyzer (HTDMA), where the width of the Gaussian average reflects particle-to-particle variability, the AFM showed higher sensitivity to particle-to-particle diversity that revealed a correlation between 3D morphology and hygroscopic growth (Figure 1B, D). The bulk ensemble data from the HTDMA on the other hand, returned one "representative" average value (Figure 1C). Overall, the water uptake response from substrate-deposited particles measured with the AFM is in agreement with the HTDMA results collected for airborne particles. In other words, the presence of the substrate for the AFM measurements did not alter the water uptake property of the aerosol particle. Following the successful demonstration of the AFM ability to quantitatively measure water uptake of individual particles as a function of RH prompted us to further develop the technique to study important physical-chemical properties of aerosols: physical phase state and surface tension. Both have strong significance to how aerosols affect the climate and also evolve due to the variation of atmospheric RH that modulates the water uptake and subsequently changes the concentration and viscosity.

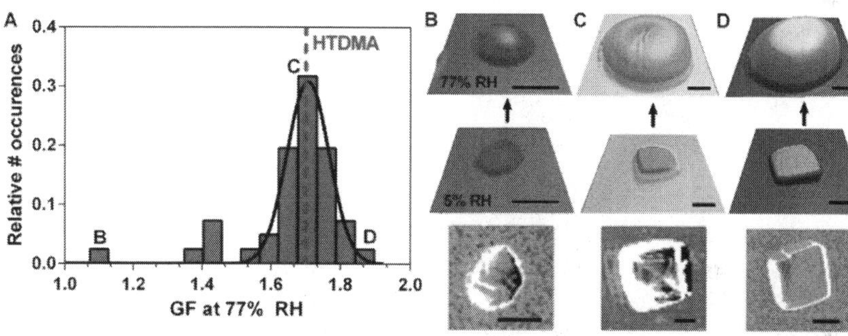

Figure 1. (A) Volume growth factor (GF) at 77% RH distribution measured for 40 individual nonanoic acid/NaCl particles. The data was fit to a Gaussian distribution. The red dashed line labels the GF value determined with HTDMA. Letters correspond to particles shown in (B)–(D). (B–D) The middle row of images shows dehydrated particles at 5% RH and corresponding deliquesced liquid droplets at 77% RH are shown above. The bottom row is phase images. (B) corresponds to the extreme left side of the distribution of GFs, while (C) corresponds to the most probable GF value and is representative of the majority of particles. (D) corresponds to the extreme right side of the GFs distribution. All scale bars are 500 nm. Reprinted with permission from (Reference (26)). Copyright (2016) American Chemical Society.

In this chapter, we review the latest works that have further developed the AFM techniques to directly measure atmospherically relevant properties, such as phase state and surface tension of individual particles. Here, sucrose is used as an ideal model system because of its access to all phase states at subsaturated RH (below 100%) and its quantified relationship between RH, solute concentration, and viscosity (*27, 28*). Following a brief introduction of each property of interest, we show two methodologies that, with minimal modeling and assumptions, are then used to interpret the force profile data collected from AFM. The two methodologies are called nanoindentation and nano-Wilhelmy methods, which use the AFM probe to directly interact with the sucrose particle, and record the particle response to the applied mechanical force. The response is then used to both qualitatively and quantitatively determine the phase state and surface tension at varying RH. We conclude by suggesting future research directions from this established work that involve applying these methodologies to highly complex nascent aerosol systems, to further improve our understanding of how aerosols influence our world by using novel microscopy techniques such as the AFM.

Discussion

Assessment of the Particle Phase States with Hygroscopic Growth

Particle phase state can dictate water uptake, determine whether particles can help form clouds, and control the chemical reactivity with gas phase molecules in the atmosphere (*17*, *29–34*). This is especially important to understand for submicrometer-sized particles that can last a much longer time (up to weeks) in the atmosphere than supermicrometer sized aerosols (*35*). However, the small particle size has been limiting experimental capabilities to directly measure the phase state until very recently. Here, we review the latest AFM technique that directly identifies the phase state of submicrometer-sized particles, with minimal modeling and assumptions used (*36*). The AFM nanoindentation technique uses a sharp (typically several to tens of nanometers) tip that presses into a substrate deposited particle with a specific amount of applied force and records the resulting particle response to establish mechanical equilibrium as a function of RH. The quantitative analysis is used to interpret the force profile or force spectroscopy, which is a collection of force *versus* tip-sample separation data, to quantify the viscoelastic response distance (VRD) and relative indentation depth (RID). These two measurements are newly established means of assessing the phase state of individual sucrose particles which may vary with RH, from solid, semisolid, and liquid phase, or the corresponding viscosity range from 10^{16} to 10^{-3} Pa s, respectively.

To modulate the viscosity of sucrose, a humidity cell is used to control the RH, and thus the concentration of the particles. In hydration mode, the particle grows in size and height from the increased RH, and this growth can be directly observed using 3D AFM imaging. In Figure 2A, the AFM 3D height image of a single sucrose particle at varying RH is shown, where the left and right images are at 4% and 80% RH, respectively. Thus, by increasing the RH from 4% to 80%, the particle grows from ~500 nm to ~700 nm in diameter. To quantify the water uptake behavior, volume-equivalent growth factor (GF) is calculated by the ratio of the wet volume-equivalent diameter divided by the dry volume-equivalent diameter, recorded at ~4% RH, which is shown in Figure 2B. Single particle data show that the sucrose particle has a continuous water uptake at subsaturated RH range, which agrees well with the theoretical prediction from the aerosol inorganic-organic mixtures functional groups activity coefficients (AIOMFAC) models (*37*). The hydration step is performed to modulate the viscosity and concentration, and then to identify the phase states on a single particle basis by observing the particle response to applied force (Figure 3). The same idea is also used for the AFM-based surface tension measurements, as discussed in a later section.

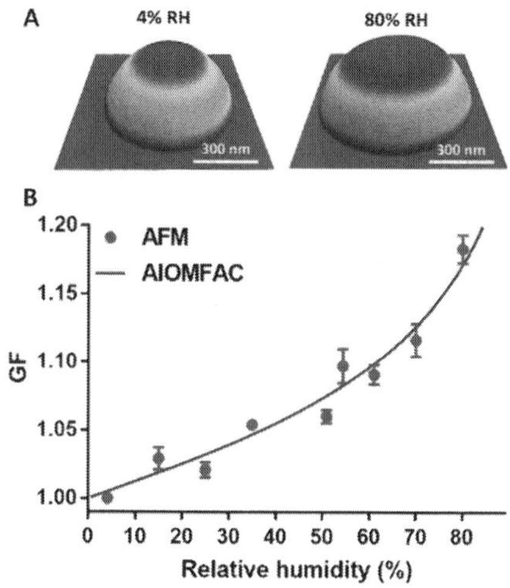

Figure 2. (A) AFM 3D height images of sucrose particle at 4% and 80% RH. As the RH is increased, the particle uptakes water and grows in size, from ~500 nm in diameter and 250 nm in height at 4% RH to ~700 nm in diameter and 300 nm in height at 80% RH. (B) AFM hydration volume-equivalent growth factor versus RH. Red circles with error bars represent average AFM growth factor data and two standard deviations. Purple line represents theoretical prediction of the GF using the AIOMFAC model (15). Reprinted with permission from (Reference (36)). Copyright (2017) American Chemical Society.

Figure 3. Representation of AFM probe applying predefined force onto a particle at varying RH. Reprinted with permission from (Reference (36)). Copyright (2017) American Chemical Society.

RH identification for the phase transition between solid and semisolid phase is not expected to be as discrete as that for the solid-to-liquid transition; instead, a smooth and relatively small change is expected. However, force spectroscopy measurements over an individual particle after 3D imaging can quantify both the phase state and surface tension. Figure 4 shows several representative examples of forces acting onto the AFM probe as a function of tip–sample separation (TSS), which is the experimentally obtainable distance between the tip and the top surface of the sucrose particle. With decreasing TSS, the probe comes down from the z-direction perpendicular to the particle surface, until there is a mechanical contact with the particle, with continued downward motion into and through the particle resulting in an indentation until a predefined maximum force is reached. After reaching the predefined maximum force, the AFM probe reverses in direction and retracts away from the particle and towards its original starting position above the particle. Within the force profiles data, the phase transition between solid and semisolid phase state was determined by observing the viscoelastic response distance, or the hysteresis distance between the approach and retract force curves in the contact region, from 7% to 19% RH (Figure 4A). Increasing hysteresis indicates that the sucrose particle has undergone transition from purely elastic to viscoelastic material, which shows both viscous and elastic behaviors when being deformed. This is normalized at the 0 nN force position on the force profile. Data show that for the sucrose particle at 7% RH, the VRD is 0.2 ± 0.1 nm, and shows a small increase to 0.4 ± 0.3 nm at 15% RH. At 19% RH, the VRD is 0.6 ± 0.2 nm, which is now statistically different relative to the expected uncertainty in the VRD value (~ 0.2 nm). Thus, the phase transition between solid and semisolid can be determined to be at ~18% RH, which corresponds to the viscosity value of $10^{11.2}$ Pa s. This agrees well with the currently accepted value of 10^{12} Pa s from multiple sources in the literature (*38*, *39*).

The transition from semisolid to liquid can also be determined from the force profiles data, by observing the change in the indentation depth with varying RH. At 54% RH, the approach force curve does not show significant indentation from the observed negative shallow slope with continued decrease in the TSS with increase in force (Figure 4B). Upon increasing the RH to 61%, however, the approach curve shows a drastically different response with the AFM tip indenting completely through the particle at the given maximum applied force, and touching the substrate underneath the particle. This is evident from the approach curve with the slope becoming almost perpendicular after the initial contact, thus indicating a phase transition from semisolid to liquid. Therefore, this type of the data can be used as a qualitative means to predict the semisolid to liquid transition, which should be at around 60% RH, or viscosity of $10^{2.5}$ Pa s for sucrose (*39*).

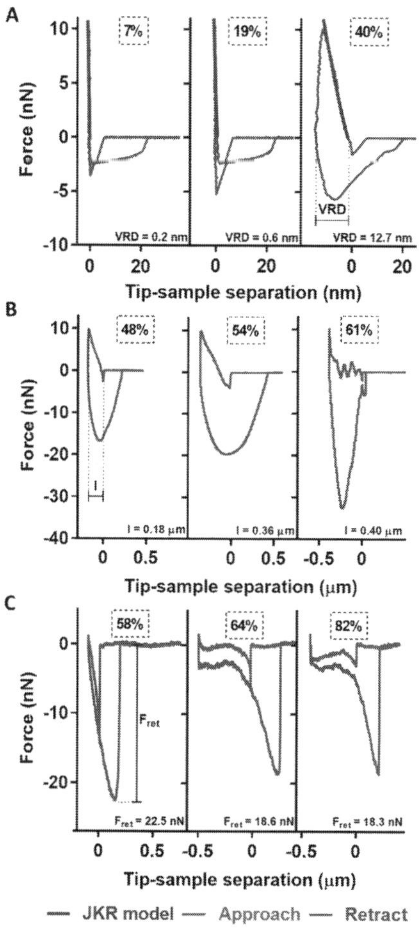

Figure 4. Force versus tip–sample separation plots for selected RH ranging from 7% to 82% of a sucrose particle. Tip sample separation of 0 indicates the initial position of AFM tip contact with the surface of the particle. The evolution of the force profiles from low to high RH is shown from left to right, using x-offset. Red lines indicate approach to the sucrose particle, and blue lines retract away from the particle. The purple lines indicate the JKR model fit in the contact region. (A) Force profiles at 7%, 19%, and 40% RH collected at the maximum applied force of 10 nN. The corresponding viscoelastic response distance (VRD) is specified below. (B) Force profiles at 48%, 54%, and 61% RH collected at the maximum applied force of 10 nN. The corresponding indentation depth (I) is specified below. (C) Force profiles at 58%, 64%, and 82% RH collected at the maximum applied force of 1 nN. The retention force (F_{ret}) used to calculate surface tension is specified below. Reprinted with permission from (Reference (36)). Copyright (2017) American Chemical Society.

To further accurately assess the semisolid to liquid phase transition of sucrose, the quantitative measure of relative indentation depth at varying RH is developed and employed. This unitless value at a particular RH is defined as the following:

$$RID = \frac{I}{H} \tag{1}$$

where I is the indentation depth that represents total distance that the AFM probe moves in the z-direction from the top surface of the particle to the depth indented into the particle and H is the maximum height of the particle. A low relative indentation depth value would thus represent a dehydrated and mechanically stiff particle that resists the AFM tip to indent significantly into the particle, and vice versa (Figure 4B, C). The result is shown in Figure 5A, where the combination of both viscoelastic response distance and relative indentation depth at varying RH and viscosity are shown. Due to high viscosity from 7% to 34% RH range, the measured RID is small at ~0.04. RID does not significantly increase until 44% RH, where beyond this RH value, the sucrose viscosity becomes low enough that the AFM probe makes an appreciable amount of indentation into the particle in comparison to its maximum height. Further increase to 47% RH, RID is approximately 0.18, and leading up to the previously proposed transition point at 60% RH, the RID becomes 0.98 and the particle now behaves like a liquid droplet. This transition, in comparison to the solid and semisolid phase state, is not discrete but continuous. Increase in RH beyond 60% shows RID plateauing to the value of 1, where the indentation depth and particle height becomes equal to each other. Further study was designed to determine whether experimental parameters such as applied force affects the relative indentation depth measurements, which indicated that the phase state measurements from the AFM are unaffected by varying applied force, as long as the maximum force value of 5 nN and higher is used (Figure 5B).

Surface Tension Assessment of Liquid Phase Droplets

To theoretically predict cloud formation by atmospheric aerosols, κ-Köhler theory and compressed film theory are often used (*40*, *41*). A key factor in both, however, is the surface tension of the aerosol droplet, where the present assumption is a fixed value equal to that of pure water (72.8 mN/m at 25 °C). However, aerosols often contain surface-active compounds that can significantly decrease the surface tension of the aerosol droplet, thus challenging the validity of the constant surface tension assumption (*42*, *43*). These models are expected to be significantly improved in their accuracy if direct surface tension measurements of relevant aerosol particles are provided to establish accurate and relevant surface tension value ranges for nascent particles with diverse chemical compositions and origin.

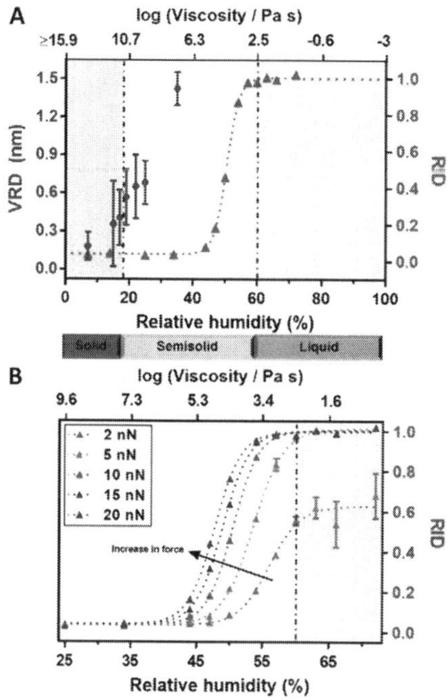

Figure 5. (A) AFM viscoelastic response distance (VRD, left, blue circles) and relative indentation depth (RID, right, red triangles) versus RH and corresponding viscosity collected with 10 nN of maximum applied force. The error bars represent two standard deviations for the VRD and RID, although sometimes smaller than the symbol. The RH–viscosity relationship is taken from Song et al. The red dotted line is shown for clarity and represents the fit using a four-parameter logistic sigmoidal function. Color bars indicate the various phase states of the sucrose particle at given RH from solid, semisolid, and liquid. The dashed black lines represent phase transition points determined by the force profiles of sucrose. The phase transition from solid to semisolid states occurred at ~18% RH and $10^{11.2}$ Pa s, while the semisolid to liquid state transition occurred at ~60% RH and $10^{2.5}$ Pa s. (B) RID versus RH and corresponding viscosity from 25% to 70% RH with varying maximum applied forces from 2 to 20 nN. The colored dotted lines are shown for clarity and represent the fit using a four-parameter logistic sigmoidal function. Reprinted with permission from (Reference (36)). Copyright (2017) American Chemical Society.

Here, we review the latest AFM technique that directly measures the surface tension of submicrometer-sized particles (*43, 44*). With a constant radius AFM nanoneedle, nano-Wilhelmy method is used to probe the substrate-deposited particle with the nanoneedle, which measures the retention force required to break the formed meniscus between the droplet and the nanoneedle as a function of RH (Figure 6). The approach is nearly identical to the Wilhelmy plate method, which uses a macro-sized needle to immerse into the liquid, and measure the force required to break the meniscus. This force is used to quantify the surface tension.

The general expression for surface tension is defined as force over unit length for a bulk solution. Extending this to the nano-Wilhelmy method applicable for this methodology, the equation for surface tension is expressed as the following:

$$\sigma = \frac{F_{ret}}{2\pi r} \tag{2}$$

where F_{ret} is the retention force and r is radius of the nanoneedle. Similar to phase state measurements, this methodology is repeated after 3D particle imaging as a function of increasing RH. As RH increases, the particle uptakes more water and becomes more dilute, decreasing the solute concentration which then changes the surface tension. The force spectroscopy is used to measure the retention force (Figure 4C) for the RH range where sucrose particle is liquid in phase.

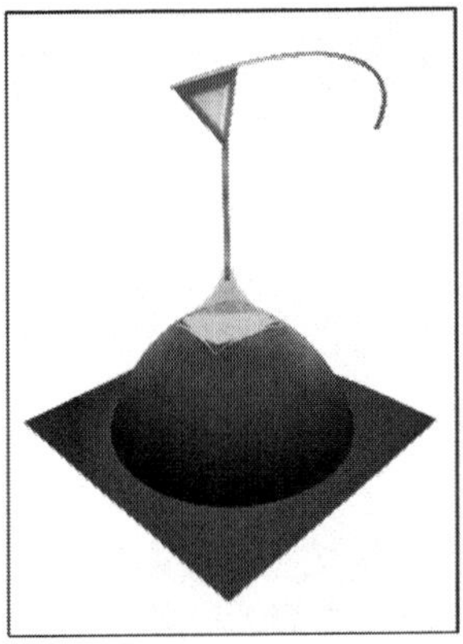

Figure 6. Representation of AFM nanoneedle probe applying predefined force onto a particle and measuring the retention force from the meniscus. Reprinted with permission from (Reference (43)). Copyright (2017) American Chemical Society.

As an example, AFM 3D height image of a glucose particle at ~20% RH is shown (Figure 7A); sucrose also displays similar rounded morphology. The surface tension result for sucrose is shown in Figure 7B, where single particle surface tension measurements are shown in blue circles with error bars, indicating two standard deviations. The red dashed line indicates surface tension prediction from bulk solution surface tension measurements. The bulk surface tension data collected with the conventional Wilhelmy plate method, used to measure the surface tension value of liquids approximately 4 mL in volume. Within the experimentally identified RH range, there is a good overlap between the AFM and bulk surface tension data, despite the individual particles being submicrometer in size. The good overlap is also seen even below ~70% RH, the solubility point of sucrose, when the particle is becoming supersaturated in concentration. However, if the RH is to go even lower, enough to undergo a phase transition from liquid to semisolid below 60% RH, the measured retention force now has a strong viscosity contribution that cannot be decoupled, and the resultant surface tension will deviate in comparison to the bulk trend line.

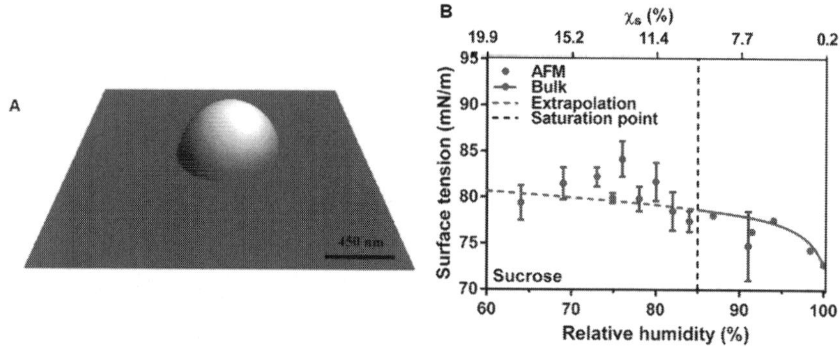

Figure 7. Representative AFM 3D height image at ~20% RH of glucose particle with height of ~350 nm (A). Surface tension vs RH (bottom axis) and corresponding solute mole percentage (top horizontal axis) for ~600 nm diameter sucrose (blue circles with error bars) (B). Average AFM surface tension data and two standard deviations. (red circles) Bulk solution surface tension measurements. Because of the low standard deviation, the error bars for the bulk data are smaller than the size of the symbol. Red solid line is the bulk trend line. Red dashed line is the extrapolation of the bulk trend line above the saturation point, which is indicated by the vertical black dashed line. Reprinted with permission from (Reference (43)). Copyright (2017) American Chemical Society.

Conclusions

In this review, we assessed new AFM methodologies that directly measure as a function of relative humidity, atmospherically relevant physical-chemical properties including the phase state and surface tension, of individual submicrometer-sized substrate-deposited particles. Complementary AFM nanoindentation and nano-Wilhelmy methodologies were used to collect forces acting on the AFM tip *versus* tip-sample separation profiles at varying RH that were then interpreted to quantify the two properties at subsaturated RH. For qualitative and quantitative assessment of the phase states, AFM methods revealed distinct changes in the phase state within viscosity ranges from 10^{16} to 10^{-3} Pa s for sucrose. Three different phase states of an individual sucrose particle corresponding to solid, semisolid, and liquid states were accessed and quantified. In the future, this approach can be further developed to determine the particle phase states of other atmospherically relevant chemical systems, at varying mixing states and increased complexity. In this regard, nascent aerosol particles collected on a substrate may be characterized for their phase states at a specific RH. AFM was used to also directly measure the surface tension of individual submicrometer sucrose particles at ambient temperature and varying RH conditions. The single particle AFM data overlap well with the bulk surface tension values when the particle phase state is liquid. In the future, applying this methodology to measure the surface tension of more complex particles, such as nascent aerosols of unknown composition, would provide further insights into the extent of particle-to-particle surface tension variation for aerosols of different origins, ultimately improving our understanding of the factors that facilitate cloud formation at a specific RH. With further developmental progress, this may allow us to elucidate the multiphase role of various aerosols on the climate and atmosphere.

Acknowledgments

This work was funded by the National Science Foundation through the Center for Aerosol Impacts on Chemistry of the Environment under grant no. CHE 1305427. Any opinions, findings, and conclusions or recommendations expressed in this material are those of the authors and do not necessarily reflect the views of the National Science Foundation.

References

1. Haywood, J.; Boucher, O. Estimates of the direct and indirect radiative forcing due to tropospheric aerosols: A review. *Rev. Geophys.* **2000**, *38*, 513–543.
2. Fuentes, E.; Coe, H.; Green, D.; de Leeuw, G.; McFiggans, G. Laboratory-generated primary marine aerosol via bubble-bursting and atomization. *Atmos. Meas. Tech.* **2010**, *3*, 141–162.
3. Jacobson, M. Z. Global direct radiative forcing due to multicomponent anthropogenic and natural aerosols. *J. Geophys. Res.-Atmos.* **2001**, *106*, 1551–1568.
4. de Leeuw, G.; Andreas, E. L.; Anguelova, M. D.; Fairall, C. W.; Lewis, E. R.; O'Dowd, C.; Schulz, M.; Schwartz, S. E. Production Flux of Sea Spray Aerosol. *Rev. Geophys.* **2011**, *49*, RG2001.
5. Calvo, A. I.; Alves, C.; Castro, A.; Pont, V.; Vicente, A. M.; Fraile, R. Research on aerosol sources and chemical composition: Past, current and emerging issues. *Atmos. Res.* **2013**, *120-121*, 1–28.
6. Fuzzi, S.; Baltensperger, U.; Carslaw, K.; Decesari, S.; Denier van der Gon, H.; Facchini, M. C.; Fowler, D.; Koren, I.; Langford, B.; Lohmann, U.; Nemitz, E.; Pandis, S.; Riipinen, I.; Rudich, Y.; Schaap, M.; Slowik, J. G.; Spracklen, D. V.; Vignati, E.; Wild, M.; Williams, M.; Gilardoni, S. Particulate matter, air quality and climate: lessons learned and future needs. *Atmos. Chem. Phys.* **2015**, *15*, 8217–8299.
7. Saukko, E.; Lambe, A. T.; Massoli, P.; Koop, T.; Wright, J. P.; Croasdale, D. R.; Pedernera, D. A.; Onasch, T. B.; Laaksonen, A.; Davidovits, P.; Worsnop, D. R.; Virtanen, A. Humidity-dependent phase state of SOA particles from biogenic and anthropogenic precursors. *Atmos. Chem. Phys.* **2012**, *12*, 7517–7529.
8. Song, M. J.; Liu, P. F. F.; Hanna, S. J.; Zaveri, R. A.; Potter, K.; You, Y.; Martin, S. T.; Bertram, A. K. Relative humidity-dependent viscosity of

secondary organic material from toluene photo-oxidation and possible implications for organic particulate matter over megacities. *Atmos. Chem. Phys.* **2016**, *16*, 8817–8830.

9. Koop, T.; Bookhold, J.; Shiraiwa, M.; Poschl, U. Glass transition and phase state of organic compounds: dependency on molecular properties and implications for secondary organic aerosols in the atmosphere. *Phys. Chem. Chem. Phys.* **2011**, *13*, 19238–19255.
10. Virtanen, A.; Joutsensaari, J.; Koop, T.; Kannosto, J.; Yli-Pirila, P.; Leskinen, J.; Makela, J. M.; Holopainen, J. K.; Poschl, U.; Kulmala, M.; Worsnop, D. R.; Laaksonen, A. An amorphous solid state of biogenic secondary organic aerosol particles. *Nature* **2010**, *467*, 824–827.
11. Baustian, K. J.; Wise, M. E.; Jensen, E. J.; Schill, G. P.; Freedman, M. A.; Tolbert, M. A. State transformations and ice nucleation in amorphous (semi-)solid organic aerosol. *Atmos. Chem. Phys.* **2013**, *13*, 5615–5628.
12. Perraud, V.; Bruns, E. A.; Ezell, M. J.; Johnson, S. N.; Yu, Y.; Alexander, M. L.; Zelenyuk, A.; Imre, D.; Chang, W. L.; Dabdub, D.; Pankow, J. F.; Finlayson-Pitts, B. J. Nonequilibrium atmospheric secondary organic aerosol formation and growth. *Proc. Natl. Acad. Sci. U.S.A.* **2012**, *109*, 2836–2841.
13. Schill, G. P.; Tolbert, M. A. Heterogeneous Ice Nucleation on Simulated Sea-Spray Aerosol Using Raman Microscopy. *J. Phys. Chem. C* **2014**, *118*, 29234–29241.
14. Wang, B. B.; O'Brien, R. E.; Kelly, S. T.; Shilling, J. E.; Moffet, R. C.; Gilles, M. K.; Laskin, A. Reactivity of Liquid and Semisolid Secondary Organic Carbon with Chloride and Nitrate in Atmospheric Aerosols. *J. Phys. Chem. A* **2015**, *119*, 4498–4508.
15. Hodas, N.; Zuend, A.; Mui, W.; Flagan, R. C.; Seinfeld, J. H. Influence of particle-phase state on the hygroscopic behavior of mixed organic-inorganic aerosols. *Atmos. Chem. Phys.* **2015**, *15*, 5027–5045.
16. Tong, H. J.; Reid, J. P.; Bones, D. L.; Luo, B. P.; Krieger, U. K. Measurements of the timescales for the mass transfer of water in glassy aerosol at low relative humidity and ambient temperature. *Atmos. Chem. Phys.* **2011**, *11*, 4739–4754.
17. Shiraiwa, M.; Ammann, M.; Koop, T.; Poschl, U. Gas uptake and chemical aging of semisolid organic aerosol particles. *Proc. Natl. Acad. Sci. U.S.A.* **2011**, *108*, 11003–11008.
18. Cappa, C. D.; Lovejoy, E. R.; Ravishankara, A. R. Evidence for liquid-like and nonideal behavior of a mixture of organic aerosol components. *Proc. Natl. Acad. Sci. U.S.A.* **2008**, *105*, 18687–18691.
19. DeMott, P. J.; Hill, T. C. J.; Petters, M. D.; Bertram, A. K.; Tobo, Y.; Mason, R. H.; Suski, K. J.; McCluskey, C. S.; Levin, E. J. T.; Schill, G. P.; Boose, Y.; Rauker, A. M.; Miller, A. J.; Zaragoza, J.; Rocci, K.; Rothfuss, N. E.; Taylor, H. P.; Hader, J. D.; Chou, C.; Huffman, J. A.; Poschl, U.; Prenni, A. J.; Kreidenweis, S. M. Comparative measurements of ambient atmospheric concentrations of ice nucleating particles using multiple immersion freezing methods and a continuous flow diffusion chamber. *Atmos. Chem. Phys.* **2017**, *17*, 11227–11245.

20. DeMott, P. J.; Hill, T. C. J.; McCluskey, C. S.; Prather, K. A.; Collins, D. B.; Sullivan, R. C.; Ruppel, M. J.; Mason, R. H.; Irish, V. E.; Lee, T.; Hwang, C. Y.; Rhee, T. S.; Snider, J. R.; McMeeking, G. R.; Dhaniyala, S.; Lewis, E. R.; Wentzell, J. J. B.; Abbatt, J.; Lee, C.; Sultana, C. M.; Ault, A. P.; Axson, J. L.; Martinez, M. D.; Venero, I.; Santos-Figueroa, G.; Stokes, M. D.; Deane, G. B.; Mayol-Bracero, O. L.; Grassian, V. H.; Bertram, T. H.; Bertram, A. K.; Moffett, B. F.; Franc, G. D. Sea spray aerosol as a unique source of ice nucleating particles. *Proc. Natl. Acad. Sci. U.S.A.* **2016**, *113*, 5797–5803.
21. Woodward, X.; Kostinski, A.; China, S.; Mazzoleni, C.; Cantrell, W. Characterization of Dust Particles' 3D Shape and Roughness with Nanometer Resolution. *Aerosol Sci. Tech.* **2015**, *49*, 229–238.
22. Chou, C.; Formenti, P.; Maille, M.; Ausset, P.; Helas, G.; Harrison, M.; Osborne, S. Size distribution, shape, and composition of mineral dust aerosols collected during the African Monsoon Multidisciplinary Analysis Special Observation Period 0: Dust and Biomass-Burning Experiment field campaign in Niger, January 2006. *J. Geophys. Res.-Atmos.* **2008**, *113*, D00C10.
23. Helas, G.; Andreae, M. O. Surface features on Sahara soil dust particles made visible by atomic force microscope (AFM) phase images. *Atmos. Meas. Tech.* **2008**, *1*, 1–8.
24. Li, W. J.; Shao, L. Y.; Zhang, D. Z.; Ro, C. U.; Hu, M.; Bi, X. H.; Geng, H.; Matsuki, A.; Niu, H. Y.; Chen, J. M. A review of single aerosol particle studies in the atmosphere of East Asia: morphology, mixing state, source, and heterogeneous reactions. *J. Clean. Prod.* **2016**, *112*, 1330–1349.
25. Barkay, Z.; Teller, A.; Ganor, E.; Levin, Z.; Shapira, Y. Atomic force and scanning electron microscopy of atmospheric particles. *Microsc. Res. Tech.* **2005**, *68*, 107–114.
26. Morris, H. S.; Estillore, A. D.; Laskina, O.; Grassian, V. H.; Tivanski, A. V. Quantifying the Hygroscopic Growth of Individual Submicrometer Particles with Atomic Force Microscopy. *Anal. Chem.* **2016**, *88*, 3647–3654.
27. Song, Y. C.; Haddrell, A. E.; Bzdek, B. R.; Reid, J. P.; Barman, T.; Topping, D. O.; Percival, C.; Cai, C. Measurements and Predictions of Binary Component Aerosol Particle Viscosity. *J. Phys. Chem. A* **2016**, *120*, 8123–8137.
28. Bones, D. L.; Reid, J. P.; Lienhard, D. M.; Krieger, U. K. Comparing the mechanism of water condensation and evaporation in glassy aerosol. *Proc. Natl. Acad. Sci. U.S.A.* **2012**, *109*, 11613–11618.
29. Kuwata, M.; Martin, S. T. Phase of atmospheric secondary organic material affects its reactivity. *Proc. Natl. Acad. Sci. U.S.A.* **2012**, *109*, 17354–17359.
30. Renbaum-Wolff, L.; Grayson, J. W.; Bateman, A. P.; Kuwata, M.; Sellier, M.; Murray, B. J.; Shilling, J. E.; Martin, S. T.; Bertram, A. K. Viscosity of alpha-pinene secondary organic material and implications for particle growth and reactivity. *Proc. Natl. Acad. Sci. U.S.A.* **2013**, *110*, 8014–8019.
31. Murray, B. J.; Wilson, T. W.; Dobbie, S.; Cui, Z. Q.; Al-Jumur, S. M. R. K.; Mohler, O.; Schnaiter, M.; Wagner, R.; Benz, S.; Niemand, M.; Saathoff, H.; Ebert, V.; Wagner, S.; Karcher, B. Heterogeneous nucleation of ice particles on glassy aerosols under cirrus conditions. *Nat. Geosci.* **2010**, *3*, 233–237.

32. Mikhailov, E.; Vlasenko, S.; Martin, S. T.; Koop, T.; Poschl, U. Amorphous and crystalline aerosol particles interacting with water vapor: conceptual framework and experimental evidence for restructuring, phase transitions and kinetic limitations. *Atmos. Chem. Phys.* **2009**, *9*, 9491–9522.
33. Shiraiwa, M.; Berkemeier, T.; Schilling-Fahnestock, K. A.; Seinfeld, J. H.; Poschl, U. Molecular corridors and kinetic regimes in the multiphase chemical evolution of secondary organic aerosol. *Atmos. Chem. Phys.* **2014**, *14*, 8323–8341.
34. Berkemeier, T.; Huisman, A. J.; Ammann, M.; Shiraiwa, M.; Koop, T.; Poschl, U. Kinetic regimes and limiting cases of gas uptake and heterogeneous reactions in atmospheric aerosols and clouds: a general classification scheme. *Atmos. Chem. Phys.* **2013**, *13*, 6663–6686.
35. Gong, S. L.; Barrie, L. A.; Blanchet, J. P. Modeling sea-salt aerosols in the atmosphere .1. Model development. *J. Geophys. Res.-Atmos.* **1997**, *102*, 3805–3818.
36. Lee, H. D.; Ray, K. K.; Tivanski, A. V. Solid, Semisolid, and Liquid Phase States of Individual Submicrometer Particles Directly Probed Using Atomic Force Microscopy. *Anal. Chem.* **2017**, *89*, 12720–12726.
37. Ovadnevaite, J.; Zuend, A.; Laaksonen, A.; Sanchez, K. J.; Roberts, G.; Ceburnis, D.; Decesari, S.; Rinaldi, M.; Hodas, N.; Facchini, M. C.; Seinfeld, J. H.; Dowd, C. O. Surface tension prevails over solute effect in organic-influenced cloud droplet activation. *Nature* **2017**, *546*, 637–641.
38. Koop, T.; Bookhold, J.; Shiraiwa, M.; Poschl, U. Glass transition and phase state of organic compounds: dependency on molecular properties and implications for secondary organic aerosols in the atmosphere. *Phys. Chem. Chem. Phys.* **2011**, *13*, 19238–55.
39. Power, R. M.; Simpson, S. H.; Reid, J. P.; Hudson, A. J. The transition from liquid to solid-like behaviour in ultrahigh viscosity aerosol particles. *Chem. Sci.* **2013**, *4*, 2597–2604.
40. Petters, M. D.; Kreidenweis, S. M. A single parameter representation of hygroscopic growth and cloud condensation nucleus activity. *Atmos. Chem. Phys.* **2007**, *7*, 1961–1971.
41. Ruehl, C. R.; Davies, J. F.; Wilson, K. R. An interfacial mechanism for cloud droplet formation on organic aerosols. *Science* **2016**, *351*, 1447–1450.
42. Ruehl, C. R.; Wilson, K. R. Surface Organic Mono layers Control the Hygroscopic Growth of Submicrometer Particles at High Relative Humidity. *J. Phys. Chem. A* **2014**, *118*, 3952–3966.
43. Lee, H. D.; Estillore, A. D.; Morris, H. S.; Ray, K. K.; Alejandro, A.; Grassian, V. H.; Tivanski, A. V. Direct Surface Tension Measurements of Individual Sub-Micrometer Particles Using Atomic Force Microscopy. *J. Phys. Chem. A* **2017**, *121*, 8296–8305.
44. Morris, H.; Grassian, V.; Tivanski, A. Humidity-dependent surface tension measurements of individual inorganic and organic submicrometre liquid particles. *Chem. Sci.* **2015**, *6*, 3242–3247.

Chapter 13

Molecular Characterization of Atmospheric Brown Carbon

Alexander Laskin,*,[1] Peng Lin,[1] Julia Laskin,[1] Lauren T. Fleming,[2] and Sergey Nizkorodov[2]

[1]Department of Chemistry, Purdue University, West Lafayette, Indiana 47906, United States
[2]Department of Chemistry, University of California, Irvine, California 92697, United States
***E-mail: alaskin@purdue.edu; phone: +1-765-494-5243**

Light-absorbing organic aerosol, commonly known as brown carbon (BrC), is a significant contributor to radiative forcing of the Earth's climate and also is of potential toxicological concern. Understanding the environmental effects of BrC, its sources, formation, and aging processes requires molecular-level speciation of its chromophores and characterization of their light-absorption properties. This chapter highlights recent analytical chemistry developments and applications in the area of molecular characterization of BrC that provided first insights into the diverse composition and properties of its common chromophores. We present chemical analysis of BrC reported in a number of case studies associated with emissions from biomass burning and anthropogenic sources producing secondary organic aerosols. The results highlight major classes of organic BrC differing in structures, polarities, and volatilities. Understanding their chemical identity requires applications of multi-modal complementary separation and ionization approaches in combination with high resolution mass spectrometry. Overall, these studies allow deciphering the BrC absorbance and studying its atmospheric evolution with respect to relative contributions from different classes of chromophores such as aromatic carboxylic acids, nitro-phenols; substituted, heterocyclic, and pure polycyclic aromatic hydrocarbons. These studies also provide a glimpse into the complex atmospheric evolution of BrC as a result of photooxidation, photolysis and different levels of acidity. Samples of BrC materials discussed in this chapter were obtained in FIREX 2016 biomass burning experiments performed at the U.S. Forest Service Fire Science Laboratory in Missoula, MT.

Introduction

A significant and highly variable fraction of anthropogenic atmospheric aerosol absorbs solar radiation resulting in reduced visibility on regional scales and also affecting global climate forcing. "Black carbon" (BC), the most efficient light-absorbing aerosol (*1*), is composed of highly carbonized graphitic-like particulates generated by incomplete combustion of fuels. BC absorbs solar radiation over a broad spectral range from the ultra-violet (UV) into far infra-red (IR). Effects of BC on the environment are particularly significant in the areas heavily impacted by either large-scale forest fires or combustion of fossil fuels in densely populated developing countries. In addition to BC, certain components of organic aerosol (OA) absorb efficiently in the UV/Vis range – these species are collectively called "brown carbon" (BrC) (*2*).

The importance of BrC absorption on regional-to-global scales has been highlighted in a number of atmospheric modeling studies (*3*, *4*) and has been confirmed by the satellite (*5*) and ground (*6–8*) observations. In particular, regional effects of BrC over major areas of biomass burning and biofuel combustion are substantial, where BrC and BC become equivalently important light-absorbing materials in the atmosphere (*3*). In these regions, the BrC warming effect is comparable to cooling by non-absorbing OA (−0.1 to −0.4 W m^{-2}). However, because of large variability in their chemical composition, concentrations and optical properties of specific chromophores, it is not clear to what extent the warming effect of BrC and cooling effect of OA may cancel each other (*9*). Inherent to their chemical complexity, the sources of BrC are not well understood; even less is known about its transformations resulting from atmospheric ageing. Most of the earlier reports (*2*) attributed BrC to primary organic aerosols (POA) in the emissions from biomass burning and combustion of fossil fuels. More recent studies (*10*, *11*) indicate that secondary organic aerosol (SOA) formed from the oxidation of selected natural and anthropogenic volatile organic compounds (VOC) also may contribute to BrC.

Even small fractions of strong chromophores may determine the overall absorption of light by the BrC material (*12*). Because of the low concentrations of light-absorbing molecules in complex organic mixtures present in the aerosol, identification of BrC chromophores is a challenging task. The identification and structural characterization of BrC chromophores require applications of highly sensitive molecular characterization approaches capable of detecting both strongly and weakly absorbing species in complex organic mixtures (*10*). This chapter provides a brief summary of the most recent advances in the molecular-level studies of BrC materials conducted in our groups.

Separation and Analysis of BrC Chromophores

A non-targeted detection of a priori unknown light absorbing components (chromophores) within complex environmental mixtures is the major challenge in the chemical characterization of BrC. To address this challenge experimentally, we use a multi-stage analytical platform that combines high performance liquid chromatography (HPLC), photodiode array (PDA) spectrophotometry, and high-resolution mass spectrometry (HRMS) assisted with soft ionization methods (*13–19*). The ultimate success of the entire method critically depends on the performance at each of its stages. First, the extent of HPLC separation is crucial to the subsequent analysis of BrC chromophores eluting from the column. Besides the natural variability in the chromophores' composition between different BrC samples, HPLC separation is affected by many factors, such as stationary phase and operation temperature of the column, pH of the analyte mixture, flow rate, chemical composition of the mobile phase, and other parameters. Selection of the best performing chromatographic column and development of robust separation protocols were addressed in our initial studies employing laboratory generated proxies of BrC. In these studies, BrC materials were produced through: 1) reactions of methylglyoxal (MG) and ammonium sulfate (AS) (*13*) and 2) OH/NO_x-induced photo-oxidation of toluene (*14*). We conducted systematic studies to evaluate performance of several HPLC columns for separation of BrC chromophores. We confirmed that the reverse-phase C18 column provides most practical separation of organics in complex real-world samples (*15–17*), where a broad range of chemically diverse BrC chromophores are typically present. However, for separation of BrC chromophores belonging to the same chemical class, columns with specially tailored stationary phases may provide a better performance. For instance, best separation of reduced nitrogen chromophores formed in the MG+AS reacting system was achieved using the SM-C18 column (*13*), which combines ion exchange capability with the more commonly used reverse-phase interactions.

Figure 1 illustrates the HPLC-PDA chromatogram of BrC chromophores identified in a sample of biomass burning organic aerosol (BBOA) collected from controlled burns of sagebrush biofuel during Fire Influence on Regional and Global Environments Experiment (FIREX) campaign in 2016 (*20*) at the U.S. Forest Service Fire Science Laboratory in Missoula, MT. The Y-axis of the plot shows wavelength of UV-vis spectra and the heatmap colors indicate absorption intensities. The chromatogram reveals abundant well-separated chromophores at retention times (RT) between 10 and 100 min (*17*). A substantial fraction of their light absorption lies in the visible spectral range > 390nm. The chemical composition of BrC chromophores is diverse and includes different classes of organic molecules that cannot be ionized using one common ionization mechanism. To address this limitation, their composition was investigated in four separate experiments utilizing electrospray ionization (ESI) and atmospheric pressure photoionization (APPI) sources operated in positive (+) and negative (-) modes, as illustrated in the figure legends. Elemental formulas and plausible structures of the BrC chromophores were then identified through correlative analysis of the combined HPLC-PDA and HPLC-HRMS records.

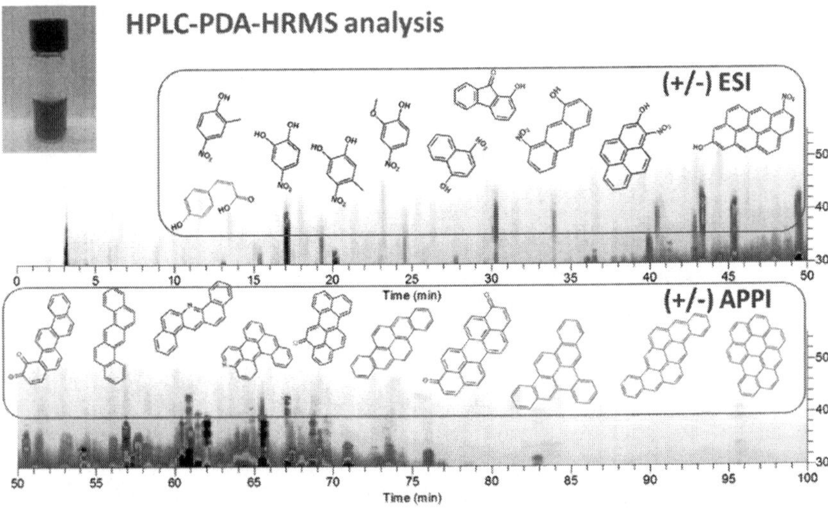

Figure 1. HPLC-PDA chromatogram of a selected BBOA sample (17). The x-axis is a retention time (RT). The y-axis and heatmap refer to the wavelength and intensity of the UV-vis absorption spectra, respectively. The molecular structures denote BrC chromophores identified in HPLC-HRMS experiments interfaced with ESI and APPI sources. Polar compounds eluted at earlier RT and are detected mostly in the ESI-HRMS experiments; non-polar compounds eluted later and are detected mostly in the APPI-HRMS experiments.

ESI-HRMS experiments are particularly sensitive to polar molecules such as carboxylic acids, nitro-organics, organo-sulfates, etc., while APPI-HRMS is the method for chemical analysis of non- and low-polarity compounds with extensive network of π-bonds such as PAH, N- and O-heterocyclic compounds. Figure 2 illustrates differences in the direct infusion mass spectra of the same BBOA sample acquired with ESI and APPI sources, each operated in positive-ion and negative-ion modes, respectively (*17*). Remarkably, HRMS spectra obtained using ESI+, ESI-, APPI+ and APPI- are distinctly different and only minor overlap was observed between them. These results are attributed to the differences between the ionization mechanisms. Specifically, ESI generates ions through desolvation of charged droplets, in which polar analyte molecules that possess either acidic or basic functionalities efficiently compete for charge. In contrast, APPI ionizes non-polar molecules through electron detachment or charge transfer from photo-ionized dopant ions purposely added to the sample. As illustrated by Figure 2, distinctly different classes of compounds were identified in each of the modes. It follows that repeated HRMS experiments employing different ionization modes are required for the comprehensive characterization of a broad range of organic compounds present in BBOA samples (*17*).

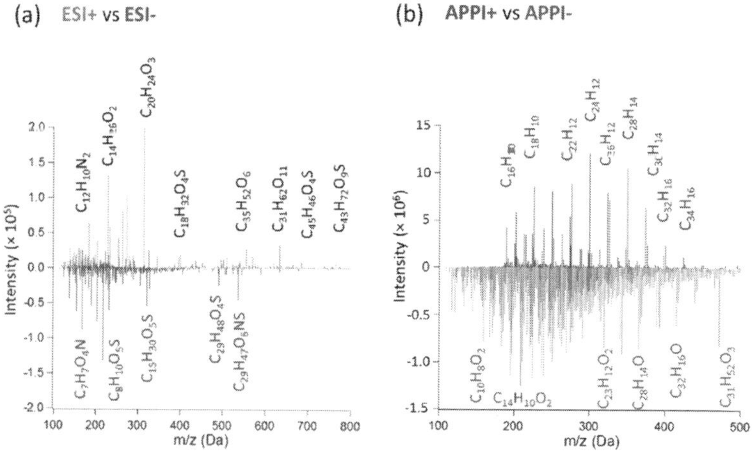

Figure 2. Mass spectra of solvent extractable BBOA compounds detected in positive and negative modes of the direct infusion ESI-HRMS (panel a) and APPI-HRMS (panel b) experiments (17). Reproduced with permission from ref. (17). Copyright 2018, ACS.

High mass resolving power and high mass accuracy are essential for assigning elemental compositions to individual BBOA components, while MS^n fragmentation experiments can provide additional insights into their structural characterization (*21*, *22*). Although minimal mass resolution required for unambiguous formula assignments depends on the complexity of the BBOA analytes, we showed that good-quality assignments of the most common $C_xH_yO_zN_{0\text{-}3}S_{0\text{-}2}$ formulas can be routinely obtained from HRMS data recorded at 100,000 *m/Δm* mass resolving power (*13*, *14*, *17*, *19*, *23–26*). Furthermore, accurate mass measurements combined with calculated element-specific isotope distributions can reveal the presence of less common species, such as Mg, Al, Ca, Cr, Mn, Fe, Ni, Cu, Zn, and Ba-containing metal-organic components of BBOA (*27*).

Different configurations of HPLC-PDA-HRMS analytical platform were used in a number of studies where a variety of BrC chromophores were characterized in the samples of ambient BBOA (*15–17*, *19*, *28–31*) and lab-generated proxies (*13*, *14*, *18*, *32–34*) of light-absorbing organic aerosol. Several classes of BrC chromophores such as alkyl-phenols, methoxy-phenols, nitro-phenols, substituted benzoic acids, imidazole- and furan-based oligomers, quinones, PAHs, nitro-PAHs, N- and O-heterocyclic compounds and others have been identified. Beyond this, it has been shown that charge transfer complexes of supramolecular aggregates contribute additionally to the overall optical properties of BrC (*35*).

Comparison of the BrC chromophores between different BBOA samples indicates that a large fraction of them are biofuel-specific, while certain chromophores also appear repeatedly among different samples, which may suggest their more common nature across certain types of biofuels. In nearly all BBOA samples studied so far, we found that the total absorbance by 20-25 individual BrC chromophores accounted for 40-60% of the overall absorbance in the wavelength range of 300–500 nm. Relative contributions of the different types of BrC chromophores may be estimated by combining them into broader groups based on their chemical properties, such as similarity in molecular composition and estimated volatility, as illustrated by Figure 3. Such a grouping may provide practical input for future atmospheric process models to simulate sources, composition and transformations of BrC.

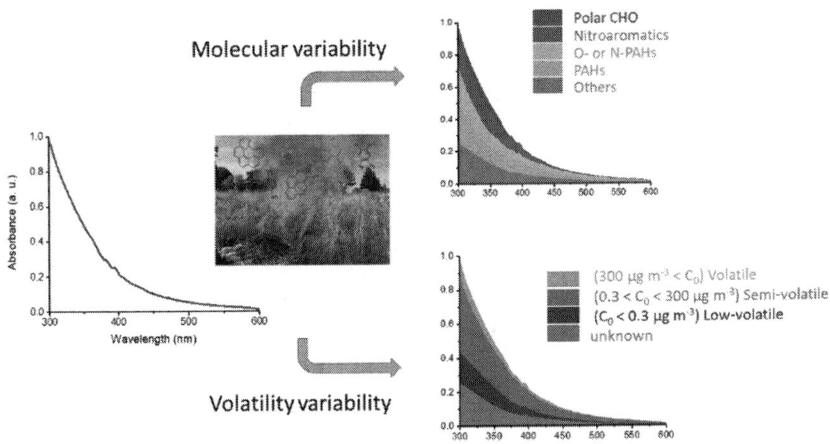

Figure 3. Relative contributions to the overall BrC absorption of a selected BBOA sample by individual chromophores grouped based on their molecular composition (upper plot) and estimated volatility (lower plot).

Processes Affecting BrC Composition and Transformations

BrC contains both water-soluble and low-solubility organic carbon, which must be considered when designing the experimental protocol for chemical analysis. The water-soluble fraction of BrC is usually below 70%, while nearly 90% of BrC can be extracted into organic solvents, such as acetonitrile and methanol (*16*). The water-insoluble fraction of BrC has greater absorption per unit of mass than the water-soluble fraction (*36*). In BBOA, the relative abundances of water-soluble and water-insoluble fractions as well as partitioning of their light-absorption properties are strongly affected by the temperature of the biomass burning processes, fuel type and its moisture content.

Light-absorption properties of water-soluble BrC are pH dependent because its chromophores exist in both neutral and ionized (protonated/deprotonated) forms and their relative abundances are defined by their pK_a and the pH of the solution. For example, UV-vis absorption by the nitro-phenol based chromophores shifts toward longer wavelengths (red shift) when they are deprotonated at neutral pH (*37*). Figure 4 illustrates the pH effect in the overall light absorption of the

BBOA sample with abundant presence of the nitro-aromatic chromophores (*16*). In contrast, chromophores containing moieties of benzoic acid and its derivatives exhibit blue shift when deprotonated (*38*). Therefore, although aerosol acidity plays an important role in modulating the optical properties of BrC, the combined effect of pH may not be always the same for all BrC materials.

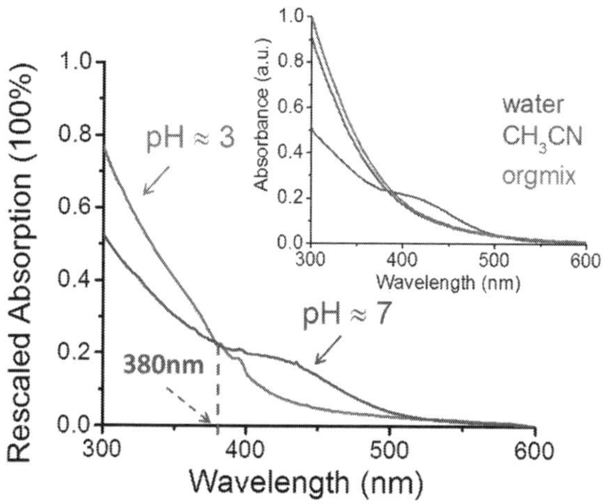

Figure 4. UV−vis spectra of water-soluble BrC fraction extracted from BBOA sample and measured at different pH conditions. The inset illustrates the UV−vis spectra of BrC extracted with water and organic solvents from the same sample (16). Reproduced with permission from ref. (16). Copyright 2016, ACS.

Heterogeneous gas-particle and condensed phase reactions of aerosols during atmospheric aging also modify the composition and light absorption of BrC chromophores. Notably, BrC components containing aromatic molecules with various oxygenated groups present in fresh BBOA (attributed to both thermal decomposition of biofuel in BBOA and to photooxidation of aromatic VOC in anthropogenic SOA) may be converted into stronger-absorbing nitro-aromatic species as a result of their reactions with various forms of nitrogen oxides (*39*, *40*). Our studies showed that evaporation-driven (*41*, *42*) and particle-phase condensation (*18*, *43*, *44*) reactions forming oligomer species may generate strong BrC chromophores resulting in further 'browning' of aged aerosol. Oxidation of nitro-aromatic compounds by OH initially makes them more light absorbing, but further oxidation destroys network of carbon-carbon double bonds in chromophores and therefore results in less absorbing BrC components (*45*, *46*).

Direct photolysis of chromophores by solar radiation also modifies light absorption properties of BrC. We observed that photo-bleaching is the common trend, and overall BrC becomes less absorbing upon exposure to sunlight (*47–49*). However, photolysis rates of individual chromophores were found to be different by orders of magnitude, depending on their specific molecular composition. For example, Figure 5 shows a fragment of HPLC-PDA chromatogram for lodge pole pine BBOA collected during the FIREX campaign. A portion of the filter sample was irradiated by 300 nm radiation from a UV light-emitting diode, while the remainder of the filter was left in the dark. Both photolyzed and unphotolyzed samples were analyzed by the HPLC-PDA-HRMS methods. The major chromophores, tentatively assigned as salicylic acid ($C_7H_6O_3$, RT=11.94 min), veratraldehyde ($C_7H_6O_3$, RT= 14.49 min), and coniferaldehyde ($C_{10}H_{10}O_3$, RT=18.45 min) diminish in abundance but remain visible in the chromatogram. Other BrC chromophores are almost completely destroyed by the prolonged exposure to 300 nm radiation. Such an analysis provides insights into which BrC chromophores are photolabile and which are photo-resistant.

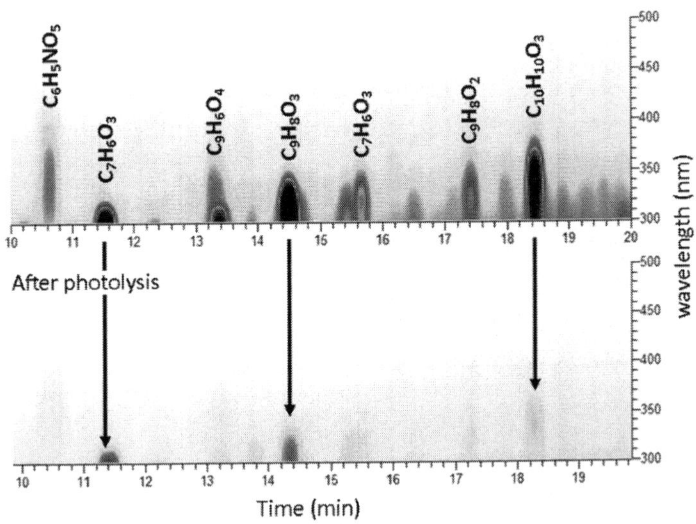

Figure 5. Upper Panel: HPLC-PDA chromatogram for lodge pole pine BBOA with tentative elemental formulas assigned for each light-absorbing compound. Lower Panel: HPLC-chromatogram for the sample after direct photolysis of the particulate matter on the filter.

We need to emphasize that photobleaching is not the only possible effect of UV irradiation on BrC. Figure 6 illustrates a more complex case where 2,4-dinitrophenol (2,4-DNP) chromophores absorbing in the UV range decompose under photolysis, but then their photolysis products undergo secondary reactions to form a new dimer product absorbing in the visible range (*48*). Similar observations were made for nitrophenols undergoing aqueous OH oxidation (*45*, *46*), in which the absorption coefficient of the mixture first increased and then decreased during the oxidation process. These experiments emphasize the complicated role of photochemical aging in determining the light absorption properties of BrC samples.

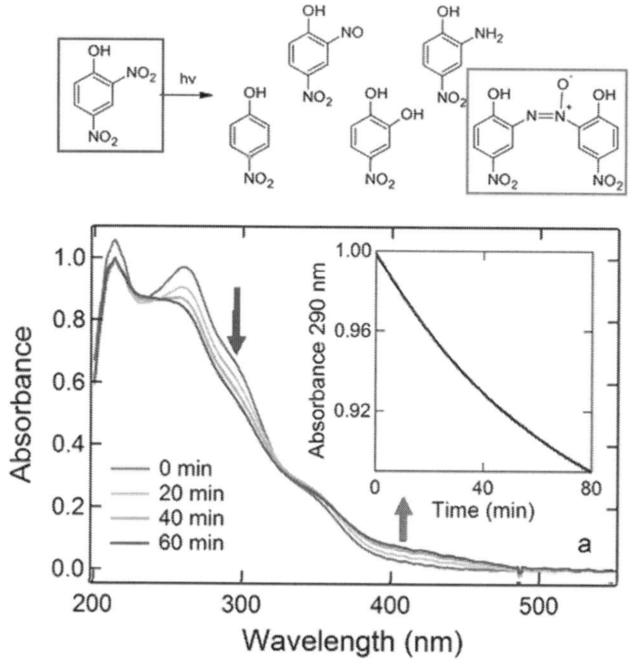

Figure 6. Upper panel: Products identified in photo-degradation of 2,4-DNP (marked by blue frame). Brown frame indicates a dimer product absorbing in the visible range. Lower panel: UV-vis absorption spectra recorded during the photo-degradation of 2,4-DNP. The blue arrow and the inset show the decay of the 290 nm peak of 2,4-DNP and the brown arrow shows buildup of the 400-450 nm absorbance attributed to the dimer product (48).

To conclude, the initial phenomenological information on the composition and transformations of BrC chromophores characteristic of BBOA and anthropogenic SOA has been formulated based on the studies highlighted above. Studies performed so far indicate that BrC chromophores have complex and diverse composition and that their molecular identity may depend on the primary source. On the other hand, several common BrC constituents also have been identified. These include nitro-aromatics, PAH, N- and O-heterocyclic compounds, and their derivatives. Several studies have examined atmospheric stability of common chromophores along with mechanisms and lifetimes of transformation under atmospherically relevant conditions. Chemical diversity of BrC chromophores and their low abundance in aerosol samples present a challenge to their successful characterization. Creative applications of multi-modal complementary mass spectrometry techniques interfaced with separation, ionization and spectroscopic platforms are essential to obtain a predictive understanding of the relationship between the chemical composition and optical properties of BrC. Because BrC is a highly dynamic system, its composition and optical properties at the emission source and later downwind may be substantially different. A general trend of the decaying BrC light absorption properties upon its atmospheric ageing has been reported by majority of the published studies. However, certain instances of the new chromophore formation in aged BrC were also noted.

Despite all these recent research advances, the overall level of our understanding is still insufficient for predicting the properties of BrC in the atmosphere. Future studies are needed to enable practical classification of broader BrC materials for practical parameterization in the next-generation atmospheric and climate models. Although atmospheric models cannot capture the complexity of the chemical composition and optical properties of BrC in its entirety, detailed understanding of the BrC chemistry will facilitate the development of simplified model of the BrC properties and establish parameters necessary for its adequate description in atmospheric models.

Acknowledgments

We acknowledge support by the U.S. Department of Commerce, National Oceanic and Atmospheric Administration through Climate Program Office's AC4 program, awards NA16OAR4310101 and NA16OAR4310102. The HPLC/PDA/ESI-HRMS measurements were performed at the Environmental Molecular Sciences Laboratory (EMSL), a national scientific user facility located at PNNL, and sponsored by the Office of Biological and Environmental Research of the U.S. DOE.

References

1. Bond, T. C.; Doherty, S. J.; Fahey, D. W.; Forster, P. M.; Berntsen, T.; DeAngelo, B. J.; Flanner, M. G.; Ghan, S.; Kaercher, B.; Koch, D.; Kinne, S.; Kondo, Y.; Quinn, P. K.; Sarofim, M. C.; Schultz, M. G.; Schulz, M.; Venkataraman, C.; Zhang, H.; Zhang, S.; Bellouin, N.; Guttikunda, S. K.; Hopke, P. K.; Jacobson, M. Z.; Kaiser, J. W.; Klimont, Z.; Lohmann, U.; Schwarz, J. P.; Shindell, D.; Storelvmo, T.; Warren, S. G.; Zender, C. S. Bounding the role of black carbon in the climate system: A scientific assessment. *J. Geophys. Res. - Atmos.* **2013**, *118*, 5380–5552.
2. Andreae, M. O.; Gelencser, A. Black carbon or brown carbon? The nature of light-absorbing carbonaceous aerosols. *Atmos. Chem. Phys.* **2006**, *6*, 3131–3148.
3. Feng, Y.; Ramanathan, V.; Kotamarthi, V. R. Brown carbon: a significant atmospheric absorber of solar radiation. *Atmos. Chem. Phys.* **2013**, *13*, 8607–8621.
4. Wang, X.; Heald, C. L.; Ridley, D. A.; Schwarz, J. P.; Spackman, J. R.; Perring, A. E.; Coe, H.; Liu, D.; Clarke, A. D. Exploiting simultaneous observational constraints on mass and absorption to estimate the global direct radiative forcing of black carbon and brown carbon. *Atmos. Chem. Phys.* **2014**, *14*, 10989–11010.
5. Jethva, H.; Torres, O. Satellite-based evidence of wavelength-dependent aerosol absorption in biomass burning smoke inferred from Ozone Monitoring Instrument. *Atmos. Chem. Phys.* **2011**, *11*, 10541–10551.

6. Arola, A.; Schuster, G.; Myhre, G.; Kazadzis, S.; Dey, S.; Tripathi, S. N. Inferring absorbing organic carbon content from AERONET data. *Atmos. Chem. Phys.* **2011**, *11*, 215–225.
7. Bahadur, R.; Praveen, P. S.; Xu, Y.; Ramanathan, V. Solar absorption by elemental and brown carbon determined from spectral observations. *Proc. Natl. Acad. Sci. U.S.A.* **2012**, *109*, 17366–17371.
8. Wang, L.; Li, Z.; Tian, Q.; Ma, Y.; Zhang, F.; Zhang, Y.; Li, D.; Li, K.; Li, L. Estimate of aerosol absorbing components of black carbon, brown carbon, and dust from ground-based remote sensing data of sun-sky radiometers. *J. Geophys. Res. - Atmos.* **2013**, *118*, 6534–6543.
9. Chung, C. E.; Ramanathan, V.; Decremer, D. Observationally constrained estimates of carbonaceous aerosol radiative forcing. *Proc. Natl. Acad. Sci. U.S.A.* **2012**, *109*, 11624–11629.
10. Laskin, A.; Laskin, J.; Nizkorodov, S. A. Chemistry of Atmospheric Brown Carbon. *Chem. Rev.* **2015**, *115*, 4335–4382.
11. Moise, T.; Flores, J. M.; Rudich, Y. Optical Properties of Secondary Organic Aerosols and Their Changes by Chemical Processes. *Chem. Rev.* **2015**, *115*, 4400–4439.
12. Laskin, J.; Laskin, A.; Nizkorodov, S. A.; Roach, P. J.; Eckert, P.; Gilles, M. K.; Wang, B.; Lee, J.; Hu, Q. Molecular Selectivity of Brown Carbon Chromophores. *Environ. Sci. Technol.* **2014**, *48*, 12047–12055.
13. Lin, P.; Laskin, J.; Nizkorodov, S. A.; Laskin, A. Revealing Brown Carbon Chromophores Produced in Reactions of Methylglyoxal with Ammonium Sulfate. *Environ. Sci. Technol.* **2015**, *49*, 14257–14266.
14. Lin, P.; Liu, J.; Shilling, J. E.; Kathmann, S. M.; Laskin, J.; Laskin, A. Molecular Characterization of Brown Carbon (BrC) Chromophores in Secondary Organic Aerosol Generated From Photo-Oxidation of Toluene. *Phys. Chem. Chem. Phys.* **2015**, *17*, 23312–23325.
15. Lin, P.; Aiona, P. K.; Li, Y.; Shiraiwa, M.; Laskin, J.; Nizkorodov, S. A.; Laskin, A. Molecular Characterization of Brown Carbon in Biomass Burning Aerosol Particles. *Environ. Sci. Technol.* **2016**, *50*, 11815–11824.
16. Lin, P.; Bluvshtein, N.; Rudich, Y.; Nizkorodov, S. A.; Laskin, J.; Laskin, A. Molecular Chemistry of Atmospheric Brown Carbon Inferred from a Nationwide Biomass Burning Event. *Environ. Sci. Technol.* **2017**, *51*, 11561–11570.
17. Lin, P.; Fleming, L.; Nizkorodov, S.; Laskin, J.; Laskin, A. Comprehensive Characterization of Brown Carbon in Biomass Burning Organic Aerosols by High Resolution Mass Spectrometry with Electrospray and Atmospheric Pressure Photoionization. *Anal. Chem.* **2018**, submitted.
18. Lavi, A.; Lin, P.; Bhaduri, B.; Carmieli, R.; Laskin, A.; Rudich, Y. Characterization of Light-Absorbing Oligomers from Phenolic Compounds and Fe(III). *ACS Earth Space Chem.* **2017**, *1*, 637–646.
19. Fleming, L. T.; Lin, P.; Laskin, A.; Laskin, J.; Weltman, R.; Edwards, R. D.; Arora, N. K.; Yadav, A.; Meinardi, S.; Blake, D. R.; Pillarisetti, A.; Smith, K. R.; Nizkorodov, S. A. Molecular composition of particulate matter emissions from dung and brushwood burning household cookstoves in Haryana, India. *Atmos. Chem. Phys.* **2018**, *18*, 2461–2480.

20. Selimovic, V.; Yokelson, R. J.; Warneke, C.; Roberts, J. M.; de Gouw, J.; Reardon, J.; Griffith, D. W. T. Aerosol optical properties and trace gas emissions by PAX and OP-FTIR for laboratory-simulated western US wildfires during FIREX. *Atmos. Chem. Phys.* **2018**, *18*, 2929–2948.
21. Nizkorodov, S. A.; Laskin, J.; Laskin, A. Molecular chemistry of organic aerosols through the application of high resolution mass spectrometry. *Phys. Chem. Chem. Phys.* **2011**, *13*, 3612–3629.
22. Laskin, J.; Laskin, A.; Nizkorodov, S. Mass Spectrometry Analysis in Atmospheric Chemistry. *Anal. Chem.* **2018**, *90*, 166–189.
23. Blair, S. L.; MacMillan, A. C.; Drozd, G. T.; Goldstein, A. H.; Chu, R. K.; Pasa-Tolic, L.; Shaw, J. B.; Tolic, N.; Lin, P.; Laskin, J.; Laskin, A.; Nizkorodov, S. A. Molecular Characterization of Organosulfur Compounds in Biodiesel and Diesel Fuel Secondary Organic Aerosol. *Environ. Sci. Technol.* **2017**, *51*, 119–127.
24. O'Brien, R. E.; Laskin, A.; Laskin, J.; Rubitschun, C. L.; Surratt, J. D.; Goldstein, A. H. Molecular characterization of S- and N-containing organic constituents in ambient aerosols by negative ion mode high-resolution Nanospray Desorption Electrospray Ionization Mass Spectrometry: CalNex 2010 field study. *J. Geophys. Res. - Atmos.* **2014**, *119*, 12706–12720.
25. O'Brien, R. E.; Nguyen, T. B.; Laskin, A.; Laskin, J.; Hayes, P. L.; Liu, S.; Jimenez, J. L.; Russell, L. M.; Nizkorodov, S. A.; Goldstein, A. H. Probing molecular associations of field-collected and laboratory-generated SOA with nano-DESI high-resolution mass spectrometry. *J. Geophys. Res. - Atmos.* **2013**, *118*, 1042–1051.
26. Tao, S.; Lu, X.; Levac, N.; Bateman, A. P.; Nguyen, T. B.; Bones, D. L.; Nizkorodov, S. A.; Laskin, J.; Laskin, A.; Yang, X. Molecular Characterization of Organosulfates in Organic Aerosols from Shanghai and Los Angeles Urban Areas by Nanospray-Desorption Electrospray Ionization High-Resolution Mass Spectrometry. *Environ. Sci. Technol.* **2014**, *48*, 10993–11001.
27. Chang-Graham, A. L.; Profeta, L. T. M.; Johnson, T. J.; Yokelson, R. J.; Laskin, A.; Laskin, J. Case Study of Water-Soluble Metal Containing Organic Constituents of Biomass Burning Aerosol. *Environ. Sci. Technol.* **2011**, *45*, 1257–1263.
28. Samburova, V.; Connolly, J.; Gyawali, M.; Yatavelli, R. L. N.; Watts, A. C.; Chakrabarty, R. K.; Zielinska, B.; Moosmuller, H.; Khlystov, A. Polycyclic aromatic hydrocarbons in biomass-burning emissions and their contribution to light absorption and aerosol toxicity. *Sci. Total Environ.* **2016**, *568*, 391–401.
29. Kitanovski, Z.; Grgic, I.; Vermeylen, R.; Claeys, M.; Maenhaut, W. Liquid chromatography tandem mass spectrometry method for characterization of monoaromatic nitro-compounds in atmospheric particulate matter. *J. Chromatogr. A* **2012**, *1268*, 35–43.
30. Desyaterik, Y.; Sun, Y.; Shen, X.; Lee, T.; Wang, X.; Wang, T.; Collett, J. L., Jr. Speciation of "brown" carbon in cloud water impacted by agricultural biomass burning in eastern China. *J. Geophys. Res. - Atmos.* **2013**, *118*, 7389–7399.

31. Budisulistiorini, S. H.; Riva, M.; Williams, M.; Chen, J.; Itoh, M.; Surratt, J. D.; Kuwata, M. Light-Absorbing Brown Carbon Aerosol Constituents from Combustion of Indonesian Peat and Biomass. *Environ. Sci. Technol.* **2017**, *51*, 4415–4423.
32. Kitanovski, Z.; Čusak, A.; Grgić, I.; Claeys, M. Chemical characterization of the main products formed through aqueous-phase photonitration of guaiacol. *Atmos. Meas. Tech.* **2014**, *7*, 2457–2470.
33. Montoya-Aguilera, J.; Horne, J. R.; Hinks, M. L.; Fleming, L. T.; Perraud, V.; Lin, P.; Laskin, A.; Laskin, J.; Dabdub, D.; Nizkorodov, S. A. Secondary organic aerosol from atmospheric photooxidation of indole. *Atmos. Chem. Phys.* **2017**, *17*, 11605–11621.
34. Lin, Y. H.; Budisulistiorini, H.; Chu, K.; Siejack, R. A.; Zhang, H. F.; Riva, M.; Zhang, Z. F.; Gold, A.; Kautzman, K. E.; Surratt, J. D. Light-Absorbing Oligomer Formation in Secondary Organic Aerosol from Reactive Uptake of Isoprene Epoxydiols. *Environ. Sci. Technol.* **2014**, *48*, 12012–12021.
35. Phillips, S. M.; Smith, G. D. Light Absorption by Charge Transfer Complexes in Brown Carbon Aerosols. *Environ. Sci. Technol. Lett.* **2014**, *1*, 382–386.
36. Bergstrom, R. W.; Pilewskie, P.; Russell, P. B.; Redemann, J.; Bond, T. C.; Quinn, P. K.; Sierau, B. Spectral absorption properties of atmospheric aerosols. *Atmos. Chem. Phys.* **2007**, *7*, 5937–5943.
37. Cornard, J. P.; , Rasmiwetti; Merlin, J. C. Molecular structure and spectroscopic properties of 4-nitrocatechol at different pH: UV-visible, Raman, DFT and TD-DFT calculations. *Chem. Phys.* **2005**, *309*, 239–249.
38. Guo, H. B.; He, F.; Gu, B. H.; Liang, L. Y.; Smith, J. C. Time-Dependent Density Functional Theory Assessment of UV Absorption of Benzoic Acid Derivatives. *J. Phys. Chem. A* **2012**, *116*, 11870–11879.
39. Iinuma, Y.; Böge, O.; Gräfe, R.; Herrmann, H. Methyl-Nitrocatechols: Atmospheric Tracer Compounds for Biomass Burning Secondary Organic Aerosols. *Environ. Sci. Technol.* **2010**, *44*, 8453–8459.
40. Barzaghi, P.; Herrmann, H. Kinetics and mechanisms of reactions of the nitrate radical (NO3) with substituted phenols in aqueous solution. *Phys. Chem. Chem. Phys.* **2004**, *6*, 5379–5388.
41. Nguyen, T. B.; Laskin, A.; Laskin, J.; Nizkorodov, S. A. Brown carbon formation from ketoaldehydes of biogenic monoterpenes. *Faraday Discuss.* **2013**, *165*, 473–494.
42. Nguyen, T. B.; Lee, P. B.; Updyke, K. M.; Bones, D. L.; Laskin, J.; Laskin, A.; Nizkorodov, S. A. Formation of nitrogen- and sulfur-containing light-absorbing compounds accelerated by evaporation of water from secondary organic aerosols. *J. Geophys. Res. - Atmos.* **2012**, *117*, D01207.
43. Laskin, J.; Laskin, A.; Roach, P. J.; Slysz, G. W.; Anderson, G. A.; Nizkorodov, S. A.; Bones, D. L.; Nguyen, L. Q. High-Resolution Desorption Electrospray Ionization Mass Spectrometry for Chemical Characterization of Organic Aerosols. *Anal. Chem.* **2010**, *82*, 2048–2058.
44. Aiona, P. K.; Lee, H. J.; Lin, P.; Heller, F.; Laskin, A.; Laskin, J.; Nizkorodov, S. A. A Role for 2-Methyl Pyrrole in the Browning of

4-Oxopentanal and Limonene Secondary Organic Aerosol. *Environ. Sci. Technol.* **2017**, *51*, 11048–11056.

45. Zhao, R.; Lee, A. K. Y.; Huang, L.; Li, X.; Yang, F.; Abbatt, J. P. D. Photochemical processing of aqueous atmospheric brown carbon. *Atmos. Chem. Phys.* **2015**, *15*, 6087–6100.
46. Hems, R. F.; Abbatt, J. P. D. Aqueous Phase Photo-oxidation of Brown Carbon Nitrophenols: Reaction Kinetics, Mechanism, and Evolution of Light Absorption. *ACS Earth Sci. Chem.* **2018**, *2*, 225–234.
47. Aiona, P. K.; Lee, H. J.; Leslie, R.; Lin, P.; Laskin, A.; Laskin, J.; Nizkorodov, S. A. Photochemistry of Products of the Aqueous Reaction of Methylglyoxal with Ammonium Sulfate. *ACS Earth Sci. Chem.* **2017**, *1*, 522–532.
48. Hinks, M. L.; Brady, M. V.; Lignell, H.; Song, M.; Grayson, J. W.; Bertram, A. K.; Lin, P.; Laskin, A.; Laskin, J.; Nizkorodov, S. A. Effect of viscosity on photodegradation rates in complex secondary organic aerosol materials. *Phys. Chem. Chem. Phys.* **2016**, *18*, 8785–8793.
49. Lee, H. J.; Aiona, P. K.; Laskin, A.; Laskin, J.; Nizkorodov, S. A. Effect of Solar Radiation on the Optical Properties and Molecular Composition of Laboratory Proxies of Atmospheric Brown Carbon. *Environ. Sci. Technol.* **2014**, *48*, 10217–10226.

Chapter 14

Absorption Spectroscopy of Black and Brown Carbon Aerosol

Christopher D. Zangmeister* and James G. Radney

Chemical Sciences Division, Material Measurement Laboratory, 100 Bureau Dr., Mail Stop 8320, Gaithersburg, Maryland 20899, United States
***E-mail: cdzang@nist.gov**

Understanding the absorption spectra of aerosolized carbonaceous materials in the atmosphere enables their identification and offers improved parameterizations for use in radiative forcing models and aerosol remote sensing. This Chapter will focus on: 1) the parameters that influence aerosol absorption, 2) quantitative measurements of mass-normalized aerosol absorption spectra, and 3) how these developments have been applied to atmospherically relevant systems. We present measured mass-normalized absorption spectra of two families of absorbing carbonaceous aerosol: strongly absorbing and weakly wavelength dependent black carbon (BC) aerosol generated from liquid and gas fueled flames and allotropes of carbon, and weakly absorbing and strongly wavelength dependent brown carbon (BrC) derived from biomass burning.

Introduction

Absorption of light by suspended micro- and nano-meter size particles in the atmosphere (aerosol) directly exerts a positive radiative forcing (i.e. warming) relative to non-absorbing particles (*1*). Radiative forcing calculations (*2*) and satellite remote sensing measurements (*3*) require high-quality input data with spectroscopic measurements serving as foundational inputs. The heterogeneity of atmospheric aerosol provides a challenge in understanding the factors that influence their optical properties such as chemical composition (e.g. refractive index), particle size and morphology (e.g. lacey aggregates versus spheres, particles embedded in a coating material versus bare, etc.). One method of overcoming this challenge is to perform measurements under well-controlled laboratory conditions with aerosolized materials of known size (electrical mobility), mass and/or chemical composition and morphology to investigate each factor independently. This Chapter will give an introduction to the general principles of aerosol light absorption and how each factor affects particle optical properties. It will also use laboratory-based measurements of the material's chemical composition, size, and morphology to demonstrate the impact of how particle characteristics impact absorption individually or in concert.

Carbonaceous aerosols are a diverse class of materials observed in the atmosphere. Their light absorption properties are correlated to the material's chemical composition and can be spectroscopically classified into 3 broad categories: non-absorbing and absorbing organic carbonaceous species and highly absorbing nearly elemental carbon.

Organic carbonaceous aerosol is formed from the condensation of complex mixtures of organic compounds resulting from natural processes such as biomass burning or gas-phase reactions of organic materials to form particles that range in size from less than 1 nm to greater than 1 μm, see Figure 1. Absorption by organic carbonaceous aerosols in the optical atmospheric transmission window is wavelength dependent with very low absorption at long wavelengths (near-IR) and increasing sharply toward the shorter wavelengths (UV), giving them a faint yellow or brown-like color, and the often used nomenclature, brown carbon (BrC) (*4*).

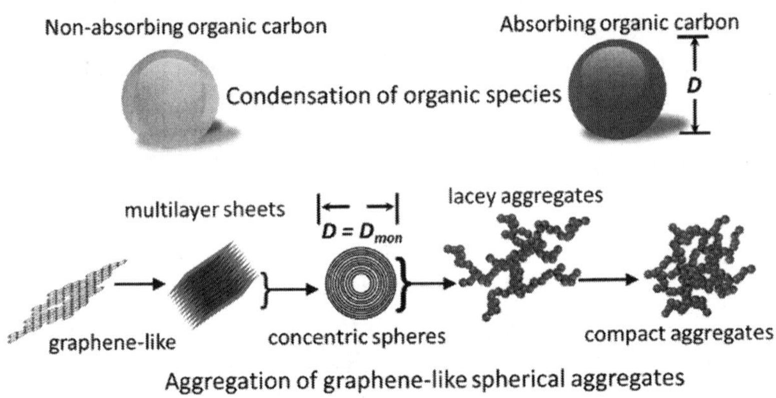

Figure 1. Classification of carbonaceous aerosol. Top shows condensation of organic carbonaceous species into non-absorbing and absorbing aerosol, brown carbon (BrC). Bottom shows schematic of aggregation of graphene-like spherical monomers to form black carbon (BC). Definitions of D and D_{mon} used in the text are shown for BrC and BC particles.

Highly absorbing carbonaceous aerosol, or black carbon (BC), is produced during the incomplete combustion of carbonaceous fuels. These particles consist of stacked graphene-like sheets arranged into concentric multilayered nanoscale spheres that aggregate into lacey structures. The diameter of individual monomers (D_{mon}) ranges between 10 nm and 100 nm, with linear dimensions spanning from 30 nm for individual aggregates of a few monomers up to a few μm for freshly formed particles consisting of multiple aggregates, see Figure 1. In rare instances, superaggregates with linear dimensions > 20 μm up to a few mm have been observed (*5, 6*). Aggregates may also subsequently become compacted upon interaction with gaseous and/or condensed phase materials. The graphene-like composition makes these materials highly absorbing across the visible spectral region with only a modest wavelength dependence (*7, 8*).

The magnitude of aerosol absorption per unit mass is a function of the primary, secondary, and tertiary structures of an aerosol (*9–13*), see Figure 2. The intrinsic (primary) absorption strength of a material is a function of its chemical composition (i.e. refractive index); for graphitic and graphenic materials, this is related to the extent of sp^2-bonding, surface imperfections and sheet lateral dimension. Absorption can be enhanced or dampened depending upon the atomistic arrangement of materials into secondary structure(s), such as spherical monomers or single- and multi-layered sheets. For BC, monomers are aggregated into an open, lacey tertiary structure, also referred to as nanosphere soot, or ns-soot (*14*).

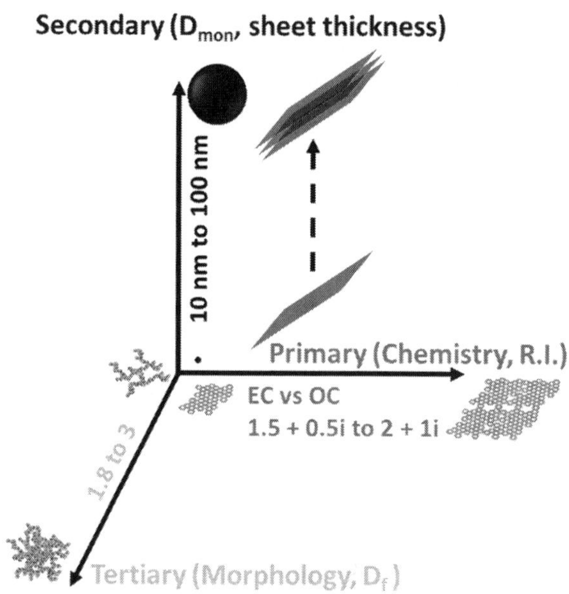

Figure 2. Aerosol absorption properties are a function of primary (refractive index), secondary (atomistic arrangement), and tertiary structures (particle morphology).

Refractive Indices, Coefficients, Cross-Sections and Efficiencies

The primary parameter influencing the optical properties of any material is the complex refractive index (n) given by

$$n = m + ik \tag{1}$$

where the real (m) and imaginary (k) components relate to the velocity and attenuation of light in the material, respectively. In the macroscopic limit (i.e. bulk), the absorption coefficient ($\alpha_{\text{abs,bulk}}$) of a material is related to the imaginary component of the refractive index and the wavelength of light (λ) through (*15*)

$$\alpha_{\text{abs,bulk}} = \frac{4\pi k}{\lambda} \tag{2}$$

The Beer-Lambert Law is used to relate the absorption coefficient (α_{abs}) to the fractional change in light intensity (dI) due to absorption per unit penetration distance in an absorbing medium (dl)

$$\frac{dI}{I} = \alpha_{\text{abs}} dl \tag{3}$$

with the absorption coefficient being dependent upon the number concentration of absorbers (N) and their absorption cross-section (C_{abs})

$$\alpha_{\text{abs}} = N C_{\text{abs}} \tag{4}$$

Integrating Eq. 3 yields the absorption optical depth (τ_{abs}). With respect to atmospheric observations, τ_{abs} is typically calculated from an arbitrary altitude (z) to the top of the atmosphere (z_{TOA})

$$\tau_{\text{abs}} = \int_z^{z_{\text{TOA}}} \alpha_{\text{abs}} dz = \int_z^{z_{\text{TOA}}} N C_{\text{abs}} dz \tag{5}$$

The absorption coefficient (α_{abs}), as defined in Eqs. 3 and 5, is reported in units of inverse length (m^{-1}), whereas ome studies use units of Mm^{-1} or cm^{-1}, representing part-per-million and percent losses, respectively. In the solution phase, Eq. 5 is commonly written as the absorbance (A), instead of optical depth, using the base 10 logarithm, molar absorptivity (ε) and molar concentration (c)

$$A = \varepsilon c l \tag{6}$$

For aerosol measurements, α_{abs} in Eq. 5 is often replaced with the extinction coefficient (α_{ext}) as the total optical depth (τ) contains contributions from both scattering (α_{scat}) and absorption (α_{abs}) such that

$$\alpha_{\text{ext}} = \alpha_{\text{scat}} + \alpha_{\text{abs}} \tag{7}$$

and

$$\tau_{\text{ext}} = \int_z^{z_{\text{TOA}}} \alpha_{\text{ext}} dz = \int_z^{z_{\text{TOA}}} \alpha_{\text{scat}} dz + \int_z^{z_{\text{TOA}}} \alpha_{\text{abs}} dz \tag{8}$$

since scattering increases with particle diameter (D) to the sixth power, α_{scat} approaches 0 for decreasing particle diameters and $\alpha_{\text{ext}} = \alpha_{\text{abs}}$ in the molecular and small particle limit (*16*).

The size parameter (x) is often used to relate a characteristic dimension of particle size to the wavelength of light:

$$x = \frac{\pi D}{\lambda} \tag{9}$$

For spherical particles, the physical diameter (D) is the characteristic dimension, while monomer diameter (D_{mon}) can be used for aggregated flame generated particles, see Figure 1. For non-spherical particles, the surface area or volume can be used instead. Particle optics are scale invariant, so particles with the same refractive index and x will behave similarly; e.g. nanometer-sized particles will interact with light in the visible (λ = 400 nm to 750 nm) similar to millimeter-sized particles and millimeter wavelength light (*17*, *18*).

The absorption efficiency (Q_{abs}) represents the probability that a photon incident on the physical cross-section of a particle will be absorbed and is expressed as the ratio of the absorption cross-section (C_{abs}) to the physical cross-section (C_{phys}):

$$Q_{abs} = \frac{C_{abs}}{C_{phys}} = \frac{4C_{abs}}{\pi D^2} \tag{10}$$

Similar quantities can be calculated for scattering (Q_{scat}) and extinction (Q_{ext}). Figure 3 shows these values as a function of size parameter for a brown carbon sphere ("tar ball") with a refractive index of $n = 1.84 + 0.21i$ (*19*). As seen in the plot, three regions of these values are evident: 1) a linearly increasing region at small x for all quantities, 2) an enhanced/peaking region at intermediate x and 3) a nearly constant region at large x. These three regions roughly correspond to the 1) Rayleigh regime, 2) Mie regime and 3) geometric regime where particle circumferences are 1) significantly smaller than the wavelength of light (x << 1), 2) approximately equal to the wavelength of light ($\sim 1 \leq x \leq \sim 10$, bound by grey lines in Figure 3) and 3) significantly larger than the wavelength of light (x >> 10), respectively. The Q values peak in the Mie regime from internal resonances enhancing scattering and absorption where the particle circumference and wavelength of light are comparable.

Spectroscopic efficiencies can be calculated explicitly from Maxwell's equations and are converted into cross-sections through Eq. 10. For spherical particles, Mie theory represents an analytical solution to Maxwell's equations assuming homogeneous spherical particles. For non-spherical particles, more complex algorithms such as the discrete-dipole approximation (DDA) (*20*) or the multiple sphere T-matrix algorithm (*21*) must be used. In the limit of aggregated particles composed of N_{mon} spherical monomers, the Rayleigh-Debye-Gans theory (RDG) has been used and assumes that multiple scattering by, and interactions between, monomers are negligible. First, the absorption cross-section of an individual monomer is calculated using Mie theory, and the aggregate cross-section is computed by multiplying the individual cross-section by N_{mon} (*7*, *22–24*). Although RDG has seen significant use in the literature, the absolute accuracy remains uncertain (*25–28*).

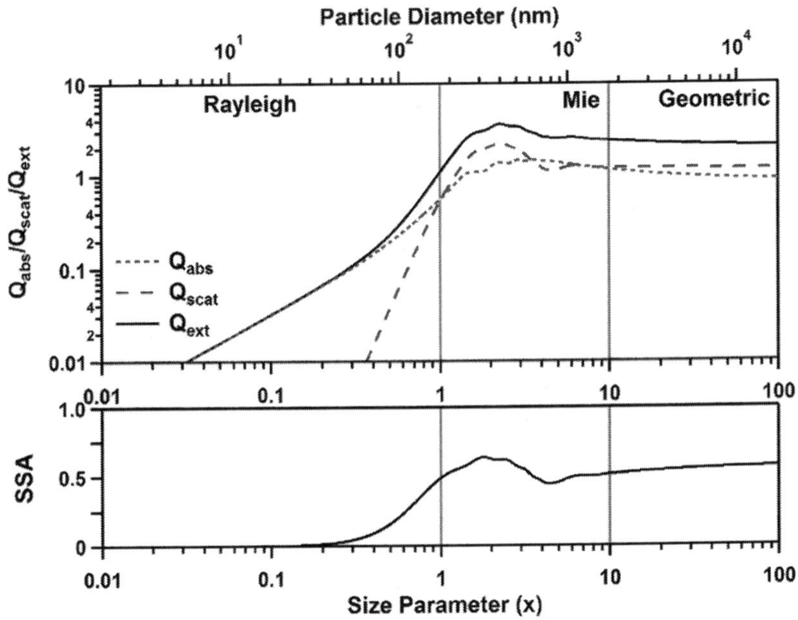

Figure 3. Plot of absorption, scattering and extinction efficiencies (Q_{abs}, Q_{scat} and Q_{ext}, respectively, top) and single-scattering albedo (SSA, bottom) as a function of size parameter (x, bottom axis) calculated using Mie theory assuming spherical particles with $n = 1.84 + 0.21i$ at $\lambda = 550$ nm (19). For reference, the corresponding particle diameters are shown on the top axis. Grey lines shown at $x = 1$ and 10 approximate the bounds of the Mie regime.

Aerosol spectroscopic measurements are often normalized to the average particle mass (m_p) or the mass concentration (M, μg m^{-3}) of aerosol particles in a distribution. For absorption (*MAC*)

$$MAC = \frac{C_{abs}}{m_p} = \frac{\alpha_{abs}}{Nm_p} = \frac{\alpha_{abs}}{M} \tag{11}$$

Mass-normalized scattering and extinction (*MSC* and *MEC*) are calculated similarly but using the corresponding scattering and extinction values, respectively. The *MAC*, *MSC* and *MEC* were calculated using Mie theory and are shown in Figure 4 as a function of size parameter (x) for spherical particles with a refractive index of $n = 1.95 + 0.79i$ at $\lambda = 550$ nm and a mass density (ρ) of 1.8 g cm^{-3}, typical values for BC (*29–31*). Grey lines at $x = 1$ and 10 approximate the bounds of the Mie regime and are shown only for reference.

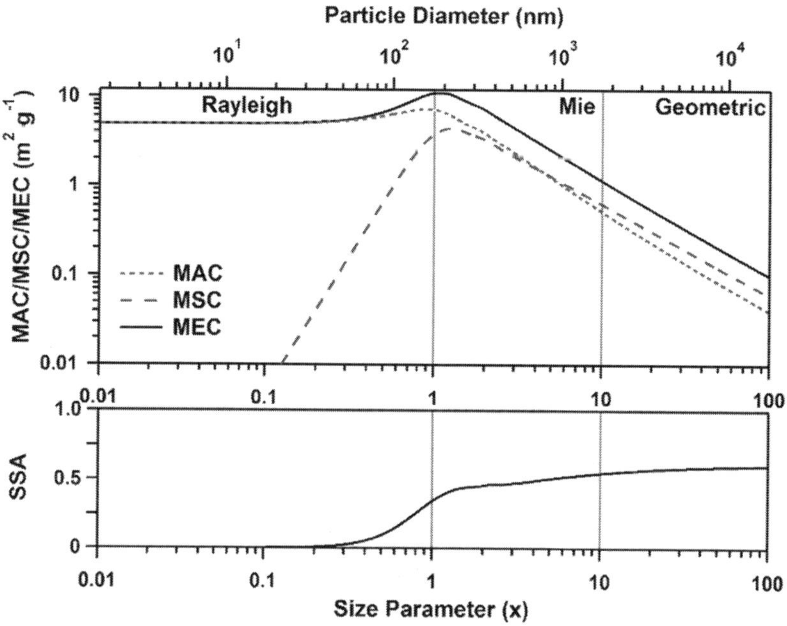

Figure 4. Plot MAC, MSC and MEC (top) and single scattering albedo (SSA, bottom) as a function of size parameter (x, bottom axis) calculated using Mie theory assuming spherical particles with a refractive index of (n = 1.95 + 0.79i) (7) and a mass density of 1.8 g cm⁻³ (i.e. BC) (29–31). See discussion in text for definitions of MAC, MSC and MEC. Grey lines shown at x = 1 and 10 approximate the bounds of the Mie regime.

Importantly, all 3 forms of MAC in Eq. 11 have the same units ($m^2\ g^{-1}$), although they are not equivalent. The first (C_{abs}/m_p) and second forms (α_{abs}/Nm_p) are equivalent (see Eq. 4), and represent the mass-normalized absorption cross-section (a.k.a. mass absorption cross-section) while the third form (α_{abs}/M) represents the mass absorption coefficient. In terms of more familiar solution-phase spectroscopy,

$$A = MAC * M * l \tag{12}$$

In the dilute limit, absorbance scales linearly with M and MAC represents the mass-normalized absorption cross-section as this value is independent of the amount of absorbing material present. Upon saturation where A becomes concentration dependent, MAC now represents the mass absorption coefficient. Translated into aerosol spectroscopy, where absorption cross-sections are size-dependent, the mass-normalized absorption cross-section represents the window where the absorption cross-section is a direct function of particle mass (volume) – e.g. Rayleigh regime – which corresponds to particles < ≈10 nm in diameter in Figure 4. The mass-normalized absorption cross-section can also apply to a single cross-section and mass combination at larger x, or D. For example, from Figure 4, the mass-normalized absorption cross-section of a 300 nm diameter black carbon particle with a refractive index of $n = 1.95 + 0.79i$ and $\rho =$ of 1.8 g cm^{-3} ($m_p = 25.6$ fg), is 4.53 m^2 g^{-1}. Varying any of these parameters would concomitantly alter the mass-normalized absorption cross-section.

The mass absorption coefficient, third form of Eq. 11 ($MAC = \alpha_{abs}/M$), is the absorption coefficient normalized to the mass concentration of aerosol particles ($M = \int m_p dN$). Unlike the mass-normalized absorption cross-section, the mass absorption coefficient is dependent upon the distribution of particles measured; i.e. it accounts for size effects. Placing this framework of mass-normalized absorption cross-sections and mass absorption coefficients relative to actual measurements of aerosol particles, it stands to reason that a distribution of particles is always measured. Even particles size-selected by a differential mobility analyzer with an output geometric standard deviation (σ_g) ≤ 1.05 still have a size distribution. Thus, for practical purposes *MAC* is defined as the mass absorption coefficient when discussing optical properties. In instances referring to the mass-normalized absorption cross-section, it should be used explicitly enumerated to avoid confusion.

In the small particle limit ($x << 1$), the mass-normalized absorption cross-section and mass absorption coefficient can be equal. From Figure 4, for $n = 1.95 + 0.79i$ at $\lambda = 550$ nm and $\rho = 1.8$ g cm^{-3} the mass-normalized absorption cross-section is 4.85 m^2 g^{-1}. At particle diameters of ≈ 7 nm this value increases to 4.86 m^2 g^{-1}, indicating that the particles are no longer in the Rayleigh limit and the mass absorption coefficient is the more accurate term.

Figures 3 and 4, show the size dependence of Q_{abs} and *MAC* at a single wavelength, respectively. Organic materials formed in biomass burning may have a wavelength dependent refractive index, impacting the measured absorption cross-section as a function of *D*. Figure 5a, shows the mass absorption coefficients (*MAC*) for humic acid (a brown carbonaceous material optically similar to those found in the atmosphere) (*32*) for a dilute aqueous solution measured using a UV-visible spectrophotometer (green line) and as a function of particle size for 250 nm (black circles), 500 nm (red squares) and 750 nm (blue triangles) diameter particles measured using photoacoustic spectroscopy. The lines represent the best fit to the measured data using a power law (e.g. absorption Ångström exponent, see discussion below). These data show that both the magnitude and shape of the absorption spectra are size dependent. By taking the ratio of the particle *MAC* values to that of the solution (Figure 5b), it is observed that the ratios are also size dependent. Similar to Figure 3, at the shorter wavelengths (e.g. $\lambda = 405$ nm), the *MAC* initially increases in going from the molecular limit (e.g. solution) to 250 nm and decreases at ≥ 250 nm. For the longer wavelengths, the maximum ratio of *MAC* values relative to the solution occurs with larger particle sizes with the 500 nm particles having the maximum ratio for 500 nm $\leq \lambda \leq 575$ nm and the 750 nm particles having the maximum ratio at $\lambda > 575$ nm. These spectral observations can be rationalized by the size parameter (x) at each wavelength. For a given refractive index, particles in the Mie regime will possess higher *MAC* values due to absorption enhancement from internal resonances. For strongly absorbing particles, maximum *MAC* values will be achieved with $x \approx 1$; see Figure 4. For weakly absorbing particles, like those shown in Figure 3 and 5, the maximum *MAC* occurs at larger values of x.

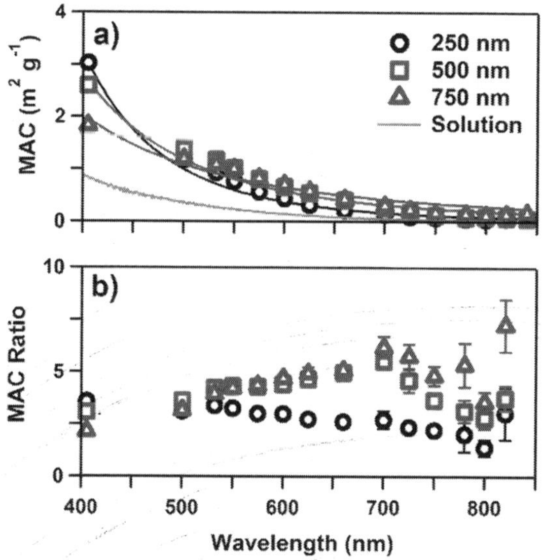

Figure 5. a) Plot of the mass absorption coefficient (MAC) for humic acid in aqueous solution (green line) and for particles with diameters of 250 nm (black circles), 500 nm (red squares) and 750 nm (blue triangles). b) Ratio of particle MAC values to those of the solution absorption spectrum.

Following the discussion of *MAC*, the *MSC* and *MEC* are defined as the mass scattering coefficient and mass extinction coefficient, respectively. From Figure 4, *MSC* does not approach a constant, mass-independent value (scattering scales as D^6 in the Rayleigh limit), thereby precluding it from being defined as a mass-normalized scattering cross-section. *MEC* does, however, approach a constant value in the Rayleigh limit since $MEC \approx MAC$ allowing for the use of a mass-normalized extinction cross-section. For simplicity, *MEC* is defined as the mass extinction coefficient for the same reasons the *MAC* is defined as the mass absorption coefficient.

Note that in some publications the term mass absorption efficiency (E_{abs} or *MAE* in units of m^2 g^{-1}) *(33–36)* is used to describe the mass-normalized absorption cross-section or the mass absorption coefficient. Scattering and extinction efficiencies have been defined similarly. Regardless of the proper use of E_{abs}, it should be noted that efficiencies are unitless (see absorption efficiency, Eq. 10), and may lead to confusion.

Single-Scattering Albedo (*SSA*)

Published studies often report the single-scattering albedo (*SSA* or ϖ_0) which represents the ratio of scattering to extinction

$$SSA = \frac{MSC}{MEC} = 1 - \frac{MAC}{MEC} \qquad (13)$$

The *SSA* is only applicable in the optically thin limit where it can be assumed that photons interact with particles a single time. Eq. 13 defines *SSA* based upon the mass-dependent coefficients, but any ratio of scattering to extinction can be used (e.g. $\alpha_{scat}/\alpha_{ext}$, C_{scat}/C_{ext}, etc.). Similarly, the co-albedo is the ratio of absorption to extinction or 1 – *SSA*. The *SSA* represents the dominant intensive parameter for determining aerosol direct radiative forcing (*37*). For non-absorbing particles where the imaginary component of the refractive index is zero ($k = 0i$), the *SSA* = 1 and is independent of particle size. When $k > 0i$ the *SSA* is dependent upon the complex refractive index, particle size (see lower panels of Figures 3 and 4) and wavelength. In the Rayleigh limit, *SSA* tends towards 0 with decreasing particle size (independent of the magnitude of n), but at different rates (scattering and absorption are proportional to D^6 and D^3, respectively) (*15*, *38*). Large particle approximations have been derived as a function of the imaginary component of the refractive index and size parameters (kx) (*37*), but may have limited applicability to particles suspended in the atmosphere with more modest values of x. For particles between these two limits, measurement of the actual value (or calculation if the refractive index is known) likely remains the best method for quantifying *SSA* rather than using assumed values.

Wavelength Dependence and Ångström Exponents

In 1929, Anders Ångström published his seminal work "On the Atmospheric Transmission of Sun Radiation and on Dust in the Air" (*39*) linking changes in the solar transmission of Earth's atmosphere to the presence of suspended particulates. In its original form, the extinction Ångström exponent (*EAE*) was defined as

$$\frac{\tau_\lambda}{\tau_{\lambda_0}} = \left(\frac{\lambda}{\lambda_0}\right)^{-EAE} \tag{14}$$

where τ_λ is the aerosol optical depth at wavelength λ and $\tau_{\lambda 0}$ is the optical depth at a reference wavelength λ_0. This original derivation of *EAE* was intended to be applied to particles dominated by scattering where the absorption component was negligible. From this work, Ångström concluded that the exponent has "a very marked dependence on the size of the scattering particles, but not, in the case of non-absorbing materials such as MgO, ZnO, SbO on the substance itself" (*39*). This relationship is confirmed by the well-known λ^{-4} scattering dependence for particles in the Rayleigh regime (assuming wavelength independent refractive indices) (*15*, *38*, *40*, *41*). Note, increasing particle sizes result in lower scattering Ångström exponent (*SAE*) values. Continuing upon the idea of wavelength independent refractive indices, the absorption of light by particles in the Rayleigh limit follows a λ^{-1} trend yielding an absorption Ångström exponent (*AAE*) of 1 (*16*). While the historical definition of Ångström exponents (scattering and absorption) was based upon the optical depth, in practice any optical parameter can be used (e.g. for absorption α_{abs}, C_{abs}, *MAC*, etc.) where for *MAC*

$$\frac{MAC_\lambda}{MAC_{\lambda_0}} = \left(\frac{\lambda}{\lambda_0}\right)^{-AAE} \tag{15}$$

Implicit in this mathematical representation of *AAE* is that *MAC* is a continuous function of λ with enough spectral coverage allowing the *AAE* to be calculated from a fit of multiple values. In practice, many measurements of aerosol optical properties are performed at only a few discrete wavelengths spanning the UV to near-IR. In these instances, it is usually preferable to use an explicit form of the *AAE* where for *MAC*

$$AAE = -\frac{\ln(MAC_{\lambda 1}) - \ln(MAC_{\lambda 2})}{\ln(\lambda_1) - \ln(\lambda_2)} \quad (16)$$

with the explicit forms of the Ångström exponents for extinction and scattering being defined similarly. These discrete calculations allow for approximation of the spectral curvature between wavelengths without any *a priori* knowledge on the absolute shape of the spectral dependence.

Another fundamental assumption of the Ångström exponent is that the spectral dependence of absorption/scattering can be well-defined by a single power law expression (*41*, *42*). To test this, Figure 6 shows the simulated *MAC* and *AAE* of spherical particles using Mie theory as a function of the imaginary component of the refractive index (D = 1 nm, m = 1.77 and ρ = 1.65 g cm^{-3}) and as a function of particle diameters (D) at constant refractive index (n = 1.77 + 0.8i and ρ = 1.65 g cm^{-3}), corresponding to carbon black particles measured in You, et al. (2017) (*43*) while D = 1 nm and ρ = 1.65 g cm^{-3} corresponds to a spherical particle that mimics a single fullerene (C_{60}) molecule (*44*). For the 1 nm diameter particle (Figure 6a), *MAC* increases monotonically with k while the spectral shape is invariant; *AAE* = 1.0 (*16*, *38*). Increasing particle size at a constant refractive index (Figure 6b) yields significantly different results. At λ = 400 nm the *MAC* increases by 1 percent for particle diameters between 1 nm and 10 nm and 13 percent between 10 nm and 50 nm. For the refractive index at this wavelength, similar to BC (*7*, *43*), particle diameters ≥ 10 nm are outside of the Rayleigh regime, and *MAC* varies for larger particle diameters. The results also show the *AAE* increases from 1.0 to 1.2 for particle diameters increasing from 1 nm to 100 nm. For $D \geq$ 100 nm the spectral dependence may not be adequately captured by the *AAE* due to the transition of particle absorption from the volumetric (mass) regime to the surface absorption regime.

Figure 6 also shows the size dependence of *MAC* and *AAE* on the imaginary component of the refractive index or particle size. It has been reported and referenced that BC *MAC* is 7.5 m^2 g^{-1} with a wavelength independent refractive index leading to an *AAE* = 1.0 (*7*, *8*). As shown above, particles in the Rayleigh regime with a wavelength independent refractive index will have *AAE* = 1.0. Further, within a narrow combination of refractive index and density space, *MAC* values of 7.5 m^2 g^{-1} are reasonable. However, these values should only be applied to particles in the Rayleigh regime. As seen from Figure 6b, particles with a refractive index of 1.77 + 0.8i (comparable to that of BC) (*7*, *8*) transition out of the Rayleigh regime at particle diameters (i.e. BC aggregate monomer diameters, D_{mon}) of $D \approx$ 10 nm. Once out of the Rayleigh regime, *MAC* and *AAE* are no longer constant but a function of monomer size.

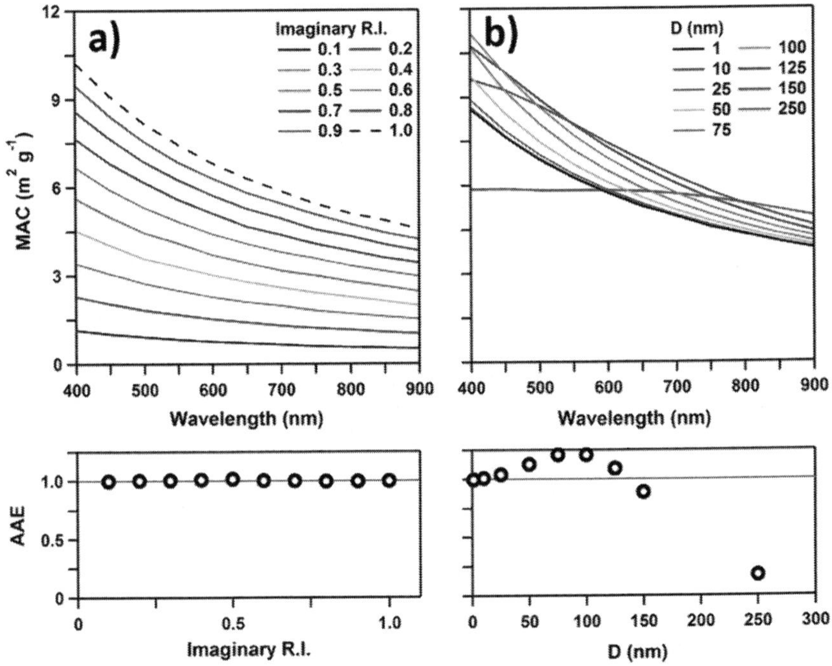

Figure 6. a) MAC and AAE as a function of imaginary refractive index for particles with a 1 nm diameter (D) and real refractive index component = 1.77. b) MAC and AAE as a function of D at constant refractive index (1.77 + 0.8i).

Laboratory-Based Absorption Measurements of Carbonaceous Aerosol

Laboratory-based spectroscopic measurements using materials of known size (electrical mobility), mass and/or chemical and morphological composition serves as a foundation for the calculation of aerosol positive radiative forcing. Using the parameters that impact aerosol absorption detailed above gives context to the factors that influence measured absorption data.

In this section carbonaceous aerosol absorption is presented using well-characterized aerosol of two types: highly absorbing black carbon-like particles formed during flaming, high-temperature combustion and weakly absorbing brown carbon organic aerosols formed during smoldering low-temperature combustion of biomass. As described above, the size range of the particles and their morphologies may affect the measured aerosol absorption. For BC-like aerosol, the particles studied were composed of aggregated monomers with diameters (D_{mon}) on the order of 10 nm to 40 nm. For BrC aerosol, individual spherical particles with diameters more than an order of magnitude larger than the BC monomers were produced and studied, impacting measured cross-sections. The aerosol was selected by both electrical mobility diameter (D_m) and mass (m_p) prior to absorption measurements (*45*). Aerosol absorption was measured as a function of wavelength (8 points) to construct absorption spectra between $\lambda = 500$ nm and 840 nm. This enabled the determination of *MAC* and *AAE* using Eqs. 11 and 15, respectively.

Figure 7 shows the absorption spectra of three types of aerosol generated *in situ* from flames (blue diamonds) and three commercially available materials: two BC-mimics (black circles) and one carbon allotrope (crumpled graphene sheets, red squares). The corresponding TEM images are shown to the right of the *MAC* plots for each material with data tabulated in Table 1.

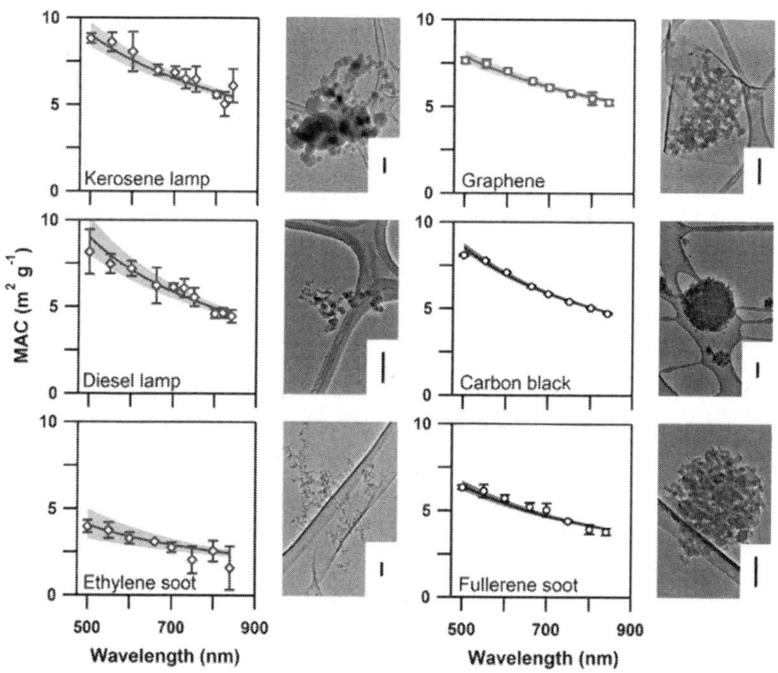

Figure 7. Plot of mass absorption coefficients (MAC) for the measured particles; corresponding D_m and ρ_{eff} are shown in in Table 1. All plots have the same abscissa and ordinate axis ranges. Blue diamonds are flame-generated particles in the laboratory, red squares are carbon allotropes and black circles are commercially available materials. Solid line is AAE determined using Eq. 15. Shaded areas represent fit uncertainties. Corresponding TEM images are shown to the right of each plot. Scale bars are 200 nm.

Table 1. Measured electrical mobility diameter (D_m), effective density (ρ_{eff}), mean monomer diameter (D_{mon}), MAC (λ = 550 nm), and AAE for all particle types. Uncertainties shown in p arenthesis are 2σ. Green (red) text indicates species does (does not) meet definition for BC from Bond and Bergstrom (2006) (*7*).

Species	D_m (nm)	ρ_{eff}[a] (g cm^{-3})	D_{mon} (nm)	*MAC* (λ = 550 nm)	*AAE*
Ethylene soot	250	0.18	17 (3)[b]	3.8 (0.5)[c]	1.0 (0.3)
Kerosene lamp	700	0.12	42 (12)[b]	8.6 (0.6)[b]	1.0 (0.1)
Diesel lamp	700	0.12	26 (8)[b]	7.5 (0.6)[b]	1.3 (0.2)
Carbon black	250	0.77	27 (7)[b]	7.78 (0.06)[b]	1.11 (0.03)
Graphene	450	0.51	--[c]	7.5 (0.1)[b]	0.74 (0.05)
Fullerene soot	350	0.74	--[c]	6.1 (0.4)[b]	0.92 (0.05)

[a] $\rho_{eff} = 6m_p/D_m^3$. [b] Meets BC definition, see Ref. (*7*). [c] Does not meet BC definition, see Ref. (*7*).

From Figure 7, it is evident that the absorption spectrum of highly absorbing carbonaceous aerosol is variable. As stated above, highly absorbing aerosol from flames are often referred to as BC, with a given *MAC* (7.5 m^2 g^{-1} at λ = 550 nm). The definition of BC includes four measurable properties (*8*): 1) composed of aggregates of small carbon spherules, 2) $MAC \geq 5$ m^2 g^{-1} at λ = 550 nm, 3) refractory with a vaporization temperature near 4000 K, and 4) insoluble in water and common organic solvents. Only three of the six samples shown in Figure 7 meet both the morphological and spectroscopic definitions of BC (particles generated from kerosene and diesel simple wick lamps and carbon black), see green text in Table 1, although the water solubility of carbon black negates its inclusion. Perhaps most surprising, particles generated from an ethylene-fueled flame did not meet the spectroscopic definition for BC (*46*). Conversely, crumpled graphene nanosheets and fullerene soot met the spectroscopic definition, but did not meet the morphological definition of BC.

As shown in Figure 2, in addition to being dependent on material refractive index and D_{mon}, *MAC* may also be a function of particle morphology (*13*). It can be envisioned that particles with a lacey, open morphology – e.g. freshly-emitted flame-generated particles – with small D_{mon} enable light to interrogate the entire particle resulting in absorption that scales linearly with particle mass (volume). In contrast, particles with a compacted morphology may only be in the volume absorption regime (constant *MAC* and *AAE)* at very small D_{mon} and transition to the Mie or geometric absorption regimes with either increasing D_{mon} or mobility diameter (D_m); for these spherical particles $D_m \approx$ particle diameter (D). Figure 8 tests this hypothesis by measuring the change (Δ) in *MAC* and *AAE* as a function of D_m between $\lambda = 532$ nm and 780 nm of multilayered crumpled graphene-like sheets (formed from thermally reduced graphene oxide, rGO) (*47*), compacted flame-generated particles (carbon black) and freshly-formed, lacey particles generated from an ethylene flame with $D_{mon} \approx 20$ nm. The *MAC* of the crumpled sheets and compacted spherical particles decreases nearly monotonically with D_m, consistent with geometric absorption even for the smallest particles measured ($D_m = 150$ nm). A similar dependence was measured for the change in *AAE* with D_m for crumpled and compacted particles, see Figure 8b. For particles generated from an ethylene flame, the *MAC* and *AAE* are constant within measurement uncertainty for particles up to $D_m = 550$ nm and typical of particles in the volumetric absorption regime. Notably, based upon differences in D_{mon} between carbon black and ethylene soot at this wavelength ($\lambda = 532$ nm), the *MAC* should vary by ≈ 1.5 % assuming a constant refractive index (see Figure 6b). This is significantly less than the ≈ 40 % reduction in *MAC* observed for carbon black observed for D_m between 150 nm and 500 nm. These data confirm that particle morphology may impact the spectral properties of highly-absorbing carbonaceous particles. For flame-generated particles where D_{mon} is small and the morphology is consistent with an open, lacey structure (fractal dimension, $D_f \approx 1.8$) it is reasonable to assume that the particle absorption scales directly with particle mass and *MAC* and *AAE* are invariant. The data also illustrate that this assumption is invalid for particles that fall outside of this limited parameter space.

Low temperature combustion (e.g. smoldering) produces primarily viscous liquid particles formed from the condensation of gaseous organic compounds with a significantly lower, wavelength-dependent, imaginary component of the refractive index when compared to BC (*48–50*). This results in a strongly wavelength dependent absorption spectrum with weak or negligible absorption in the near-IR that increases towards the UV (*4*). The wavelength dependence of BrC may introduce large uncertainty into radiative forcing calculations (*50*) with predictions of BrC's climate impact ranging from net-neutral[1] to net-positive (*51, 52*).

Figure 9 shows the *MAC* spectra of BrC aerosol generated from smoldering smoke generated by six North American wood species. Shaded regions represent the 16th and 84th percentiles (i.e. 1σ). Solid and dashed lines represent *AAE* fits to the median and 16th and 84th percentiles, respectively, using Eq. 15. Values are shown in Table 2 with percentiles in parenthesis. The measured *MAC* at $\lambda = 500$ nm is up to 200 times lower for BrC compared to aerosol generated from flames. All data was well described by a single *AAE* with the exception of Baldcypress, potentially indicating the presence of more complex wavelength dependencies (*41, 42*). Point-by-point comparison of the individual absorption spectra in Figure 9 reveals that Oak, Western redcedar and Blue spruce are statistically similar ($p > 0.05$) while Hickory, Mesquite and Baldcypress were each distinct ($p < 0.01$) suggesting that, for smoldering biomass, *MAC* and spectral shape can be species dependent (*53, 54*).

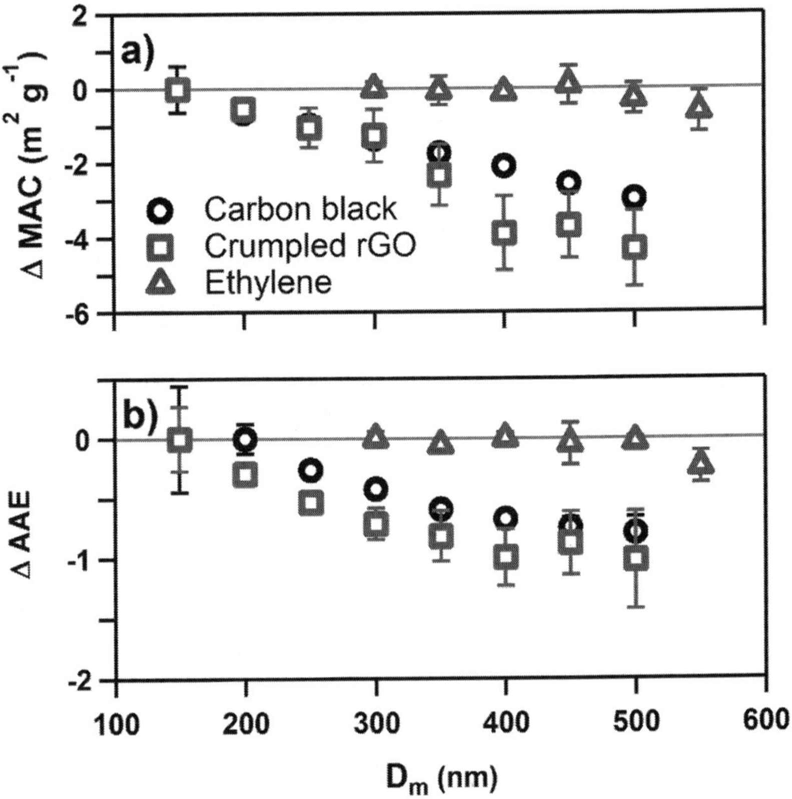

Figure 8. a) Change (Δ) in the mass absorption coefficient (MAC) at λ = 532 nm and b) change (Δ) in absorption Ångström exponent (AAE) between λ = 532 nm and 780 nm as function of particle mobility diameter (D_m) for compact carbon black (black circles), crumpled rGO (red squares), and fresh, lacey particles from ethylene flame (blue triangles). Error bars are 1 standard deviation of a minimum of 3 replicate measurements.

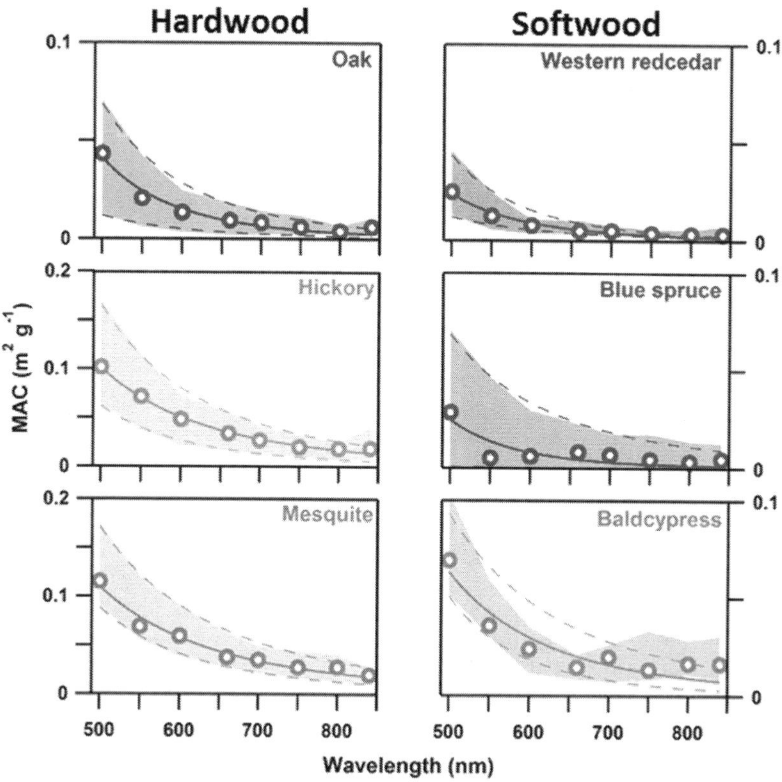

Figure 9. Measured mass absorption coefficients (MAC) as a function of wavelength for smoldering wood particles from six species of wood. Open circles and shaded areas represent the median MAC values and the 16th and 84th percentiles (i.e. 1σ) at each wavelength, respectively; solid and dashed lines represent AAE fits to the median and 16th and 84th percentile values, respectively. Adapted with permission from reference (55). Copyright 2017, ACS Publications.

Table 2. Measured effective density (ρ_{eff}), median *MAC* and *AAE* for D_m = 750 nm smoldering wood particles. Table adapted from Radney, et al. (2017) (*55*). Adapted with permission from reference (*55*). Copyright 2017, ACS Publications.

Species	ρ_{eff} *(g cm*$^{-3}$*)*	*Median MAC (x* 10^{-2} m^2 g^{-1}*)*[a]	*AAE*[a]
Oak[b]	1.41 (0.04)	2.5 (0.76/4.3)	5.5 (4.8/4.7)
Hickory[b]	1.44 (0.05)	7.0 (4.0/11)	3.8 (4.0/4.7)
Mesquite[b]	1.51 (0.03)	7.9 (6.0/12)	3.5 (3.5/4.2)
Western redcedar[c]	1.36 (0.03)	1.4 (0.81/2.6)	5.2 (5.6/4.3)
Blue spruce[c]	1.35 (0.02)	1.4 (0.039/4.7)	6.2 (3.9/3.4)
Baldcypress[c]	1.35 (0.01)	4.3 (3.0/6.6)	4.1 (3.6/5.5)

[a] λ = 550 nm value calculated from *AAE* fit (Eq. 2) of average values spanning λ = 500 nm to 840 nm. Values in parenthesis are the values from fits of the 16th and 84th percentile data. [b] Hardwood. [c] Softwood.

Conclusions

This chapter focused on the introduction of absorption spectroscopy for carbonaceous aerosol in the atmosphere and the relevant parameters that affect aerosol optical properties. Using these parameters gives better insight into reported aerosol absorption cross-sections. As shown, the material refractive index, particle size (D_{mon} for aggregates or D for individual particles) and morphology (lacey versus compact) may impact both absorption strength and spectral shape. These same parameters can also impact extinction, scattering and derived values such as *SSA*. With this in mind, it is important to highlight that care must be given when reporting aerosol optical properties and that additional data regarding size, mass, and morphology should also be included to enable quantitative comparisons between studies as many of these quantities are only applicable to the distribution(s) for which they were measured or calculated.

Definition of Variables, Terms and Units Used

A = absorbance
AAE = absorption Ångström exponent
BC = black carbon
BrC = brown carbon
C_{abs} = absorption cross-section (m^2)
C_{phys} = physical cross-section (m^2)
D = particle diameter (nm)
D_f = fractal dimension
D_m = particle mobility diameter (nm)
D_{mon} = diameter of monomers (nm)
l = propagation distance (m)

k = imaginary component of refractive index
m = real component of refractive index
m_p= average particle mass (g)
M = mass concentration (μg m^{-3})
MAC = mass absorption coefficient (m^2 g^{-1})
MEC = mass extinction coefficient (m^2 g^{-1})
MSC = mass scattering coefficient (m^2 g^{-1})
n = complex refractive index
N_{mon} = number of monomers in an aggregate
Q_{abs} = absorption efficiency
Q_{ext} = extinction efficiency
Q_{scat} = scattering efficiency
SSA = single scatter albedo, ϖ_0
x = size parameter
z = altitude (m)
α_{Abs} = absorption coefficient (m^{-1})
λ = wavelength (nm)
ρ = density (g cm^{-3})
ρ_{eff} = effective density (= $6m_p/D_m^3$, g cm^{-3})

References

1. Boucher, O.; Randall, D.; Artaxo, P.; Bretherton, C.; Feingold, G.; Forster, P.; Kerminen, V.-M.; Kondo, Y.; Liao, H.; Lohmann, U.; Rash, P.; Satheesh, S. K.; Sherwood, S.; Stevens, B.; Zhang, X.-Y. Clouds and Aerosols. In *Climate Change 2013: The Physical Science Basis. Contribution of Working Group I to the Fifth Assessment Report of the Intergovernmental Panel on Climate Change*; Fuzzi, S., Penner, J., Ramaswamy, V., Stubenrauch, C., Eds.; Cambridge University Press: Cambridge, United Kingdom, and New York, NY, USA, 2013.
2. Burkholder, J. B.; Abbatt, J. P. D.; Barnes, I.; Roberts, J. M.; Melamed, M. L.; Ammann, M.; Bertram, A. K.; Cappa, C. D.; Carlton, A. G.; Carpenter, L. J.; Crowley, J. N.; Dubowski, Y.; George, C.; Heard, D. E.; Herrmann, H.; Keutsch, F. N.; Kroll, J. H.; McNeill, V. F.; Ng, N. L.; Nizkorodov, S. A.; Orlando, J. J.; Percival, C. J.; Picquet-Varrault, B.; Rudich, Y.; Seakins, P. W.; Surratt, J. D.; Tanimoto, H.; Thornton, J. A.; Tong, Z.; Tyndall, G. S.; Wahner, A.; Weschler, C. J.; Wilson, K. R.; Ziemann, P. J. The Essential Role for Laboratory Studies in Atmospheric Chemistry. *Environ. Sci. Technol.* **2017**, *51*, 2519–2528.
3. Seinfeld, J. H.; Bretherton, C.; Carslaw, K. S.; Coe, H.; DeMott, P. J.; Dunlea, E. J.; Feingold, G.; Ghan, S.; Guenther, A. B.; Kahn, R.; Kraucunas, I.; Kreidenweis, S. M.; Molina, M. J.; Nenes, A.; Penner, J. E.; Prather, K. A.; Ramanathan, V.; Ramaswamy, V.; Rasch, P. J.; Ravishankara, A. R.; Rosenfeld, D.; Stephens, G.; Wood, R. Improving our fundamental understanding of the role of aerosol–cloud interactions in the climate system. *Proc. Natl. Acad. Sci. U.S.A.* **2016**, *113*, 5781–5790.

4. Andreae, M. O.; Gelencser, A. Black carbon or brown carbon? The nature of light-absorbing carbonaceous aerosols. *Atmos. Chem. Phys.* **2006**, *6*, 3131–3148.
5. Chakrabarty, R. K.; Beres, N. D.; Moosmüller, H.; China, S.; Mazzoleni, C.; Dubey, M. K.; Liu, L.; Mishchenko, M. I. Soot superaggregates from flaming wildfires and their direct radiative forcing. *Sci. Rep.* **2014**, *4*, 5508.
6. Liu, P.; Heinson, W. R.; Chakrabarty, R. K. Fractal scaling of soot packing density across five size decades. *Aerosol Sci. Technol.* **2017**, *51*, 879–886.
7. Bond, T. C.; Bergstrom, R. W. Light absorption by carbonaceous particles: An investigative review. *Aerosol Sci. Technol.* **2006**, *40*, 27–67.
8. Bond, T. C.; Doherty, S. J.; Fahey, D. W.; Forster, P. M.; Berntsen, T.; DeAngelo, B. J.; Flanner, M. G.; Ghan, S.; Kärcher, B.; Koch, D.; Kinne, S.; Kondo, Y.; Quinn, P. K.; Sarofim, M. C.; Schultz, M. G.; Schulz, M.; Venkataraman, C.; Zhang, H.; Zhang, S.; Bellouin, N.; Guttikunda, S. K.; Hopke, P. K.; Jacobson, M. Z.; Kaiser, J. W.; Klimont, Z.; Lohmann, U.; Schwarz, J. P.; Shindell, D.; Storelvmo, T.; Warren, S. G.; Zender, C. S. Bounding the role of black carbon in the climate system: A scientific assessment. *J. Geophys. Res. Atmos.* **2013**, *118*, 5380–5552.
9. García Fernández, C.; Picaud, S.; Devel, M. Calculations of the mass absorption cross sections for carbonaceous nanoparticles modeling soot. *J. Quant. Spectrosc. Radiat. Transfer* **2015**, *164*, 69–81.
10. Langlet, R.; Vanacharla, M. R.; Picaud, S.; Devel, M. Bottom-up multi-step approach to study the relations between the structure and the optical properties of carbon soot nanoparticles. *J. Quant. Spectrosc. Radiat. Transfer* **2009**, *110*, 1615–1627.
11. Moulin, F.; Devel, M.; Picaud, S. Optical properties of soot nanoparticles. *J. Quant. Spectrosc. Radiat. Transfer* **2008**, *109*, 1791–1801.
12. Scarnato, B. V.; Vahidinia, S.; Richard, D. T.; Kirchstetter, T. W. Effects of internal mixing and aggregate morphology on optical properties of black carbon using a discrete dipole approximation model. *Atmos. Chem. Phys.* **2013**, *13*, 5089–5101.
13. Liu, L.; Mishchenko, M. I.; Patrick Arnott, W. A study of radiative properties of fractal soot aggregates using the superposition T-matrix method. *J. Quant. Spectrosc. Radiat. Transfer* **2008**, *109*, 2656–2663.
14. Buseck, P. R.; Adachi, K.; Gelencsér, A.; Tompa, É.; Pósfai, M. ns-Soot: A Material-Based Term for Strongly Light-Absorbing Carbonaceous Particles. *Aerosol Sci. Technol.* **2014**, *48*, 777–788.
15. Bohren, C. F.; Huffman, D. R. *Absorption and scattering of light by small particles*; Wiley: New York, 1983.
16. Moosmüller, H.; Arnott, W. P. Particle optics in the Rayleigh regime. *J. Air Waste Manage. Assoc.* **2009**, *59*, 1028–31.
17. Chýlek, P.; Srivastava, V.; Pinnick, R. G.; Wang, R. T. Scattering of electromagnetic waves by composite spherical particles: experiment and effective medium approximations. *Appl. Opt.* **1988**, *27*, 2396–2404.
18. Mishchenko, M. I. Scale invariance rule in electromagnetic scattering. *J. Quant. Spectrosc. Radiat. Transfer* **2006**, *101*, 411–415.

19. Hoffer, A.; Tóth, A.; Nyirő-Kósa, I.; Pósfai, M.; Gelencsér, A. Light absorption properties of laboratory-generated tar ball particles. *Atmos. Chem. Phys.* **2016**, *16*, 239–246.
20. Draine, B. T.; Flatau, P. J. Discrete-Dipole Approximation For Scattering Calculations. *J. Opt. Soc. Am. A* **1994**, *11*, 1491–1499.
21. Mackowski, D. W.; Mishchenko, M. I. A multiple sphere T-matrix Fortran code for use on parallel computer clusters. *J. Quant. Spectrosc. Radiat. Transfer* **2011**, *112*, 2182–2192.
22. Sorensen, C. M. Light Scattering by Fractal Aggregates: A Review. *Aerosol Sci. Technol.* **2001**, *35*, 648–687.
23. Dobbins, R. A.; Megaridis, C. M. Absorption and scattering of light by polydisperse aggregates. *Appl. Opt.* **1991**, *30*, 4747–4754.
24. Köylü, U. O.; Faeth, G. M. Optical Properties of Overfire Soot in Buoyant Turbulent Diffusion Flames at Long Residence Times. *J. Heat Transfer* **1994**, *116*, 152–159.
25. Iskander, M. F.; Chen, H. Y.; Penner, J. E. Resonance optical absorption by fractal agglomerates of smoke aerosols. *Atmos. Environ. Part A* **1991**, *25*, 2563–2569.
26. Fuller, K. A. Scattering and absorption cross sections of compounded spheres. II. Calculations for external aggregation. *J. Opt. Soc. Am. A* **1995**, *12*, 881–892.
27. Farias, T. L.; Köylü, Ü. Ö.; Carvalho, M. G. Range of validity of the Rayleigh–Debye–Gans theory for optics of fractal aggregates. *Appl. Opt.* **1996**, *35*, 6560–6567.
28. Mackowski, D. W. Electrostatics analysis of radiative absorption by sphere clusters in the Rayleigh limit: application to soot particles. *Appl. Opt.* **1995**, *34*, 3535–3545.
29. Mullins, J.; Williams, A. The optical properties of soot: a comparison between experimental and theoretical values. *Fuel* **1987**, *66*, 277–280.
30. Wu, J. S.; Krishnan, S. S.; Faeth, G. M. Refractive Indices at Visible Wavelengths of Soot Emitted From Buoyant Turbulent Diffusion Flames. *J. Heat Transfer* **1997**, *119*, 230–237.
31. Park, K.; Kittelson, D.; Zachariah, M.; McMurry, P. Measurement of Inherent Material Density of Nanoparticle Agglomerates. *J. Nanopart. Res.* **2004**, *6*, 267–272.
32. Graber, E. R.; Rudich, Y. Atmospheric HULIS: How humic-like are they? A comprehensive and critical review. *Atmos. Chem. Phys.* **2006**, *6*, 729–753.
33. Moosmüller, H.; Arnott, W. P.; Rogers, C. F.; Chow, J. C.; Frazier, C. A.; Sherman, L. E.; Dietrich, D. L. Photoacoustic and filter measurements related to aerosol light absorption during the Northern Front Range Air Quality Study (Colorado 1996/1997). *J. Geophys. Res. Atmos.* **1998**, *103*, 28149–28157.
34. Srinivas, B.; Rastogi, N.; Sarin, M. M.; Singh, A.; Singh, D. Mass absorption efficiency of light absorbing organic aerosols from source region of paddy-residue burning emissions in the Indo-Gangetic Plain. *Atmos. Environ.* **2016**, *125B*, 360–370.
35. Chakrabarty, R. K.; Gyawali, M.; Yatavelli, R. L. N.; Pandey, A.; Watts, A. C.; Knue, J.; Chen, L. W. A.; Pattison, R. R.; Tsibart, A.; Samburova, V.;

Moosmüller, H. Brown carbon aerosols from burning of boreal peatlands: microphysical properties, emission factors, and implications for direct radiative forcing. *Atmos. Chem. Phys.* **2016**, *16*, 3033–3040.

36. Caponi, L.; Formenti, P.; Massabó, D.; Di Biagio, C.; Cazaunau, M.; Pangui, E.; Chevaillier, S.; Landrot, G.; Andreae, M. O.; Kandler, K.; Piketh, S.; Saeed, T.; Seibert, D.; Williams, E.; Balkanski, Y.; Prati, P.; Doussin, J. F. Spectral- and size-resolved mass absorption efficiency of mineral dust aerosols in the shortwave spectrum: a simulation chamber study. *Atmos. Chem. Phys.* **2017**, *17*, 7175–7191.
37. Moosmüller, H.; Sorensen, C. M. Small and large particle limits of single scattering albedo for homogeneous, spherical particles. *J. Quant. Spectrosc. Radiat. Transfer* **2018**, *204*, 250–255.
38. Moosmüller, H.; Chakrabarty, R. K.; Arnott, W. P. Aerosol light absorption and its measurement: A review. *J. Quant. Spectrosc. Radiat. Transfer* **2009**, *110*, 844–878.
39. Ångström, A. On the Atmospheric Transmission of Sun Radiation and on Dust in the Air. *Geografiska Annaler* **1929**, *11*, 156–166.
40. Moosmüller, H.; Chakrabarty, R. K. Technical Note: Simple analytical relationships between Ångström coefficients of aerosol extinction, scattering, absorption, and single scattering albedo. *Atmos. Chem. Phys.* **2011**, *11*, 10677–10680.
41. Moosmüller, H.; Chakrabarty, R. K.; Ehlers, K. M.; Arnott, W. P. Absorption Ångström coefficient, brown carbon, and aerosols: basic concepts, bulk matter, and spherical particles. *Atmos. Chem. Phys.* **2011**, *11*, 1217–1225.
42. Utry, N.; Ajtai, T.; Filep, Á.; Dániel Pintér, M.; Hoffer, A.; Bozoki, Z.; Szabó, G. Mass specific optical absorption coefficient of HULIS aerosol measured by a four-wavelength photoacoustic spectrometer at NIR, VIS and UV wavelengths. *Atmos. Environ.* **2013**, *69*, 321–324.
43. You, R.; Radney, J. G.; Zachariah, M. R.; Zangmeister, C. D. Measured Wavelength-Dependent Absorption Enhancement of Internally Mixed Black Carbon with Absorbing and Nonabsorbing Materials. *Environ. Sci. Technol.* **2016**, *50*, 7982–7990.
44. Zangmeister, C. D.; Radney, J. G.; You, R.; Lunny, E. M.; Jacobson, A.; Okumura, M.; Zachariah, M. R. Measured In-situ Mass Specific Absorption Spectra for Nine Forms of Highly-absorbing Carbonaceous Aerosol. *Carbon* **2018**, *136*, 85–93.
45. Radney, J. G.; Ma, X.; Gillis, K. A.; Zachariah, M. R.; Hodges, J. T.; Zangmeister, C. D. Direct measurements of mass-specific optical cross sections of single component aerosol mixtures. *Anal. Chem.* **2013**, *85*, 8319–8325.
46. Radney, J. G.; You, R.; Ma, X.; Conny, J. M.; Zachariah, M. R.; Hodges, J. T.; Zangmeister, C. D. Dependence of Soot Optical Properties on Particle Morphology: Measurements and Model Comparisons. *Environ. Sci. Technol.* **2014**, *48*, 3169–3176.
47. Ma, X.; Zachariah, M. R.; Zangmeister, C. D. Reduction of Suspended Graphene Oxide Single Sheet Nanopaper: The Effect of Crumpling. *J. Phys. Chem. C* **2013**, *117*, 3185–3191.

48. Saleh, R.; Hennigan, C. J.; McMeeking, G. R.; Chuang, W. K.; Robinson, E. S.; Coe, H.; Donahue, N. M.; Robinson, A. L. Absorptivity of brown carbon in fresh and photo-chemically aged biomass-burning emissions. *Atmos. Chem. Phys.* **2013**, *13*, 7683–7693.
49. Martinsson, J.; Eriksson, A. C.; Nielsen, I. E.; Malmborg, V. B.; Ahlberg, E.; Andersen, C.; Lindgren, R.; Nyström, R.; Nordin, E. Z.; Brune, W. H.; Svenningsson, B.; Swietlicki, E.; Boman, C.; Pagels, J. H. Impacts of Combustion Conditions and Photochemical Processing on the Light Absorption of Biomass Combustion Aerosol. *Environ. Sci. Technol.* **2015**, *49*, 14663–14671.
50. Chakrabarty, R. K.; Moosmüller, H.; Chen, L. W. A.; Lewis, K.; Arnott, W. P.; Mazzoleni, C.; Dubey, M. K.; Wold, C. E.; Hao, W. M.; Kreidenweis, S. M. Brown carbon in tar balls from smoldering biomass combustion. *Atmos. Chem. Phys.* **2010**, *10*, 6363–6370.
51. Jacobson, M. Z. Effects of biomass burning on climate, accounting for heat and moisture fluxes, black and brown carbon, and cloud absorption effects. *J. Geophys. Res. Atmos.* **2014**, *119*, 8980–9002.
52. Feng, Y.; Ramanathan, V.; Kotamarthi, V. R. Brown carbon: a significant atmospheric absorber of solar radiation? *Atmos. Chem. Phys.* **2013**, *13*, 8607–8621.
53. Levin, E. J. T.; McMeeking, G. R.; Carrico, C. M.; Mack, L. E.; Kreidenweis, S. M.; Wold, C. E.; Moosmüller, H.; Arnott, W. P.; Hao, W. M.; Collett, J. L., Jr.; Malm, W. C. Biomass burning smoke aerosol properties measured during Fire Laboratory at Missoula Experiments (FLAME). *J. Geophys. Res.* **2010**, *115*, D18210.
54. Saleh, R.; Robinson, E. S.; Tkacik, D. S.; Ahern, A. T.; Liu, S.; Aiken, A. C.; Sullivan, R. C.; Presto, A. A.; Dubey, M. K.; Yokelson, R. J.; Donahue, N. M.; Robinson, A. L. Brownness of organics in aerosols from biomass burning linked to their black carbon content. *Nat. Geosci.* **2014**, *7*, 647–650.
55. Radney, J. G.; You, R.; Zachariah, M. R.; Zangmeister, C. D. Direct In Situ Mass Specific Absorption Spectra of Biomass Burning Particles Generated from Smoldering Hard and Softwoods. *Environ. Sci. Technol.* **2017**, *51*, 5622–5629.

Modeling Multiphase Chemistry

Chapter 15

Modeling Heterogeneous Oxidation of NO_x, SO_2 and Hydrocarbons in the Presence of Mineral Dust Particles under Various Atmospheric Environments

Myoseon Jang* and Zechen Yu

Department of Environmental Engineering Sciences, University of Florida, Gainesville, Florida 32611, United States
***E-mail: mjang@ufl.edu**

Airborne mineral dust particles are known to significantly promote atmospheric oxidation of SO_2, NO_x, and hydrocarbons *via* the heterogeneously photocatalytic process under ambient sunlight. However, simulation of this process is not fully taken into account by current models. This study streamlines the process of developing an atmospheric mineral dust chemistry model and simulates the influence of air–suspended mineral dust particles on the formation of sulfate, nitrate, and secondary organic aerosol under various environments. The model encompasses partitioning processes and reactions in multiple phases including gas phase, inorganic–salted aqueous phase (non–dust phase), and dust phase. The reaction of adsorbed chemical species occurs *via* two major paths: autoxidation in the open air and photocatalytic mechanisms in the presence of UV light. Photocatalytic rate constants of tracers on dust surfaces are derived from the integration of the combinational product of the dust absorbance spectrum and wave–dependent actinic flux for the full range of wavelengths from the light source. The photocatalytic ability of dust particles is diversified according to sources and metal compositions in the model. The buffering capacity of dust particles to react to acids is also used in the model to predict the dynamic chemical characteristics of dust particles. The hygroscopicity and buffering capacities of both the fresh dust particles and the aged particles are applied in the model to predict the influence of humidity on partitioning processes and heterogeneous chemistry. The model predicted concentrations of sulfate and nitrate are then compared with outdoor chamber data obtained under natural sunlight. Systematic, in-depth dust chemistry can provide a platform for predicting the formation of nitrate, sulfate, and secondary organic aerosol at regional or global scales during dust events.

1. Introduction

Despite numerous studies regarding the increased oxidation of SO_2 and NO_2 due to mineral dust particles, the underlying heterogeneous chemical processes remain uncertain, hampering the model prediction of sulfate and nitrate formation on regional and global scales. Figure 1 illustrates the fate of airborne dust particles and their atmospheric aging. This section will discuss where the airborne dust particles come from, what their compositions are, and how they impact the environment.

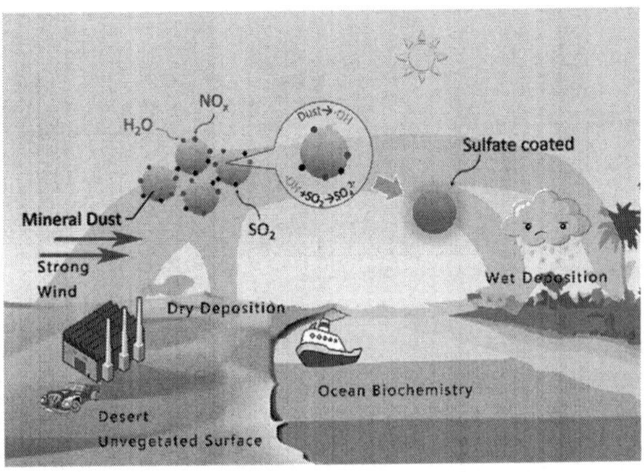

Figure 1. The schematic of heterogeneous oxidation of SO_2 and NO_x in the presence of mineral dust particles.

1.1. Dust Composition

Mineral dust usually contains aluminosilicates, calcium species, and miscellaneous metal oxides. The typical Scanning Electron Microscope (SEM)/Energy Dispersive X-ray (EDX) analysis of particles has been used to obtain the elemental compositions of mineral dust (*1–6*). Figure 2 shows the elemental composition of Arizona Test Dust (ATD) (*4*), Gobi Desert Dust (GDD) (*4*), Sahara Dust (SD) (*1*) and Australia Dust (AD) (*2*). The types of elements in the four different mineral dust particles are similar, though the fraction of alkaline ions and metals varies with sources. For example, alkaline ions such as Ca and Mg are present in higher concentration in GDD particles than those in other types of dust. Additionally, the percent composition of the element Ti in GDD is higher than in ATD but lower than in SD. The surface area of dust particles also varies with sources and influences their heterogeneous chemistry. For example, the BET surface areas of GDD and ATD particles in Figure 2 were respectively 39.6 and 47.4 $m^2\ g^{-1}$.

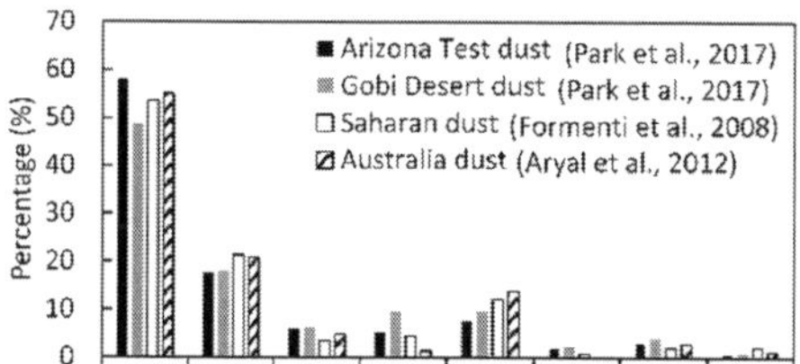

	Si	Al	K	Ca	Fe	Na	Mg	Ti
■ Arizona Test dust	57.9	17.6	5.9	5.4	7.6	1.9	3	0.7
▒ Gobi Desert dust	48.7	18	6.4	9.6	9.6	2.4	4.2	1
□ Saharan dust	53.5	21.4	3.6	4.6	12.3	0.9	2.1	2.3
▨ Australia dust	55.04	20.80	4.81	1.42	13.80	N.A.	2.81	1.33

Figure 2. Element fractions in various dust sources.

The composition of dust particles affects their ability to modulate heterogeneous chemistry of tracers. For example, it is recognized that clays in mineral dust aerosols actively affect heterogeneously nucleating ice, even at temperatures warmer than required for homogeneous nucleation (*7*). Clays, whose major components are phyllosilicates (i.e. montmorillonite, kaolinite, illite, hectorite, and talc), are very efficient in this respect. The quantity and composition of conducting metal oxides also influence the photocatalytic activities of dust particles and consequently affect the heterogeneous chemistry of tracers. For example, the high content of titanium oxides in dust may increase the photocatalytic uptake coefficients of tracers (*8*). In addition, dust's capacity to buffer acidic species is closely related to the content of alkaline carbonates. Dust compositions also vary depending upon geological environments. For example, coastal dust particles contain a large amount of Na compared with those of other source regions due to the mixing with sea salts (*9*).

1.2. Impact of Mineral Dust Particles

Mineral dust particles affect the climate by directly modulating the radiative budget due to the absorption and scattering of both short and long-wave radiation (*10, 11*). They also have an important indirect effect on the climate due to their microphysical interactions with clouds. Hansell, et al. (*12*) reported that dust's longwave warming effect counters more than 50% of dust's shortwave cooling effect. In general, the warming effect of greenhouse gases is global but the radiative impact of dust is regional. For example, dust's radiative impact, which is associated with its warming effects, measures from 2.3 to 20 watts m^{-2} of radiative forcing in Zhangye, China near the Gobi Desert area; in comparison, the warming effect by greenhouse gases measures around 2 watts m^{-2}. Thus, the influence of dust on the longwave radiation associated with climate effects is significant. Additionally, dust particles have been shown to act as efficient ice nuclei (IN) (*13–15*) and cloud condensation nuclei (CCN) (*16, 17*). Mineral dust deposited on snow can also influence snow's ability to reflect sunlight, causing it to melt faster (*18*).

Mineral dust particles can fertilize ocean surfaces because they provide nutrients to ocean plankton through long-range transport and deposition processes. For example, in the Mediterranean region, Saharan dust is an important source of nutrients for phytoplankton and other aquatic biosystems. However, Saharan dust adversely affects some ecosystems because it carries the fungus *Aspergillus sydowii* as well as others (*19*). Since 1970, dust outbreaks have worsened due to periods of drought in Africa (*20*), leading to a decline in the health of coral reefs across the Caribbean and Florida. Additionally, dust particles can carry infectious diseases such as bacterial meningitis and the diseases associated with micro-organisms transported in desert dust (i.e. Coccidioidomycosis) (*21*). The health impact of heavy metal dust exposure is also a global concern (*22–25*) due to the long-range transport of dust particles. In addition to the effect of dust itself, airborne sand dust in East Asia contains chemicals, metals, microorganisms, and ions from urban or industrial pollutant emissions across many regions (*26–28*). Watanabe, et al. (*29*) reported that the decrease in pulmonary function may be more severe when the levels of both sand dust particles and air pollution aerosols are high. In the future, it may be necessary to study the synergistic health effects of sand dust particles and air pollution aerosols in East Asia.

2. Heterogeneous Chemistry of Atmospheric Tracers on Mineral Dust Particle Surfaces

With an average lifetime of up to several weeks, mineral particles can be transported over large distances downwind from the source. Dust particles can act as an important sink for atmospheric trace gases such as NO_x, SO_2, O_3, and organics. Most laboratory studies (*30–32*) have been limited to certain metal oxides (i.e. TiO_2 and Fe_2O_3) and the uptake of NO_2 and SO_2 without the presence of radiation. As noted in the recent chamber study by Park and Jang (*33*) the reactive uptake coefficient $(\gamma_{SO_4^{2-}})$ of SO_2 in the presence of dry ATD particles increased by one order of magnitude (1.16×10^{-6}) in an indoor chamber with light integrated from UV–A and UV–B lamps compared to that from autoxidation (1.15×10^{-7}) without a light source. Dupart, et al. (*34*) also observed a significant enhancement of the uptake rate of NO_2 on ATD dust particles using UV–A irradiation. The field study by Dupart, et al. (*35*) observed an increase in conversion of SO_2 to sulfate by photooxidation chemistry during dust events. In the following sections, the mechanisms underlying the heterogeneous chemistry of atmospherically important tracers will be discussed. Figure 3 illustrates the simplified mechanisms of heterogeneous oxidation of major pollutants (O_3, SO_2, NO_2, and hydrocarbons). The reaction of adsorbed tracers occurs *via* two major pathways: autoxidation in the open air (no need of a light source) and photocatalytic mechanisms under UV light.

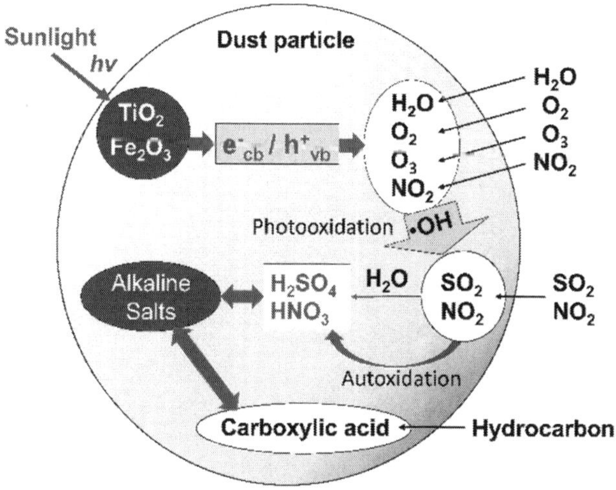

Figure 3. Mechanisms for heterogeneous oxidation of SO_2 and NO_x on airborne dust particles.

2.1. Autoxidation

Autoxidation of tracers on dust is an oxidation process *via* the reaction of an absorbed tracer molecule with an oxygen molecule. For example, the autoxidation of SO_2 on dust particles (denoted as "d") can be represented as the first order reaction (assuming a constant concentration of oxygen at 2×10^5 ppm).

$$SO_2(d) + O_2(g) \rightarrow SO_4^{2-}(d) \qquad k_{auto} \tag{1}$$

In the dark condition, the formation of sulfate is primarily from autoxidation of SO_2. For comparison with other studies, the uptake coefficient $(\gamma_{X_i,auto})$ of tracer X_i on dust is estimated. The relationship between k_{auto} and $\gamma_{X_i,auto}$ is written as (*36*)

$$k_{auto} = \frac{\omega_{X_i}\gamma_{X_i,auto}}{4} S_{dust} \tag{2}$$

where ω_{X_i} is the mean molecular velocity (m s^{-1}) and S_{dust} is the surface area (cm^2 m^{-3}) of dust particles. It should be noted that the calculation of the uptake coefficient using BET surface area will be different from the geometric surface area.

2.2. Photoactivation

Semiconducting metal oxides (i.e. TiO_2, Fe_2O_3, and their mixture with non-semiconducting metal oxides such as Al_2O_3 (*37*)) in dust particles could act as photocatalysts that yield electron (e^-_{cb} from the conduction band, cb)–hole (h^+_{vb} from the valence band, vb) pairs and produce strong oxidants (i.e. superoxide radical anions (O_2^-) and OH radicals) (*34, 38–43*). For example, oxygen molecules adsorbed on metal oxides receive an electron and produce highly reactive O_2^-

radicals. Water molecules adsorbed on dust also react with h^+_{vb} forming a hydroxyl radical (OH). The reactions are shown as follows:

$$O_2(d) + e^-_{cb} \rightarrow O_2^{-}\bullet(d) \quad (3)$$

$$e^-_{cb}(d) + O_2(d) + H^+(d) \rightarrow HO_2\bullet(d) \quad (4)$$

$$2HO_2\bullet(d) \rightarrow H_2O_2(d) + O_2(d) \quad (5)$$

$$H_2O(d) + h^+_{vb}(d) \rightarrow H^+(d) + \bullet OH(d) \quad (6)$$

$$2H_2O(d) + 2h^+_{vb}(d) \rightarrow 2H^+(d) + H_2O_2(d) \quad (7)$$

$$H_2O_2(d) + h\upsilon \rightarrow 2HO\bullet(d) \quad (8)$$

OH radicals enable rapid oxidation of tracers, such as SO_2 and NO_2, on the surface of dust particles (*33*, *34*, *44*, *45*). However, most studies on the photochemistry of dust have been limited to qualitative analyses and lack kinetic mechanisms that can be linked to a predictive model.

2.3. Heterogeneous Chemistry of Ozone

The heterogeneous uptake of ozone on the surface of metal oxides results in O_3 decomposition, forming a surface-bound atomic oxygen and an oxygen molecule. Usher, et al. (*46*) observed that O_3 increased the conversion of sulfite/bisulfite to sulfate/bisulfate on the surface of metal oxides (Al_2O_3) using FTIR spectroscopy. In the presence of O_3, the oxidation of NO_2 to form nitric acid is also enhanced on dust particles *via* the pathway of the formation of N_2O_5 followed by its hydrolysis.

$$NO_2(d) + O_3(d) \rightarrow NO_3(d) + O_2(d) \quad (9)$$

$$NO_2(d) + NO_3(d) \rightarrow N_2O_5(d) \quad (10)$$

$$N_2O_5(d) \rightarrow NO_2(d) + NO_3(d) \quad (11)$$

$$N_2O_5(d) + H_2O(d) \rightarrow 2HNO_3(d) \quad (12)$$

In the presence of UV light, O_3 is efficiently decomposed on the surface of dust due to the interaction of O_3 with e^-_{cb}, and is thereby able to form an OH radical.

$$O_3(d) + e^-_{cb}(d) \rightarrow O_3^{-}\bullet(d) + O_2(d) \quad (13)$$

$$O_3^{-}\bullet(d) + H^+(d) \rightarrow HO_3\bullet(d) \quad (14)$$

$$HO_3\bullet(d) \rightarrow O_2(d) + \bullet OH(d) \quad (15)$$

This OH radical can oxidize tracers adsorbed or absorbed on dust surfaces. Using a global three-dimensional model, Dentener et al. (*47*) estimated that O_3 levels close to a dust source are reduced by more than 10% due to this heterogeneous uptake process. Similarly, Wang et al. (*48*) reported that mixing ratios of O_3 were reduced by up to 3.8 ppb (~9%) in the presence of dust.

2.4. Heterogeneous Chemistry of Nitrogen Oxides

In the absence of a light source, the heterogeneous oxidation of NO_2 still occurs but is much weaker. Underwood et al. showed that NO_2 adsorption on the surfaces of metal oxides involved subsequent formation of surface nitrite (NO_2^-) ions and the production of nitrate ions *via* the reaction of two NO_2^- or the reaction of NO_2^- and gas phase NO_2 under dark conditions (*49*). As discussed in section 2.3, nitrate can form from the hydrolysis of N_2O_5. Liu et al. (*50*) and Ma & Liu (*51*) reported the formation of dinitrogen tetroxide (N_2O_4), a critical oxidant, on the surface of metal oxides showing that NO_x promotes SO_2 oxidation but that the formation of nitrite ion was inhibited by SO_2.

In the presence of UV light, NO_2 oxidation is enhanced on the surface of mineral dust particles. Ndour, et al. (*8*) found the production of HONO from NO_2 autoxidation was negligible, but HONO formation from photooxidation of NO_2 was significant in the presence of dust particles such as mixed TiO_2-SiO_2, Saharan dust, or Arizona Test Dust.

$$NO_2(d) + e^-_{cb}(d) \rightarrow NO_2^-(d) \quad (16)$$

$$NO_2^-(d) + H^+(d) \rightarrow HONO(d) \quad (17)$$

$$HONO(d) + h\nu \rightarrow NO(d) + \bullet OH(d) \quad (18)$$

$$NO_2(d) + \bullet OH(d) \rightarrow HNO_3(d) \quad (19)$$

The OH radical produced from the photolysis of HONO can oxidize NO_2 and increase nitrate formation. When Ndour, et al. (*8*) applied dust heterogeneous chemistry to three-dimensional modeling, they found that the photochemistry of dust may reduce the NO_2 level up to 37% and ozone up to 5% during a dust event in the free troposphere.

2.5. Heterogeneous Chemistry of Sulfur Dioxide

The laboratory studies characterizing the heterogeneous chemistry of SO_2 have focused mostly on autoxidation. Zhang et al. found that SO_2 was heterogeneously oxidized by active oxygen (O_2^-) or hydroxyl ions (OH^-), which were produced from O_2 or H_2O adsorbed on the surface of metal oxides (i.e. Al_2O_3) (*32*). The oxidation of SO_2 can also be promoted by the heterogeneous chemistry of atmospheric oxidants like O_3 (*46, 52–55*), NO_x (*50, 51, 56, 57*), and H_2O_2 (*58, 59*). For example, sulfite (SO_3^{2-}) produced on the surfaces of dust particles due to the solubility of SO_2 in the dust aqueous phase is heterogeneously oxidized to sulfate by exposure to ozone (*53, 54*). H_2O_2 adsorbed on the dust particle can produce OH radicals and oxidize adsorbed SO_2 (*59*). As shown in section 2.2, some metal oxides (i.e. photoactive semiconductor materials) in mineral dust particles are known to generate e^-_{cb}-h^+_{vb} pairs under UV light and create powerful surface oxidants (i.e. OH radicals) (*35, 60, 61*). The heterogeneous photooxidation of SO_2 occurs by oxidation with these OH radicals forming SO_3. The resulting SO_3 is hydrated and forms H_2SO_4 (*4, 42, 62, 63*).

3. Chemical and Physical Parameters To Model Dust's Heterogeneous Chemistry

Both dust's characteristics and meteorological parameters can significantly influence the heterogeneous chemistry of tracers. In this section, we will discuss how these properties are linked to the model parameters in the Atmospheric Mineral Aerosol Reaction (AMAR) model.

3.1. Buffering Capacity of Dust Particles

Calcium-rich dust particles can buffer an atmospheric acid, which condenses or heterogeneously forms on dust surfaces, and increase pH values of precipitation (*4, 64–68*). Dust's capacity to buffer acids is complex due to various reactions occurring on the dust surface:

$$CaCO_3(d) + H_2SO_4(d) \rightarrow CaSO_4(d) + H_2O(d) + CO_2 \quad (20)$$

$$M_2O_3(d) + 3H_2SO_4(d) \rightarrow M_2(SO_4)_3(d) + 3H_2O(d) \quad \text{(M: Al and Fe)} \quad (21)$$

$$CaCO_3(d) + HNO_3(d) \rightarrow Ca(NO_3)_2(d) + H_2O(d) + CO_2 \quad (22)$$

$$M_2O_3(d) + 3HNO_3(d) \rightarrow M(NO_3)_3(d) + 3H_2O(d) \quad (23)$$

In general, calcium nitrate ($Ca(NO_3)_2$) produced in Eq. 22 is much more hygroscopic than calcium carbonate. The buffering capacity of dust particles is closely related to the maximum nitrate concentration in the absence of sulfuric acid and semivolatile carboxylic acids (RCOOH), which can deplete nitrate.

$$Ca(NO_3)_2(d) + H_2SO_4(d) \rightarrow CaSO_4(d) + H_2O(d) + 2HNO_3 \quad (24)$$

$$CaCO_3(d) + RCOOH(d) \rightarrow Ca(RCOO)_2(d) + H_2O(d) + CO_2 \quad (25)$$

A recent study by Yu and Jang (*68*) reported that the measured buffering capacity of GDD (0.02 $\mu g\ \mu g^{-1}$) was significantly higher than that of ATD (0.011 $\mu g\ \mu g^{-1}$). In general, the dust originating from Chinese loess contains a high fraction of $CaCO_3$ (*28*). The inclusion of the buffering capacity of mineral dust into the model is essential in order to predict the formation of alkaline nitrate salts and the water content in dust phases.

3.2. Hygroscopicity of Dust

Hygroscopicity of mineral dust is another important parameter because the dust water content influences gas-dust partitioning of tracers, the production of oxidants *via* photocatalytic heterogeneous chemistry, and reactions of tracers adsorbed on dust particles. Hygroscopicity of mineral dust varies with dust sources and the status of dust aging. The inorganic salts and metal oxides in dust particles absorb water *via* a thermodynamic equilibrium process and form a thin layer of a water film on the dust surface (*67*, *69–71*). Figure 4 illustrates the hygroscopicity of two different mineral dust: ATD and GDD. The water content of dust particles was measured by the FTIR for %RH levels ranging from 10 to 85. No clear phase transitions nor obvious differences between the hydration and dehydration processes were observed because the hygroscopicity of dust particles is influenced by a variety of chemical species.

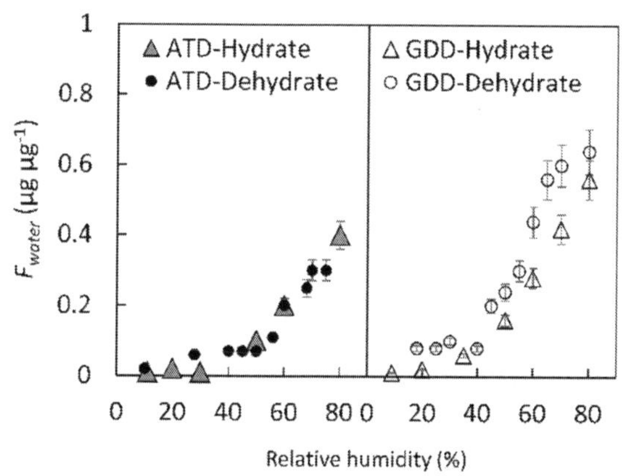

Figure 4. The measured water mass normalized by the dry dust mass using FTIR data for fresh ATD and fresh GDD. (Adapted with permission from reference (5). Copyright 2017, ACS Publications.)

In the recent study by Yu and Jang (*68*) a mathematical equation for the dust-phase water content (F_{water}, μg μg^{-1}), which is defined as the water mass normalized by the dry dust mass, was semiempirically derived using FTIR data. In their approach, F_{water} was calculated by the three major contributors (Figure 5): (1) The low amount of water in authentic dust mostly originates from metal oxides and alkaline carbonates. Additionally, sulfate salts formed in dust are generally not hygroscopic and treated in the same way as alkaline carbonates. (2) The hygroscopicity of nitrate salts (Eq. 22) is much higher than that in fresh dust particles. (3) An excess amount of acidic sulfate can be neutralized by ammonia on dust surfaces, forming the ammonium sulfate system, and the acidic sulfate's hygroscopicity can be estimated using the conventional inorganic thermodynamic model (*72–74*). The difference in model parameters for hygroscopicity between ATD and GDD is insignificant. The most influential variable that impacts the hygroscopicity of aged dust is the amount of nitrate, which is closely related to the buffering capacity of dust.

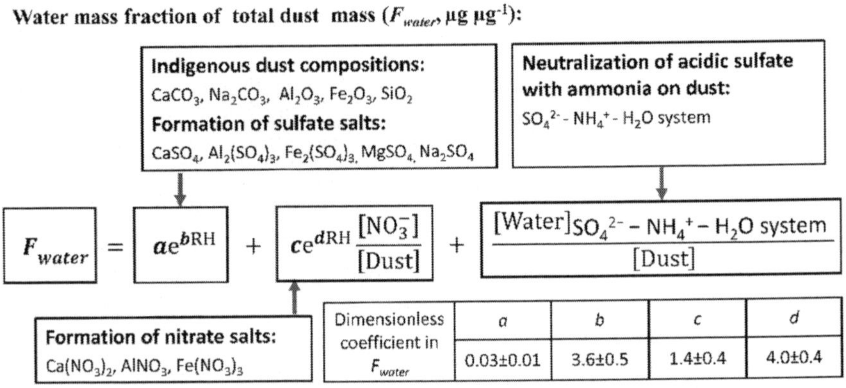

Figure 5. The calculation of the water mass (F_{water}, µg µg^{-1}) normalized by the dry dust mass. The hygroscopic parameters of GDD are sourced from work by Yu and Jang (68)

The formation of nitric acid on dust is faster than that of sulfuric acid because NO_x concentrations are generally higher than SO_2 in ambient air, and the reaction of NO_2 with OH radicals is also faster than that of SO_2 (by one order of magnitude). Thus, mineral dust in the ambient air is typically aged by NO_x first forming nitrate salts. In the presence of SO_2, carbonate in fresh dust and nitrate in aged dust are further depleted by the sulfate formation. Figure 6 illustrates how the hygroscopicity of mineral dust dynamically modulates as dust ages.

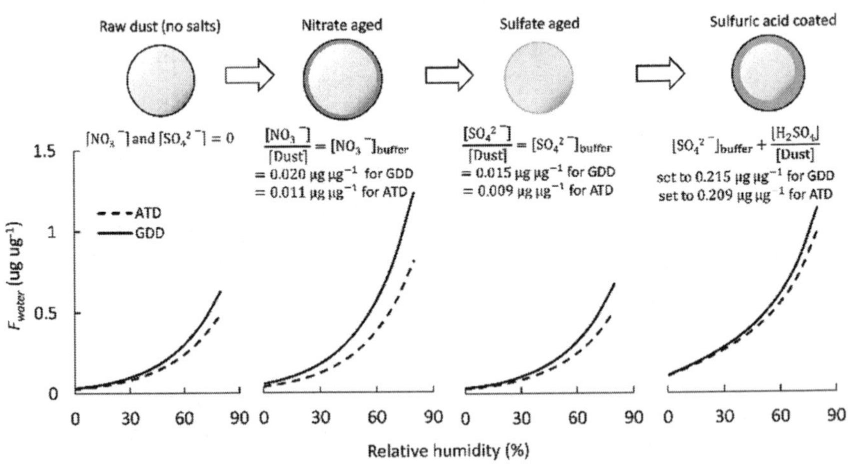

Figure 6. The hygroscopicity of fresh dust; the aged dust with nitric acid; the aged dust with sulfuric acid; and the dust coated with sulfuric acid.

The upper boundary of F_{water} at a given dust type is higher in the fully titrated mineral dust with nitric acid and is significantly affected by buffering capacity. For example, the water content in the aged GDD particles with nitric acid is almost 40% higher than in aged ATD because of the higher buffering capacity of GDD. Under the dust event, both nitrate and sulfate could possibly coexist on the dust surface, and dust particles can be partially or fully titrated by both inorganic acids and organic acids.

3.3. Photoactivation of Dust Particles

The value of the photolysis rate constant (j_i) of compound i is typically determined by actinic flux ($I(\lambda)$, quanta cm^{-2} s^{-1} nm^{-1}), the absorption cross section ($\sigma(\lambda)$, cm^2), and the quantum yield ($\phi(\lambda)$) at each wavelength range (λ, nm):

$$j_i = \int_{\lambda 1}^{\lambda 2} I(\lambda)\sigma(\lambda)\phi(\lambda)d\lambda \tag{26}$$

In order to use the preexisting structure of the photolysis rate constant, the integration of wavelength–dependent actinic flux with the photocatalytic activity of mineral dust is also needed. In the AMAR model recently derived by Yu et al. (*67*) the photoactivation rate constant on dust particles ($j_{[ATD]}$ (s^{-1})) was introduced to produce an e^-_{cb}–h^+_{vb} pair, which was also dependent on both the actinic flux originating from the light source and the photocatalytic property of dust particles. In order to deal with $\sigma(\lambda)\times\phi(\lambda)$, the mass absorption cross section of dust particles (MAC_{ATD}, m^2 g^{-1}) was estimated in the model. MAC_{ATD} was determined using the absorption coefficient (b_{ATD}, m^{-1}) of Arizona Test Dust, a reference dust, and the particle concentration (m_{ATD}, g m^{-3}):

$$MAC_{ATD} = \frac{b_{ATD}}{m_{ATD}} \tag{27}$$

b_{ATD} is calculated from the absorbance of a dust filter sample (Abs_{ATD}) measured using a reflective UV–visible spectrometer,

$$b_{ATD} = \frac{Abs_{ATD}\, A_{filter}}{f\, V} \ln(10) \tag{28}$$

where A_{filter} (7.85×10^{-5}, m^2) is the area of the filter sample and V (m^3) is the total volume of air passing through the filter during the sampling period. To eliminate the absorbance caused by filter material scattering, a correction factor (f = 1.4845) is coupled into Eq. (28) (*75*). Both $\sigma(\lambda)$ and $\varphi(\lambda)$ cannot be directly measured because of complexity in the amount of photoactive conducting matter in dust particles and the irradiation processes of the e^-_{cb}–h^+_{vb} pairs. The MAC_{ATD} of dust particles originates from light–absorbing matter (i.e. metal oxides and metal sulfate). The MAC_{ATD} spectrum is adjusted using the TiO_2 absorption spectrum in order to account for the conducting matter in dust particles (*76*) and applied to $\sigma(\lambda) \times \varphi(\lambda)$. Figure 7 illustrates the activation process of mineral dust by UV light with the MAC_{ATD} spectrum of Arizona Test Dust and the $\sigma(\lambda) \times \varphi(\lambda)$ spectrum applied to the model.

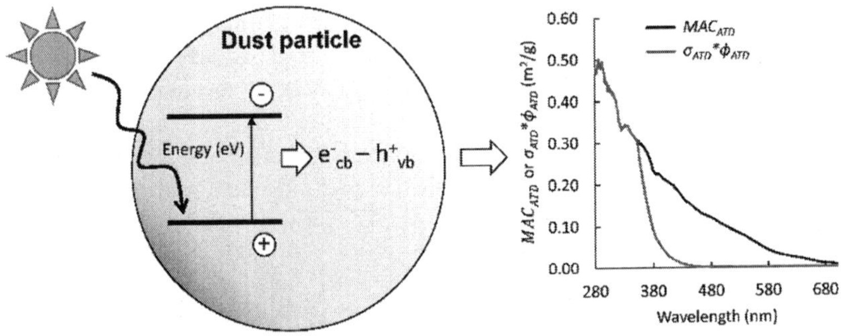

Figure 7. Photoactivation of mineral dust particles to produce electron-hole pairs.

As shown in the composition of dust sourced from different locations (Figure 2 in section 1.1), the amount of conducting matter in Gobi Desert Dust is lower than that of Saharan dust but higher than that of Arizona dust. Collectively, heterogeneous chemistry under UV light is affected by the metal compositions of mineral dust. The recent study by Park et al. (*4*) observed that the reactive uptake coefficient of SO_2 in the presence of the Gobi Desert dust was 2 to 2.5 times greater than in the presence of Arizona Test Dust.

3.4. Humidity and Temperature Effects on Heterogeneous Chemistry

The Henry's law constant of a compound can be applied to calculate the gas-particle partitioning coefficient of tracers. When temperature increases, the Henry's law constant generally decreases. Moreover, temperature influences the mobility of the compound in particles. Lower temperature leads to slower movement of a molecule due to the high viscosity of the medium. Thus, the temperature may alter the reaction rate constant of tracers in the dust phase, but this is not well studied yet.

Humidity influences the amount of dust phase water, and subsequently affects gas-particle partitioning of tracers and chemistry. As discussed in section 3.2, the dust phase water content is determined by the hygroscopicity of dust particles under varying humidity. Humidity modulates the production of the surface OH radical because the electron-hole pairs created in the photoactivation process reacts with a water molecule to generate surface oxidants such as OH radicals.

4. Modeling Heterogeneous Chemistry of Tracers on Dust

Heterogeneous chemistry of tracers on dust surfaces has been traditionally approached using an uptake coefficient based on laboratory data (*8*, *30*, *31*, *57*, *77–79*). However, this uptake coefficient is limited to the heterogeneous reaction of tracers under dark conditions or a simple mechanism controlled by a pseudo 1st-order reaction. In the ambient environment, the reactive uptake of tracers *via* the heterogeneous photocatalytic process on dust particles cannot be treated by a single uptake coefficient because it originates from multi-step reactions including 2nd order reactions. Furthermore, dust particles dynamically age, influencing partitioning and reaction rates. Therefore, an explicit kinetic model, which benefits from accurately simulating heterogeneous chemistry, can predict the change in atmospheric compositions on regional and global scales during the atmospheric processes of dust events. In the following section, the AMAR model, which is derived using explicit kinetic mechanisms (*67*), will be described and demonstrated to predict the heterogeneous oxidation of tracers.

4.1. Kinetic Mechanisms of Heterogeneous Chemistry in the Dust Model

Figure 8 illustrates the model skeleton for the heterogeneous chemistry of tracers in the presence of mineral dust. In order to predict the atmospheric fate of tracers in the presence of dust, the model needs to include chemistry in the three phases: gas phase, aqueous aerosol phase, and dust phase. The reaction of SO_2 with an OH radical in the gas phase ultimately produces sulfuric-acid vapor. The sulfuric acid vapor, with volatility as low as 10^{-9} mmHg in ambient humidity, is involved in nucleation to form sulfuric-acid seeds, hygroscopic aerosols (*80–82*). Except in the large presence of dust particles, the formation of sulfuric acid seeded aqueous aerosol is unavoidable. The explicit mechanisms of the aqueous–phase chemistry that occurs in inorganic salted aqueous aerosol (SO_4^{2-}–NH_4^+–H_2O system without dust) to form $SO_4^{2-}{}_{aq}$ and $NO_3^-{}_{aq}$ can be accounted for in the preexisting model constructed by Liang and Jacobson (*83*)

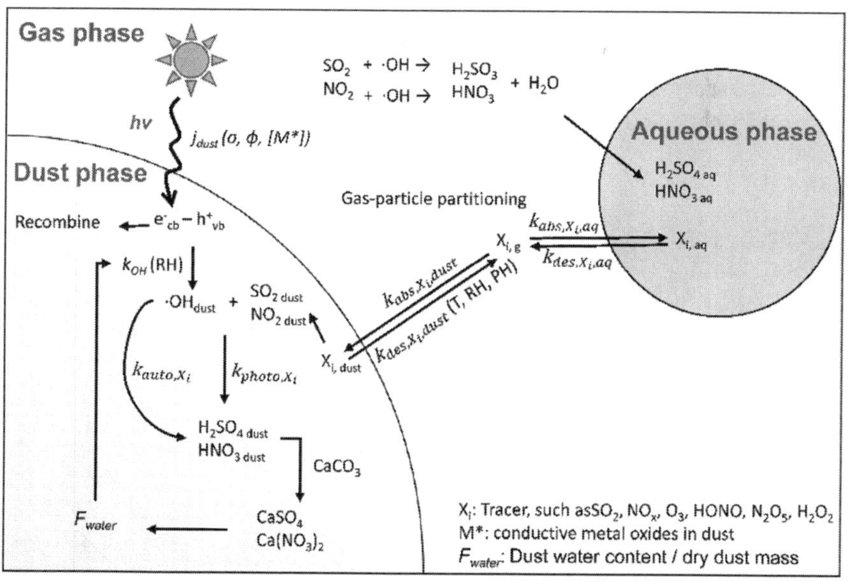

Figure 8. The schematic of explicit kinetic mechanisms for heterogeneous chemistry in mineral dust particles.

Under urban environments (relative humidity higher than 20%), dust particles are typically coated with multiple layers of water (*69*, *70*, *84*). Thus, the assumption of absorptive partitioning of tracers onto the water layer is reasonable and allows us to use the Henry's law constant for the partitioning process between the gas and dust phase. For example, the gas-dust partitioning constant (K_{dust,X_i}, m^3 m^{-2}) of tracer X_i is defined as,

$$K_{dust,X_i} = \frac{[X_i]_{dust}}{[X_i]_g A_{Dust}} \quad (29)$$

where A_{dust} (m^2 m^{-3}) is the concentration of geometric surface for dust particles. The partitioning process of tracer X_i is kinetically calculated using absorption and desorption processes as follows,

$$X_{i,gas} + A_{Dust} \rightarrow X_{i,dust} + A_{Dust} \qquad k_{abs,X_i,dust} \quad (30)$$

$$X_{i,dust} \rightarrow X_{i,gas} \qquad k_{des,X_i,dust} \quad (31)$$

where $k_{abs,X_i,dust}$ (m^3 m^{-2} s^{-1}) and $k_{des,X_i,dust}$ (s^{-1}) are the absorption rate constant and the desorption rate constant, respectively. At equilibrium, the absorption rate (Eq. 30) equals the desorption rate (Eq. 31). Thus, K_{d,SO_2} is expressed as:

$$K_{dust,X_i} = \frac{k_{abs,X_i,dust}}{k_{des,X_i,dust}}. \quad (32)$$

The characteristic time to reach equilibrium is as short as the order of 10^{-6} s. Typically, the adsorption and desorption rates are larger and faster than the reaction rates of tracers with oxidants in particle phase. In the model, the absorption rate constants are set at large numbers to simulate a fast partitioning process. The gas-dust partitioning constant is estimated first and then applied to Eq. 32 to determine the desorption rate constant. The desorption rate constant is dependent on temperature and humidity (F_{water}, Figure 5). The temperature dependence of $k_{des,X_i,dust}$ is estimated with an analog of the Henry's law constant,

$$k_{des,X_i,dust} = k_0 \exp\left(-\frac{\Delta H_{X_i}/R}{T}\right) / (F_{water}). \quad (33)$$

When the chemical species is dissociable in aqueous phase, the desorption rate constant is also affected by the acidity of particles. For example, the desorption rate constant of SO_2 that forms H_2SO_3 is equilibrated with HSO_3^- and SO_3^{2-}, and can be affected by acidity. The desorption rate constant is rewritten as shown below,

$$k_{des,SO_2,dust} = 1 \times 10^9 \exp\left(-\frac{3100}{T}\right) / (F_{water}(1 + \frac{Ka_{SO_2}}{[H^+]})). \quad (34)$$

The dust phase reactions are divided into two major aspects: autoxidation and photooxidation. As shown in Figure 8, e^-_{cb}–h^+_{vb} pairs are produced *via* the photoactivation of dust particles. These $e^-_{cb}h^+_{vb}$ pairs can be deactivated through a recombination process, or react with a water molecule or an oxygen molecule to form a strong oxidant, an OH radical, on the surface of dust as expressed below:

$$e^-_{cb}\text{–}h^+_{vb} + O_2(d) \rightarrow \cdot OH(d) \qquad k_{OH} = 1 \times 10^{-22} \exp(2.3RH) \quad (35)$$

k_{OH} (cm^3 molecules^{-1} s^{-1}) is the reaction rate constant to form the OH radical on dust and is a function of humidity. The tracer absorbed on the surface of mineral dust can be oxidized via the reaction with this OH radical. In the model, k_{auto,X_i} is the rate constant for autoxidation of species X_i and k_{photo,X_i} is the reaction rate constant for the photooxidation of X_i which is processed *via* the mediated OH radical (Figure 8). The production of e^-_{cb}–h^+_{vb} pairs is estimated using the fraction of the mass of effective conductive metals to total dust mass ([M*], ug ug^{-1}):

$$M^* + hv \rightarrow M^* + e^-_{cb}\text{–}h^+_{vb} \qquad j_{dust} \qquad (36)$$

In the model, the [M*] value of ATD is selected as reference dust and applied to scale [M*] for different types of dust particles based on laboratory data. For example, the photo-degradation rate of an organic dye (i.e. malachite green) impregnated on different dust filter samples was measured using the online reflective UV-Visible spectrometer. The [M*] value of GDD is 2.55 times higher than that of ATD. This finding is in accord with the results of Park, et al. (*4*).

4.2. Simulation of Heterogeneous Oxidation of SO_2 and NO_x

Figure 9 shows that the AMAR model simulation agrees with chamber observations in a recent study. Figure 10 represents the simulation of nitrate and sulfate formation in the presence of the fresh GDD using the explicit heterogeneous chemistry mechanisms (AMAR) integrated with a box model given a typical urban environment. Several important findings that resulted from this simulation are summarized as follows:

(1) In the presence of fresh dust particles, the formation of nitrate occurs earlier than that of sulfate because the oxidation rate of NO_x is faster than that of SO_2 in both gas phase and particle phase.
(2) The nitrate concentration on dust particles at high humidity (wet aerosol) is slightly higher than that estimated from the buffering capacity of dust particles (section 3.1) because nitric acid can partition onto the water layer associated with wet alkaline nitrate salt on dust.
(3) Dust particles are completely neutralized mostly by nitric acid in the morning under the simulation conditions shown in Figure 10.
(4) In the morning when humidity is high (before 8 AM in Figure 10(a)), the formation of hygroscopic alkaline nitrate salts rapidly increases and thus, the water fraction (F_{water}) in dust particles also increases.
(5) With 20 ppb of SO_2, nitrate is partially depleted in the simulation above (Figure 10). If more SO_2 is introduced into the system, nitrate will be further depleted.
(6) The water content of dust particles drops in the late morning because of the decrease in humidity as well as the formation of alkaline sulfate, which is less hygroscopic than nitrate salts.
(7) Overall, heterogeneous photooxidation is the most important mechanism (57%) for forming sulfate in the presence GDD in this simulation. Autoxidation is less significant than photooxidation but it is still

more significant (31%) than non-dust chemistry (gas phase oxidation + aqueous chemistry) (12%) in the daytime. In the dark period, autoxidation continuously adds sulfate to the system.

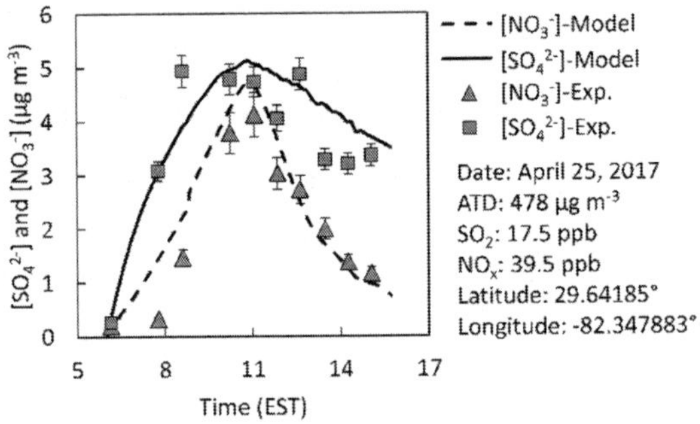

Figure 9. AMAR-model simulation of sulfate and nitrate formation against chamber data obtained on April 25, 2017 in the presence of ATD particles using the Atmospheric PHotochemical Outdoor Reaction (UF-APHOR) chamber located in the University of Florida.

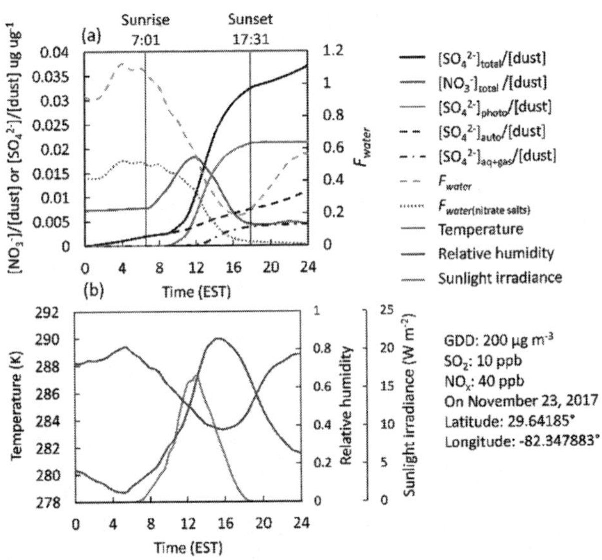

Figure 10. (a) Simulation of sulfate and nitrate formation in the presence of GDD particles under the typical urban environmental condition. (b) The ambient data for sunlight, temperature, and humidity are obtained on November 23, 2017 at Gainesville, Florida (latitude/longitude: 29.64185°/–82.347883°).

Figure 11 shows model results for the impact of dust characteristics on the heterogeneous chemistry of SO_2 in the presence of three different dust particles: ATD, GDD, and the GDD pre-aged with the oxidation of NO_2. GDD has both a higher photoactivation ability and greater buffering capacity than ATD, and thus yields higher sulfate formation. As shown in Figure 11 (a)-(c), sulfate formation increases from ATD to GDD, but the sulfate formation with GDD is similar to that with pre-aged GDD with NO_2. Considering the insensitivity of SO_2 oxidation to buffering capacity, the most critical factor in increased sulfate formation is the photoactivation capacity of dust particles. Under the ambient conditions within our simulation (see Figure 10 (b)), %RH drops to about 40 during the daytime and significantly reduces dust phase water content (Figure 4). In order to simulate the impact of humidity on sulfate formation, the prediction under ambient humidity (Figure 11 (b)) is compared to that at RH=55% (Figure 11 (d)). Humidity considerably affects sulfate formation (Figure 11 (b) vs (d)) but the impact of aging is small (Figure 11 (b) vs. (c) or Figure 11 (d) vs. (e)). Although not shown here, both the preexisting chamber studies (*4*, *33*) and the previous model simulations (*67*) showed that the oxidation of SO_2 was suppressed by NO_x because SO_2 and NO_2 competed for the consumption of OH radicals.

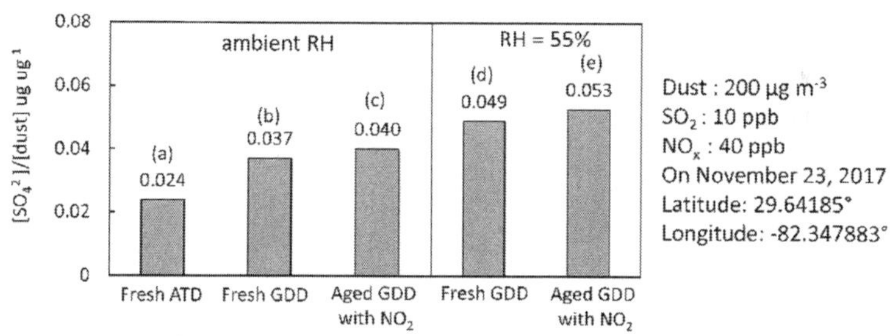

Figure 11. The formation of sulfate from the photooxidation of SO_2 in the presence of different dust particles (i.e. ATD, GDD and pre-aged GDD particles with nitrate) under the two different metrological environments (ambient conditions of Figure 10 (b) and fixed humidity at 55%). The photoactivation capability and the buffering capacity of GDD and aged GDD are scaled to reference dust, ATD.

4.3. Heterogeneous Chemistry of Hydrocarbons

The impact of mineral dust particles on hydrocarbon oxidation is not understood well yet. Lederer, et al. (*85*) reported that the oxidation of d-limonene on the surface of mineral dust formed d-limonene epoxide and dihydroxy derivatives *via* hydrolysis. However, the uptake coefficient of d-limonene *via* autoxidation was insignificant after consideration of terpene concentrations in ambient air. The most important aspect of the heterogeneous chemistry of hydrocarbons is the accommodation of semivolatile oxygenated products on preexisting dust particles. Semivolatile organics are produced from the atmospheric oxidation of some hydrocarbons in the gas phase. These semivolatile organics on the dust surface can compete with inorganic tracers (i.e. SO_2 and NO_2) for the consumption of surface OH radicals. Thus, the organic coating on dust particles may modify the production of sulfate and nitrate. The coating of secondary organic aerosol (SOA) on mineral dust particles could vary with the consumption of hydrocarbon and the concentration of dust particles (*86*, *87*).

Figure 12 illustrates our recent data that show the formation of the SOA produced from the photooxidation of α-pinene in the presence of NO_x (VOC ppbC/NO_x ppb = 6.9) and GDD using the outdoor chamber. Compared to the organic matter (OM) formed in the presence of silica particles, the OM produced with GDD was significantly greater, suggesting that authentic dust particles can be effectively coated by SOA. In Figure 12, the maximum nitrate concentration appears in the early stage but never reaches the maximum buffering capacity (7.8×10^{-4} gm^{-2}). One possible explanation for such nitrate depletion is the formation of carboxylate salts *via* the reaction of carboxylic acids with carbonate or nitrate salts. The enrichment of carboxylate salt was also observed in the formation of SOA in the presence of sea salt aerosol (*71*).

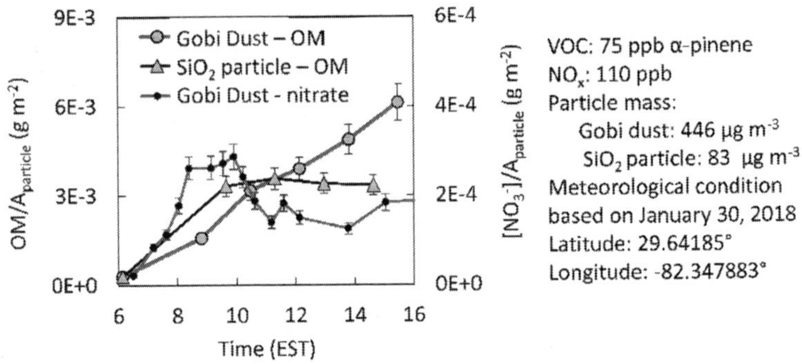

Figure 12. The formation of organic matter (OM normalized by the surface area) and nitrate from the photooxidation of α-pinene in the presence of NO_x and GDD particles (or SiO_2 particles) using the UF-APHOR chamber. The nitrate on SiO_2 is negligible.

Carboxylate salts can also influence the hygroscopicity of mineral dust particles. Additionally, adsorbed organic compounds can react with dust-phase OH radicals and influence the oxidation rate of NO_2 and SO_2. Currently, little research has been conducted regarding the interaction of mineral dust with organic compounds. The incorporation of the chemical interactions of mineral dust with hydrocarbons and inorganic tracers into the dust-phase heterogeneous chemistry model should be conducted in the future.

5. Conclusions

The primary components which are necessary to simulate the heterogeneous chemistry of atmospheric tracers (SO_2 and NO_2) in the presence of mineral dust were discussed in this chapter. The heterogeneous oxidation of tracers on dust surfaces is affected by both meteorological variables (i.e. temperature, humidity, and sunlight (section 3.4) and dust characteristics (i.e. buffering capacity, photoactivation ability, and hygroscopicity (sections 3.1-3.3). For NO_2 oxidation, the maximum concentration of nitrate is limited by the buffering capacity of dust particles. However, buffering capacity cannot limit the formation of sulfate from the heterogeneous oxidation of SO_2. Across different authentic mineral dust particles, the most important parameters that enhance tracers' photooxidation is the dust's photoactivation ability, which is catalyzed by conductive metal oxides. Hence, it is necessary to scale the photocatalytic ability of different dust particles. In our study, the photoactivation ability of different dust types (ATD and GDD) was scaled using the kinetic rate constant of a photo-degradation of the surrogate compound (section 3.3) on dust particles. In the future, the photoactivation ability of diverse dust particles should be connected to dust metal compositions.

Through atmospheric processes, the hygroscopicity of mineral dust particles dynamically changes. The dust particles aged with nitrate become hygroscopic and can also become deliquesced at low humidity (<20%). The large amount of dust phase water further enhances the gas-dust partitioning of tracers and heterogeneous reactions. However, in the presence of SO_2, nitrate is depleted by sulfuric acid forming alkaline sulfate salts. The AMAR model allows us to simulate how dust aerosol's life cycle affects the long-distance transport of SO_2. For example, the concentration of SO_2 in urban areas in the vicinity of dust sources can quickly drop due to heterogeneous reactions and thus, the amount of SO_2 transported from the polluted areas to other locations could be less.

Mineral dust is the primary component of the total atmospheric aerosol burden, while sea salt aerosol is the most widely distributed natural aerosols. Unlike dust particles, sea salt aerosols are non-light-absorbing and highly hygroscopic. However, sea salt aerosol can still produce powerful radicals such as the chloride radical *via* heterogeneous chemistry (*88*). For example, sodium chloride, the most abundant component in sea salt aerosol, can react with atmospheric pollutants such as N_2O_5 to form a photodissociable $ClNO_2$ (Cl radical and NO_2). Additionally, the transition metal ions (i.e., Fe^{3+}) in sea salt aerosols can considerably enhance the uptake of ozone to form photodissociable Cl_2 (*89*) and they can also oxidize absorbed SO_2 leading to increased sulfate concentrations (*90*).

Acknowledgments

The authors acknowledge the award from the National Institute of Metrological Sciences (KMA2018-00512), the Scholarship from the Fulbright Scholar (from the USA to Mongolia), and the award from the Ministry of Science and ICT, the Ministry of Environment, the Ministry of Health and Welfare (2017M3D8A1090654).

References

1. Formenti, P.; Rajot, J. L.; Desboeufs, K.; Caquineau, S.; Chevaillier, S.; Nava, S.; Gaudichet, A.; Journet, E.; Triquet, S.; Alfaro, S.; Chiari, M.; Haywood, J.; Coe, H.; Highwood, E. Regional variability of the composition of mineral dust from western Africa: Results from the AMMA SOP0/DABEX and DODO field campaigns. *J. Geophys. Res.: Atmos.* **2008**, *113*, D00C13.

2. Aryal, R.; Kandel, D.; Acharya, D.; Chong, M. N.; Beecham, S. Unusual Sydney dust storm and its mineralogical and organic characteristics. *Environ. Chem.* **2012**, *9*, 537–546.
3. Chen, Y.; Paytan, A.; Chase, Z.; Measures, C.; Beck, A. J.; Sanudo-Wilhelmy, S. A.; Post, A. F. Sources and fluxes of atmospheric trace elements to the Gulf of Aqaba, Red Sea. *J. Geophys. Res.: Atmos.* **2008**, *113*, D05306.
4. Park, J.; Jang, M.; Yu, Z. Heterogeneous Photo-oxidation of SO_2 in the Presence of Two Different Mineral Dust Particles: Gobi and Arizona Dust. *Environ. Sci. Technol.* **2017**, *51*, 9605–9613.
5. Yadav, S.; Rajamani, V. Geochemistry of aerosols of northwestern part of India adjoining the Thar Desert. *Geochim. Cosmochim. Acta* **2004**, *68*, 1975–1988.
6. Vlasenko, A.; Sjogren, S.; Weingartner, E.; Gaggeler, H. W.; Ammann, M. Generation of submicron Arizona test dust aerosol: Chemical and hygroscopic properties. *Aerosol Sci. Technol.* **2005**, *39*, 452–460.
7. Gallavardin, S.; Lohmann, U.; Cziczo, D. Analysis and differentiation of mineral dust by single particle laser mass spectrometry. *Int. J. Mass Spectrom.* **2008**, *274*, 56–63.
8. Ndour, M.; D'Anna, B.; George, C.; Ka, O.; Balkanski, Y.; Kleffmann, J.; Stemmler, K.; Ammann, M. Photoenhanced uptake of NO_2 on mineral dust: Laboratory experiments and model simulations. *Geophys. Res. Lett.* **2008**, *35*, L05812.
9. Niimura, N.; Okada, K.; Fan, X. B.; Kai, K.; Arao, K.; Shi, G. Y.; Takahashi, S. Formation of Asian dust-storm particles mixed internally with sea salt in the atmosphere. *J. Meteorol. Soc. Jpn.* **1998**, *76*, 275–288.
10. Tegen, I. Modeling the mineral dust aerosol cycle in the climate system. *Quat. Sci. Rev.* **2003**, *22*, 1821–1834.
11. Reddy, M. S.; Boucher, O.; Balkanski, Y.; Schulz, M. Aerosol optical depths and direct radiative perturbations by species and source type. *Geophys. Res. Lett.* **2005**, *32*, L12803.
12. Hansell, R. A.; Tsay, S. C.; Hsu, N. C.; Ji, Q.; Bell, S. W.; Holben, B. N.; Welton, E. J.; Roush, T. L.; Zhang, W.; Huang, J.; Li, Z. Q.; Chen, H. An assessment of the surface longwave direct radiative effect of airborne dust in Zhangye, China, during the Asian Monsoon Years field experiment (2008). *J. Geophys. Res.: Atmos.* **2012**, *117*, D00K39.
13. DeMott, P. J.; Sassen, K.; Poellot, M. R.; Baumgardner, D.; Rogers, D. C.; Brooks, S. D.; Prenni, A. J.; Kreidenweis, S. M. African dust aerosols as atmospheric ice nuclei. *Geophys. Res. Lett.* **2003**, *30*, L07808.
14. Crawford, I.; Bower, K. N.; Choularton, T. W.; Dearden, C.; Crosier, J.; Westbrook, C.; Capes, G.; Coe, H.; Connolly, P. J.; Dorsey, J. R.; Gallagher, M. W.; Williams, P.; Trembath, J.; Cui, Z.; Blyth, A. Ice formation and development in aged, wintertime cumulus over the UK: observations and modelling. *Atmos. Chem. Phys.* **2012**, *12*, 4963–4985.
15. Chen, J. P.; Hazra, A.; Levin, Z. Parameterizing ice nucleation rates using contact angle and activation energy derived from laboratory data. *Atmos. Chem. Phys.* **2008**, *8*, 7431–7449.

16. Rosenfeld, D.; Rudich, Y.; Lahav, R. Desert dust suppressing precipitation: a possible desertification feedback loop. *Proc. Natl. Acad. Sci. U. S. A.* **2001**, *98*, 5975–5980.
17. Kluser, L.; Holzer-Popp, T. Relationships between mineral dust and cloud properties in the West African Sahel. *Atmos. Chem. Phys.* **2010**, *10*, 6901–6915.
18. Gautam, R.; Hsu, N. C.; Lau, W. K. M.; Yasunari, T. J. Satellite observations of desert dust-induced Himalayan snow darkening. *Geophys. Res. Lett.* **2013**, *40*, 988–993.
19. Schlesinger, P.; Mamane, Y.; Grishkan, I. Transport of microorganisms to Israel during Saharan dust events. *Aerobiologia* **2006**, *22*, 259–273.
20. Prospero, J. M.; Lamb, P. J. African droughts and dust transport to the Caribbean: climate change implications. *Science (Washington, DC, United States)* **2003**, *302*, 1024–1027.
21. Kirkland, T. N.; Fierer, J. Coccidioidomycosis: a reemerging infectious disease. *Emerging Infect. Dis.* **1996**, *2*, 192–199.
22. Wang, J.; Hu, Z. M.; Chen, Y. Y.; Chen, Z. L.; Xu, S. Y. Contamination characteristics and possible sources of PM10 and PM2.5 in different functional areas of Shanghai, China. *Atmos. Environ.* **2013**, *68*, 221–229.
23. de Miranda, R. M.; de Fatima Andrade, M.; Fornaro, A.; Astolfo, R.; de Andre, P. A.; Saldiva, P. Urban air pollution: a representative survey of PM(2.5) mass concentrations in six Brazilian cities. *Air Qual., Atmos. Health* **2012**, *5*, 63–77.
24. Aldabe, J.; Elustondo, D.; Santamaria, C.; Lasheras, E.; Pandolfi, M.; Alastuey, A.; Querol, X.; Santamaria, J. M. Chemical characterisation and source apportionment of PM2.5 and PM10 at rural, urban and traffic sites in Navarra (North of Spain). *Atmos. Res.* **2011**, *102*, 191–205.
25. Keil, D. E.; Buck, B.; Goossens, D.; Teng, Y.; Pollard, J.; McLaurin, B.; Gerads, R.; DeWitt, J. Health effects from exposure to atmospheric mineral dust near Las Vegas, NV, USA. *Toxicol. Rep.* **2016**, *3*, 785–795.
26. Chan, C. C.; Chuang, K. J.; Chen, W. J.; Chang, W. T.; Lee, C. T.; Peng, C. M. Increasing cardiopulmonary emergency visits by long-range transported Asian dust storms in Taiwan. *Environ. Res.* **2008**, *106*, 393–400.
27. Onishi, K.; Kurosaki, Y.; Otani, S.; Yoshida, A.; Sugimoto, N.; Kurozawa, Y. Atmospheric transport route determines components of Asian dust and health effects in Japan. *Atmos. Environ.* **2012**, *49*, 94–102.
28. Choi, J. C.; Lee, M.; Chun, Y.; Kim, J.; Oh, S. Chemical composition and source signature of spring aerosol in Seoul, Korea. *J. Geophys. Res.: Atmos.* **2001**, *106*, 18067–18074.
29. Watanabe, M.; Noma, H.; Kurai, J.; Kato, K.; Sano, H.; Tatsukawa, T.; Nakazaki, H.; Yamasaki, A.; Shimizu, E. Association between pulmonary function and daily levels of sand dust particles assessed by light detection and ranging in schoolchildren in western Japan: A panel study. *Allergol. Int.* **2016**, *65*, 56–61.
30. Goodman, A. L.; Li, P.; Usher, C. R.; Grassian, V. H. Heterogeneous uptake of sulfur dioxide on aluminum and magnesium oxide particles. *J. Phys. Chem. A* **2001**, *105*, 6109–6120.

31. Usher, C. R.; Al-Hosney, H.; Carlos-Cuellar, S.; Grassian, V. H. A laboratory study of the heterogeneous uptake and oxidation of sulfur dioxide on mineral dust particles. *J. Geophys. Res.: Atmos.* **2002**, *107*, 4713.
32. Zhang, X.; Zhuang, G.; Chen, J.; Wang, Y.; Wang, X.; An, Z.; Zhang, P. Heterogeneous reactions of sulfur dioxide on typical mineral particles. *J. Phys. Chem. B* **2006**, *110*, 12588–12596.
33. Park, J. Y.; Jang, M. Heterogeneous photooxidation of sulfur dioxide in the presence of airborne mineral dust particles. *RSC Adv.* **2016**, *6*, 58617–58627.
34. Dupart, Y.; Fine, L.; D'Anna, B.; George, C. Heterogeneous uptake of NO_2 on Arizona Test Dust under UV-A irradiation: An aerosol flow tube study. *Aeolian Res.* **2014**, *15*, 45–51.
35. Dupart, Y.; King, S. M.; Nekat, B.; Nowak, A.; Wiedensohler, A.; Herrmann, H.; David, G.; Thomas, B.; Miffre, A.; Rairoux, P.; D'Anna, B.; George, C. Mineral dust photochemistry induces nucleation events in the presence of SO_2. *Proc. Natl. Acad. Sci. U. S. A.* **2012**, *109*, 20842–20847.
36. Schwartz, S. E. In *Chemistry of Multiphase Atmospheric Systems*; Springer: Berlin, Heidelberg, 1986; Vol. 6, pp 415–471.
37. Anderson, C.; Bard, A. J. Improved photocatalytic activity and characterization of mixed TiO_2/SiO_2 and TiO_2/Al_2O_3 materials. *J. Phys. Chem. B* **1997**, *101*, 2611–2616.
38. Linsebigler, A. L.; Lu, G. Q.; Yates, J. T. Photocatalysis on TiO_2 Surfaces - Principles, Mechanisms, and Selected Results. *Chem. Rev.* **1995**, *95*, 735–758.
39. Hoffmann, M. R.; Martin, S. T.; Choi, W. Y.; Bahnemann, D. W. Environmental Applications of Semiconductor Photocatalysis. *Chem. Rev.* **1995**, *95*, 69–96.
40. Thompson, T. L.; Yates, J. T., Jr. Surface science studies of the photoactivation of TiO_2--new photochemical processes. *Chem. Rev.* **2006**, *106*, 4428–4453.
41. Cwiertny, D. M.; Young, M. A.; Grassian, V. H. Chemistry and photochemistry of mineral dust aerosol. *Annu. Rev. Phys. Chem.* **2008**, *59*, 27–51.
42. Chen, H.; Nanayakkara, C. E.; Grassian, V. H. Titanium dioxide photocatalysis in atmospheric chemistry. *Chem. Rev.* **2012**, *112*, 5919–5948.
43. Colmenares, J. C.; Luque, R. Heterogeneous photocatalytic nanomaterials: prospects and challenges in selective transformations of biomass-derived compounds. *Chem. Soc. Rev.* **2014**, *43*, 765–778.
44. Nanayakkara, C. E.; Pettibone, J.; Grassian, V. H. Sulfur dioxide adsorption and photooxidation on isotopically-labeled titanium dioxide nanoparticle surfaces: roles of surface hydroxyl groups and adsorbed water in the formation and stability of adsorbed sulfite and sulfate. *Phys. Chem. Chem. Phys.* **2012**, *14*, 6957–6966.
45. Gankanda, A.; Coddens, E. M.; Zhang, Y.; Cwiertny, D. M.; Grassian, V. H. Sulfate formation catalyzed by coal fly ash, mineral dust and iron(iii) oxide: variable influence of temperature and light. *Environ. Sci.: Processes Impacts* **2016**, *18*, 1484–1491.

46. Usher, C. R.; Michel, A. E.; Stec, D.; Grassian, V. H. Laboratory studies of ozone uptake on processed mineral dust. *Atmos. Environ.* **2003**, *37*, 5337–5347.
47. Dentener, F. J.; Carmichael, G. R.; Zhang, Y.; Lelieveld, J.; Crutzen, P. J. Role of mineral aerosol as a reactive surface in the global troposphere. *J. Geophys. Res.: Atmos.* **1996**, *101*, 22869–22889.
48. Wang, K.; Zhang, Y.; Nenes, A.; Fountoukis, C. Implementation of dust emission and chemistry into the Community Multiscale Air Quality modeling system and initial application to an Asian dust storm episode. *Atmos. Chem. Phys.* **2012**, *12*, 10209–10237.
49. Underwood, G. M.; Song, C. H.; Phadnis, M.; Carmichael, G. R.; Grassian, V. H. Heterogeneous reactions of NO2 and HNO3 on oxides and mineral dust: A combined laboratory and modeling study. *J. Geophys. Res.: Atmos.* **2001**, *106*, 18055–18066.
50. Liu, C.; Ma, Q.; Liu, Y.; Ma, J.; He, H. Synergistic reaction between SO_2 and NO_2 on mineral oxides: a potential formation pathway of sulfate aerosol. *Phys. Chem. Chem. Phys.* **2012**, *14*, 1668–1676.
51. Ma, Q.; Liu, Y.; He, H. Synergistic effect between NO_2 and SO_2 in their adsorption and reaction on gamma-alumina. *J. Phys. Chem. A* **2008**, *112*, 6630–6635.
52. Li, L.; Chen, Z. M.; Zhang, Y. H.; Zhu, T.; Li, J. L.; Ding, J. Kinetics and mechanism of heterogeneous oxidation of sulfur dioxide by ozone on surface of calcium carbonate. *Atmos. Chem. Phys.* **2006**, *6*, 2453–2464.
53. Ullerstam, M.; Vogt, R.; Langer, S.; Ljungström, E. The kinetics and mechanism of SO_2 oxidation by O_3 on mineral dust. *Phys. Chem. Chem. Phys.* **2002**, *4*, 4694–4699.
54. Usher, C. R.; Al-Hosney, H.; Carlos-Cuellar, S.; Grassian, V. H. A laboratory study of the heterogeneous uptake and oxidation of sulfur dioxide on mineral dust particles. *J. Geophys. Res.: Atmos.* **2002**, *107*, ACH 16-1–ACH 16-9.
55. Wu, L. Y.; Tong, S. R.; Wang, W. G.; Ge, M. F. Effects of temperature on the heterogeneous oxidation of sulfur dioxide by ozone on calcium carbonate. *Atmos. Chem. Phys.* **2011**, *11*, 6593–6605.
56. He, H.; Wang, Y.; Ma, Q.; Ma, J.; Chu, B.; Ji, D.; Tang, G.; Liu, C.; Zhang, H.; Hao, J. Mineral dust and NO_x promote the conversion of SO_2 to sulfate in heavy pollution days. *Sci. Rep.* **2014**, *4*, 4172.
57. Ullerstam, M.; Johnson, M. S.; Vogt, R.; Ljungstrom, E. DRIFTS and Knudsen cell study of the heterogeneous reactivity of SO_2 and NO_2 on mineral dust. *Atmos. Chem. Phys.* **2003**, *3*, 2043–2051.
58. Hua, W.; Chen, Z. M.; Jie, C. Y.; Kondo, Y.; Hofzumahaus, A.; Takegawa, N.; Chang, C. C.; Lu, K. D.; Miyazaki, Y.; Kita, K.; Wang, H. L.; Zhang, Y. H.; Hu, M. Atmospheric hydrogen peroxide and organic hydroperoxides during PRIDE-PRD'06, China: their concentration, formation mechanism and contribution to secondary aerosols. *Atmos. Chem. Phys.* **2008**, *8*, 6755–6773.
59. Huang, L.; Zhao, Y.; Li, H.; Chen, Z. Kinetics of Heterogeneous Reaction of Sulfur Dioxide on Authentic Mineral Dust: Effects of Relative Humidity and Hydrogen Peroxide. *Environ. Sci. Technol.* **2015**, *49*, 10797–805.

60. Usher, C. R.; Michel, A. E.; Grassian, V. H. Reactions on mineral dust. *Chem. Rev.* **2003**, *103*, 4883–4939.

61. George, C.; D'Anna, B.; Herrmann, H.; Weller, C.; Vaida, V.; Donaldson, D. J.; Bartels-Rausch, T.; Ammann, M., Emerging Areas in Atmospheric Photochemistry. In *Atmospheric and Aerosol Chemistry*; McNeill, F. V., Ariya, A. P., Eds.; Springer: Berlin, Heidelberg, 2014; pp 1−53.

62. Jayne, J. T.; Poschl, U.; Chen, Y. M.; Dai, D.; Molina, L. T.; Worsnop, D. R.; Kolb, C. E.; Molina, M. J. Pressure and temperature dependence of the gas-phase reaction of SO_3 with H_2O and the heterogeneous reaction of SO_3 with H_2O/H_2SO_4 surfaces. *J. Phys. Chem. A* **1997**, *101*, 10000–10011.

63. Loerting, T.; Liedl, K. R. Toward elimination of discrepancies between theory and experiment: The rate constant of the atmospheric conversion of SO_3 to H_2SO_4. *Proc. Natl. Acad. Sci. U. S. A.* **2000**, *97*, 8874–8878.

64. Roda, F.; Bellot, J.; Avila, A.; Escarre, A.; Pinol, J.; Terradas, J. Saharan Dust and the Atmospheric Inputs of Elements and Alkalinity to Mediterranean Ecosystems. *Water, Air, Soil Pollut.* **1993**, *66*, 277–288.

65. Cao, J. J.; Lee, S. C.; Zhang, X. Y.; Chow, J. C.; An, Z. S.; Ho, K. F.; Watson, J. G.; Fung, K.; Wang, Y. Q.; Shen, Z. X. Characterization of airborne carbonate over a site near Asian dust source regions during spring 2002 and its climatic and environmental significance. *J. Geophys. Res.: Atmos.* **2005**, *110*, D03203.

66. Ooki, A.; Uematsu, M. Chemical interactions between mineral dust particles and acid gases during Asian dust events. *J. Geophys. Res.: Atmos.* **2005**, *110*, D03201.

67. Yu, Z.; Jang, M.; Park, J. Modeling atmospheric mineral aerosol chemistry to predict heterogeneous photooxidation of SO_2. *Atmos. Chem. Phys.* **2017**, *17*, 10001–10017.

68. Yu, Z.; Jang, M. Simulation of Heterogeneous Photooxidation of SO_2 and NO_x in the presence of Gobi Desert Dust Particles under Ambient Sunlight. *Atmos. Chem. Phys. Discuss.* **2018**, in review.

69. Tang, I. N.; Fung, K. H. Hydration and Raman scattering studies of levitated microparticles: $Ba(NO_3)_2$, $Sr(NO_3)_2$, and $Ca(NO_3)_2$. *J. Phys. Chem.* **1997**, *106*, 1653–1660.

70. Al-Abadleh, H. A.; Grassian, V. H. FT-IR study of water adsorption on aluminum oxide surfaces. *Langmuir* **2003**, *19*, 341–347.

71. Beardsley, R.; Jang, M.; Ori, B.; Im, Y.; Delcomyn, C. A.; Witherspoon, N. Role of sea salt aerosols in the formation of aromatic secondary organic aerosol: yields and hygroscopic properties. *Environ. Chem.* **2013**, *10*, 167–177.

72. Clegg, S. L.; Brimblecombe, P.; Wexler, A. S. Thermodynamic model of the system H^+-NH_4^+-SO_4^{2-}-NO_3^--H_2O at tropospheric temperatures. *J. Phys. Chem. A* **1998**, *102*, 2137–2154.

73. Wexler, A. S.; Clegg, S. L. Atmospheric aerosol models for systems including the ions H^+, NH_4^+, Na^+, SO_4^{2-}, NO_3^-, Cl^-, Br^-, and H_2O. *J. Geophys. Res.: Atmos.* **2002**, *107*, 4207.

74. Clegg, S.; Wexler, A. S. Densities and Apparent Molar Volumes of Atmospherically Important Electrolyte Solutions. 2. The Systems H^+–

HSO_4^-– SO_4^{2-}– H_2O from 0 to 3 mol kg^{-1} as a Function of Temperature and H^+– NH_4^+– HSO_4^-– SO_4^{2-}– H_2O from 0 to 6 mol kg^{-1} at 25° C Using a Pitzer Ion Interaction Model, and NH_4HSO_4– H_2O and $(NH_4)_3H(SO_4)_2$– H_2O over the Entire Concentration Range. *J. Phys. Chem. A* **2011**, *115*, 3461 3471.

75. Zhong, M.; Jang, M. Light absorption coefficient measurement of SOA using a UV-Visible spectrometer connected with an integrating sphere. *Atmos. Environ.* **2011**, *45*, 4263–4271.
76. Reyes-Coronado, D.; Rodríguez-Gattorno, G.; Espinosa-Pesqueira, M.; Cab, C.; De Coss, R.; Oskam, G. Phase-pure TiO_2 nanoparticles: anatase, brookite and rutile. *Nanotechnology* **2008**, *19*, 145605.
77. Underwood, G. M.; Song, C. H.; Phadnis, M.; Carmichael, G. R.; Grassian, V. H. Heterogeneous reactions of NO_2 and HNO_3 on oxides and mineral dust: A combined laboratory and modeling study. *J. Geophys. Res.: Atmos.* **2001**, *106*, 18055–18066.
78. Li, H. J.; Zhu, T.; Zhao, D. F.; Zhang, Z. F.; Chen, Z. M. Kinetics and mechanisms of heterogeneous reaction of NO_2 on $CaCO_3$ surfaces under dry and wet conditions. *Atmos. Chem. Phys.* **2010**, *10*, 463–474.
79. Adams, J. W.; Rodriguez, D.; Cox, R. A. The uptake of SO_2 on Saharan dust: a flow tube study. *Atmos. Chem. Phys.* **2005**, *5*, 2679–2689.
80. Marti, J. J.; Jefferson, A.; Cai, X. P.; Richert, C.; McMurry, P. H.; Eisele, F. H_2SO_4 vapor pressure of sulfuric acid and ammonium sulfate solutions. *J. Geophys. Res.: Atmos.* **1997**, *102*, 3725–3735.
81. Eisele, F. L.; McMurry, P. H. Recent progress in understanding particle nucleation and growth. *Philos. Trans. R. Soc., B* **1997**, *352*, 191–201.
82. Finlayson-Pitts, B. J.; Pitts, J. N., Jr. Acid Deposition: Formation and Fates of Inorganic and Organic Acids in the Troposphere. In *Chemistry of the Upper and Lower Atmosphere*; Academic Press: San Diego, 2000; pp 294–348.
83. Liang, J. Y.; Jacobson, M. Z. A study of sulfur dioxide oxidation pathways over a range of liquid water contents, pH values, and temperatures. *J. Geophys. Res.: Atmos.* **1999**, *104*, 13749–13769.
84. Gustafsson, R. J.; Orlov, A.; Badger, C. L.; Griffiths, P. T.; Cox, R. A.; Lambert, R. M. A comprehensive evaluation of water uptake on atmospherically relevant mineral surfaces: DRIFT spectroscopy, thermogravimetric analysis and aerosol growth measurements. *Atmos. Chem. Phys.* **2005**, *5*, 3415–3421.
85. Lederer, M. R.; Staniec, A. R.; Coates Fuentes, Z. L.; Van Ry, D. A.; Hinrichs, R. Z. Heterogeneous Reactions of Limonene on Mineral Dust: Impacts of Adsorbed Water and Nitric Acid. *J. Phys. Chem. A* **2016**, *120*, 9545–9556.
86. Liu, C.; Chu, B. W.; Liu, Y. C.; Ma, Q. X.; Ma, J. Z.; He, H.; Li, J. H.; Hao, J. M. Effect of mineral dust on secondary organic aerosol yield and aerosol size in alpha-pinene/NO_x photo-oxidation. *Atmos. Environ.* **2013**, *77*, 781–789.
87. Wang, G. H.; Cheng, C. L.; Meng, J. J.; Huang, Y.; Li, J. J.; Ren, Y. Q. Field observation on secondary organic aerosols during Asian dust storm periods:

Formation mechanism of oxalic acid and related compounds on dust surface. *Atmos. Environ.* **2015**, *113*, 169–176.

88. Rossi, M. J. Heterogeneous reactions on salts. *Chem. Rev.* **2003**, *103*, 4823–4882.
89. Sadanaga, Y.; Hirokawa, J.; Akimoto, H. Formation of molecular chlorine in dark condition: Heterogeneous reaction of ozone with sea salt in the presence of ferric ion. *Geophys. Res. Lett.* **2001**, *28*, 4433–4436.
90. Hoppel, W.; Pasternack, L.; Caffrey, P.; Frick, G.; Fitzgerald, J.; Hegg, D.; Gao, S.; Ambrusko, J.; Albrechcinski, T. Sulfur dioxide uptake and oxidation in sea-salt aerosol. *J. Geophys. Res.: Atmos.* **2001**, *106*, 27575–27585.

Chapter 16

Progress and Problems in Modeling Chemical Processing in Cloud Droplets and Wet Aerosol Particles

Barbara Ervens*

University of Colorado Cooperative Institute for Research in Environmental Sciences (CIRES) at the NOAA Earth System Research Laboratory (ESRL) Chemical Sciences Division, 325 Broadway, Boulder, Colorado 80305, United States

***E-mail: barbara.ervens@noaa.gov**

Water plays an important role in the atmosphere in form of cloud, fog or rain droplets and associated with aerosol particles because unique chemical reactions can occur in the aqueous medium. These reactions often lead to different products and take place on different time scales than in the surrounding air. Examples include the oxidation of sulfur dioxide (SO_2) to sulfate (SO_4^{2-}), the major contributor to 'acid rain', or the aqueous formation of secondary organic aerosol mass (aqSOA). Implementing chemical aqueous phase processes into large-scale models is a great challenge as often the resolution of such models is much coarser than the size of individual clouds and/or than the spatial and temporal scales, on which particle composition and water content change. This chapter summarizes the basics and most recent findings in the formation of sulfate and organic aerosol mass and how they are implemented in models. As many oxidation reactions of organic are initiated by OH radical, its source and sink reactions in the aqueous phase are also discussed.

Introduction

Chemical and physical processes in the atmosphere occur on very different spatial scales – from nanometers to 1000s of kilometers. Numerical models that describe the present atmosphere or make predictions of our future atmosphere need to include all these processes. The representation of clouds represents a particular challenge in some models as clouds often have smaller sizes than the grid boxes within large-scale (global, climate) models. Clouds and aerosol-cloud-interactions have the largest uncertainty in current estimates of changes in radiative forcing according to the most recent report of the Intergovernmental Panel on Climate Change (*1*).

Aerosol particles affect clouds and fog in non-linear ways by acting as condensation nuclei and, thus, impacting cloud properties, such as chemical composition of droplets, drop number concentration, drop size distributions, and evolution of precipitation. Species dissolved in cloud and aerosol water, either from cloud condensation nuclei or taken up from the gas phase, undergo chemical reactions in the aqueous phase that can lead to non-volatile compounds, such as sulfate or secondary organic aerosol (aqSOA) that subsequently contribute to ambient aerosol particle loading (Figure 1). Chemical processing in cloud droplets modifies the initial aerosol particles in terms of their physical and chemical properties, e.g., particle size, chemical composition and morphology. These properties affect particle interaction with incoming radiation by scattering/absorbing (aerosol direct effect) and their impact on consecutive cloud cycles (aerosol indirect effects).

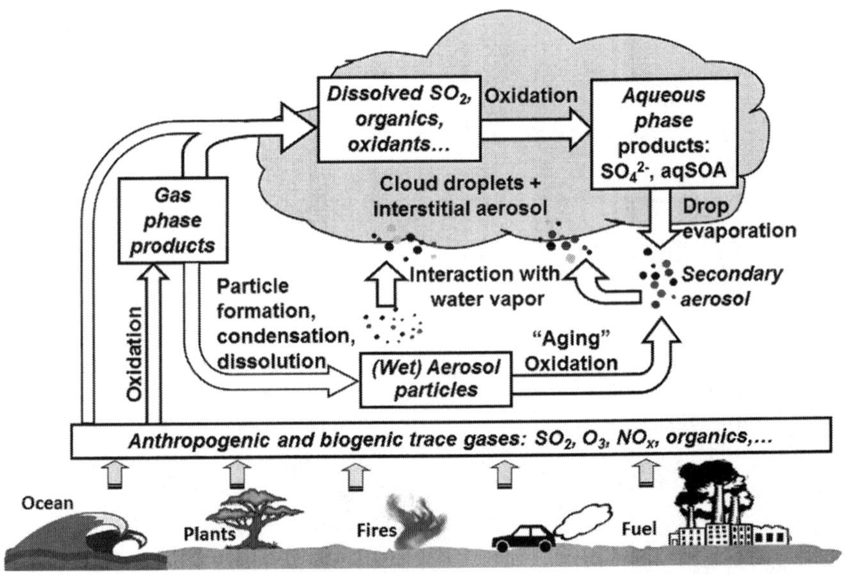

Figure 1. Schematic of aerosol mass formation in clouds and wet aerosol

Recently, it has been suggested that chemical processes in the aqueous phase of aerosol particles, i.e. outside or in the interstitial spaces of clouds, might lead to aqSOA mass. While the liquid water content of aerosol particles is smaller by several orders of magnitude than that of cloud droplets, higher solute concentrations in the former might lead to enhanced reaction rates and, thus, also to efficient aqSOA product formation.

Many of these chemical processes are initiated by photochemically-formed oxidants such as the OH radical. Its sinks and sources in the aqueous phase are not fully constrained and, therefore, there are large uncertainties in the prediction of the OH concentration levels and reaction rates, in particular for reactions of organics. As OH cannot be directly measured in cloud droplets and aerosol particles, estimates of its concentration levels largely rely on model studies.

The atmospheric multiphase system is very complex as detailed in other chapters of this book and explicit chemical mechanisms include hundreds of chemical reactions of inorganic and organic compounds. The current chapter describes findings on such processes and how they may be simplified for model implementation. The focus is set on the aqueous phase formation of sulfate and secondary organic aerosol (aqSOA) mass in clouds and wet aerosol particles and processes that affect OH levels in cloud and aerosol water.

Aerosol Mass Formation in Clouds and Aqueous Particles

Sulfate Formation in the Aqueous Phase

Sulfate accounts for ~50% of the particulate matter on a global scale, with fractions of up to ~70% regionally (*2*). It is mostly formed from anthropogenic (SO_2) and to a smaller extent from biogenic (H_2S, DMS) precursors. Many model studies agree that in-cloud oxidation might contribute significantly (~60 - 90%) to the total sulfate budget (*3, 4*).

The one-step oxidation of SO_2 by OH radicals to H_2SO_4 in the gas phase is relatively well constrained. More uncertainties exist for SO_2 oxidation in the aqueous phase. Due to the great dominance of the aqueous phase pathway, the gas phase oxidation pathway may be even neglected in models (*5*). While it is well established that ozone (O_3) and hydrogen peroxide (H_2O_2) are the main oxidants in the aqueous phase, complex pH dependencies make the prediction of sulfate formation rates difficult. Small contributions of sulfate are also ascribed to the oxidation of sulfur (IV) by molecular oxygen (O_2), catalyzed by transition metal ions (Fe^{3+}, Mn^{2+}). These pathways are included in many regional and global models, either explicitly or in form of parameterizations (*6*). Parameterizations simplify the distribution of newly formed sulfate mass to the initial aerosol population and the partitioning of SO_2, H_2O_2 and O_3 between the gas and aqueous phases. In some cases, thermodynamic equilibrium is implied based on Henry's law whereas in other models, kinetic approaches are applied that consider the time-dependent uptake of trace gases into individual cloud droplets. Whereas in the former approach, a bulk liquid water content (LWC) can be assumed, kinetic uptake is described as a function of droplet size and, thus, a cloud droplet size (distribution) is inferred.

Many regional and global models assume one or two single drop sizes depending on scenarios with smaller droplets (i.e. more aerosol particles) in continental air as compared to marine air e.g., (*7, 8*). A recent theoretical study on sulfate formation in clouds has confirmed previous findings that sulfate formation scales with the total liquid water content (*9*) if it is dominated by the reaction of sulfur (IV) and H_2O_2. It has been shown over a wide range of cloud properties that the assumption of a random single cloud droplet size leads to the same predicted sulfate mass as the assumption of a full cloud droplet size distribution (*10*).

At pH > ~5, sulfur (IV) oxidation by ozone, which is greatly dependent on pH, might become dominant (depending on the ratio of H_2O_2 and O_3). Under these conditions, when significant gradients in pH values between smallest and largest droplets are present, drop size dependence of sulfate formation needs to be considered (*11*).

Several model intercomparisons for sulfate formation have been performed on various scales. For example, in a recent model intercomparison, two global models with an identical aerosol module were used to simulate sulfate formation in East Asia (*12*). The great predicted differences in sulfate loadings were mostly ascribed to differences in cloud properties (LWC, cloud fraction), which affect sulfate loading both through formation and its wet deposition.

Even though deliquesced sea-salt particles contain much less water than cloud droplets, their high pH value might lead to efficient sulfate formation from O_3 oxidation in the marine, cloud-free boundary layer (*13*). In addition to these 'traditional' sulfate formation pathways in the aqueous phase, it has been recently shown that methane sulfonic acid (MSA) can be directly oxidized by OH radicals within or on deliquesced particles with a reactive uptake coefficient of $\gamma = 0.05 \pm 0.03$ (*14*). MSA is an oxidation product of dimethyl sulfide, which is a major source of sulfur in the marine boundary layer. MSA oxidation by OH was previously only considered in the gas phase as a sulfate source (*15*), and the new study (*14*) suggests that oxidation in the aerosol phase might be similarly important to gas phase processes.

The recent great interest in chemical reactions of organics in aerosol water (cf next section on 'Secondary organic aerosol formation in the aqueous phase (aqSOA)) has also led to the discovery of the formation of organosulfur compounds in aerosol that might eventually result in sulfate. One of these pathways is the formation of Criegee intermediates that can act as oxidants for sulfur(IV) leading to efficient sulfate formation in cloud-free biogenic scenarios as shown in a regional model study (*16–19*). Similarly, Criegee intermediate formation from gasoline vehicle exhaust at a relative humidity (RH) of 50% was also suggested to efficiently form sulfate (*20*). The role of this pathway in sulfate formation and new particle formation has not been fully evaluated yet due to the lack of a comprehensive set of chemical parameters (*21*, *22*).

The strong haze formation in Beijing, China, has been explained by the photocatalytical formation of sulfate on dust particles (*23*). These processes are suggested to occur on dust surfaces initiated by oxidants such as OH and ozone. Concurrent high NO_x and SO_2 levels in Beijing lead to enhanced nitrate and sulfate formation. This hygroscopic secondary aerosol mass takes up water where further sulfate formation can take place (*24*). Because these pathways of sulfate formation at RH < 100% are not fully implemented in models, most models cannot predict sulfate formation under cloud-free conditions in various air masses and compare the efficiency of these pathways to in-cloud or gas phase formation on global or regional scales.

Secondary Organic Aerosol Formation in the Aqueous Phase (aqSOA)

Organic aerosol comprises up to 90% of the total ambient aerosol mass. Depending on the region, a large fraction of the mass is secondary organic aerosol, i.e. formed by chemical reactions of gases that result in low volatility products that form new particles or add to the condensed phase of particles (*25*). Traditionally, SOA formation is described as absorption of low-volatility or semivolatile gases that condense on pre-existing particles (*26–28*).

Many laboratory, field and model studies suggest that not only sulfate, but also secondary organic aerosol mass, can be formed in cloud droplets (aqSOA). First ideas on this topic were published in the early 2000's, when oxalate formation from acetylene and ethene was postulated to occur in marine clouds (*29*). The facts that (i) oxalic acid/oxalate is ubiquitous in aerosol particles (e.g., (*30*, *31*)), which have an average lifetime of several days or a few weeks, and (ii) no gas phase reaction is known that results in oxalic acid, suggest that it is formed in the aqueous phase. While oxalate can be considered a tracer of cloud-processing, it only contributes a few percent to total aerosol mass with loadings of few 10's of nanograms m^{-3}.

Following these initial studies, many laboratory experiments explored the formation of oxalate and similar compounds, such as the formation of glyoxylic and pyruvic acids (*32*, *33*). From this, data chemical mechanisms were developed. These mechanisms were implemented into process models and could successfully explain enhanced oxalate concentrations in cloud-processed air (*34*). However, under particular conditions, when high iron concentrations are present, as encountered for example in ship plumes, oxalate might be efficiently depleted due to the formation and subsequent photolysis of iron-oxalato complexes that oxidizes oxalate to CO_2 (*35*).

Several recent review articles have summarized evidence and current knowledge of aqSOA formation in clouds based on laboratory, model and field studies (*36–39*). Global model studies have attempted to quantify the role of aqSOA formed in clouds in the atmosphere. As many reaction parameters and mechanisms are still unknown, these model studies come to very different conclusions regarding the importance of aqSOA. For example, assuming that glyoxal, a precursor of oxalate, is taken up with the same reactive uptake coefficient on cloud droplets and on aqueous aerosol ($\gamma = 2.9 \cdot 10^{-3}$), it was found that aqSOA formation in clouds might double the predicted water-soluble organic carbon loading over North America (*40*). The great sensitivity of cloud aqSOA to cloud properties was highlighted in another global model study, which translates into non-linear relationships between predicted aqSOA mass and cloud LWC (*41*). In this latter study, a great range of aqSOA mass (13.1 - 46.8 Tg/yr) was predicted depending on the assumed chemical mechanism of aqSOA formation. Relatively good agreement between predicted and observed oxalate concentrations was shown for many locations world-wide; however, the relative contribution of oxalate to total organic aerosol was greatly overestimated as the model generally underpredicts total water-soluble organic carbon (*7*).

Organics in clouds are mostly oxidized by the OH radical (*42*). The direct uptake of OH is assumed to be one of the main OH sources in cloud droplets (*42*, *43*). As the OH radical is only moderately soluble (K_H = 30 M atm^{-1} (*44*)) but highly reactive, OH is often consumed very quickly near the drop surface and does not reach its thermodynamic equilibrium concentration throughout the droplet. Consequently, a larger drop surface area allows more OH to be taken up and more organics to be oxidized to form aqSOA products. This leads to relatively higher aqSOA formation rates in smaller droplets due to their larger surface-to-volume ratio (*9*). This trend is supported by the analysis of fog droplet residuals in the Indo Gangetic Plains where a higher oxygen-to-carbon (O/C) ratio was found in small droplets (*45*). The O/C ratio is a measure of the oxidation state of organics in aerosol. Aqueous phase processes lead to highly oxygenated products with oxalate having the highest possible O/C ratio in an organic compound (O/C = 2, carbon oxidation number = 3). Accompanying model studies showed that OH-driven chemistry can reproduce these trends. However, the OH concentration due to direct uptake from the gas phase was not sufficient to explain the high O/C ratios; only if additional OH sources (Fenton reaction, cf Section 'OH radical') were included, the O/C ratio could be reproduced.

Including detailed drop size distributions in models is often unfeasible due to computational burden and/or due to the lack of detailed measurements. Instead, simplifications are needed that take into account the apparent surface-dependence of OH reactions. A comprehensive analysis of size-dependent, OH-driven aqSOA formation in clouds showed that for wide parameter ranges of cloud droplet distributions and liquid water contents, the effective radius of the cloud droplet distributions (total droplet volume / total droplet surface area) can be used to parameterize aqSOA formation rates in clouds (*10*). Using this parameter results in the same aqSOA formation rates (within ± 30%) as size-resolved calculations, but it is computationally much more efficient as only one drop size class has to be considered instead of multiple ones.

Cloud droplets represent a fairly diluted aqueous phase with solute concentrations on the microliter level. Aerosol particles are much more highly concentrated as the soluble fraction of the aerosol particles dissolve and might reach their solubility limit (molar concentrations). Due to these high concentrations, the aerosol water phase cannot be considered dilute and, thus, basic principles for ideal solutions do not apply. Several studies have shown that reaction rate constants might be enhanced or lowered in solutions of high ionic strength (*46–48*). In addition, since Henry's law does not describe non-ideal solutions, the solubility of organic species might be enhanced (*49*) or reduced (*50*). These facts make modeling chemical processes in aerosol water more difficult and complex than in cloud water as many of the parameters have not been systematically investigated as a function of ionic strength and different solutes. In addition, it has been shown that the viscosity of the condensed particle phase might lead to slower reactant diffusion within the particle depending on the ambient RH and water content of the particle (*51*).

Several laboratory studies have shown that at high concentrations of organic compounds in the aqueous phase, the formation of products with a higher molecular weight than the initial reactants is favored (*52–57*). Based on the study by Renard et al. (*58*), a chemical mechanism for the formation of oligomers of methyl vinyl ketone, a major oxidation product of isoprene, was developed (*59*). Unknown rate constants were fit based on the temporal profiles of intermediates and products identified in laboratory experiments. Non-idealities in the aqueous solution were inherently included in these rate constants. Applying this mechanism in a process model with atmospheric conditions suggested that oligomerization of methyl vinyl ketone and similar compounds via OH reactions is likely not a major contributor to aqSOA in the atmosphere, unless the solubility of methylvinyl ketone into aerosol water is much higher than assumed (K_H = 41 M atm^{-1}) and/or a much higher concentration of α,β-unsaturated carbonyl compounds is present in aerosol water than identified up to date. Both possibilities seem unlikely as for nearly all organics a salting-out effect, i.e. lower solubility than in ideal solutions, has been observed in salt solutions (*50*) and structurally similar compounds have not been identified in aerosols at great quantities.

The more soluble glyoxal ($K_H = 3 \cdot 10^5$ M atm^{-1}) has been shown to initiate more efficient aqSOA formation in the aerosol aqueous phase. Unlike all other identified aerosol organics, glyoxal shows a salting-in effect into solutions of common aerosol solutes (e.g., ammonium sulfate). This salting-in effect translates into an apparent effective Henry's law constant of $K_H \sim 10^8$ M atm^{-1} (*60, 61*). Glyoxal can reversibly form dimers but also other compounds such as imidazoles from the reaction with ammonium (NH_4^+) (*62*) or organosulfates (*63, 64*).

Implementing aqSOA formation from glyoxal into process models leads to ambiguous conclusions: while in Mexico City a significant fraction of observed SOA mass was ascribed to glyoxal reactions (*65, 66*), a different model only predicted very low contributions in Los Angeles (*67*). As even the sources of glyoxal in the gas phase are uncertain, models predict very different SOA levels from glyoxal (0 – 0.8 μg m^{-3} SOA) depending on the gas phase mechanism (*68*). Imidazole formation might be a great contributor to organic nitrogen from biomass burning (*69*). Uncertainties associated with the salting-in effect of glyoxal and the simultaneous salting-out effect of other organics and consequences for predicted aqSOA formation are discussed based on a regional model (*70*).

Non-oxidative pathways have been predicted as important for SOA formation in aerosol water. These pathways include the formation of isoprene epoxy diols (IEPOX) in combination with sulfate (cf previous Section 'Sulfate formation in the aqueous phase') that lead to increases in SOA mass (*71, 72*). IEPOX-derived SOA compounds have been identified in various field experiments that focused on biogenic aerosol sources (*68, 73, 74*). Process model studies of IEPOX formation in the Southeast US led to reasonable agreement between observations and model results (*75*). Trends of reduced NO_x and SO_2 with predicted SOA formation have been observed in the Southeast US as the inorganics are precursors for sulfate and nitrate which, in turn, enhance the hygroscopicity and water content of aerosol particles and, thus, allow aqSOA formation in the aqueous phase (*76, 77*). Similar feedbacks of inorganics to aerosol water content and acidity have been previously suggested and it was proposed that the strong reductions in SO_2 concentrations in the US will lead to a decrease in aqSOA formation (*78, 79*). Related connections between anthropogenic sulfate and aqSOA formation were also suggested for Southeast Asia and the Indo Gangetic Plains (*80*).

Observational evidence for aqSOA formation that was not initiated by photochemical processes has been found in the Po Valley, Italy, where a clear correlation of water-soluble organic carbon and aerosol water content was found during periods of low or no photochemical activity (*81*). Tracer compounds, including imidazoles were identified, together with high O/C ratios that all pointed to aqueous phase processing. These measurements were made during stagnating air, conditions under which aqSOA precursors accumulate and dissolve in aerosol water.

This summary of aqSOA processes and discrepancies in findings from aqSOA models shows the large need for more laboratory studies that constrain the underlying chemical mechanisms. Process model studies for ambient conditions are needed in order to identify uncertainties and sensitivities not only to chemical but also microphysical parameters.

OH Radical

The OH radical is the main oxidant both in the atmospheric gas and aqueous phases. It oxidizes a wide variety of inorganic and organic compounds. Its sources in the gas phase are relatively well constrained. However, a recent model intercomparison showed large discrepancies in predicted gas phase OH concentrations (*82*). In this intercomparison, largest discrepancies resulted from different gas phase mechanisms and methane, O_3, and CO concentrations. The presence of clouds only played a minor role.

The reduction of gas phase OH in the presence of clouds has been shown in models several decades ago (*83*). The fraction of OH in clouds is < 1% but yet OH levels in the presence of clouds are smaller by 50% or more. These trends have been also found in observations (*84*). The presence of cloud water results in the separation of HO_2 and NO, which are the two main precursors of OH radical in the gas phase. While HO_2 is highly soluble ($K_{H,\,eff} > 10^4$ M atm^{-1} at pH = 4) and readily taken up by the droplets, the much less soluble NO ($K_H \sim 10^{-3}$ M atm^{-1}) remains in the gas phase.

The drop-size dependence of OH concentrations in cloud droplets due this surface-limitation was discussed in a model study (cf also previous section) (*9*). In the same model study, observational data suggested a surface dependence of oxalate formation in clouds. As the OH radical is the main oxidant for most organic compounds in the aqueous phase, the drop-size dependence of OH translates also into a drop-size dependence of all OH-initiated aqSOA formation.

In addition to the direct uptake from the gas phase, the main known OH sources in the aqueous phase are Fenton reactions (reaction of Fe^{2+} and Cu^+ with H_2O_2) and direct photolyses of NO_3^- and H_2O_2 (*42*). Both the oxidation state and the dissolved fraction of iron has to be known in cloud and aerosol water, in order to accurately model Fenton chemistry. The fact that iron can be complexed by organics, making it unavailable for Fenton chemistry, may lead to a reduced efficiency of OH formation (*85*). Such interactions might be the reason that the role of Fenton chemistry is overestimated for the production of aqueous OH in models (*86*). However, it should be also noted that the number fraction of particles where iron is found is rather low (mostly in the dust coarse mode) so that these reactions might not be efficient in the majority of particles.

While some organics, such as oxalate and similar compounds, might suppress Fenton reactions, other oxidized compounds, including organic hydroperoxides from the oxidation of small carbonyl compounds, might act as an OH source in the aqueous phase by direct photolysis (*87*, *88*). Such compounds seem abundant in water-soluble SOA as it was found that the decay of SOA material in the aqueous phase efficiently produces OH radicals (*89*). These OH production pathways from organic compounds are not comprehensively implemented in models yet as complete sets of reaction parameters are not available. Under conditions when they represent a significant OH source in the aqueous phase, the deviations of OH concentration from Henry's law constants and gradients between drop surface and drop center might be smaller than predicted based on current models (*9*).

There is no direct measurement of OH available in real cloud droplets. Thus, modeled OH(aq) concentrations have to be compared to indirectly derived levels. Utilizing the well-constrained loss rates of benzoic acid as a measure of OH, concentrations of $\sim 10^{-15}$ M in cloud water were estimated (*90*). The main sinks for OH in the aqueous phase are generally organics. However, only a small fraction of all organics in the aqueous phase can be identified on a molecular level (*91*). As it is neither practical for computational reasons nor possible due to the lack of experimental data to include all organic sinks for aqueous OH in models, it is suggested to use a general rate constant of $(3.8 \pm 1.9) \cdot 10^{8}$ M^{-1} s^{-1} to describe the loss of OH due to reactions with dissolved water-soluble organic carbon, a parameter that is routinely measured in cloud and aerosol samples. This general rate constant is in agreement with weighted rate constants that can be derived from structure-reactivity relationships, which were developed for OH reactions with organics including a wide variety of structures and functional groups (*92–95*).

Aqueous Phase Processing in the Future Atmosphere

The water content of clouds is mostly determined by the cooling rate and by meteorological and dynamical conditions. Under the assumption of a constant liquid water content, more numerous particles lead to more and smaller cloud droplets (first indirect effect) (*96*). Thus, with increasing aerosol loading as observed above Asia, the surface-to-volume ratio of cloud droplets might become larger (i.e. droplets are smaller) which might trigger more efficient OH-driven aqSOA formation. Additionally, higher global temperature will change the global distribution of clouds (*97*) and decrease cloudiness (*98*, *99*). Thus, a straightforward prediction of trends in aqSOA formation in clouds cannot be given.

The water content of aerosol particles is much smaller than that of clouds. Particles are usually in thermodynamic equilibrium with the surrounding water vapor and, thus, an increase in particle number and/or mass results in an increase in particle water. The efficient reduction of sulfur emissions in the US and Europe has led to less sulfate aerosol and consequently to a smaller aqueous reaction volume where particle reactions leading to aqSOA can occur (*79*, *100*). In addition, higher temperature leads to evaporation of particle constituents such as ammonium nitrate (*101*, *102*) or organics. Less particle mass will lead to less aerosol water where chemical reactions can occur. The evaporation of organic aerosol might be partially offset by higher biogenic emissions that can act as precursors for SOA. However, higher temperature will also increase reaction rates. This might lead to faster SO_2 conversion and sulfate formation. The expected trend is ambiguous because SOA mass might be formed faster while fragmentation of low volatility compounds to more volatile compounds that escape to the gas phase might be accelerated. Thus, the consequence of temperature increase for net SOA formation cannot be easily estimated.

The formation processes of organic aerosol largely determine the radiative forcing of aerosol in the atmosphere as different mechanisms lead to the growth of particles in different sections of the aerosol population (*103*). Thus, models should not only target the reproduction of total observed aerosol mass but also the distribution of the mass throughout the aerosol populations of particle size, mixing state, and morphology determines their interaction with radiation.

References

1. Stocker, T. F.; Qin, D.; Plattner, G. K.; Tignor, M.; Allen, S. K.; Boschung, J.; Nauels, A.; Xia, Y.; Bex, V.; Midgley, P. M. *Climate Change 2013: The Physical Science Basis. Contribution of Working Group I to the Fifth Assessment Report of the Intergovernmental Panel on Climate Change*; Intergovernmental Panel on Climate Change: United Kingdom and New York, 2013; p 1535.
2. Zhang, Q.; Jimenez, J. L.; Canagaratna, M. R.; Allan, J. D.; Coe, H.; Ulbrich, I.; Alfarra, M. R.; Takami, A.; Middlebrook, A. M.; Sun, Y. L.; Dzepina, K.; Dunlea, E.; Docherty, K.; DeCarlo, P. F.; Salcedo, D.; Onasch, T.; Jayne, J. T.; Miyoshi, T.; Shimono, A.; Hatekayama, S.; Takegawa, N.; Kondo, Y.; Schneider, J.; Drewnick, F.; Borrmann, S.; Weiner, S.; Demerijan, K.; Williams, P.; Bower, K.; Bahreini, R.; Cottrell, L.; Griffin, R. J.; Rautiainen, J.; Sun, J. Y.; Zhang, Y. M.; Worsnop, D. R. Ubiquity and dominance of oxygenated species in organic aerosols in anthropogenically-influenced Northern Hemisphere midlatitudes. *Geophys. Res. Lett.* **2007**, *34*, L13801.
3. Faloona, I. Sulfur processing in the marine atmospheric boundary layer: A review and critical assessment of modeling uncertainties. *Atmos. Environ.* **2009**, *43*, 2841–2854.
4. Barth, M. C.; Rasch, P. J.; Kiehl, J. T.; Benkovitz, C. M.; Schwartz, S. E. Sulfur chemistry in the National Center for Atmospheric Research Community Climate Model: Description, evaluation, features, and sensitivity to aqueous chemistry. *J. Geophys. Res. - Atmos.* **2000**, *105*, 1387–1415.
5. Cautenet, S.; Lefeivre, B. Contrasting behavior of gas and aerosol scavenging in convective rain: A numerical and experimental study in the African equatorial forest. *J. Geophys. Res. - Atmos.* **1994**, *99*, 13013–13024.
6. Ervens, B. Modeling the Processing of Aerosol and Trace Gases in Clouds and Fogs. *Chem. Rev.* **2015**, *115*, 4157–4198.

7. Myriokefalitakis, S.; Tsigaridis, K.; Mihalopoulos, N.; Sciare, J.; Nenes, A.; Segers, A.; Kanakidou, M. In-cloud oxalate formation in the global troposphere: a 3-D modeling study. *Atmos. Chem. Phys.* **2011**, *11*, 5761–5782.
8. Roelofs, G.; Lelieveld, J.; Ganzeveld, L. Simulation of global sulfate distribution and the influence on effective cloud drop radii with a coupled photochemistry sulfur cycle model. *Tellus B* **1998**, *50*, 224–242.
9. Ervens, B.; Sorooshian, A.; Lim, Y. B.; Turpin, B. J. Key parameters controlling OH-initiated formation of secondary organic aerosol in the aqueous phase (aqSOA). *J. Geophys. Res. - Atmos.* **2014**, *119*, 3997–4016.
10. McVay, R.; Ervens, B. A microphysical parameterization of aqSOA and sulfate formation in clouds. *Geophys. Res. Lett.* **2017**, *44*, 7500–7509.
11. Seinfeld, J. H.; Pandis, S. N. *Atmospheric Chemistry and Physics - From air pollution to climate change*, 2nd ed.; John Wiley & Sons, Inc.: Hoboken, NJ, 2006; p 1225.
12. Goto, D.; Nakajima, T.; Dai, T.; Takemura, T.; Kajino, M.; Matsui, H.; Takami, A.; Hatakeyama, S.; Sugimoto, N.; Shimizu, A.; Ohara, T. An evaluation of simulated particulate sulfate over East Asia through global model intercomparison. *J. Geophys. Res: Atmos.* **2015**, *120*, 2014JD021693.
13. Sievering, H.; Boatman, J.; Galloway, J.; Keene, W.; Kim, Y.; Luria, M.; Ray, J. Heterogeneous sulfur conversion in sea-salt aerosol particles: the role of aerosol water content and size distribution. *Atmos. Environ. A* **1991**, *25*, 1479–1487.
14. Mungall, E. L.; Wong, J. P. S.; Abbatt, J. P. D. Heterogeneous Oxidation of Particulate Methanesulfonic Acid by the Hydroxyl Radical: Kinetics and Atmospheric Implications. *ACS Earth Space Chem.* **2018**, *2*, 48–55.
15. Saltzman, E. S.; Savoie, D. L.; Zika, R. G.; Prospero, J. M. Methane sulfonic acid in the marine atmosphere. *J.Geophys. Res.: Oceans* **1983**, *88*, 10897–10902.
16. Sarwar, G.; Fahey, K.; Kwok, R.; Gilliam, R. C.; Roselle, S. J.; Mathur, R.; Xue, J.; Yu, J.; Carter, W. P. L. Potential impacts of two SO2 oxidation pathways on regional sulfate concentrations: Aqueous-phase oxidation by NO2 and gas-phase oxidation by Stabilized Criegee Intermediates. *Atmos. Environ.* **2013**, *68*, 186–197.
17. Sarwar, G.; Simon, H.; Fahey, K.; Mathur, R.; Goliff, W. S.; Stockwell, W. R. Impact of sulfur dioxide oxidation by Stabilized Criegee Intermediate on sulfate. *Atmos. Environ.* **2014**, *85*, 204–214.
18. Huang, H.-L.; Chao, W.; Lin, J. J.-M. Kinetics of a Criegee intermediate that would survive high humidity and may oxidize atmospheric SO_2. *Proc. Natl. Acad. Sci. U.S.A.* **2015**, *112*, 10857–10862.
19. Kim, P. S.; Jacob, D. J.; Fisher, J. A.; Travis, K.; Yu, K.; Zhu, L.; Yantosca, R. M.; Sulprizio, M. P.; Jimenez, J. L.; Campuzano-Jost, P.; Froyd, K. D.; Liao, J.; Hair, J. W.; Fenn, M. A.; Butler, C. F.; Wagner, N. L.; Gordon, T. D.; Welti, A.; Wennberg, P. O.; Crounse, J. D.; St. Clair, J. M.; Teng, A. P.; Millet, D. B.; Schwarz, J. P.; Markovic, M. Z.; Perring, A. E. Sources, seasonality, and trends of southeast US aerosol: an integrated analysis of

surface, aircraft, and satellite observations with the GEOS-Chem chemical transport model. *Atmos. Chem. Phys.* **2015**, *15*, 10411–10433.

20. Liu, T.; Wang, X.; Hu, Q.; Deng, W.; Zhang, Y.; Ding, X.; Fu, X.; Bernard, F.; Zhang, Z.; Lü, S.; He, Q.; Bi, X.; Chen, J.; Sun, Y.; Yu, J.; Peng, P.; Sheng, G.; Fu, J. Formation of secondary aerosols from gasoline vehicle exhaust when mixing with SO2. *Atmos. Chem. Phys.* **2016**, *16*, 675–689.
21. Newland, M. J.; Rickard, A. R.; Vereecken, L.; Muñoz, A.; Ródenas, M.; Bloss, W. J. Atmospheric isoprene ozonolysis: impacts of stabilised Criegee intermediate reactions with SO_2, H_2O and dimethyl sulfide. *Atmos. Chem. Phys.* **2015**, *15*, 9521–9536.
22. Li, J.; Ying, Q.; Yi, B.; Yang, P. Role of stabilized Criegee Intermediates in the formation of atmospheric sulfate in eastern United States. *Atmos. Environ.* **2013**, *79*, 442–447.
23. Yu, Z.; Jang, M.; Park, J. Modeling atmospheric mineral aerosol chemistry to predict heterogeneous photooxidation of SO2. *Atmos. Chem. Phys.* **2017**, *17*, 10001–10017.
24. Ma, J.; Chu, B.; Liu, J.; Liu, Y.; Zhang, H.; He, H. NOx promotion of SO2 conversion to sulfate: An important mechanism for the occurrence of heavy haze during winter in Beijing. *Environ. Pollut.* **2018**, *233*, 662–669.
25. Kanakidou, M.; Seinfeld, J. H.; Pandis, S.; Barnes, I.; Dentener, F. J.; Facchini, M. C.; van Dingenen, R.; Ervens, B.; Nenes, A.; Nielsen, C. J.; Swietlicki, E.; Putaud, J. P.; Balkanski, Y.; C. E., S. F.; Hjorth, J.; Moortgat, G.; Winterhalter, R.; Myhre, C. E. L.; Tsigaridis, K.; Vignati, E.; Stephanou, E.; Wilson, J. Organic aerosol and global climate modelling: A review. *Atmos. Chem. Phys.* **2005**, *5*, 1–70.
26. Pankow, J. F.; Seinfeld, J. H.; Asher, W. E.; Erdakos, C. B. Modeling the Formation of Secondary Aerosol. 1. Application of Theoretical Principles to Measurements Obtained in the α-Pinene/,β-Pinene/,Sabinene/,Δ3 Carene/, and Cyclohexene/Ozone Systems. *Environ. Sci. Technol.* **2001**, *35*, 1164–1172.
27. Pankow, J. F. An Absorption model of the gas/aerosol partitioning involved in the formation of secondary organic aerosol. *Atmos. Environ.* **1994**, *28*, 189–193.
28. Donahue, N. M.; Robinson, A. L.; Stanier, C. O.; Pandis, S. N. Coupled partitioning, dilution and chemical aging of semivolatile organics. *Environ. Sci. Technol.* **2006**, *40*, 2635–2643.
29. Warneck, P. In-cloud chemistry opens pathway to the formation of oxalic acid in the marine atmosphere. *Atmos. Environ.* **2003**, *37*, 2423–2427.
30. Mochida, M.; Umemoto, N.; Kawamura, K.; Uematsu, M. Bimodal size distribution of C_2-C_4 dicarboxylic acids in the marine aerosols. *Geophys. Res. Lett.* **2003**, *30*, 1672–1676.
31. Kawamura, K.; Sakaguchi, F. Molecular distributions of water soluble dicarboxylic acids in marine aerosols over the Pacific Ocean including tropics. *J. Geophys. Res. - Atmos.* **1999**, *104*, 3501–3509.
32. Carlton, A. G.; Turpin, B. J.; Altieri, K. E.; Reff, A.; Seitzinger, S.; Lim, H.; Ervens, B. Atmospheric oxalic acid and SOA production from glyoxal:

Results of aqueous photooxidation experiments. *Atmos. Environ.* **2007**, *41*, 7588–7602.

33. Lim, H.; Carlton, A. G.; Turpin, B. J. Isoprene Forms Secondary Organic Aerosol through Cloud Processing: Model Simulations. *Environ. Sci. Technol.* **2005**, *39*, 4441–4446.
34. Wonaschuetz, A.; Sorooshian, A.; Ervens, B.; Chuang, P. Y.; Feingold, G.; Murphy, S. M.; de Gouw, J.; Warneke, C.; Jonsson, H. H. Aerosol and gas re-distribution by shallow cumulus clouds: An investigation using airborne measurements. *J. Geophys. Res. - Atmos.* **2012**, *117*, D17202.
35. Sorooshian, A.; Wang, Z.; Coggon, M. M.; Jonsson, H. H.; Ervens, B. Observations of Sharp Oxalate Reductions in Stratocumulus Clouds at Variable Altitudes: Organic Acid and Metal Measurements During the 2011 E-PEACE Campaign. *Environ. Sci. Technol.* **2013**, *47*, 7747–7756.
36. McNeill, V. F. Aqueous Organic Chemistry in the Atmosphere: Sources and Chemical Processing of Organic Aerosols. *Environ. Sci. Technol.* **2015**, *49*, 1237–1244.
37. Ervens, B.; Turpin, B. J.; Weber, R. J. Secondary organic aerosol formation in cloud droplets and aqueous particles (aqSOA): a review of laboratory, field and model studies. *Atmos. Chem. Phys.* **2011**, *11*, 11069–11102.
38. Shrivastava, M.; Cappa, C. D.; Fan, J.; Goldstein, A. H.; Guenther, A. B.; Jimenez, J. L.; Kuang, C.; Laskin, A.; Martin, S. T.; Ng, N. L.; Petaja, T.; Pierce, J. R.; Rasch, P. J.; Roldin, P.; Seinfeld, J. H.; Shilling, J.; Smith, J. N.; Thornton, J. A.; Volkamer, R.; Wang, J.; Worsnop, D. R.; Zaveri, R. A.; Zelenyuk, A.; Zhang, Q. Recent advances in understanding secondary organic aerosol: Implications for global climate forcing. *Rev. Geophys.* **2017**, *55*, 509–559.
39. Lim, Y. B.; Tan, Y.; Perri, M. J.; Seitzinger, S. P.; Turpin, B. J. Aqueous chemistry and its role in secondary organic aerosol (SOA) formation. *Atmos. Chem. Phys.* **2010**, *10*, 10521–10539.
40. Fu, T.; Jacob, D. J.; Heald, C. L. Aqueous-phase reactive uptake of dicarbonyls as a source of organic aerosol over eastern North America. *Atmos. Environ.* **2009**, *43*, 1814–1822.
41. Lin, G.; Sillman, S.; Penner, J. E.; Ito, A. Global modeling of SOA: the use of different mechanisms for aqueous phase formation. *Atmos. Chem. Phys.* **2014**, *14*, 5451–5475.
42. Ervens, B.; George, C.; Williams, J. E.; Buxton, G. V.; Salmon, G. A.; Bydder, M.; Wilkinson, F.; Dentener, F.; Mirabel, P.; Wolke, R.; Herrmann, H. CAPRAM2.4 (MODAC mechanism): An extended and condensed tropospheric aqueous phase mechanism and its application. *J. Geophys. Res.* **2003**, *108*, 4426.
43. Herrmann, H.; Tilgner, A.; Barzaghi, P.; Majdik, Z.; Gligorovski, S.; Poulain, L.; Monod, A. Towards a more detailed description of tropospheric aqueous phase organic chemistry: CAPRAM 3.0. *Atmos. Environ* **2005**, *39*, 4351–4363.
44. Hanson, D. R.; Burkholder, J. B.; Howard, C. J.; Ravishankara, A. R. Measurement of hydroxyl and hydroperoxy radical uptake coefficients on water and sulfuric acid surfaces. *J. Phys. Chem.* **1992**, *96*, 4979–4985.

45. Chakraborty, A.; Ervens, B.; Gupta, T.; Tripathi, S. N. Characterization of organic residues of size-resolved fog droplets and their atmospheric implications. *J. Geophys. Res: Atmos.* **2016**, *121*, 4317–4332.
46. Tursic, J.; Grgic, I.; Podkrajsek, B. Influence of ionic strength on aqueous oxidation of SO_2 catalysed by manganese. *Atmos. Environ.* **2003**, *37*, 2589–2595.
47. Bao, Z.; Barker, J. R. Temperature and Ionic Strength Effects of some Reactions Involving Sulfate Radical (SO4-(aq)). *J. Phys. Chem. A* **1996**, *100*, 9780–9787.
48. Herrmann, H. Kinetics of aqueous phase reactions relevant for atmospheric chemistry. *Chem. Rev.* **2003**, *103*, 4691–4716.
49. Kampf, C. J.; Waxman, E. M.; Slowik, J. G.; Dommen, J.; Pfaffenberger, L.; Praplan, A. P.; Prévôt, A. S. H.; Baltensperger, U.; Hoffmann, T.; Volkamer, R. Effective Henry's Law Partitioning and the Salting Constant of Glyoxal in Aerosols Containing Sulfate. *Environ. Sci. Technol.* **2013**, *47*, 4236–4244.
50. Wang, C.; Lei, Y. D.; Endo, S.; Wania, F. Measuring and Modeling the Salting-out Effect in Ammonium Sulfate Solutions. *Environ. Sci. Technol.* **2014**, *48*, 13238–13245.
51. Pajunoja, A.; Hu, W.; Leong, Y. J.; Taylor, N. F.; Miettinen, P.; Palm, B. B.; Mikkonen, S.; Collins, D. R.; Jimenez, J. L.; Virtanen, A. Phase state of ambient aerosol linked with water uptake and chemical aging in the southeastern US. *Atmos. Chem. Phys.* **2016**, *16*, 11163–11176.
52. Tan, Y.; Lim, Y. B.; Altieri, K. E.; Seitzinger, S. P.; Turpin, B. J. Mechanisms leading to oligomers and SOA through aqueous photooxidation: insights from OH radical oxidation of acetic acid and methylglyoxal. *Atmos. Chem. Phys.* **2012**, *12*, 801–813.
53. Tan, Y.; Carlton, A. G.; Seitzinger, S. P.; Turpin, B. J. SOA from methylglyoxal in clouds and wet aerosols: Measurement and prediction of key products. *Atmos. Environ.* **2010**, *44*, 5218–5226.
54. Altieri, K. E.; Seitzinger, S. P.; Carlton, A. G.; Turpin, B. J.; Klein, G. C.; Marshall, A. G. Oligomers formed through in-cloud methylglyoxal reactions: Chemical composition, properties, and mechanisms investigated by ultra-high resolution FT-ICR mass spectrometry. *Atmos. Environ.* **2008**, *42*, 1476–1490.
55. Renard, P.; Reed Harris, A. E.; Rapf, R. J.; Ravier, S.; Demelas, C.; Coulomb, B.; Quivet, E.; Vaida, V.; Monod, A. Aqueous Phase Oligomerization of Methyl Vinyl Ketone by Atmospheric Radical Reactions. *J. Phys. Chem. C* **2014**, *14*, 29421–29430.
56. Yasmeen, F.; Sauret, N.; Gal, J.-F.; Maria, P.-C.; Massi, L.; Maenhaut, W.; Claeys, M. Characterization of oligomers from methylglyoxal under dark conditions: a pathway to produce secondary organic aerosol through cloud processing during nighttime. *Atmos. Chem. Phys.* **2010**, *10*, 3803–3812.
57. Zhang, X.; Chen, Z. M.; Zhao, Y. Laboratory simulation for the aqueous OH-oxidation of methyl vinyl ketone and methacrolein: significance to the in-cloud SOA production. *Atmos. Chem. Phys.* **2010**, *10*, 9551–9561.

58. Renard, P.; Siekmann, F.; Salque, G.; Smaani, A.; Demelas, C.; Coulomb, B.; Vassalo, L.; Ravier, S.; Temime-Roussel, B.; Voisin, D.; Monod, A. Aqueous phase oligomerization of methyl vinyl ketone through photooxidation – Part 1: Aging processes of oligomers. *Atmos. Chem. Phys.* **2015**, *15*, 51–35.
59. Ervens, B.; Renard, P.; Tlili, S.; Ravier, S.; Clément, J. L.; Monod, A. Aqueous-phase oligomerization of methyl vinyl ketone through photooxidation – Part 2: Development of the chemical mechanism and atmospheric implications. *Atmos. Chem. Phys.* **2015**, *15*, 9109–9127.
60. Ervens, B.; Volkamer, R. Glyoxal processing by aerosol multiphase chemistry: towards a kinetic modeling framework of secondary organic aerosol formation in aqueous particles. *Atmos. Chem. Phys.* **2010**, *10*, 8219–8244.
61. Volkamer, R.; Ziemann, P. J.; Molina, M. J. Secondary organic aerosol formation from acetylene (C_2H_2): seed effect on SOA yields due to organic photochemistry in the aerosol aqueous phase. *Atmos. Chem. Phys.* **2009**, *9*, 1907–1928.
62. Noziere, B.; Dziedzic, P.; Cordova, A. Products and kinetics of the liquid-phase reaction of glyoxal catalysed by ammonium ions (NH_4^+). *J. Phys. Chem. A* **2009**, *113*, 231–237.
63. Galloway, M. M.; Chhabra, P. S.; Chan, A. W. H.; Surratt, J. D.; Flagan, R. C.; Seinfeld, J. H.; Keutsch, F. N. Glyoxal uptake on ammonium sulphate seed aerosol: reaction products and reversibility of uptake under dark and irradiated conditions. *Atmos. Chem. Phys.* **2009**, *9*, 3331–3345.
64. Yu, G.; Bayer, A. R.; Galloway, M. M.; Korshavn, K. J.; Fry, C. G.; Keutsch, F. N. Glyoxal in aqueous ammonium sulfate solutions: Products, kinetics and hydration effects. *Environ. Sci. Technol.* **2011**, *45*, 6336–6342.
65. Volkamer, R.; Jimenez, J. L.; SanMartini, F.; Dzepina, K.; Zhang, Q.; Salcedo, D.; Molina, L. T.; Worsnop, D. R.; Molina, M. J. Secondary organic aerosol formation from anthropogenic air pollution: Rapid and higher than expected. *Geophys. Res. Lett.* **2006**, *33*, L17811.
66. Waxman, E. M.; Dzepina, K.; Ervens, B.; Lee-Taylor, J.; Aumont, B.; Jimenez, J. L.; Madronich, S.; Volkamer, R. Secondary organic aerosol formation from semi- and intermediate-volatility organic compounds and glyoxal: Relevance of O/C as a tracer for aqueous multiphase chemistry. *Geophys. Res. Lett.* **2013**, *40*, 1–5.
67. Washenfelder, R. A.; Young, C. J.; Brown, S. S.; Angevine, W. M.; Atlas, E. L.; Blake, D. R.; Bon, D. M.; Cubison, M. J.; de Gouw, J. A.; Dusanter, S.; Flynn, J.; Gilman, J. B.; Graus, M.; Griffith, S.; Grossberg, N.; Hayes, P. L.; Jimenez, J. L.; Kuster, W. C.; Lefer, B. L.; Pollack, I. B.; Ryerson, T. B.; Stark, H.; Stevens, P. S.; Trainer, M. K. The glyoxal budget and its contribution to organic aerosol for Los Angeles, California, during CalNex 2010. *J. Geophys. Res: Atmos.* **2011**, *116*, D00V02.
68. Li, J.; Mao, J.; Min, K.-E.; Washenfelder, R. A.; Brown, S. S.; Kaiser, J.; Keutsch, F. N.; Volkamer, R.; Wolfe, G. M.; Hanisco, T. F.; Pollack, I. B.; Ryerson, T. B.; Graus, M.; Gilman, J. B.; Lerner, B. M.; Warneke, C.; de Gouw, J. A.; Middlebrook, A. M.; Liao, J.; Welti, A.; Henderson, B. H.; McNeill, V. F.; Hall, S. R.; Ullmann, K.; Donner, L. J.; Paulot, F.;

Horowitz, L. W. Observational constraints on glyoxal production from isoprene oxidation and its contribution to organic aerosol over the Southeast United States. *J. Geophys. Res: Atmos.* **2016**, *121*, 2016JD025331.

69. Ito, A.; Lin, G.; Penner, J. E. Global modeling study of soluble organic nitrogen from open biomass burning. *Atmos. Environ.* **2015**, *121*, 103–112.
70. Toivola, M.; Prisle, N. L.; Elm, J.; Waxman, E. M.; Volkamer, R.; Kurtén, T. Can COSMOTherm Predict a Salting in Effect? *J. Phys. Chem. A* **2017**, *121*, 6288–6295.
71. Surratt, J. D.; Chan, A. W. H.; Eddingsaas, N. C.; Chan, M.; Loza, C. L.; Kwan, A. J.; Hersey, S. P.; Flagan, R. C.; Wennberg, P. O.; Seinfeld, J. H. Reactive intermediates revealed in secondary organic aerosol formation from isoprene. *Proc. Natl. Acad. Sci. U.S.A.* **2010**, *107*, 6640–6645.
72. Worton, D. R.; Surratt, J. D.; LaFranchi, B. W.; Chan, A. W. H.; Zhao, Y.; Weber, R. J.; Park, J.-H.; Gilman, J. B.; de Gouw, J.; Park, C.; Schade, G.; Beaver, M.; Clair, J. M. S.; Crounse, J.; Wennberg, P.; Wolfe, G. M.; Harrold, S.; Thornton, J. A.; Farmer, D. K.; Docherty, K. S.; Cubison, M. J.; Jimenez, J.-L.; Frossard, A. A.; Russell, L. M.; Kristensen, K.; Glasius, M.; Mao, J.; Ren, X.; Brune, W.; Browne, E. C.; Pusede, S. E.; Cohen, R. C.; Seinfeld, J. H.; Goldstein, A. H. Observational Insights into Aerosol Formation from Isoprene. *Environ. Sci. Technol.* **2013**, *47*, 11403–11413.
73. Budisulistiorini, S. H.; Canagaratna, M. R.; Croteau, P. L.; Marth, W. J.; Baumann, K.; Edgerton, E. S.; Shaw, S. L.; Knipping, E. M.; Worsnop, D. R.; Jayne, J. T.; Gold, A.; Surratt, J. D. Real-Time Continuous Characterization of Secondary Organic Aerosol Derived from Isoprene Epoxydiols in Downtown Atlanta, Georgia, Using the Aerodyne Aerosol Chemical Speciation Monitor. *Environ. Sci. Technol.* **2013**, *47*, 5686–5694.
74. Chan, M. N.; Surratt, J. D.; Claeys, M.; Edgerton, E. S.; Tanner, R. L.; Shaw, S. L.; Zheng, M.; Knipping, E. M.; Eddingsaas, N. C.; Wennberg, P. O.; Seinfeld, J. H. Characterization and Quantification of Isoprene-Derived Epoxydiols in Ambient Aerosol in the Southeastern United States. *Environ. Sci. Technol.* **2010**, *44*, 4590–4596.
75. Budisulistiorini, S. H.; Nenes, A.; Carlton, A. G.; Surratt, J. D.; McNeill, V. F.; Pye, H. O. T. Simulating Aqueous-Phase Isoprene-Epoxydiol (IEPOX) Secondary Organic Aerosol Production During the 2013 Southern Oxidant and Aerosol Study (SOAS). *Environ. Sci. Technol.* **2017**, *51*, 5026–5034.
76. Marais, E. A.; Jacob, D. J.; Jimenez, J. L.; Campuzano-Jost, P.; Day, D. A.; Hu, W.; Krechmer, J.; Zhu, L.; Kim, P. S.; Miller, C. C.; Fisher, J. A.; Travis, K.; Yu, K.; Hanisco, T. F.; Wolfe, G. M.; Arkinson, H. L.; Pye, H. O. T.; Froyd, K. D.; Liao, J.; McNeill, V. F. Aqueous-phase mechanism for secondary organic aerosol formation from isoprene: application to the southeast United States and co-benefit of SO2 emission controls. *Atmos. Chem. Phys.* **2016**, *16*, 1603–1618.
77. Rattanavaraha, W.; Chu, K.; Budisulistiorini, S. H.; Riva, M.; Lin, Y. H.; Edgerton, E. S.; Baumann, K.; Shaw, S. L.; Guo, H.; King, L.; Weber, R. J.; Neff, M. E.; Stone, E. A.; Offenberg, J. H.; Zhang, Z.; Gold, A.; Surratt, J. D. Assessing the impact of anthropogenic pollution on isoprene-derived secondary organic aerosol formation in PM2.5 collected

from the Birmingham, Alabama, ground site during the 2013 Southern Oxidant and Aerosol Study. *Atmos. Chem. Phys.* **2016**, *16*, 4897–4914.

78. Sareen, N.; Carlton, A. G.; Surratt, J. D.; Gold, A.; Lee, B.; Lopez-Hilfiker, F. D.; Mohr, C.; Thornton, J. A.; Zhang, Z.; Lim, Y. B.; Turpin, B. J. Identifying precursors and aqueous organic aerosol formation pathways during the SOAS campaign. *Atmos. Chem. Phys.* **2016**, *16*, 14409–14420.
79. Carlton, A. G.; Turpin, B. J. Particle partitioning potential of organic compounds is highest in the Eastern US and driven by anthropogenic water. *Atmos. Chem. Phys.* **2013**, *13*, 10203–10214.
80. Bikkina, S.; Kawamura, K.; Sarin, M. Secondary Organic Aerosol Formation over Coastal Ocean: Inferences from Atmospheric Water-Soluble Low Molecular Weight Organic Compounds. *Environ. Sci. Technol.* **2017**, *51*, 4347–4357.
81. Sullivan, A. P.; Hodas, N.; Turpin, B. J.; Skog, K.; Keutsch, F. N.; Gilardoni, S.; Paglione, M.; Rinaldi, M.; Decesari, S.; Facchini, M. C.; Poulain, L.; Herrmann, H.; Wiedensohler, A.; Nemitz, E.; Twigg, M. M.; Collett, Jr, J. L. Evidence for ambient dark aqueous SOA formation in the Po Valley, Italy. *Atmos. Chem. Phys.* **2016**, *16*, 8095–8108.
82. Nicely, J. M.; Salawitch, R. J.; Canty, T.; Anderson, D. C.; Arnold, S. R.; Chipperfield, M. P.; Emmons, L. K.; Flemming, J.; Huijnen, V.; Kinnison, D. E.; Lamarque, J.-F.; Mao, J.; Monks, S. A.; Steenrod, S. D.; Tilmes, S.; Turquety, S. Quantifying the causes of differences in tropospheric OH within global models. *J. Geophys. Res: Atmos.* **2017**, *122*, 2016JD026239.
83. Lelieveld, J.; Crutzen, P. J. Influences of cloud photochemical processes on tropospheric ozone. *Nature* **1990**, *343*, 227–233.
84. Frost, G. J.; Trainer, M.; Mauldin, R. L., III; Eisele, F. L.; Prevot, A. S. H.; Flocke, S. J.; Madronich, S.; Kok, G.; Schillawski, R. D.; Baumgardner, D.; Bradshaw, J. Photochemical modeling of OH levels during the First Aerosol Characterization Experiment (ACE 1). *J. Geophys. Res. - Atmos.* **1999**, *104*, 16041–16052.
85. Hems, R. F.; Hsieh, J. S.; Slodki, M. A.; Zhou, S.; Abbatt, J. P. D. Suppression of OH Generation from the Photo-Fenton Reaction in the Presence of α-Pinene Secondary Organic Aerosol Material. *Environ. Sci. Technol. Lett.* **2017**, *4*, 439–443.
86. Bianco, A.; Passananti, M.; Perroux, H.; Voyard, G.; Mouchel-Vallon, C.; Chaumerliac, N.; Mailhot, G.; Deguillaume, L.; Brigante, M. A better understanding of hydroxyl radical photochemical sources in cloud waters collected at the puy de Dôme station – experimental versus modelled formation rates. *Atmos. Chem. Phys.* **2015**, *15*, 9191–9202.
87. Lim, Y. B.; Turpin, B. J. Laboratory evidence of organic peroxide and peroxyhemiacetal formation in the aqueous phase and implications for aqueous OH. *Atmos. Chem. Phys.* **2015**, *15*, 12867–12877.
88. Badali, K. M.; Zhou, S.; Aljawhary, D.; Antiñolo, M.; Chen, W. J.; Lok, A.; Mungall, E.; Wong, J. P. S.; Zhao, R.; Abbatt, J. P. D. Formation of hydroxyl radicals from photolysis of secondary organic aerosol material. *Atmos. Chem. Phys.* **2015**, *15*, 7831–7840.

89. Tong, H.; Arangio, A. M.; Lakey, P. S. J.; Berkemeier, T.; Liu, F.; Kampf, C. J.; Brune, W. H.; Pöschl, U.; Shiraiwa, M. Hydroxyl radicals from secondary organic aerosol decomposition in water. *Atmos. Chem. Phys.* **2016**, *16*, 1761–1771.
90. Arakaki, T.; Anastasio, C.; Kuroki, Y.; Nakajima, H.; Okada, K.; Kotani, Y.; Handa, D.; Azechi, S.; Kimura, T.; Tsuhako, A.; Miyagi, Y. A general scavenging rate constant for reaction of hydroxyl radical with organic carbon in atmospheric waters. *Environ. Sci. Technol.* **2013**, *47*, 8196–8203.
91. Herckes, P.; Valsaraj, K. T.; Collett, Jr, J. L. A review of observations of organic matter in fogs and clouds: Origin, processing and fate. *Atmos. Res.* **2013**, *132–133*, 434–449.
92. Doussin, J. F.; Monod, A. Structure–activity relationship for the estimation of OH-oxidation rate constants of carbonyl compounds in the aqueous phase. *Atmos. Chem. Phys.* **2013**, *13*, 11625–11641.
93. Luo, X.; Yang, X.; Qiao, X.; Wang, Y.; Chen, J.; Wei, X.; Peijnenburg, W. J. G. M. Development of a QSAR model for predicting aqueous reaction rate constants of organic chemicals with hydroxyl radicals. *Environ. Sci. Process. Impact* **2017**, *19*, 350–356.
94. Ervens, B.; Gligorovski, S.; Herrmann, H. Temperature dependent rate constants for hydroxyl radical reactions with organic compounds in aqueous solution. *Phys. Chem. Chem. Phys.* **2003**, *5*, 1811–1824.
95. Wang, Y.-N.; Chen, J.; Li, X.; Zhang, S.; Qiao, X. Estimation of Aqueous-Phase Reaction Rate Constants of Hydroxyl Radical with Phenols, Alkanes and Alcohols. *QSAR Comb. Sci.* **2009**, *28*, 1309–1316.
96. Twomey, S. The Influence of Pollution on the Shortwave Albedo of Clouds. *J. Atmos. Sci.* **1977**, *34*, 1149–1152.
97. Chepfer, H.; Noel, V.; Winker, D.; Chiriaco, M. Where and when will we observe cloud changes due to climate warming? *Geophys. Res. Lett.* **2014**, *41*, 2014GL061792.
98. Meleux, F.; Solmon, F.; Giorgi, F. Increase in summer European ozone amounts due to climate change. *Atmos. Environ.* **2007**, *41*, 7577–7587.
99. Sherwood, S. C.; Bony, S.; Dufresne, J.-L. Spread in model climate sensitivity traced to atmospheric convective mixing. *Nature* **2014**, *505*, 37–42.
100. Sareen, N.; Waxman, E. M.; Turpin, B. J.; Volkamer, R.; Carlton, A. G. Potential of Aerosol Liquid Water to Facilitate Organic Aerosol Formation: Assessing Knowledge Gaps about Precursors and Partitioning. *Environ. Sci. Technol.* **2017**, *51*, 3327–3335.
101. Tai, A. P. K.; Mickley, L. J.; Jacob, D. J. Correlations between fine particulate matter (PM2.5) and meteorological variables in the United States: Implications for the sensitivity of PM2.5 to climate change. *Atmos. Environ.* **2010**, *44*, 3976–3984.
102. Tai, A. P. K.; Mickley, L. J.; Jacob, D. J.; Leibensperger, E. M.; Zhang, L.; Fisher, J. A.; Pye, H. O. T. Meteorological modes of variability for fine particulate matter (PM2.5) air quality in the United States: implications for PM2.5 sensitivity to climate change. *Atmos. Chem. Phys.* **2012**, *12*, 3131–3145.

103. Zhu, J.; Penner, J. E.; Lin, G.; Zhou, C.; Xu, L.; Zhuang, B. Mechanism of SOA formation determines magnitude of radiative effects. *Proc. Natl. Acad. Sci. U.S.A.* **2017**, *114*, 12685–12690.

Chemistry and Characterization of Fires

Chapter 17

Detailed Characterization of Organic Carbon from Fire: Capitalizing on Analytical Advances To Improve Atmospheric Models

Annmarie G. Carlton,[1,*] Kelley C. Barsanti,[2] Christine Wiedinmyer,[3] and Isaac Afreh[2]

[1]Department of Chemistry, University of California, Irvine, California 92697, United States
[2]Department of Chemical & Environmental Engineering, University of California, Riverside, California 92521, United States
[3]Coperative Institute for Research in Environmental Sciences, University of Colorado, Boulder, Colorado 80309, United States
[*]E-mail: agcarlto@uci.edu

Biomass burning — including wildfires, agricultural burning and prescribed fires — injects large amounts of particulate matter (PM) and reactive trace gases to the atmosphere, affecting air quality and climate. Advances in analytical technology have allowed improvements in identification and quantification of non-methane volatile organic carbon (VOC) species in fire emissions. Application of new techniques including two-dimensional gas chromatography time-of-flight mass spectrometry (GCxGC/TOFMS) leads to unprecedented identification of total detected VOC mass and associated chemical properties. Hundreds of chemical species emitted from fires can now be identified and quantified. To preserve computational speed in chemical transport models, individual NMOCs are mapped to a smaller number of lumped surrogate species, typically grouped by reactivity with the hydroxyl radical (•OH). Yet, underlying a priori assumptions regarding mapping individual compounds to simplified chemical mechanism surrogates may inhibit atmospheric models from capitalizing on recent measurement advances. Mapping protocols often do not consider semi-volatile and solubility properties, all of which impact the fate and transport of trace species through the atmosphere. This reduction in modeled chemical complexity contributes to substantial uncertainties in simulations of biomass burning derived O_3 and PM. In this chapter we explore how a multi-dimensional physicochemical property characterization of VOC species emitted by biomass burning fire emissions and their mapping to existing gas-phase chemical mechanisms will enable more holistic and accurate descriptions of organic compounds in atmospheric models.

Introduction

Gas-phase chemical mechanisms employed in atmospheric models were developed primarily from chamber studies with the intended design to accurately predict ground-level ozone (O_3) mixing ratios in polluted urban environments (e.g., carbon bond and SAPRC approaches (*1*, *2*)). When these kinetic mechanisms are applied in models that account for emission, physical transport and chemistry, the evaluation metrics are often bias calculations relative to measurements that are indicators of poor air quality, such as O_3, fine particulate matter mass ($PM_{2.5}$) and NO_X concentrations (e.g., (*3*, *4*)). Although extensively evaluated for O_3 predictive skill, evaluations of VOCs, a key ozone precursor are relatively sparse in comparison. The extent to which these mechanisms and their application in atmospheric models accurately describe the fate and transport of organic trace species, and in particular, their oxidation products and subsequent formation of secondary organic aerosol (SOA) is unclear.

Model evaluation for VOC prediction is typically assessed through proxies such as O_3 formation potential (*5*) and total VOC reactivity (*6*). VOC reactivity is defined as the potential of an air mass to form organic peroxy radicals and is quantified as the sum of individual organic species concentration times their respective reaction rate constant with the hydroxyl radical: $\sum k_{OH,i}[VOC]_i$, where *i* is each individual VOC species. In laboratory and field studies, instantaneous (*6*) or integrated VOC-reactivity (*7*) is evaluated for the predicted bulk VOCs and compared to overall measured VOC loss (e.g., (*8*)). Episodically, when detailed and specific gas-phase organic measurements are available, more comprehensive model evaluation for VOCs is possible and may reveal that proxy analysis of VOCs can masks compensating errors among individual VOCs or VOC classes (*9*). For example, Steiner et al. find that oxygenated VOCs (oVOCs), such as aldehyde and formaldehyde, dominate total VOC-reactivity in rural areas, but are so poorly constrained that more comprehensive oVOC measurements are necessary before such evaluations are reliable (*10*). Doraiswamy mapped individual VOC hourly measurements to six lumped species in the carbon bond 4 (CB4) chemical mechanism and evaluated model performance at sixteen sites in the northeast U.S. Ozone predictions were within ±15% of measurements, yet VOCs, with the exception of isoprene, were underpredicted (*11*). This demonstrates modeling systems can predict ozone well even if precursors are not accurately described, suggesting model predictions are sometimes right for the wrong reasons.

VOC species are aggregated in atmospheric models because explicit representation of the tens of thousands of organic gas phase species present in the Earth's atmosphere (*12*) is not computationally feasible. VOCs are "lumped" (*13*) in one dimension, usually by hydroxyl radical reactivity (i.e., k_{OH}), an important parameter for ozone production. Reaction rate constants are engineered to the mechanism and represent a weighted average for individual compounds assigned to the same lumped species. Empirical values are fit based on an ability to predict O_3 in the original smog chamber experiment. This hinders expansion of the gas phase chemical mechanism to represent newly discovered organic chemistry (*14*), in particular for SOA. Isoprene is a notable exception to organic species aggregation due to its large emission flux to the atmosphere (*15*), high hydroxyl radical reactivity and contribution to ozone (*16, 17*). Unique representation for isoprene facilitates compound-specific evaluation during intensive observation periods. Isoprene is often not predicted well (i.e., factor of 2 difference compared to measurements) (*18, 19*). In the rural Ozarks, home to the 'isoprene volcano' (*20*), predictions of ambient isoprene mixing ratios in two air quality simulations, each with the same gas phase chemistry and meteorology but different underlying biogenic emission models, are biased relative to isoprene measurements by up to a factor of five, yet ozone compares well with observations in both simulations (*21*). This is consistent with other findings that suggest compensating errors within atmospheric models whereby O_3 is well predicted even when key precursors are not. Expanding organic gas-phase photochemistry to include new VOCs, semi-volatile organic carbon (SVOC), water-soluble organic carbon (WSOC), and other SOA-forming precursors and pathways is desirable but difficult because it requires some 'unlumping' and possibly extrapolation beyond the original smog chamber experiments upon which the mechanisms were developed that may not have included the targeted compound(s).

Biomass burning VOC emissions are substantial (*22*) and impact air quality through O_3 (*23*) and $PM_{2.5}$ by both direct particle emissions and SOA formation (*24*). The simplified and lumped representation of VOCs is a challenge to accurate SOA prediction, in particular for fires, because of substantial emissions of chemically complex SOA precursors not considered in the original ozone mechanism design experiments (*25–27*). Hatch et al. recently measured 38 different monoterpene isomers in biomass burning emissions and 11 sesquiterpenes (*25*). In the SAPRC gas phase chemical mechanism, as employed in the Community Multiscale Air Quality (CMAQ) model, sesquiterpenes are implemented as a single species in all phases. Gas-phase oxidation of sesquiterpenes (SESQ) to generate SOA-forming species that occurs via reaction with ozone does not impact ozone (i.e., $SESQ + O_3 \rightarrow \Delta SESQ + O_3$) (*28*) in order to maintain integrity of the entire chemical mechanism. Lumping of VOCs in the gas-phase extends to aggregation of SOA formation chemistry and properties. For example, in CMAQ, mass-weighted partitioning parameters (i.e., k and α (*29*)) are calculated from a single year's annual emission totals over the United States of β-caryophyllene and α-humulene, and are used to describe SOA formation for a single sesquiterpene model species. SOA from the monoterpenes α-pinene,

Δ^3-carene, sabinene, limonene is represented as a single species in CMAQ also with mass-weighted parameters based on one year's annual emission totals (*28*). Recently, more explicit gas-phase and SOA chemistry was implemented in CMAQ for β-pinene when it was separated from the other monoterpenes after discovery that it exhibits anomalous NO_3 radical chemistry compared to the other lumped monoterpenes. An intensive period of detailed atmospheric chemical composition during the Southern Oxidant and Aerosol Study (SOAS) (*30*) provided a unique opportunity to test and evaluate implementation of independent β-pinene chemistry (*31*). Greater chemical complexity for organic, SOA-forming species is required for proper representation of biomass burning emissions impacts on air quality, and this requires accurate prediction of semi-volatile partitioning (*32*), uptake and condensed phase chemistry, (e.g. (*33*),) and precursors including VOCs and oVOCs.

Computational numerical constraints persist despite the pressing need for increased chemical complexity. Pankow and Barsanti (2009) proposed that organic species could be mapped in two dimensions, carbon number and polarity, to properly describe partitioning, particle phase state, water uptake and other chemical properties important for SOA formation, while maintaining a computationally efficient framework (*34*). We explore the diversity of chemical properties for individual SOA-forming species that are mapped to the same lumped species in the gas phase chemical mechanism SAPRC for biomass burning emissions analyzed using two-dimensional gas chromatography with time-of-flight mass spectrometry (GC x GC-ToFMS) (*25*) during the Fire Lab at Missoula Experiment (FLAME-4) experiment (*35*). Henry's Law solubility is discussed as a potential surrogate for polarity. Atmospheric aerosol liquid water is abundant (*36*), affects gas-to-particle partitioning (*37*), and gas-to-water solubility is a chemical property used in both wet and dry deposition calculations (formulations for both employ Henry's Law). Henry's Law is not applicable for atmospheric partitioning calculations to non-ideal solutions, such as wet particles (*38*), but is useful as an index and appropriate for many deposition flux calculations. Deposition velocity is calculated as an inverse function of flux resistance due to turbulence and the surface (*39*), and most of the Earth's surface is water. Brownian motion for dry deposition of gases is usually ignored in atmospheric models. Specifically terms related to surface and stomatal resistance, calculated every time step have a Henry's Law dependence (*39*). We discuss potential 2-dimensional mapping approaches for efficient implementation of expanded and coherent organic species chemistry in atmospheric models.

Methods

Species Categories and Chemical Properties

Gas-phase non-methane organic compounds (NMOCs) were collected using adsorption-thermal desorption (ATD) cartridges during biomass burning studies as part of FLAME-4. The fuels and fire conditions of the controlled burns in the US Forest Service Fire Science Laboratory are discussed in detail by

Stockwell et al. (*35*). Detailed description of the measurements, GCxGC-ToF analysis, organic compound speciation and emission factor calculations are based on data collected and analyzed by Hatch et al., 2015. A subset of organic gas-phase species that may form SOA were assigned to preliminary SAPRC-16 lump categories (http://www.cert.ucr.edu/~carter/SAPRC/16). SAPRC identification is subject to change and this analysis is intended to explore the distribution in chemical properties for individual species grouped together for atmospheric modeling. Henry's Law constants were estimated with the HENRYWIN™ model in the Estimation Programs Interface (EPI) Suite™ (https://www.epa.gov/tsca-screening-tools/epi-suitetm-estimation-program-interface#copyright) and estimated values are based on the bond contribution method (*40*). Estimated vapor pressure values are calculated using the group contribution method at 298.15K in the UManSysProp online facility (http://umansysprop.seaes.manchester.ac.uk/tool/vapour_pressure) by uploading SMILES strings as text file (*41*, *42*).

Results and Discussion

A wide array of organic chemical species in biomass burning emissions were identified with GCxGC-ToF (Table 1) during FLAME-4, and mapped to SAPRC-16 gas phase species for which SOA parameterizations exist. There is distribution in the number of carbons (nC) for measured chemicals lumped to species ALK5, ARO1, ARO2 and XYNL (Figure 1a). Single values are assigned in atmospheric models to describe the chemical properties of semi-volatile oxidation products of VOC species that condense to form SOA from these parent compounds. For example, the number of carbons assigned to the semi-volatile species that form SOA from ALK5 in the CMAQ model is 8 (*28*). There is little overlap with the actual measured constituents of biomass burning plumes from FLAME-4 assigned to that lumped category. The median value for nC of those molecules when mapped to the chemical mechanism is 9 carbon atoms per compound. Carbon present in individual species with nC greater than their lumped model species (e.g., nC > 8 for compounds mapped to ALK5) is not accounted for in CMAQ and that carbon is 'lost' in the modeling system at the time of emission. For individual species with a number of carbons less than the nC assigned to its lumped species (e.g., compounds mapped to ALK5 for which nC < 8) the mass amount of carbon created within the modeling system is higher than from the actual emission. Analogies for hydrogen and oxygen atoms also exist and this is inferred in the distribution of molecular weight in Figure 1b. There is also variability in the vapor pressures of individual VOCs, in particular for species mapped to ALK5 that vary by a factor greater than 4000 (Figure 2). While these compounds do not directly partition to the condensed phase to form SOA (only a fraction of their oxidation products condense), SOA yields for the alkanes increase with increasing carbon number (*43*). Benzene and isoprene are notable exceptions to lumping approaches and there is no distribution because those species represent single, individual compounds in the mechanism.

Table 1. SOA Precursors: Lumped and Measured Species

SOA precursor	*Lump species*	*Lump Definition*	*Measured Species*
long chain alkanes	ALK5	alkanes and other non-aromatic compounds that react only with OH and k_{OH} > $1x10^4$ ppm^{-1} min^{-1}	1,1,3-trimethyl cyclohexane; 1,4-dimethyl cyclohexane; ethyl cyclohexane; methyl cyclohexane, butyl cyclopentane; ethyl cyclopentane; propyl cyclopentane; propyl cyclopentane; decane; dodecane; heptane, 2,3-dimethyl heptane; 2,6- dimethyl heptane; 2-methyl heptane; 3-methyl heptane; 2,3,-dimethyl hexane; 3-methyl heptane; 2,3-dimethyl hexane; 2,5-dimethyl hexane; nonane, 2-methyl nonane; 2-methyl octane; 3-methyl octane; patchulane; pentadecane; tetradecane; tridecane; undecane
high-yield aromatics	ARO1	aromatics with k_{OH} < $2x10^4$ ppm^{-1} min^{-1}	1,3-dimethyl-butyl benzene; butyl benzene; hexyl benzene; isopropyl benzene; propyl benzene; pentyl benzene; propyl; ethylbenzene; isobutylbenzene; p-cymene; toluene
low-yield aromatics	ARO2	aromatics with k_{OH} > $2x10^4$ ppm^{-1} min^{-1}	1,2,3,5-tetramethylbenzene; 1,2-dimethyl-3-ethylbenzene; 1,3-dimethyl-4-ethyl-benzene; 1,2,3,4-tetramethyl benzene, 1,2,3-trimethyl benzene, 1,2,4-trimethyl benzene; 1,3,5-trimethyl benzene; 1,4-diethyl-benzene; 1-ethyl-2-methyl-benzene; 1-methyl-2-propyl-benzene; 1-methyl-3-propyl-benzene; 1-methyl-4-propyl-benzene; 2-ethyl-1,3-dimethyl-benzene; 2-ethyl-1,4-dimethyl-benzene; 4-ethyl-1,2-dimethyl-benzene; heptyl benzene; biphenyl; indane; m-cymene; naphthalene; 1,3-dimethyl-naphthalene; 1,6-dimethyl naphthalene; 1-ethyl naphthalene; 1-methyl naphthalene; 2-methyl naphthalene; o-cymene; o-xylene
benzene	BENZ	benzene	benzene

Continued on next page.

Table 1. (Continued). SOA Precursors: Lumped and Measured Species

SOA precursor	*Lump species*	*Lump Definition*	*Measured Species*
isoprene	ISOP	isoprene	isoprene
xylenes	XYNL	xylenes	2,5-dimethyl phenol; o-guaiacol; phenol; 2,6-dimethyl salicyladehyde

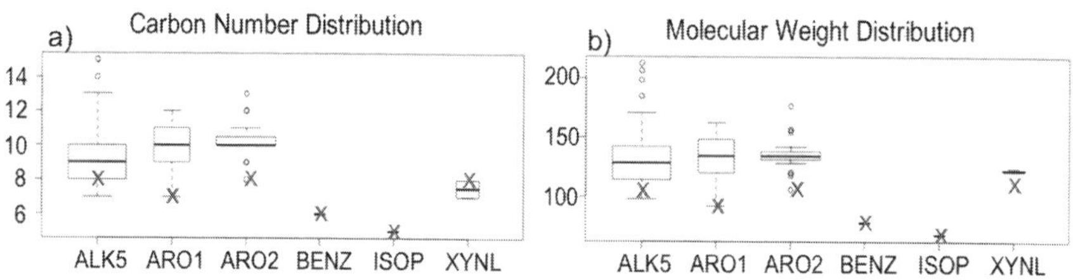

Figure 1. Distribution in a) carbon number (nC) and b) molecular weight (g gmole^{-1}) of individual chemical species measured in biomass burning and assigned to SAPRC-16 lumped species. Measured values are shown as median (black line), 25th and 75th percentiles (box edges), and 10th and 90th percentiles (whiskers). Outliers are circles. Assigned CMAQ values are identified as red "x" and inferred from Carlton et al., 2010 and current subroutines available via GitHub (https://github.com/USEPA/CMAQ).

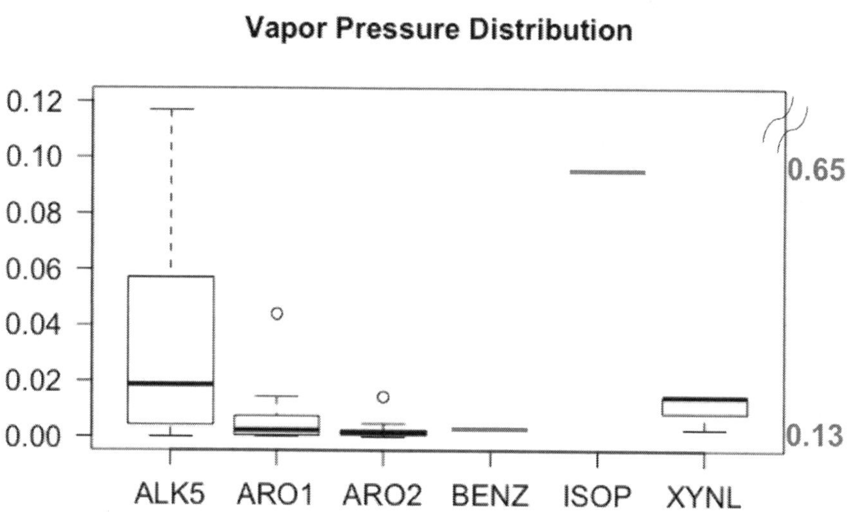

Figure 2. Distribution in vapor pressure (atm) values for individual chemical species measured during FLAME-4 and assigned to SAPRC-16 lumped species. Black line is median, box is 25th and 75th percentiles, whiskers are 10th and 90th percentile. Circles are outliers.

Dry deposition is a continual process and dominates physical removal of species to the land surface over wet deposition, which is highly precipitation-dependent and episodic. Wet *and* dry deposition in atmospheric models is estimated for individual species as a function of their Henry's Law values in uni- and bidirectional deposition calculations (*39*). The distribution in Henry's Law values for individual lumped species spans orders of magnitude (Figure 3) and outliers are not shown because they are off-scale. The range in Henry's Law values for ALK5 is the smallest for the species considered here, yet there is more than a factor of 3000 difference between the highest and lowest values in the distribution: 7890 – 2 atm m^3 mol^{-1}. For ARO1, the distribution is 7890 – 105 atm m^3 mol^{-1}, a factor of 75 different. The maximum Henry's Law value for an individual chemical species assigned to ARO2 is 77,100 atm m^3 mol^{-1}, more than 600 times the minimum value of 116 atm m^3 mol^{-1} in the distribution (Figure 3). Planned model sensitivity simulations include mass-weighted Henry's Law values for lumped species based on emission instead of hard a priori assignment. It should be noted however, that neither the average or median Henry's Law value for biomass burning species identified and assigned to ARO1 or ARO2 during FLAME-4 can describe the full bimodal distribution (Figure 4), suggesting a single value will not suffice. Model predictions of subsequent impacts, such as deposition, are poorly described. This suggests that accurate and full accounting for biogeochemical cycling cannot be accomplished with current atmospheric modeling approaches that have been optimized for O_3 predictive skill.

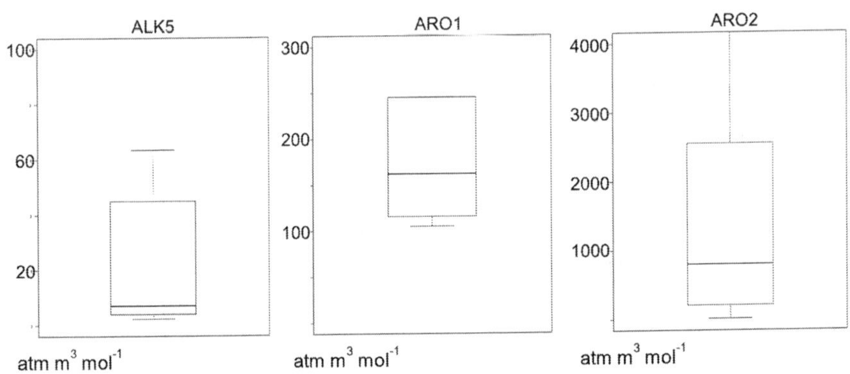

Figure 3. Distribution in Henry's Law values for individual chemical species measured during FLAME-4 and assigned to SAPRC-16 lumped species ALK5, ARO1 and ARO2. Black line is median, box is 25th and 75th percentiles, whiskers are 10th and 90th percentile. Circles are outliers.

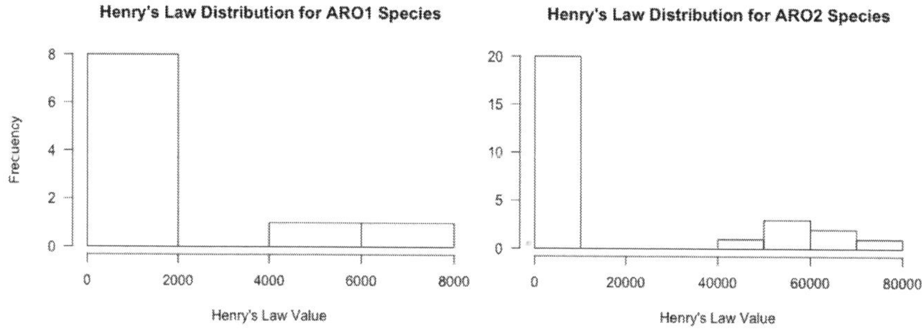

Figure 4. Bimodal Distribution in Henry's Law values for individual chemical species measured during FLAME-4 and assigned to SAPRC-16 lumped species ARO1 and ARO2.

Currently emitted aromatics are speciated to ARO1 and ARO2 depending on comparison of their reaction rate constant with the hydroxyl radical to a threshold value. For $k_{OH} < 2x10^4$ ppm^{-1} min^{-1} species are assigned to ARO1, when k_{OH} is greater aromatic species are assigned to ARO2 (Table 1). Novel 2-dimensional binning for organic species that takes into account k_{OH} *and* Henry's Law may better describe O_3 photochemistry while improving deposition estimates without increasing the number of transported species, a key determinant of model simulation time. This is an area of future study.

Organic compounds in Earth's atmosphere oxidize to CO_2 unless deposited. Current approaches in atmospheric models are insufficient to fully study the fate and transport of organic carbon emitted to the atmosphere because carbon is not conserved and deposition values are not well constrained. The distribution in chemical properties of identified chemical species in biomass burning plumes is diverse and deposition measurements are sparse. This hinders efforts to accurately predict the fate and transport of organic carbon in the Earth's atmosphere.

References

1. Gery, M. W.; Whiten, G. Z.; Killus, J. P.; Dodge, M. C. A photochemical kinetics mechanism for urban and regional scale computer modeling. *J. Geophys. Res.* **1989**, *94*, 12,925–12,956.
2. Carter, W. P. L. *Implementation of the SAPRC-99 chemical mechanism into the models-3 framework*; U.S. Environmental Protection Agency: 2000.
3. Eder, B.; Kang, D. W.; Mathur, R.; Pleim, J.; Yu, S. C.; Otte, T.; Pouliot, G. A performance evaluation of the National Air Quality Forecast Capability for the summer of 2007. *Atmos. Environ.* **2009**, *43*, 2312–2320.
4. Henneman, L. R. F.; Liu, C.; Hu, Y. T.; Mulholland, J. A.; Russell, A. G. Air quality modeling for accountability research: Operational, dynamic, and diagnostic evaluation. *Atmos. Environ.* **2017**, *166*, 551–565.
5. Finlayson-Pitts, B. J.; Pitts, J. N. Tropospheric Air Pollution: Ozone, Airborne Toxics, Polycyclic Aromatic Hydrocarbons, and Particles. *Science* **1997**, *276*, 1045–1051.

6. Hakami, A.; Bergin, M. S.; Russell, A. G. Ozone formation potential of organic compounds in the eastern United States: A comparison of episodes, inventories, and domains. *Environ. Sci. Technol.* **2004**, *38*, 6748–6759.
7. Middleton, P.; Stockwell, W. R.; Carter, W. P. L. Aggregation and analysis of volatile organic-compound emissions For regional modeling. *Atmos. Environ. Part A* **1990**, *24*, 1107–1133.
8. Kovacs, T. A.; Brune, W. H.; Harder, H.; Martinez, M.; Simpas, J. B.; Frost, G. J.; Williams, E.; Jobson, T.; Stroud, C.; Young, V.; Fried, A.; Wert, B. Direct measurements of urban OH reactivity during Nashville SOS in summer 1999. *J. Environ. Monit.* **2003**, *5*, 68–74.
9. Stroud, C. A.; Morneau, G.; Makar, P. A.; Moran, M. D.; Gong, W.; Pabla, B.; Zhang, J.; Bouchet, V. S.; Fox, D.; Venkatesh, S.; Wang, D.; Dann, T. OH-reactivity of volatile organic compounds at urban and rural sites across Canada: Evaluation of air quality model predictions using speciated VOC measurements. *Atmos. Environ.* **2008**, *42*, 7746–7756.
10. Steiner, A. L.; Cohen, R. C.; Harley, R. A.; Tonse, S.; Millet, D. B.; Schade, G. W.; Goldstein, A. H. VOC reactivity in central California: comparing an air quality model to ground-based measurements. *Atmos. Chem. Phys.* **2008**, *8*, 351–368.
11. Doraiswamy, P.; Hogrefe, C.; Hao, W.; Henry, R. F.; Civerolo, K.; Ku, J. Y.; Sistla, G.; Schwab, J. J.; Demerjian, K. L. A diagnostic comparison of measured and model-predicted speciated VOC concentrations. *Atmos. Environ.* **2009**, *43*, 5759–5770.
12. Goldstein, A. H.; Galbally, I. E. Known and unexplored organic constituents in the earth's atmosphere. *Environ. Sci. Technol.* **2007**, *41*, 1514–1521.
13. Makar, P. A.; Stockwell, W. R.; Li, S. M. Gas-phase chemical mechanism compression strategies: Treatment of reactants. *Atmos. Environ.* **1996**, *30*, 831–842.
14. Carter, W. P. L. Development of a database for chemical mechanism assignments for volatile organic emissions. *J. Air Waste Manage. Assoc.* **2015**, *65*, 1171–1184.
15. Guenther, A.; Karl, T.; Harley, P.; Wiedinmyer, C.; Palmer, P. I.; Geron, C. Estimates of global terrestrial isoprene emissions using MEGAN (Model of Emissions of Gases and Aerosols from Nature). *Atmos. Chem. Phys.* **2006**, *6*, 3181–3210.
16. Chameides, W. L.; Lindsay, R. W.; Richardson, J.; Kiang, C. S. The Role Of Biogenic Hydrocarbons In Urban Photochemical Smog - Atlanta As A Case-Study. *Science* **1988**, *241*, 1473–1475.
17. Karl, T.; Guenther, A.; Yokelson, R. J.; Greenberg, J.; Potosnak, M.; Blake, D. R.; Artaxo, P. The tropical forest and fire emissions experiment: Emission, chemistry, and transport of biogenic volatile organic compounds in the lower atmosphere over Amazonia. *J. Geophys. Res.* **2007**, *112*, D18302.
18. Warneke, C.; de Gouw, J. A.; Del Negro, L.; Brioude, J.; McKeen, S.; Stark, H.; Kuster, W. C.; Goldan, P. D.; Trainer, M.; Fehsenfeld, F. C.; Wiedinmyer, C.; Guenther, A. B.; Hansel, A.; Wisthaler, A.; Atlas, E.; Holloway, J. S.; Ryerson, T. B.; Peischl, J.; Huey, L. G.; Hanks, A. T. C.

Biogenic emission measurement and inventories determination of biogenic emissions in the eastern United States and Texas and comparison with biogenic emission inventories. *J. Geophys. Res.* **2010**, *115*, D00F18.

19. Kang, D. W.; Aneja, V. P.; Mathur, R.; Ray, J. D. Nonmethane hydrocarbons and ozone in three rural southeast United States national parks: A model sensitivity analysis and comparison to measurements. *J. Geophys. Res.* **2003**, *108*, 4604.
20. Wiedinmyer, C.; Greenberg, J.; Guenther, A.; Hopkins, B.; Baker, K.; Geron, C.; Palmer, P. I.; Long, B. P.; Turner, J. R.; Petron, G.; Harley, P.; Pierce, T. E.; Lamb, B.; Westberg, H.; Baugh, W.; Koerber, M.; Janssen, M. Ozarks Isoprene Experiment (OZIE): Measurements and modeling of the "isoprene volcano". *J. Geophys. Res.* **2005**, *110*, D1837.
21. Carlton, A. G.; Baker, K. Photochemical modeling of the Ozark isoprene volcano: MEGAN, BEIS and their impacts on air quality predictions. *Environ. Sci. Technol.* **2011**, *45*, 4438–4445.
22. Akagi, S. K.; Yokelson, R. J.; Wiedinmyer, C.; Alvarado, M. J.; Reid, J. S.; Karl, T.; Crounse, J. D.; Wennberg, P. O. Emission factors for open and domestic biomass burning for use in atmospheric models. *Atmos. Chem. Phys.* **2011**, *11*, 4039–4072.
23. Jaffe, D. A.; Wigder, N. L. Ozone production from wildfires: A critical review. *Atmos. Environ.* **2012**, *51*, 1–10.
24. Bian, Q. J.; Jathar, S. H.; Kodros, J. K.; Barsanti, K. C.; Hatch, L. E.; May, A. A.; Kreidenweis, S. M.; Pierce, J. R. Secondary organic aerosol formation in biomass-burning plumes: theoretical analysis of lab studies and ambient plumes. *Atmos. Chem. Phys.* **2017**, *17*, 5459–5475.
25. Hatch, L. E.; Luo, W.; Pankow, J. F.; Yokelson, R. J.; Stockwell, C. E.; Barsanti, K. C. Identification and quantification of gaseous organic compounds emitted from biomass burning using two-dimensional gas chromatography-time-of-flight mass spectrometry. *Atmos. Chem. Phys* **2015**, *15*, 1865–1899.
26. Sakulyanontvittaya, T.; Duhl, T.; Wiedinmyer, C.; Helmig, D.; Matsunaga, S.; Potosnak, M.; Milford, J.; Guenther, A. Monoterpene and sesquiterpene emission estimates for the United States. *Environ. Sci. Technol.* **2008**, *42*, 1623–1629.
27. Barsanti, K. C.; Carlton, A. G.; Chung, S. H. Analyzing experimental data and model parameters: implications for predictions of SOA using chemical transport models. *Atmos. Chem. Phys.* **2013**, *13*, 12073–12088.
28. Carlton, A. G.; Bhave, P. V.; Napelenok, S. L.; Edney, E. O.; Sarwar, G.; Pinder, R. W.; Pouliot, G. A.; Houyoux, M. Treatment of secondary organic aerosol in CMAQv4.7. *Environ. Sci. Technol.* **2010**, 8553–8560.
29. Pankow, J. An Absorption Model of Gas/Particle Partitioning of Organic Compounds in the Atmosphere. *Atmos. Environ.* **1994**, *28*, 185–188.
30. Carlton, A. G.; de Gouw, J.; Jimenez, J. L.; Ambrose, J. L.; Attwood, A. R.; Brown, S.; Baker, K. R.; Brock, C.; Cohen, R. C.; Edgerton, S.; Farkas, C. M.; Farmer, D.; Goldstein, A. H.; Gratz, L.; Guenther, A.; Hunt, S.; Jaegle, L.; Jaffe, D. A.; Mak, J.; McClure, C.; Nenes, A.; Nguyen, T. K.; Pierce, J. R.; de Sa, S.; Selin, N. E.; Shah, V.; Shaw, S.; Shepson, P.

B.; Song, S. J.; Stutz, J.; Surratt, J. D.; Turpin, B. J.; Warneke, C.; Washenfelder, R. A.; Wennberg, P. O.; Zhou, X. L. Synthesis Of The Southeast Atmosphere Studies: Investigating Fundamental Atmospheric Chemistry Questions. *Bull. Am. Meteorol. Soc.* **2018**, *99*, 547–567.

31. Pye, H. O. T.; Luecken, D. J.; Xu, L.; Boyd, C. M.; Ng, N. L.; Baker, K. R.; Ayres, B. R.; Bash, J. O.; Baumann, K.; Carter, W. P. L.; Edgerton, E.; Fry, J. L.; Hutzell, W. T.; Schwede, D. B.; Shepson, P. B. Modeling the Current and Future Roles of Particulate Organic Nitrates in the Southeastern United States. *Environ. Sci. Technol.* **2015**, *49*, 14195–14203.
32. Liang, C.; Pankow, J. F.; Odum, J. R.; Seinfeld, J. H. Gas/particle partitioning of semivolatile organic compounds to model inorganic, organic, and ambient smog aerosols. *Environ. Sci. Technol.* **1997**, *31*, 3086–3092.
33. Ervens, B.; Turpin, B. J.; Weber, R. J. Secondary organic aerosol formation in cloud droplets and aqueous particles (aqSOA): a review of laboratory, field and model studies. *Atmos. Chem. Phys.* **2011**, *11*, 11069–11102.
34. Pankow, J. F.; Barsanti, K. C. The carbon number-polarity grid: a means to manage the complexity of the mix of organic compounds when modeling atmospheric organic particulate matter. *Atmos. Environ.* **2009**, *43*, 2829–2835.
35. Stockwell, C. E.; Yokelson, R. J.; Kreidenweis, S. M.; Robinson, A. L.; DeMott, P. J.; Sullivan, R. C.; Reardon, J.; Ryan, K. C.; Griffith, D. W. T.; Stevens, L. Trace gas emissions from combustion of peat, crop residue, domestic biofuels, grasses, and other fuels: configuration and Fourier transform infrared (FTIR) component of the fourth Fire Lab at Missoula Experiment (FLAME-4). *Atmos. Chem. Phys.* **2014**, *14*, 9727–9754.
36. Carlton, A. G.; Turpin, B. J. Particle partitioning potential of organic compounds is highest in the Eastern US and driven by anthropogenic water. *Atmos. Chem. Phys.* **2013**, *13*, 10203–10214.
37. Jathar, S. H.; Mahmud, A.; Barsanti, K. C.; Asher, W. E.; Pankow, J. F.; Kleeman, M. J. Water uptake by organic aerosol and its influence on gas/particle partitioning of secondary organic aerosol in the United States. *Atmos. Environ.* **2016**, *129*, 142–154.
38. Sareen, N.; Waxman, E. M.; Turpin, B. J.; Volkamer, R.; Carlton, A. G. Potential of Aerosol Liquid Water to Facilitate Organic Aerosol Formation: Assessing Knowledge Gaps about Precursors and Partitioning. *Environ. Sci. Technol.* **2017**, *51*, 3327–3335.
39. Pleim, J. E.; Bash, J. O.; Walker, J. T.; Cooter, E. J. Development and evaluation of an ammonia bidirectional flux parameterization for air quality models. *J.Geophys. Res.* **2013**, *118*, 3794–3806.
40. U. S. EPA, Science Advisory Board (SAB). *Review of the Estimation Programs Interface Suite (EPI Suite)*; 2007. Available: https://www.epa.gov/tsca-screening-tools/epi-suitetm-estimation-program-interface (accessed December 13, 2017).
41. Barley, M. H.; McFiggans, G. The critical assessment of vapour pressure estimation methods for use in modelling the formation of atmospheric organic aerosol. *Atmos. Chem. Phys.* **2010**, *10*, 749–767.

42. O'Meara, S.; Booth, A. M.; Barley, M. H.; Topping, D.; McFiggans, G. An assessment of vapour pressure estimation methods. *Phys. Chem. Chem. Phys.* **2014**, *16*, 19453–19469.
43. Lim, Y. B.; Ziemann, P. J. Effects of Molecular Structure on Aerosol Yields from OH Radical-Initiated Reactions of Linear, Branched, and Cyclic Alkanes in the Presence of NOx. *Environ. Sci. Technol.* **2009**, *43*, 2328–2334.

Chapter 18

Understanding Composition, Formation, and Aging of Organic Aerosols in Wildfire Emissions via Combined Mountain Top and Airborne Measurements

Q. Zhang,[1,*] S. Zhou,[1] S. Collier,[1] D. Jaffe,[2] T. Onasch,[3] J. Shilling,[4] L. Kleinman,[5] and A. Sedlacek[5]

[1]Department of Environmental Toxicology, University of California, Davis, California 95616, United States
[2]School of Science and Technology, University of Washington, Bothell, Washington 98011, United States
[3]Aerodyne Research Inc., Billerica, Massachusetts 01821, United States
[4]Pacific Northwest National Laboratory, Richland, Washington 99352, United States
[5]Brookhaven National Laboratory, Upton, New York 11973, United States
[*]E-mail: dkwzhang@ucdavis.edu; tel.: 530-752-5779

Wildfires are an important source of global aerosols that have pronounced impacts on regional air quality and global climate. A major component of wildfire emissions is organic, including primary organic aerosols (POA) and large amounts of gases that can be oxidized to form secondary organic aerosols (SOA). The environmental impacts of wildfire emissions are strongly correlated with the chemical, optical, and microphysical properties of biomass-burning organic aerosols (BBOA), which are dependent, in a complex manner, on combustion and atmospheric aging processes. Here we discuss results from recent field and laboratory studies aimed at understanding the chemical composition, formation, and evolution of BBOA in wildfire emissions. Comprehensive analysis of a large number of fresh and aged (up to days of aging) fire plumes originating in the Pacific Northwest region of the United States reveals that the enhancement of BBOA mass concentrations relative to the enhancement of the sum of CO and CO_2 in wildfire smokes is primarily driven by combustion efficiency, whereas the composition and properties of the BBOA are strongly influenced by atmospheric oxidation and other aging processes. These results shed light on organic aerosol formation and transformation in biomass burning emissions.

Introduction

Biomass burning (BB) is one of the largest sources of trace gases and primary carbonaceous aerosols on a global scale (*1–5*). Wildfires, in particular, are a highly variable component of BB emissions, and typically an "uncontrollable" source of aerosols that can severely impair downwind air quality and threaten human health (*6–10*). In the western U.S., for instance, wildfire activities have been frequently tied to haze episodes at various receptor sites and elevated background aerosol concentrations during summertime (*9–14*). BB emissions also represent a large source of uncertainty in model simulations of the climate effects of aerosols. The Earth's radiation budget is influenced by BB aerosols directly, due to the primary light scattering organic carbon (OC) component and the absorbing black carbon (BC) and brown carbon (BrC) components, and indirectly, due to their role as cloud condensation nuclei (CCN) and ice nuclei (*5, 15–18*). However, the indirect radiative forcing of BB aerosols remains poorly understood and the direct forcing of BB aerosols is highly uncertain, since the strong positive radiative forcing from the absorbing components is offset by the strong negative forcing from the scattering components (*19, 20*). Furthermore, the seasonal and regional climate effects of BB aerosols vary widely due to substantial temporal and spatial variability of BB activities and smoke properties (*19–21*).

Organic compounds are a major component of BB emissions, which include primary organic aerosols (POA) and volatile organic compounds (VOCs) that can be oxidized to form secondary organic aerosols (SOA) (*22, 23*). A thorough understanding of the concentration and properties of biomass-burning organic aerosols (BBOA) and how they change in the atmosphere since emission is necessary for achieving a quantitative assessment of the environmental impacts of fires (*24, 25*). However, the emission, formation, and transformation of BBOA are highly complex and large knowledge gaps currently remain. For example, the emission rate and properties of POA from BB depend on fuel types, burning conditions, and season (*26–29*). There are contradictory findings about the importance of BB emissions as a source of SOA – laboratory results have shown substantial SOA forming potential in BB emissions (*27, 30–34*) whereas field studies frequently observed little to no net enhancement of SOA in BB plumes, even as BBOA clearly age and evolve chemically during transport (*35–41*). There is abundant evidence showing that the composition, optical properties, and hygroscopicity of aerosols can change substantially in BB smoke subjected to photooxidation and aqueous-phase reactions (*42–46*), suggesting that atmospheric aging processes play important roles in modulating the impacts that BB has on climate and air quality. In addition, recent studies have observed the formation and persistence of tar balls, which represent organic particles characterized by spherical shape, high-viscosity, and low volatility, in BB emission plumes (*47–49*). These observations underscore the atmospheric lifetimes of BBOA and their subsequent spatial and temporal impacts (*47–49*). Unraveling the multiphase chemical processes that affect the abundance and properties of BBOA in the atmosphere is thus crucial for improving model parameterizations and predictions, as well as for developing air pollution control strategies.

Knowledge about BB emissions and aging processes can be gained from laboratory experiments and measurements conducted in the field. Laboratory experiments, which can be performed in a controlled manner, have greatly enriched our knowledge on BBOA behavior under various burning conditions and simulated aging (*27, 30–33, 37, 50, 51*). However, the highly complex nature of real-world fires on both the micro- and macro-physical level is impossible to fully represent in lab settings. The realistic behaviors of BB aerosols as they are emitted, transported, and processed in the atmosphere can only be characterized through field studies. Ambient measurements are crucial for attaining a process-level understanding of the formation and chemical transformations of OA within BB plumes and for evaluating model simulations and performance.

Measurements of Wildfire Plumes in the Western U.S.

In this chapter, we discuss recent findings on the chemical characteristics of wildfire emissions and atmospheric aging of BB aerosols, focusing on results from the 2013 Biomass Burning Observation Project (BBOP) – a field campaign sponsored by the US Department of Energy (DOE). BBOP strategically combined aircraft-based measurements with mountain top observations to investigate the near-field and regional evolution of carbonaceous aerosol properties in wildfire emissions located in the western U.S. A large suite of real-time instruments was deployed on both platforms to measure chemical and physical properties of aerosol particles, trace gases, and atmospheric conditions with high time resolution (*13, 14, 49*). These measurements produced one of the most extensive datasets on gas and particle phase emissions from wildfires in the contiguous U.S.

The BBOP mountain-top site was at Mt. Bachelor Observatory (MBO; 43.979 °N, 121.687 °W, ~ 2700 m above sea level) located on the summit of an isolated volcanic peak in central Oregon (Figure 1). MBO has almost no local source of air pollutants and its altitude enables the sampling of regional background air, including free tropospheric air at night when the planetary boundary layer shrinks (*52*). The common clean air conditions at MBO facilitate observations of polluted air masses transported from upwind sources, such as wildfires occurring within Oregon, Northern California and the surrounding Pacific Northwest region during summer months (*53*). While there are numerous studies investigating emissions and evolution of air pollutants from open biomass burning using airborne observations (*2, 12, 35–38, 40, 41, 49, 51, 54–56*), ground-based measurements are less frequently used for these purposes. However, observations from strategically situated high-elevation sites, such as MBO, can be complimentary to airborne measurements as they are able to acquire continuous data on transported plumes that arrive at different aging times downstream of the sources and from different fires with minimal local influence. Indeed, a number of studies have used MBO and other mountain-top observatories in the western U.S. to study transported air pollutants from BB sources over intermediate and long distances and the results have helped shape our current understanding of the effects that BB emissions have on air quality in downwind sites and atmospheric chemistry on a regional scale (*13*, *14*, *53*, *57–60*). Continuous ground observations, especially those made with high time resolution, have also proven to be highly valuable at understanding BB aerosols aging as functions of transport time and characterizing differences among wildfires (*13*, *14*, *49*). Another benefit of the continuous data is that they also make it possible to monitor differences between night and day chemistry of transported BB plumes, a topic that is severely understudied (*61*).

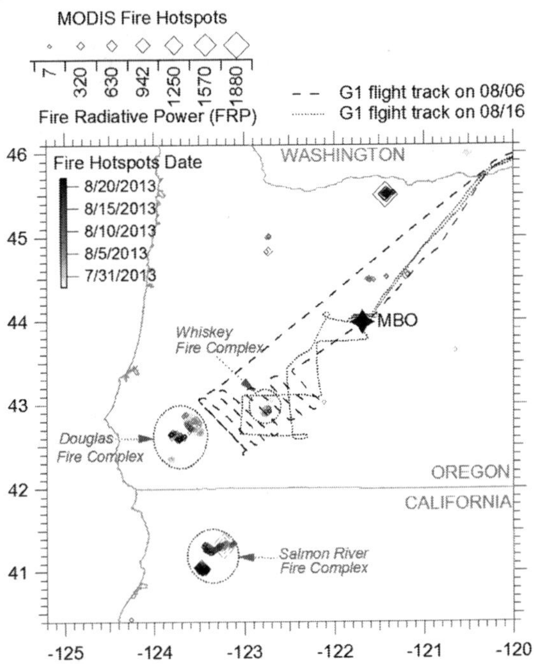

Figure 1. A map with MBO indicated by the black star. Open diamonds represent satellite fire dots colored by date and sized by fire radiation power (FRP). Persistent and large-scale fire complexes are highlighted by circles. Adapted with permission from Collier et al. (13). Copyright 2016 American Chemical Society.

During the BBOP campaign period from July 25th to August 25th 2013, various instances of strong wildfire activities occurred in the western U.S., including three large complex fires within southern Oregon and northern California (Figure 1). The complex fires persisted for weeks during the study and their emissions were transported to MBO by the prevailing westerly and southwesterly winds. The wind pattern created ideal ambient conditions for studying chemical evolution of wildfire plumes as a function of transport time – a proxy for atmospheric aging

(*13, 14*). Based on back-trajectory analysis of air mass origins (*62*), the transport times for a number of well-defined fire plumes were estimated to be ~6 – 48 h from the sources to MBO (*13*). More aged BBOA particles were sampled at MBO as well considering the time span of the fire events and the typical lifetimes (~ 2 weeks) of ambient aerosols. In addition, the DOE Gulfstream-1 (G-1) aircraft made coordinated flights on Aug. 6th and 16th to systematically sample fresh (< 1 h aging) plumes from the Whiskey Fire Complex and moderately aged plumes (~ 4 - 10 h) transported from the Salmon River and Douglas Fire Complexes, respectively (Figure 1).

Chemical Composition of BBOA in Wildfire Plumes

Submicrometer diameter particles (PM_1) during BBOP were characterized using a high-resolution time-of-flight aerosol mass spectrometer (AMS) (*63*) and a soot-particle AMS (SP-AMS) (*64*) deployed at MBO and on-board the G-1, respectively. The AMS uses an aerodynamic lens to sample PM_1 into vacuum where they are aerodynamically sized, thermally vaporized on a resistively heated tungsten vaporizer at ~ 600 °C, and chemically analyzed via 70 eV electron impact ionization (EI) time-of-flight mass spectrometry. Since aerosol species must be vaporized to be detected, the AMS only measures nonrefractory (NR) species such as organics and the ammonium salts of nitrate, sulfate and chloride (*63*). The SP-AMS samples and analyzes particles in the same manner as the AMS, except that it has an additional intracavity infrared laser vaporizer which can volatilize black carbon (BC) particles and associated coatings (*64*). The SP-AMS on board the G-1 during BBOP was operated with the tungsten vaporizer set at a constant 600 °C while the laser was alternated between on and off. For the results discussed in this chapter, only the laser off data are used; i.e., when the SP-AMS was operated identically as the MBO AMS. Thus, in the rest of the chapter, we refer to the SP-AMS data as AMS data as well.

Furthermore, the MBO AMS was operated downstream of a digitally controlled thermodenuder (TD) to characterize the volatility of the aerosols (*14*). The TD consists of a bypass line and a heated line terminating in a section with activated carbon cloth. The temperature inside the heated line was digitally controlled and programmed to cycle through 12 temperatures between 40 and 200 °C (*14*). The sample flow was switched between bypass and TD modes automatically every 5 min to alternate between ambient and heated air flow. The volatility profiles of individual aerosol species were determined by comparing the measurements between these two modes.

Anhydrous sugars from plant-based cellulose, such as levoglucosan, are a key component of POA from BB emissions and the $C_2H_4O_2^+$ and $C_3H_5O_2^+$ ions they produce through electron impact ionization have been identified as AMS tracer ions for BB (*65, 66*). It has been established that f_{60} – defined as the fraction of the signal at *m/z* 60 (mostly $C_2H_4O_2^+$) with respect to the sum of organic signals in AMS OA spectrum – is a tracer for BB influence and that an f_{60} higher than 0.3% suggests the presence of BBOA in ambient particles (*67*). Based on this f_{60} criterion, MBO was found to be impacted by transported wildfire plumes for ~ 70%

of the time during the BBOP deployment period. These BB-influenced periods were characterized by substantially increased concentrations of air pollutants such as CO (up to ~ 700 ppb), NO_y (up to ~ 6.5 ppb), and aerosols ($NR\text{-}PM_1$ up to ~ 210 μg m^{-3}) (*14*). The interstitial time periods were characterized by very clean air with an average (± 1σ) concentration of $NR\text{-}PM_1$ (= organics + sulfate + nitrate + ammonium + chloride) of only 3.7 (± 4.2) μg m^{-3}. The concentrations of CO and NO_y were very low during clean periods as well with average values of 87 ppb and 0.44 ppb, respectively (*14*).

Aerosol composition varied significantly at MBO throughout the entire campaign, due to the influences from different air masses, which included free tropospheric air, regional background air dominated by biogenic emissions, wildfire plumes, and mixtures thereof. Aerosols in wildfire plumes were predominantly organic (OA / PM_1 > 90%) and the chemical compositions of BBOA varied significantly from plume to plume, probably due to differences in fuel or burning conditions and/or varying transport times and hence varying degrees of aging. Figure 2 illustrates the changes in OA composition observed from MBO and G-1 via comparing their AMS spectra in the f_{44} vs. f_{60} space. f_{44} is defined as the fraction of *m/z* 44 signal in AMS organic spectrum. It has been established as an indicator for the average degree of oxidation of OA since oxygenated organic compounds are the primary contributors to *m/z* 44 (mostly CO_2^+). Indeed, f_{44} has been shown to correlate well with the oxygen-to-carbon (O/C) ratio of OA in various studies (*68–70*).

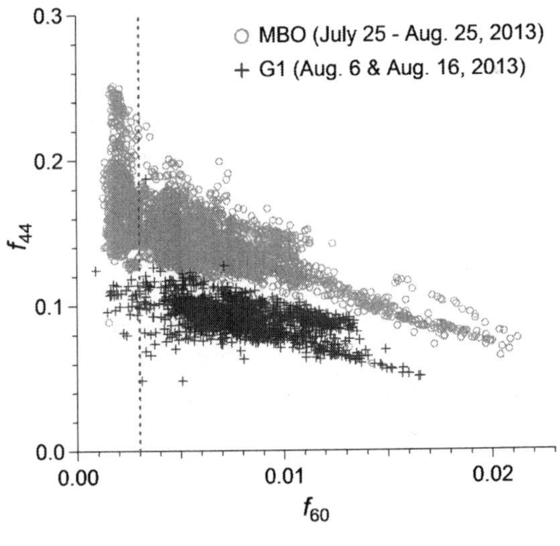

Figure 2. f_{44} vs. f_{60} in OA observed at MBO and from G-1 during BBOP. The G-1 data are from flights on Aug. 6 and Aug. 16 over the area between 42.3 ° – 45° latitude and -121 ° – -123.5° longitude. The dash line is the ambient background value of f_{60} reported in Cubison et al. (37).

The OA data showed a general progression where lower f_{60} values are associated with higher f_{44}, consistent with aging of BBOA observed previously both in laboratory studies and from airborne measurements (*27*, *37*, *40*). In addition, fire plumes sampled from G-1 had systematically lower f_{44} than those from MBO for the same f_{60}, consistent with the fact that G-1 sampled fresher and less oxidized BBOA closer to the wildfire sources. However, the f_{60} values in air masses measured from both platforms appear to overlap, despite different atmospheric aging times. Lab studies have shown that levoglucosan can react with OH radicals with an estimated atmospheric lifetime of 1 ~ 2 days under typical summertime OH concentration conditions (*71*). However, since the OH uptake can be kinetically limited by bulk diffusion of the organic reactant, the lifetime of levoglucosan in particles may increase substantially under dry conditions (up to 1 week) (*72*). Evolution of air mass histories suggest that the BB plumes sampled during BBOP mostly experienced dry air conditions since emission (*13*, *14*), which might be one of the reasons for the presence of similar anhydrous sugar content between more aged (MBO) and fresher BBOAs (G-1). In addition, combustion conditions might have also played a role in how plumes map to the $f_{44} \sim f_{60}$ space since smoldering combustion led to more gasification of unburned fuel whereas flaming combustion tends to be dominated by pyrolysis products which undergo high temperature, in-flame processing. As a result, flaming-dominated fires tend to produce BBOA with lower f_{60} compared to more smoldering fires (*13*, *73*, *74*).

In order to gain further insights into the influences of different sources and processes on OA loading and composition, the AMS data from MBO were analyzed by positive matrix factorization (PMF (*75*)) – a multivariate model that has been broadly used for aerosol source apportionment. The application of PMF on AMS mass spectral analysis has been shown to be particularly powerful at separating unique factors which allow for a quantitative interpretation of emission sources and aging processes affecting bulk OA chemistry (*76*). A total of five distinct OA factors were identified during this study, including three types of BBOA and two types of oxygenated OA (OOA) representing SOA associated with boundary layer (BL) dynamics (BL-OOA, O/C = 0.69) and more oxidized regional background OA with low-volatility (LV-OOA, O/C = 1.09), respectively (*14*). All three BBOA factors correlated well with CO (Pearson's $R > 0.8$) and showed high amplitude episodes driven by wildfire smoke. As illustrated in Figure 3e, the elevated BBOA periods were associated with high SW winds which facilitated the transport of plumes from major wildfire clusters located in SW Oregon and NW California (Figure 1). Nevertheless, the three BBOA factors had significantly different mass spectral profiles and oxygen-to-carbon ratios, showed different volatility profiles, and displayed different temporal variation trends (Figure 3), indicating major differences in extents of aging and processing pathways.

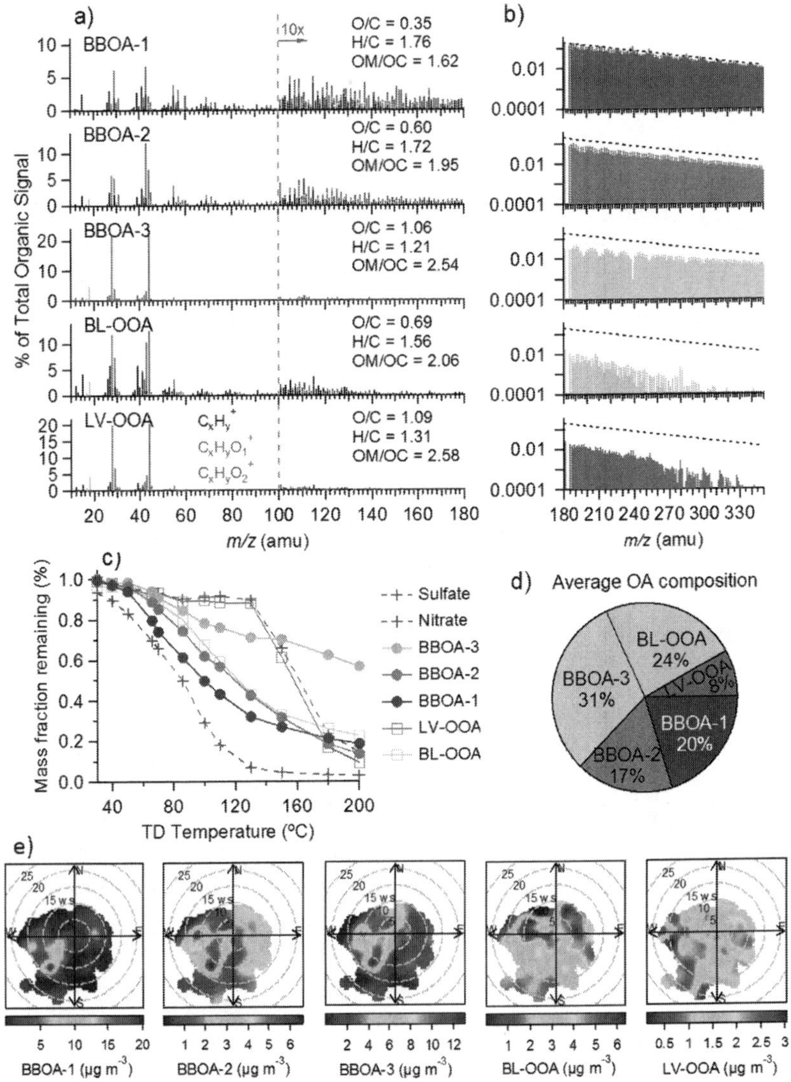

Figure 3. AMS spectra of OA factors determined at MBO at a) m/z < 180 and b) m/z > 180; c) volatility profiles of OA factors, sulfate, and nitrate; d) average composition of OA during BBOP; and e) bivariate polar plots that illustrate the variations of the concentrations of each OA factor as a function of wind speed and wind direction. The elemental ratios of each OA factor are shown in the legends of a. Modified with permission from Zhou et al. (14), Copyright 2017, the authors.

BBOA-1 had the lowest O/C (0.35) and the highest H/C (1.76) and f_{60} (2.2%) among all OA factors identified at MBO. Its spectrum showed prominent ions representing chemically-reduced organics, e.g., $C_2H_3^+$, CHO^+, $C_4H_7^+$, $C_4H_9^+$, and

$C_9H_7^+$ (Figure 3a), and a high abundance of ions larger than 100 amu ($f_{m/z>100}$ = 25%; Figure 3b). In addition, the BBOA-1 spectrum exhibited a group of peaks that differ by 14 amu at *m/z* > 180, suggesting the occurrence of molecules with hydrocarbon moieties containing different units of the CH_2 group. Figure 3e reveals a strong association of high BBOA-1 concentrations with high SW wind, suggesting that BBOA-1 was associated with plumes experiencing shorter transport times. According to TD measurements, BBOA-1 was semivolatile (Figure 3c), consistent with previous findings that a majority (50% - 80%) of the POA in BB emissions is semivolatile (*77*). Taken together, these findings suggest that BBOA-1 was primarily associated with fresher and less processed BB sources, and thus represents primary BBOA. On average, BBOA-1 comprised 20% of total OA mass at MBO (Figure 3d), indicating that fresh BB emissions exerted a significant impact on regional air masses during the 2013 fire seasons in the western U.S.

BBOA-2 was moderately oxidized (O/C = 0.60; H/C = 1.72) with a lower abundance of $C_2H_4O_2^+$ (f_{60} = 1.1%). Its polar plot displays a more dispersed pattern of sources but a hotspot in the SW direction is still visible (Figure 3e). The time series of BBOA-2 correlated tightly with tracers for secondary species such as carboxylic acids, e.g., CHO_2^+ and CO_2^+. These results suggest that BBOA-2 was more chemically processed than BBOA-1 and likely contained secondary products. This conclusion is corroborated by BBOA-2 being somewhat less volatile than BBOA-1, especially at TD temperatures under 150°C (Figure 3c).

BBOA-3 was distinctly different than BBOA-1 and BBOA-2 by being highly oxidized (O/C = 1.07; Figure 3) and containing negligible anhydrous sugars (f_{60} ~ 0). Another distinct characteristic of BBOA-3 was the very low apparent volatility: nearly 60% of its mass remained in aerosol phase after 200 °C heating in TD (Figure 3c). The BBOA-3 AMS spectrum closely resembles the LV-OOA spectrum in the low *m/z* region (Figure 3a) but there are significant differences at *m/z* > 180, which suggests a higher abundance of high molecular weight species in BBOA-3 than in LV-OOA. In addition, the temporal patterns of BBOA-3 and LV-OOA are dramatically different in that BBOA-3 correlated tightly with CO but LV-OOA had nearly no correlation with CO (*14*). Furthermore, the polar plot of BBOA-3 (Figure 3e) showed a high concentration band in the SW direction that is accompanied by wind speeds varying between 5 – 15 m s^{-1}. These results together suggest that BBOA-3 might represent SOA formed both through rapid processing near the wildfire source and during transport to MBO. An important implication of all these findings is that highly aged BBOA could appear indistinguishable from OOA from other sources (e.g., anthropogenic, biogenic, and maritime) due to mass spectral similarities (e.g., low f_{60} and high f_{44}). Given that BBOA-3 accounted for 31% of the total OA mass or 46% of the BBOA at MBO, extra caution should be paid in interpreting AMS ambient data from fire influenced regions as relying on the f_{60} criterion alone would lead to significantly underestimation of the influence of BB emissions, especially in highly aged BB plumes. Note that for the study at MBO, since BBOA-3 always coexisted with BBOA-1 and BBOA-2 (*14*), the BB-impacted time periods identified based on elevated BBOA mass concentration match well with those determined by the f_{60} criterion of Cubison et al. (*37*).

Chemical Aging of BBOA in Wildfire Plumes

The fire plumes sampled at MBO represent a range of atmospheric aging times and thus were subjected to different processes, depending on fire locations, air mass trajectories, and the meteorological conditions the plumes experienced during transport. These variations can be explored to gain process level understanding of how BBOA characteristics change due to aging and whether there is significant SOA formation in BB emissions. The strategy undertaken to investigate this is to perform back-trajectory analysis to identify case periods when emissions from persistent fire clusters were transported to the receptor site continuously. Parameters such as approximate transport time since emission, the amount of solar exposure, and the likelihood of cloud processing can be subsequently estimated by overlapping satellite fire hotspot information with air mass back-trajectories, along with information on the air mass along each trajectory such as sunlight intensity, relative humidity (RH), and temperature.

Figure 4 shows an analysis of the emissions from the Salmon River Complex fire (SRCF) in Northern California using this approach. SRCF was active during the entire month of August, 2013. Three-day back-trajectories for air masses arriving at MBO were calculated using the HYSPLIT back-trajectory model (*62, 78*). The results indicate that the air masses observed during August 14 22:00 to August 16 09:00 at MBO passed over the SRCF with estimated transport times of 8 to 11 h (Figure 4a). The observed smoke originated from a single fire complex and, thus, likely from the burning of the same type of fuel. In addition, based on measurements of CO and CO_2, the burn conditions were found to be relatively constant during this period with an average modified combustion efficiency (MCE, discussed further below) value of 0.88 (±0.03). These conditions, together with the high emissions of both gaseous and particle phase components (Figure 4b), provide a near ideal case study where atmospheric aging is likely the largest factor affecting the chemical evolution of BBOA. In order to estimate the total amount of solar radiation that the smoke plumes were exposed to during transport, cumulative solar radiation ($\sum$SR) over the transport period (from source to MBO) was calculated for each trajectory. $\sum$SR was calculated by summing the hourly solar irradiation values along the backtrajectory between the fire source and MBO and it is used as an indicator for the extent of photochemical aging assuming the plumes were optically thin.

During this SRCF case period, MBO was clearly shrouded with BB smoke as the BBOA concentration was on average more than 10 times higher than background OA concentration (~ 3.7 μg m^{-3}; Figure 4a). The average carbon oxidation state ($OS_c = 2 \times O/C - H/C$ (*79*)) of the BBOA showed a clear increasing trend with respect to $\sum$SR (Figure 4c), indicating photo-oxidative aging. The correlation between $f_{BBOA\text{-}1}$ (= BBOA-1 / total BBOA) and $\sum$SR was negative, supporting the hypothesis that BBOA-1 represents primary BBOA. Both $f_{BBOA\text{-}2}$ and $f_{BBOA\text{-}3}$ correlated positively with $\sum$SR (Figure 4c), corroborating the finding that BBOA-2 and BBOA-3 represent more aged, secondary BBOA.

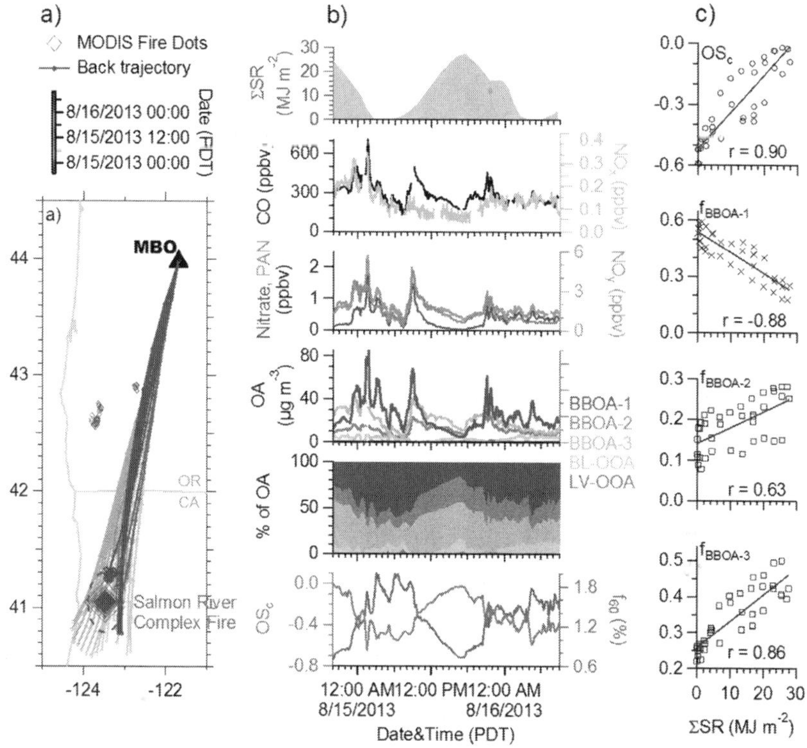

Figure 4. Observations during the SRCF case study period. a) Map that shows back-trajectories of air masses arriving at MBO colored by time of arrival and MODIS fire dots detected during Aug. 13 – 17 (sized by fire radiative power); b) temporal profiles of cumulative solar radiation (∑SR), gaseous pollutants, and aerosol parameters; and c) scatter plots of BBOA chemistry parameters vs. ∑SR. The Pearson's correlation coefficients (r) are reported on the scatter plots. Modified with permission from Zhou et al. (14), Copyright 2017, the authors.

OA and CO were found to correlate well in the wildfire plumes sampled at MBO, including during the SRCF period (Figure 5a). Since CO is regarded as a stable trace gas mainly emitted from combustion processes, it has been commonly used to account for plume dispersion and dilution during atmospheric transport/aging. The enhancement ratio of OA to CO over their respective background concentrations (i.e., ΔOA/ΔCO) is therefore a useful parameter for evaluating factors other than dilution, e.g., SOA formation or OA evaporation, that have influenced OA concentration during plume aging. A tight correlation between OA and CO suggests zero net production of SOA. To investigate the OA vs. CO relationship more closely for the SRCF case period, the air masses were classified into day-time and night-time transported plumes based on ∑SR values. ΔOA/ΔCO was determined for the plumes through linear regression fit to the OA and CO time series. OA and CO correlated tightly during this period and ΔOA/ΔCO was nearly identical in the nighttime- and daytime-plumes (Figure

5a). However, compared to daytime-plumes, OA in plumes transported during night time were less oxidized and were dominated by the fresh BBOA-1 (53%; Figure 5b). In contrast, daytime plumes were characterized by a significantly lower mass fraction of BBOA-1 (37%) with commensurate fractional increases of BBOA-2 (20%) and BBOA-3 (37%). Overall, the daytime plumes were more oxidized than the night-time plumes (average O/C = 0.66 vs. 0.55). Taken together, these observations suggest that while there was no net BBOA production with higher photochemical aging, BBOA was chemically transformed, likely due to evaporation of POA as well as oxidative processing in both gas and particle phases followed by fragmentation and volatilization.

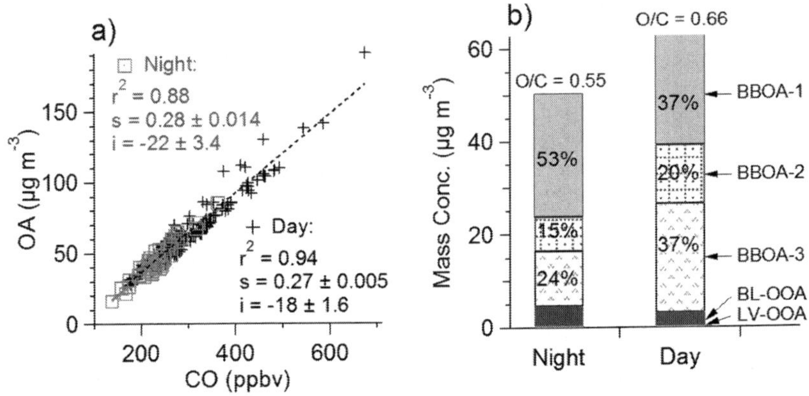

Figure 5. (a) OA vs. CO during the SRCF case study period. The orthogonal distance regression results for the two plume types are shown with the 1σ uncertainties reported for the fit slopes (s) and intercepts (i); (b) a comparison of the average concentrations of 5 OA factors (stacked) between the nighttime and daytime plumes. The average mass fractions of the BBOAs in OA in each plume type are shown. Modified with permission from Zhou et al. (14), Copyright 2017, the authors.

Impacts of Combustion Efficiency on Ambient BBOA Loading and Chemical Properties

In addition to atmospheric aging, biomass burning conditions also exert an important influence on the emissions and chemical properties of BBOA. The modified combustion efficiency (MCE) is a metric of combustion conditions in a fire. It is defined as the molar ratio of the enhanced concentration of CO_2 over background, to the sum of the enhanced concentrations of CO and CO_2: MCE = $\Delta CO_2 / (\Delta CO + \Delta CO_2)$ (*80, 81*). Higher MCE values (> 0.9) are associated with flaming combustion, higher mass fractions of black carbon (*82*), and less overall emitted PM per unit of fuel burned. Lower MCE values (< 0.9) are associated with smoldering combustion and tend to emit significantly more POA and VOCs.

In order to explore the relationships between BB aerosols and combustion conditions, Collier et al. (*13*) developed a method for analyzing ambient measurements of BB plumes made with real-time, fast instruments. The first step of the method is to identify individual plumes in which BB related gaseous and particulate species (e.g., CO, CO_2, aerosol components) are well above their corresponding background levels and show tight correlations. The criterion is set to minimize background and mixing effects on calculations of MCE and emission ratios (ER). The second step is to determine the MCE of each plume by finding the slope between CO_2 and CO using an unconstrained linear orthogonal distance regression and solving for MCE = $1/(1+\Delta CO/\Delta CO_2)$. Thirdly, the ERs of individual aerosol species relative to CO+CO_2 are determined using orthogonal distance fitting. The benefit of using the regression slope method is that it is a more objective method without the need to assume background concentrations, thus mitigates issues identified in the literature with ER calculations (*50*).

Using this method, a total of 32 plumes, 18 from MBO and 14 from G-1, were identified for the BBOP study. The calculated MCE values ranged between 0.80 and 0.99 for the MBO plumes and between 0.86 and 0.96 for the G-1 plumes. The ER of OA mass per unit mass burned (i.e., $\Delta OA/(\Delta CO+\Delta CO_2)$) showed a strong negative correlation with MCE for the plumes (Figure 6a). In addition, the values measured from both MBO and G-1 fall tightly along the same trend (Figure 6a). Since the estimated ages of the MBO BB plumes (6 – 48 h) were older than G-1 plumes (1 – 6 h), this strong agreement indicates that net changes in BBOA loadings were either slow or very similar plume to plume (i.e., independent to transport time or atmospheric aging). This finding is similar to that from the daytime and nighttime plume comparisons, where little difference was observed in $\Delta OA/\Delta CO$. One explanation is that the semivolatile BB POA evaporated during plume dilution while SOA increased in the plume with more atmospheric processing. In other words, the negligible change in the apparent ERs of BBOA with transport time for the range of MCEs encountered is likely due to the offsetting of primary BBOA losses driven by dilution and subsequent evaporation of the semi-volatile components, with the production of SOA. Another explanation for the consistency among plumes measured from the G-1 and MBO is that BBOA had undergone rapid processing near the wildfire source (e.g., < 1 h) prior to sampling by the G-1 but went through relatively little net change in mass during subsequent atmospheric transport to MBO.

As discussed above, $\Delta OA/\Delta CO$ often used to determine net formation of SOA in combustion plumes assuming that CO is a stable tracer to account for dilution effect. However, since the amount of CO emitted increases with decreasing MCE, it is important to examine the dependence of $\Delta OA/\Delta CO$ on MCE in wildfire plumes. As shown in Figure 6b, $\Delta OA/\Delta CO$ is relatively flat for various MCEs and there is high consistency between MBO and G-1 plumes with the exception of a low MCE (= 0.8) plume. Based on back-trajectory analysis, this plume appeared to be of similar age to other plumes but originated from the Douglas Complex Fire (Figure 1), which was less frequently sampled at MBO during this study. A possible reason for the larger $\Delta OA/\Delta CO$ for this plume is differences in fuel type. It is also possible that SOA formation was enhanced in this fire plume compared to the others, due to differences in emission compositions.

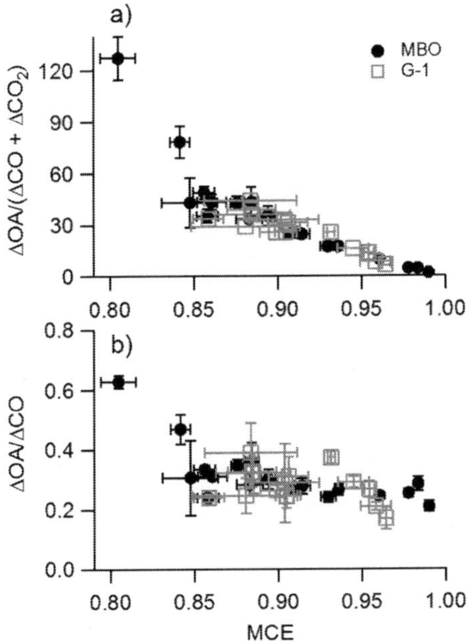

Figure 6. Variations of the enhancement ratio of a) OA mass relative to sum of $CO+CO_2$ ($\mu g\ m^{-3}$ / ppm) and b) OA mass relative to CO ($\mu g\ m^{-3}$ / ppb) as a function of MCE. Error bars represent the standard errors for each plume. Modified with permission from Collier et al. (13). Copyright 2016 American Chemical Society.

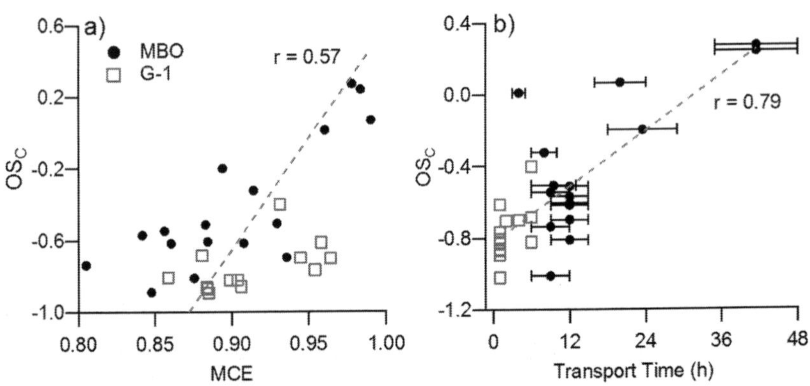

Figure 7. Correlations of average oxidation state of BBOA vs. a) MCE and b) Estimated plume transport time. The broken lines in a) and b) are determined through the orthogonal distance regression (ODR). Modified with permission from Collier et al. (13). Copyright 2016, American Chemical Society.

Ambient BBOA composition is influenced by both combustion conditions and atmospheric aging. According to Figure 7, atmospheric aging had a larger influence on the average oxidation state of BBOA as OS_C had a higher coefficient of correlation with transport time than with MCE. Note that transport times estimated based on HYSPLIT trajectories are approximate mainly due to plume injection heights, which are not available from the MODIS fire hotspots. Although transport time affected oxidation state more strongly, it appeared that burning condition, as characterized using the MCE parameter, drove the total enhancement of OA relative to the sum of CO and CO_2.

Overall, regional enhancements of organic aerosol loading relative to the total enhancements of CO and CO_2 due to wildfire emissions are mostly driven by burning efficiency and less by aging. This finding implies that bulk aerosol emission ratios of biomass burning measured at the sources can be used to estimate regional BBOA mass contributions. However, it is also evident that BBOA composition and particularly the average degree of oxidation is strongly influenced by atmospheric aging. Since aerosol microphysical and optical properties, as well as their effects on climate and human health, are strongly correlated with their chemical properties, it is important to take into account the impacts of both combustion conditions and atmospheric aging on BBOA.

Acknowledgments

This work was funded by US Department of Energy (DOE) Atmospheric Radiation Measurement (ARM) program grant DE-SC0014620.

References

1. Andreae, M. O.; Merlet, P. Emission of Trace Gases and Aerosols from Biomass Burning. *Global Biogeochem. Cycles* **2001**, *15*, 955–966.
2. Yokelson, R. J.; Christian, T. J.; Karl, T. G.; Guenther, A. The Tropical Forest and Fire Emissions Experiment: Laboratory Fire Measurements and Synthesis of Campaign Data. *Atmos. Chem. Phys.* **2008**, *8*, 3509–3527.
3. van der Werf, G. R.; Randerson, J. T.; Giglio, L.; Collatz, G. J.; Mu, M.; Kasibhatla, P. S.; Morton, D. C.; DeFries, R. S.; Jin, Y.; van Leeuwen, T. T. Global Fire Emissions and the Contribution of Deforestation, Savanna, Forest, Agricultural, and Peat Fires (1997–2009). *Atmos. Chem. Phys.* **2010**, *10*, 11707–11735.
4. Akagi, S. K.; Yokelson, R. J.; Wiedinmyer, C.; Alvarado, M. J.; Reid, J. S.; Karl, T.; Crounse, J. D.; Wennberg, P. O. Emission Factors for Open and Domestic Biomass Burning for Use in Atmospheric Models. *Atmos. Chem. Phys.* **2011**, *11*, 4039–4072.
5. Bond, T. C.; Doherty, S. J.; Fahey, D. W.; Forster, P. M.; Berntsen, T.; DeAngelo, B. J.; Flanner, M. G.; Ghan, S.; Kärcher, B.; Koch, D.; Kinne, S.; Kondo, Y.; Quinn, P. K.; Sarofim, M. C.; Schultz, M. G.; Schulz, M.; Venkataraman, C.; Zhang, H.; Zhang, S.; Bellouin, N.; Guttikunda, S. K.; Hopke, P. K.; Jacobson, M. Z.; Kaiser, J. W.; Klimont, Z.; Lohmann, U.; Schwarz, J. P.; Shindell, D.; Storelvmo, T.; Warren, S. G.; Zender, C. S. Bounding the Role of Black Carbon in the Climate System: A Scientific Assessment. *J. Geophys. Res.: Atmos.* **2013**, *118*, 5380–5552.

6. Jaffe, D. A.; Wigder, N. L. Ozone Production from Wildfires: A Critical Review. *Atmos. Environ.* **2012**, *51*, 1–10.
7. Chen, J.; Li, C.; Ristovski, Z.; Milic, A.; Gu, Y.; Islam, M. S.; Wang, S.; Hao, J.; Zhang, H.; He, C.; Guo, H.; Fu, H.; Miljevic, B.; Morawska, L.; Thai, P.; Lam, Y. F.; Pereira, G.; Ding, A.; Huang, X.; Dumka, U. C. A Review of Biomass Burning: Emissions and Impacts on Air Quality, Health and Climate in China. *Sci. Total Environ.* **2017**, *579*, 1000–1034.
8. Knorr, W.; Dentener, F.; Lamarque, J. F.; Jiang, L.; Arneth, A. Wildfire Air Pollution Hazard During the 21st century. *Atmos. Chem. Phys.* **2017**, *17*, 9223–9236.
9. Brewer, P.; Moore, T. Source Contributions to Visibility Impairment in the Southeastern and Western United States. *J. Air Waste Manage. Assoc.* **2009**, *59*, 1070–1081.
10. Jaffe, D.; Hafner, W.; Chand, D.; Westerling, A.; Spracklen, D. Interannual Variations in $PM_{2.5}$ Due to Wildfires in the Western United States. *Environ. Sci. Technol.* **2008**, *42*, 2812–2818.
11. Holden, A. S.; Sullivan, A. P.; Munchak, L. A.; Kreidenweis, S. M.; Schichtel, B. A.; Malm, W. C.; Collett, J. L., Jr. Determining Contributions of Biomass Burning and Other Sources to Fine Particle Contemporary Carbon in the Western United States. *Atmos. Environ.* **2011**, *45*, 1986–1993.
12. Liu, X.; Huey, L. G.; Yokelson, R. J.; Selimovic, V.; Simpson, I. J.; Müller, M.; Jimenez, J. L.; Campuzano-Jost, P.; Beyersdorf, A. J.; Blake, D. R.; Butterfield, Z.; Choi, Y.; Crounse, J. D.; Day, D. A.; Diskin, G. S.; Dubey, M. K.; Fortner, E.; Hanisco, T. F.; Hu, W.; King, L. E.; Kleinman, L.; Meinardi, S.; Mikoviny, T.; Onasch, T. B.; Palm, B. B.; Peischl, J.; Pollack, I. B.; Ryerson, T. B.; Sachse, G. W.; Sedlacek, A. J.; Shilling, J. E.; Springston, S.; St. Clair, J. M.; Tanner, D. J.; Teng, A. P.; Wennberg, P. O.; Wisthaler, A.; Wolfe, G. M. Airborne Measurements of Western U.S. Wildfire Emissions: Comparison with Prescribed Burning and Air Quality Implications. *J. Geophys. Res.: Atmos.* **2017**, *122*, 6108–6129.
13. Collier, S.; Zhou, S.; Onasch, T.; Jaffe, D.; Kleinman, L.; Sedlacek, A.; Briggs, N.; Hee, J.; Fortner, E.; Shilling, J. E.; Worsnop, D.; Yokelson, R. J.; Parworth, C.; Ge, X.; Xu, J.; Butterfield, Z.; Chand, D.; Dubey, M. K.; Pekour, M.; Springston, S.; Zhang, Q. Regional Influence of Aerosol Emissions from Wildfires Driven by Combustion Efficiency: Insights from the Bbop Campaign. *Environ. Sci. Technol.* **2016**, *50*, 8613–8622.
14. Zhou, S.; Collier, S.; Jaffe, D. A.; Briggs, N. L.; Hee, J.; Sedlacek Iii, A. J.; Kleinman, L.; Onasch, T. B.; Zhang, Q. Regional Influence of Wildfires on Aerosol Chemistry in the Western Us and Insights into Atmospheric Aging of Biomass Burning Organic Aerosol. *Atmos. Chem. Phys.* **2017**, *17*, 2477–2493.
15. Liu, Y.; Goodrick, S.; Heilman, W. Wildland Fire Emissions, Carbon, and Climate: Wildfire–Climate Interactions. *For. Ecol. Manage.* **2013**, *317*, 80–96.
16. Corbin, J. C.; Rehbein, P. J. G.; Evans, G. J.; Abbatt, J. P. D. Combustion Particles as Ice Nuclei in an Urban Environment: Evidence from Single-Particle Mass Spectrometry. *Atmos. Environ.* **2012**, *51*, 286–292.

17. Petters, M. D.; Carrico, C. M.; Kreidenweis, S. M.; Prenni, A. J.; DeMott, P. J.; Collett, J. L.; Moosmüller, H. Cloud Condensation Nucleation Activity of Biomass Burning Aerosol. *J. Geophys. Res.: Atmos.* **2009**, *114*, D22205.
18. Andreae, M. O.; Rosenfeld, D. Aerosol–Cloud–Precipitation Interactions. Part 1. The Nature and Sources of Cloud-Active Aerosols. *Earth-Science Reviews* **2008**, *89*, 13–41.
19. IPCC. *IPCC - Climate Change 2013: The Physical Science Basis*; 2013.
20. Randerson, J. T.; Liu, H.; Flanner, M. G.; Chambers, S. D.; Jin, Y.; Hess, P. G.; Pfister, G.; Mack, M. C.; Treseder, K. K.; Welp, L. R.; Chapin, F. S.; Harden, J. W.; Goulden, M. L.; Lyons, E.; Neff, J. C.; Schuur, E. A. G.; Zender, C. S. The Impact of Boreal Forest Fire on Climate Warming. *Science* **2006**, *314*, 1130–1132.
21. Carslaw, K. S.; Boucher, O.; Spracklen, D. V.; Mann, G. W.; Rae, J. G. L.; Woodward, S.; Kulmala, M. A Review of Natural Aerosol Interactions and Feedbacks within the Earth System. *Atmos. Chem. Phys.* **2010**, *10*, 1701–1737.
22. de Gouw, J. A.; Warneke, C.; Stohl, A.; Wollny, A. G.; Brock, C. A.; Cooper, O. R.; Holloway, J. S.; Trainer, M.; Fehsenfeld, F. C.; Atlas, E. L.; Donnelly, S. G.; Stroud, V.; Lueb, A. Volatile Organic Compounds Composition of Merged and Aged Forest Fire Plumes from Alaska and Western Canada. *J. Geophys. Res.: Atmos.* **2006**, *111*, D10303.
23. Bond, T. C.; Streets, D. G.; Yarber, K. F.; Nelson, S. M.; Woo, J.-H.; Klimont, Z. A Technology-Based Global Inventory of Black and Organic Carbon Emissions from Combustion. *J. Geophys. Res.: Atmos.* **2004**, *109*, D14203.
24. De Gouw, J.; Jimenez, J. L. Organic Aerosols in the Earth's Atmosphere. *Environ. Sci. Technol.* **2009**, *43*, 7614–7618.
25. Shrivastava, M.; Cappa, C. D.; Fan, J.; Goldstein, A. H.; Guenther, A. B.; Jimenez, J. L.; Kuang, C.; Laskin, A.; Martin, S. T.; Ng, N. L.; Petaja, T.; Pierce, J. R.; Rasch, P. J.; Roldin, P.; Seinfeld, J. H.; Shilling, J.; Smith, J. N.; Thornton, J. A.; Volkamer, R.; Wang, J.; Worsnop, D. R.; Zaveri, R. A.; Zelenyuk, A.; Zhang, Q. Recent Advances in Understanding Secondary Organic Aerosol: Implications for Global Climate Forcing. *Rev. Geophys.* **2017**, *55*, 509–559.
26. Akagi, S. K.; Burling, I. R.; Mendoza, A.; Johnson, T. J.; Cameron, M.; Griffith, D. W. T.; Paton-Walsh, C.; Weise, D. R.; Reardon, J.; Yokelson, R. J. Field Measurements of Trace Gases Emitted by Prescribed Fires in Southeastern Us Pine Forests Using an Open-Path Ftir System. *Atmos. Chem. Phys.* **2014**, *14*, 199–215.
27. Ortega, A. M.; Day, D. A.; Cubison, M. J.; Brune, W. H.; Bon, D.; de Gouw, J. A.; Jimenez, J. L. Secondary Organic Aerosol Formation and Primary Organic Aerosol Oxidation from Biomass-Burning Smoke in a Flow Reactor During Flame-3. *Atmos. Chem. Phys.* **2013**, *13*, 11551–11571.
28. Carrico, C. M.; Petters, M. D.; Kreidenweis, S. M.; Sullivan, A. P.; McMeeking, G. R.; Levin, E. J. T.; Engling, G.; Malm, W. C.; Collett, J. L. Water Uptake and Chemical Composition of Fresh Aerosols Generated in Open Burning of Biomass. *Atmos. Chem. Phys.* **2010**, *10*, 5165–5178.

29. Lapina, K.; Honrath, R. E.; Owen, R. C.; Val Martín, M.; Hyer, E. J.; Fialho, P. Late Summer Changes in Burning Conditions in the Boreal Regions and Their Implications for Nox and Co Emissions from Boreal Fires. *J. Geophys. Res.: Atmos.* **2008**, *113*, D11304.
30. Grieshop, A. P.; Donahue, N. M.; Robinson, A. L. Laboratory Investigation of Photochemical Oxidation of Organic Aerosol from Wood Fires 2: Analysis of Aerosol Mass Spectrometer Data. *Atmos. Chem. Phys.* **2009**, *9*, 2227–2240.
31. Grieshop, A. P.; Logue, J. M.; Donahue, N. M.; Robinson, A. L. Laboratory Investigation of Photochemical Oxidation of Organic Aerosol from Wood Fires 1: Measurement and Simulation of Organic Aerosol Evolution. *Atmos. Chem. Phys.* **2009**, *9*, 1263–1277.
32. Hennigan, C. J.; Miracolo, M. A.; Engelhart, G. J.; May, A. A.; Presto, A. A.; Lee, T.; Sullivan, A. P.; McMeeking, G. R.; Coe, H.; Wold, C. E.; Hao, W. M.; Gilman, J. B.; Kuster, W. C.; de Gouw, J.; Schichtel, B. A.; Collett, J. L.; Kreidenweis, S. M.; Robinson, A. L. Chemical and Physical Transformations of Organic Aerosol from the Photo-Oxidation of Open Biomass Burning Emissions in an Environmental Chamber. *Atmos. Chem. Phys.* **2011**, *11*, 7669–7686.
33. Bruns, E. A.; Krapf, M.; Orasche, J.; Huang, Y.; Zimmermann, R.; Drinovec, L.; Močnik, G.; El-Haddad, I.; Slowik, J. G.; Dommen, J.; Baltensperger, U.; Prévôt, A. S. H. Characterization of Primary and Secondary Wood Combustion Products Generated under Different Burner Loads. *Atmos. Chem. Phys.* **2015**, *15*, 2825–2841.
34. Tiitta, P.; Leskinen, A.; Hao, L.; Yli-Pirilä, P.; Kortelainen, M.; Grigonyte, J.; Tissari, J.; Lamberg, H.; Hartikainen, A.; Kuuspalo, K.; Kortelainen, A. M.; Virtanen, A.; Lehtinen, K. E. J.; Komppula, M.; Pieber, S.; Prévôt, A. S. H.; Onasch, T. B.; Worsnop, D. R.; Czech, H.; Zimmermann, R.; Jokiniemi, J.; Sippula, O. Transformation of Logwood Combustion Emissions in a Smog Chamber: Formation of Secondary Organic Aerosol and Changes in the Primary Organic Aerosol Upon Daytime and Nighttime Aging. *Atmos. Chem. Phys.* **2016**, *16*, 13251–13269.
35. Hobbs, P. V.; Sinha, P.; Yokelson, R. J.; Christian, T. J.; Blake, D. R.; Gao, S.; Kirchstetter, T. W.; Novakov, T.; Pilewskie, P. Evolution of Gases and Particles from a Savanna Fire in South Africa. *J. Geophys. Res.: Atmos.* **2003**, *108*, 8485.
36. Capes, G.; Johnson, B.; McFiggans, G.; Williams, P. I.; Haywood, J.; Coe, H. Aging of Biomass Burning Aerosols over West Africa: Aircraft Measurements of Chemical Composition, Microphysical Properties, and Emission Ratios. *J. Geophys. Res.: Atmos.* **2008**, *113*, D00C15.
37. Cubison, M. J.; Ortega, A. M.; Hayes, P. L.; Farmer, D. K.; Day, D.; Lechner, M. J.; Brune, W. H.; Apel, E.; Diskin, G. S.; Fisher, J. A.; Fuelberg, H. E.; Hecobian, A.; Knapp, D. J.; Mikoviny, T.; Riemer, D.; Sachse, G. W.; Sessions, W.; Weber, R. J.; Weinheimer, A. J.; Wisthaler, A.; Jimenez, J. L. Effects of Aging on Organic Aerosol from Open Biomass Burning Smoke in Aircraft and Laboratory Studies. *Atmos. Chem. Phys.* **2011**, *11*, 12049–12064.

38. Akagi, S. K.; Craven, J. S.; Taylor, J. W.; McMeeking, G. R.; Yokelson, R. J.; Burling, I. R.; Urbanski, S. P.; Wold, C. E.; Seinfeld, J. H.; Coe, H.; Alvarado, M. J.; Weise, D. R. Evolution of Trace Gases and Particles Emitted by a Chaparral Fire in California. *Atmos. Chem. Phys.* **2012**, *12*, 1397–1421.
39. Jolleys, M. D.; Coe, H.; McFiggans, G.; Capes, G.; Allan, J. D.; Crosier, J.; Williams, P. I.; Allen, G.; Bower, K. N.; Jimenez, J. L.; Russell, L. M.; Grutter, M.; Baumgardner, D. Characterizing the Aging of Biomass Burning Organic Aerosol by Use of Mixing Ratios: A Meta-Analysis of Four Regions. *Environ. Sci. Technol.* **2012**, *46*, 13093–13102.
40. Jolleys, M. D.; Coe, H.; McFiggans, G.; Taylor, J. W.; O'Shea, S. J.; Le Breton, M.; Bauguitte, S. J. B.; Moller, S.; Di Carlo, P.; Aruffo, E.; Palmer, P. I.; Lee, J. D.; Percival, C. J.; Gallagher, M. W. Properties and Evolution of Biomass Burning Organic Aerosol from Canadian Boreal Forest Fires. *Atmos. Chem. Phys.* **2015**, *15*, 3077–3095.
41. Forrister, H.; Liu, J.; Scheuer, E.; Dibb, J.; Ziemba, L.; Thornhill, K. L.; Anderson, B.; Diskin, G.; Perring, A. E.; Schwarz, J. P.; Campuzano-Jost, P.; Day, D. A.; Palm, B. B.; Jimenez, J. L.; Nenes, A.; Weber, R. J. Evolution of Brown Carbon in Wildfire Plumes. *Geophys. Res. Lett.* **2015**, *42*, 4623–4630.
42. Martin, M.; Tritscher, T.; Jurányi, Z.; Heringa, M. F.; Sierau, B.; Weingartner, E.; Chirico, R.; Gysel, M.; Prévôt, A. S. H.; Baltensperger, U.; Lohmann, U. Hygroscopic Properties of Fresh and Aged Wood Burning Particles. *J. Aerosol Sci.* **2013**, *56*, 15–29.
43. Engelhart, G. J.; Hennigan, C. J.; Miracolo, M. A.; Robinson, A. L.; Pandis, S. N. Cloud Condensation Nuclei Activity of Fresh Primary and Aged Biomass Burning Aerosol. *Atmos. Chem. Phys.* **2012**, *12*, 7285–7293.
44. Sumlin, B. J.; Pandey, A.; Walker, M. J.; Pattison, R. S.; Williams, B. J.; Chakrabarty, R. K. Atmospheric Photooxidation Diminishes Light Absorption by Primary Brown Carbon Aerosol from Biomass Burning. *Environ. Sci. Technol. Lett.* **2017**, *4*, 540–545.
45. Vakkari, V.; Kerminen, V.-M.; Beukes, J. P.; Tiitta, P.; van Zyl, P. G.; Josipovic, M.; Venter, A. D.; Jaars, K.; Worsnop, D. R.; Kulmala, M.; Laakso, L. Rapid Changes in Biomass Burning Aerosols by Atmospheric Oxidation. *Geophys. Res. Lett.* **2014**, *41*, 2644–2651.
46. Gilardoni, S.; Massoli, P.; Paglione, M.; Giulianelli, L.; Carbone, C.; Rinaldi, M.; Decesari, S.; Sandrini, S.; Costabile, F.; Gobbi, G. P.; Pietrogrande, M. C.; Visentin, M.; Scotto, F.; Fuzzi, S.; Facchini, M. C. Direct Observation of Aqueous Secondary Organic Aerosol from Biomass-Burning Emissions. *Proc. Natl. Acad. Sci. U.S.A.* **2016**, *113*, 10013–10018.
47. Abel, S. J.; Haywood, J. M.; Highwood, E. J.; Li, J.; Buseck, P. R. Evolution of Biomass Burning Aerosol Properties from an Agricultural Fire in Southern Africa. *Geophys. Res. Lett.* **2003**, *30*, 1783–1787.
48. Adachi, K.; Buseck, P. R. Atmospheric Tar Balls from Biomass Burning in Mexico. *J. Geophys. Res.: Atmos.* **2011**, *116*, D05204.

49. Sedlacek Iii, A. J.; Buseck, P. R.; Adachi, K.; Onasch, T. B.; Springston, S. R.; Kleinman, L. Formation and Evolution of Tar Balls from Northwestern Us Wildfires. *Atmos. Chem. Phys. Discuss.* **2018**, *2018*, 1–28.
50. Yokelson, R. J.; Burling, I. R.; Gilman, J. B.; Warneke, C.; Stockwell, C. E.; de Gouw, J.; Akagi, S. K.; Urbanski, S. P.; Veres, P.; Roberts, J. M.; Kuster, W. C.; Reardon, J.; Griffith, D. W. T.; Johnson, T. J.; Hosseini, S.; Miller, J. W.; Cocker Iii, D. R.; Jung, H.; Weise, D. R. Coupling Field and Laboratory Measurements to Estimate the Emission Factors of Identified and Unidentified Trace Gases for Prescribed Fires. *Atmos. Chem. Phys.* **2013**, *13*, 89–116.
51. May, A. A.; McMeeking, G. R.; Lee, T.; Taylor, J. W.; Craven, J. S.; Burling, I.; Sullivan, A. P.; Akagi, S.; Collett, J. L.; Flynn, M.; Coe, H.; Urbanski, S. P.; Seinfeld, J. H.; Yokelson, R. J.; Kreidenweis, S. M. Aerosol Emissions from Prescribed Fires in the United States: A Synthesis of Laboratory and Aircraft Measurements. *J. Geophys. Res.: Atmos.* **2014**, *119*, 11826–11849.
52. Weiss-Penzias, P.; Jaffe, D. A.; Swartzendruber, P.; Dennison, J. B.; Chand, D.; Hafner, W.; Prestbo, E. Observations of Asian Air Pollution in the Free Troposphere at Mount Bachelor Observatory During the Spring of 2004. *J. Geophys. Res.: Atmos.* **2006**, *111*, D10304.
53. Wigder, N. L.; Jaffe, D. A.; Saketa, F. A. Ozone and Particulate Matter Enhancements from Regional Wildfires Observed at Mount Bachelor During 2004–2011. *Atmos. Environ.* **2013**, *75*, 24–31.
54. Hecobian, A.; Liu, Z.; Hennigan, C. J.; Huey, L. G.; Jimenez, J. L.; Cubison, M. J.; Vay, S.; Diskin, G. S.; Sachse, G. W.; Wisthaler, A.; Mikoviny, T.; Weinheimer, A. J.; Liao, J.; Knapp, D. J.; Wennberg, P. O.; Kürten, A.; Crounse, J. D.; Clair, J. S.; Wang, Y.; Weber, R. J. Comparison of Chemical Characteristics of 495 Biomass Burning Plumes Intercepted by the Nasa Dc-8 Aircraft During the Arctas/Carb-2008 Field Campaign. *Atmos. Chem. Phys.* **2011**, *11*, 13325–13337.
55. Sahu, L. K.; Kondo, Y.; Moteki, N.; Takegawa, N.; Zhao, Y.; Cubison, M. J.; Jimenez, J. L.; Vay, S.; Diskin, G. S.; Wisthaler, A.; Mikoviny, T.; Huey, L. G.; Weinheimer, A. J.; Knapp, D. J. Emission Characteristics of Black Carbon in Anthropogenic and Biomass Burning Plumes over California During Arctas-Carb 2008. *J. Geophys. Res.: Atmos.* **2012**, *117*, D16302.
56. Liu, X.; Zhang, Y.; Huey, L. G.; Yokelson, R. J.; Wang, Y.; Jimenez, J. L.; Campuzano-Jost, P.; Beyersdorf, A. J.; Blake, D. R.; Choi, Y.; St. Clair, J. M.; Crounse, J. D.; Day, D. A.; Diskin, G. S.; Fried, A.; Hall, S. R.; Hanisco, T. F.; King, L. E.; Meinardi, S.; Mikoviny, T.; Palm, B. B.; Peischl, J.; Perring, A. E.; Pollack, I. B.; Ryerson, T. B.; Sachse, G.; Schwarz, J. P.; Simpson, I. J.; Tanner, D. J.; Thornhill, K. L.; Ullmann, K.; Weber, R. J.; Wennberg, P. O.; Wisthaler, A.; Wolfe, G. M.; Ziemba, L. D. Agricultural Fires in the Southeastern U.S. During Seac4rs: Emissions of Trace Gases and Particles and Evolution of Ozone, Reactive Nitrogen, and Organic Aerosol. *J. Geophys. Res.: Atmos.* **2016**, *121*, 7383–7414.

57. Hallar, A. G.; Petersen, R.; Andrews, E.; Michalsky, J.; McCubbin, I. B.; Ogren, J. A. Contributions of Dust and Biomass-Burning to Aerosols at a Colorado Mountain-Top Site. *Atmos. Chem. Phys.* **2015**, *15*, 13665–13679.
58. Timonen, H.; Jaffe, D. A.; Wigder, N.; Hee, J.; Gao, H.; Pitzman, L.; Cary, R. A. Sources of Carbonaceous Aerosol in the Free Troposphere. *Atmos. Environ.* **2014**, *92*, 146–153.
59. Laing, J. R.; Jaffe, D. A.; Hee, J. R. Physical and Optical Properties of Aged Biomass Burning Aerosol from Wildfires in Siberia and the Western US at the Mt. Bachelor Observatory. *Atmos. Chem. Phys.* **2016**, *16*, 15185–15197.
60. Briggs, N. L.; Jaffe, D. A.; Gao, H. L.; Hee, J. R.; Baylon, P. M.; Zhang, Q.; Zhou, S.; Collier, S. C.; Sampson, P. D.; Cary, R. A. Particulate Matter, Ozone, and Nitrogen Species in Aged Wildfire Plumes Observed at the Mount Bachelor Observatory. *Aerosol Air Qual. Res.* **2016**, *16*, 3075–3087.
61. Saide, P. E.; Peterson, D. A.; da Silva, A.; Anderson, B.; Ziemba, L. D.; Diskin, G.; Sachse, G.; Hair, J.; Butler, C.; Fenn, M.; Jimenez, J. L.; Campuzano-Jost, P.; Perring, A. E.; Schwarz, J. P.; Markovic, M. Z.; Russell, P.; Redemann, J.; Shinozuka, Y.; Streets, D. G.; Yan, F.; Dibb, J.; Yokelson, R.; Toon, O. B.; Hyer, E.; Carmichael, G. R. Revealing Important Nocturnal and Day-to-Day Variations in Fire Smoke Emissions through a Multiplatform Inversion. *Geophys. Res. Lett.* **2015**, *42*, 3609–3618.
62. Draxler, R. R.; Hess, G. D. An Overview of the Hysplit_4 Modeling System of Trajectories, Dispersion, and Deposition. *Aust. Meteor. Mag.* **1998**, *47*, 295–308.
63. DeCarlo, P. F.; Kimmel, J. R.; Trimborn, A.; Jayne, J.; Aiken, A. C.; Gonin, M.; Fuhrer, K.; Horvath, T.; Worsnop, D. R.; Jimenez, J. L. A Field-Deployable High-Resolution Time-of-Flight Aerosol Mass Spectrometer. *Anal. Chem.* **2006**, *78*, 8281–8289.
64. Onasch, T. B.; Trimborn, A.; Fortner, E. C.; Jayne, J. T.; Kok, G. L.; Williams, L. R.; Davidovits, P.; Worsnop, D. R. Soot Particle Aerosol Mass Spectrometer: Development, Validation, and Initial Application. *Aerosol Sci. Technol.* **2012**, *46*, 804–817.
65. Aiken, A. C.; Foy, B. d.; Wiedinmyer, C.; DeCarlo, P. F.; Ulbrich, I. M.; Wehrli, M. N.; Szidat, S.; Prevot, A. S. H.; Noda, J.; L.Wacker; Volkamer, R.; Fortner, E.; Wang, J.; Laskin, A.; Shutthanandan, V.; Zheng, J.; Zhang, R.; Paredes-Miranda, G.; Arnott, W. P.; Molina, L. T.; Sosa, G.; Querol, X.; Jimenez, J. L. Mexico City Aerosol Analysis During Milagro Using High Resolution Aerosol Mass Spectrometry at the Urban Supersite (T0) – Part 2: Analysis of the Biomass Burning Contribution and the Non-Fossil Carbon Fraction. *Atmos. Chem. Phys.* **2010**, *10*, 5315–5341.
66. Alfarra, M. R.; Prevot, A. S. H.; Szidat, S.; Sandradewi, J.; Weimer, S.; Lanz, V. A.; Schreiber, D.; Mohr, M.; Baltensperger, U. Identification of the Mass Spectral Signature of Organic Aerosols from Wood Burning Emissions. *Environ. Sci. Technol.* **2007**, *41*, 5770–5777.
67. Cubison, M. J.; Ortega, A. M.; Hayes, P. L.; Farmer, D. K.; Day, D.; Lechner, M. J.; Brune, W. H.; Apel, E.; Diskin, G. S.; Fisher, J. A.; Fuelberg, H. E.; Hecobian, A.; Knapp, D. J.; Mikoviny, T.; Riemer, D.; Sachse, G. W.; Sessions, W.; Weber, R. J.; Weinheimer, A. J.; Wisthaler, A.;

Jimenez, J. L. Effects of Aging on Organic Aerosol from Open Biomass Burning Smoke in Aircraft and Laboratory Studies. *Atmos. Chem. Phys.* **2011**, *11*, 12049–12064.

68. Aiken, A. C.; DeCarlo, P. F.; Kroll, J. H.; Worsnop, D. R.; Huffman, J. A.; Docherty, K.; Ulbrich, I. M.; Mohr, C.; Kimmel, J. R.; Sueper, D.; Sun, Y.; Zhang, Q.; Trimborn, A.; Northway, M.; Ziemann, P. J.; Canagaratna, M. R.; Onasch, T. B.; Alfarra, M. R.; Prevot, A. S. H.; Dommen, J.; Duplissy, J.; Metzger, A.; Baltensperger, U.; Jimenez, J. L. O/C and Om/Oc Ratios of Primary, Secondary, and Ambient Organic Aerosols with a High Resolution Time-of-Flight Aerosol Mass Spectrometer. *Environ. Sci. Technol.* **2008**, *42*, 4478–4485.
69. Zhang, Q.; Alfarra, M. R.; Worsnop, D. R.; Allan, J. D.; Coe, H.; Canagaratna, M. R.; Jimenez, J. L. Deconvolution and Quantification of Hydrocarbon-Like and Oxygenated Organic Aerosols Based on Aerosol Mass Spectrometry. *Environ. Sci. Technol.* **2005**, *39*, 4938–4952.
70. Alfarra, M. R.; Coe, H.; Allan, J. D.; Bower, K. N.; Boudries, H.; Canagaratna, M. R.; Jimenez, J. L.; Jayne, J. T.; Garforth, A.; Li, S.-M.; Worsnop, D. R. Characterization of Urban and Regional Organic Aerosols in the Lower Fraser Valley Using Two Aerodyne Aerosol Mass Spectrometers. *Atmos. Environ.* **2004**, *38*, 5745–5758.
71. Hennigan, C. J.; Sullivan, A. P.; Collett, J. L.; Robinson, A. L. Levoglucosan Stability in Biomass Burning Particles Exposed to Hydroxyl Radicals. *Geophys. Res. Lett.* **2010**, *37*, L09806.
72. Arangio, A. M.; Slade, J. H.; Berkemeier, T.; Pöschl, U.; Knopf, D. A.; Shiraiwa, M. Multiphase Chemical Kinetics of OH Radical Uptake by Molecular Organic Markers of Biomass Burning Aerosols: Humidity and Temperature Dependence, Surface Reaction, and Bulk Diffusion. *J. Phys. Chem. A* **2015**, *119*, 4533–4544.
73. Jolleys, M. D.; Coe, H.; McFiggans, G.; McMeeking, G. R.; Lee, T.; Kreidenweis, S. M.; Collett, J. L.; Sullivan, A. P. Organic Aerosol Emission Ratios from the Laboratory Combustion of Biomass Fuels. *J. Geophys. Res.: Atmos.* **2014**, *119*, 12850–12871.
74. Weimer, S.; Alfarra, M. R.; Schreiber, D.; Mohr, M.; Prevot, A. S. H.; Baltensperger, U. Organic Aerosol Mass Spectral Signatures from Wood-Burning Emissions: Influence of Burning Conditions and Wood Type. *J. Geophys. Res.: Atmos.* **2008**, *113*, D10304.
75. Paatero, P.; Tapper, U. Positive Matrix Factorization: A Non-Negative Factor Model with Optimal Utilization of Error Estimates of Data Values. *Environmetrics* **1994**, *5*, 111–126.
76. Zhang, Q.; Jimenez, J. L.; Canagaratna, M.; Ng, N. L.; Ulbrich, I.; Worsnop, D.; Sun, Y. L. Understanding Organic Aerosols Via Factor Analysis of Aerosol Mass Spectrometry: A Review. *Anal. BioAnal. Chem.* **2011**, *401*, 3045–3067.
77. May, A. A.; Levin, E. J. T.; Hennigan, C. J.; Riipinen, I.; Lee, T.; Collett, J. L.; Jimenez, J. L.; Kreidenweis, S. M.; Robinson, A. L. Gas-Particle Partitioning of Primary Organic Aerosol Emissions: 3. Biomass Burning. *J. Geophys. Res.: Atmos.* **2013**, *118*, 11327–11338.

78. Draxler, R. R.; Rolph, G. D. *Hysplit (Hybrid Single-Particle Lagrangian Integrated Trajectory) Model Access Via Noaa Arl Ready Website*; NOAA Air Resources Laboratory, Silver Spring, MD: Silver Spring, MD, 2015. http://Ready.Arl.Noaa.Gov/Hysplit.Php (accessed December 10, 2015).
79. Kroll, J. H.; Donahue, N. M.; Jimenez, J. L.; Kessler, S. H.; Canagaratna, M. R.; Wilson, K. R.; Altieri, K. E.; Mazzoleni, L. R.; Wozniak, A. S.; Bluhm, H.; Mysak, E. R.; Smith, J. D.; Kolb, C. E.; Worsnop, D. R. Carbon Oxidation State as a Metric for Describing the Chemistry of Atmospheric Organic Aerosol. *Nat. Chem.* **2011**, *3*, 133–139.
80. Lobert, J. M.; Scharffe, D. H.; Hao, W. M.; Kuhlbusch, T. A.; Seuwan, R.; Warneck, P.; Crutzen, P. J., Experimental Evaluation of Biomass Burning Emissions: Nitrogen and Carbon Containing Compounds. In *Global Biomass Burning: Atmospheric, Climatic and Biospheric Implications*; Levine, J. S., Ed.; MIT Press: Cambridge, MA, 1991; pp 289–304.
81. Ward, D. E.; Radke, L. F. Emission Measruements from Vegetation Fires: Acomparative Evaluation of Methods and Results. In *Fire in the Environment: The Ecological, Atmospheric, and Climatic Importance of Vegetation Fires*; Crutzen, P. J., Goldammer, J. G., Eds.; John Wiley: Chichester, U.K., 1993; pp 53–76.
82. McMeeking, G. R.; Kreidenweis, S. M.; Baker, S.; Carrico, C. M.; Chow, J. C.; Collett, J. L.; Hao, W. M.; Holden, A. S.; Kirchstetter, T. W.; Malm, W. C.; Moosmüller, H.; Sullivan, A. P.; Wold, C. E. Emissions of Trace Gases and Aerosols During the Open Combustion of Biomass in the Laboratory. *J. Geophys. Res.: Atmos.* **2009**, *114*, D19210.

Toxicity and Impacts of Aerosols

Chapter 19

Oxidative Properties of Ambient Particulate Matter - An Assessment of the Relative Contributions from Various Aerosol Components and Their Emission Sources

Vishal Verma,[*,1] Constantinos Sioutas,[2] and Rodney J. Weber[3]

[1]Department of Civil and Environmental Engineering, University of Illinois at Urbana Champaign, Urbana, Illinois 61801, United States
[2]Department of Civil and Environmental Engineering, University of Southern California, Los Angeles, California 90089, United States
[3]School of Earth and Atmospheric Sciences, Georgia Institute of Technology, Atlanta, Georgia 30332, United States
[*]E-mail: vverma@illinois.edu

The current national ambient air quality standards for $PM_{2.5}$ are based on particulate matter (PM) mass. Although epidemiological research conducted in the last few decades has associated PM mass with both respiratory and cardiovascular diseases, the heterogeneous and inconsistent nature of these associations suggests that not all components of PM are equally toxic. The capability of ambient particles to generate reactive oxygen species (ROS), also called the ROS activity or the oxidative potential is proposed as an alternative metric for relating the PM concentrations with health effects.

In this chapter, we discuss our work on the measurement of oxidative potential of ambient PM from various sampling campaigns in the United States. The core objective of this work was to identify the components of ambient PM and their emission sources, which are most responsible for inducing the ROS generation. The role of organic compounds in the oxidative potential of PM was assessed by their removal using thermodenuder and solid phase extraction techniques, while the contribution of metals was quantified with a chelation technique. A class of water-soluble organic compounds characterized by their strong hydrophobicity known as humic-like substances or HULIS, and transition metals (particularly Fe, Cu, and Mn) were identified as the major species driving the ROS generation mechanisms in ambient particles. However, our work shows that there are strong synergistic and antagonistic interactions among the HULIS components and transition metals. Limited source apportionment results revealed that biomass burning and secondary organic aerosol are the largest contributors to the oxidative potential in the southeastern United States. Further studies in this direction should help to develop useful insights on the origin of PM toxicity leading to a better assessment of the human health effects of ambient PM pollution.

Introduction

Ambient particulate matter (PM) pollution is one of the most important environmental challenges of modern society. Both short- and long-term exposure to fine PM ($PM_{2.5}$) have been linked to respiratory (*1–4*) and cardiovascular (*5–9*) health endpoints. The toxicity of ambient PM appears to be a multi-faceted phenomenon, probably involving many chemical constituents (*10*). The exact mechanisms explaining the toxic effects of PM have been a matter of debate for more than a decade; however, most of the recent studies (*11–15*) seem to indicate that the initiating step of the PM toxicity ladder is the generation of reactive oxygen species (ROS).

The most comprehensive evidence for the associations of PM concentrations with health effects comes from the epidemiological studies (*16–19*). However, the inconsistent and heterogeneous nature of these associations (i.e. spatiotemporal variability in the strength of associations) suggests that not all components of PM are equally toxic, and thus bulk $PM_{2.5}$ mass is not the best surrogate of aerosol toxicity and its health effects. Some of the recent studies have shown a better association of certain PM components [e.g. black carbon (BC), organic carbon (OC), and metals] with health endpoints than $PM_{2.5}$ mass (*15*, *20–22*). However, the limited chemical components used in the epidemiological analysis so far have been mostly driven by available measurement technology, without much consideration given to the ideal metrics suitable for assessing health impacts. Many of the components contributing to a large fraction of the PM mass could be biologically inactive. Since complete speciation of PM composition has not been possible yet, we need an integrative PM property, which combines the individual and synergistic actions of biologically relevant components to the extent possible. The capability of PM to generate ROS, also called the ROS activity or oxidative potential (OP), was proposed as one such property and several assays were developed to measure this property (*12*). While individual components may yield inconsistent results, the OP measured by these assays may be a better indicator of the aerosol toxicity. This has not been much explored in the epidemiological investigations. In epidemiological and large scale toxicological studies, there are two major challenges for testing the biological relevance of OP and its usefulness compared to the segregated chemical composition: first is the extremely time and labor intensive protocols of OP assays in comparison to other relatively easier measurement methods for PM mass and its components (e.g. EC, OC and inorganic elements); and second is the lack of sufficient information on the chemical components and emission sources contributing to the OP of the ambient particles.

This chapter primarily discusses the work conducted by the authors in the last ten years in the field of OP measurements for addressing these challenges by presenting results from various sampling campaigns in the western and southeastern United States. All of these campaigns were conducted under two research centers, i.e. Southern California Particle Center (SCPC) and Southeastern Center for Air Pollution and Epidemiology (SCAPE) funded by the United States Environmental Protection Agency (USEPA). The overall objective of SCPC was to investigate the underlying mechanisms that produce the health effects associated with exposure to PM, in relation to the sources, chemical composition and physical characteristics of PM, and with a focus on the unique urban settings of the Los Angeles air basin. Particularly, we present the results of OP measurements on the PM samples collected intermittently between Fall 2007 and Fall 2009, at an urban site near the University of Southern California (USC) in Los Angeles. We also include discussion of diesel exhaust PM samples collected under a different project (funded by California Air Resources Board) and analyzed for the OP measurement.

SCAPE was a collaborative center combining five multi-disciplinary studies performed by researchers at the Georgia Institute of Technology and Emory University. The overarching theme of SCAPE was a focus on characterizing ambient air pollution mixtures and elucidating their role in human health risks associated with air pollution. In that regard, an intensive sampling campaign was launched, which continued for over a year in the period from June 2012 to September 2013 and involved sampling at multiple sites in the southeastern US using a high-volume sampler (Hi-Vol) for filter collection and other aerosol instruments for real-time characterization of the PM chemical composition. The major findings of all these campaigns (i.e. both western and southeastern US) are already reported in various papers published over last ten years (*11*, *23–34*). This chapter summarizes those results and presents the findings from these studies to synthesize a cohesive picture on the holistic assessment of the role of aerosol ROS in health studies.

Methods

This chapter presents the data collected from many field campaigns conducted across the United States. A brief description of these campaigns is provided below.

Summer 2007: Diesel Exhaust Particles (DEPs) from Heavy-Duty Vehicles

DEPs were collected from heavy-duty vehicles tested on a chassis dynamometer at the California Air Resources Board's laboratory in Los Angeles.

Details of the test vehicle-configurations including their efficiency to control mass- and number-based PM emissions are given in the appendix (Table A1). Integrated PM samples were collected from the vehicular exhausts using a Hi-Vol.

Fall 2007: Los Angeles Air Basin during Southern California Wildfires

Sampling was conducted at the USC site, 2.5 km south of downtown Los Angeles on five different days (October 24, October 26, October 27, November 1, and November 14). Since the Los Angeles wildfires started on October 20 and continued until October 30, samples collected from October 24 to October 27 represent conditions impacted by both vehicular traffic and wildfire emissions, while November 1 and 14 samples were considered representative of typical ambient conditions in that area, affected mostly by vehicular traffic emissions.

Summer 2008: Los Angeles Air Basin

Samples were collected again at the USC site between June and August over a period of 10 consecutive weeks (excluding weekends). Time-integrated samples were collected during both "morning" (6:00-9:00 AM) and "afternoon" (11:00-14:00 PM) time periods and were grouped into three sample sets (S1, S2, and S3). There were 18 days total for the S1 sampling set, while both the S2 and S3 sets were collected for 16 days each. The "morning" period corresponds to rush hour traffic when the ambient aerosols at the sampling site are dominated by primary particles freshly emitted from vehicles on the nearby freeway. The "afternoon" period represents the mixture of primary and secondary particles undergoing physical and chemical changes (i.e., photo-oxidation, volatilization, dilution and possibly resuspension).

Fall 2009: Los Angeles Basin

This campaign was also conducted at the USC site. Sampling was conducted over a period of 6 consecutive weeks (excluding weekends during October and November), for 5-6 hrs each day. Figure 1 shows schematic of the experimental setup used in this campaign. Concentrated ultrafine particles with diameters of less than 180 nm (obtained by a versatile aerosol concentration enrichment system) were collected, both upstream and downstream of a thermodenuder (TD). As the ambient aerosol is drawn and passed through the heating section of TD, part of its volatile/semi-volatile component is evaporated. The aerosol then enters the adsorption/cooling tube, where the evaporated compounds are adsorbed onto activated charcoal placed on the walls of this section. The remaining PM (not volatilized) is collected onto the filters placed downstream of TD.

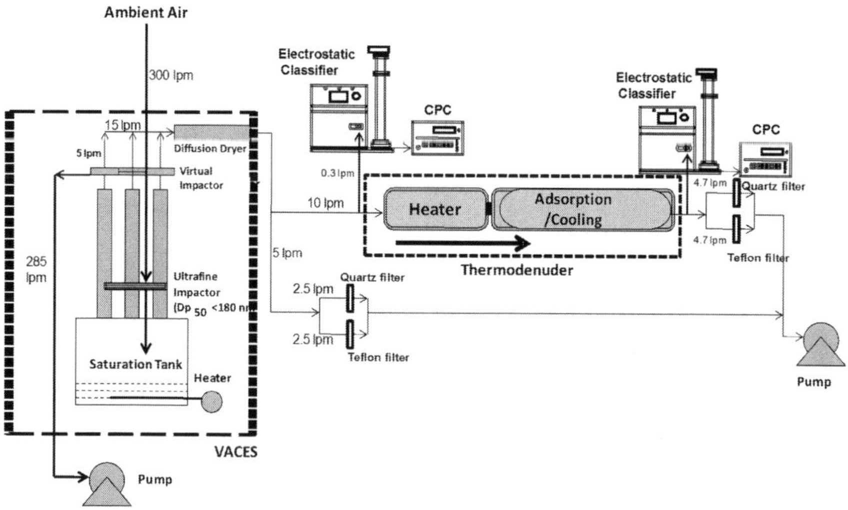

Figure 1. Experimental setup for collecting the total and non-volatile concentrated ambient particles at the sampling site in Los Angeles. Adapted with permission from reference (26). Copyright 2011, Elsevier

2012-2013: Southeast US

This was a more intensive campaign, which involved a paired sampling plan consisting of an identical set of instruments at two sites for over a year (June 2012 to September 2013). One of sites was always the Jefferson Street (JST) site, and the other was rotating between Yorkville (YRK), Georgia Tech (GT) and a roadside (RS) site in different seasons (Figure A1 in appendix). Thus, JST served as the central site, representative of the Atlanta urban background to which data from other sites was compared. RS site was in the immediate vicinity of a major freeway in Atlanta (I-75/85). GT (located on the roof of Ford Environmental Science and Engineering building) was considered the near-road site, which served as an intermediate location between the roadside (RS) and urban (JST) sites, because it is moderately impacted by the roadway emissions. All the Atlanta sites, i.e. JST, GT and RS were within few kilometers of each other. YRK, which is a site about 70 km west of Atlanta served as the rural component of this sampling network. As a contrast, measurements were also made in additional southeastern locations, i.e. Birmingham and Centreville in Alabama. Birmingham can be considered as a mixed urban-industrial site, while Centreville is the rural pair, located approximately 85km south-southwest of Birmingham. Further details of these sites are provided in previous publications (*23*, *33*).

A Hi-Vol sampler (flow rate = 1.13 m^3/min) was setup at all the sites and daily $PM_{2.5}$ were collected onto prebaked quartz filters from noon to 11 AM next day. An Aerodyne high-resolution time-of-flight aerosol mass spectrometer (HR-ToF-AMS) was also installed in parallel to the HiVol, rotating between the sites to characterize the chemical composition of non-refractory PM1 (non-light absorbing particles with diameters less than one micrometer). Positive matrix factorization

(PMF) was performed on the high-resolution organic mass spectra for resolving the organic aerosols into different factors, such as isoprene-derived OA (Isop_OA), less-oxidized oxygenated OA (LO-OOA), more-oxidized oxygenated OA (MO-OOA), hydrocarbon-like OA (HOA), cooking OA (COA), and biomass-burning OA (BBOA). Detailed discussion on the PMF procedure and results can be found in Verma et al. (*24*).

Chemical Analysis

The filters collected from all these campaigns underwent a variety of chemical and OP analyses, which are summarized in Table 1, along with the details of the specific methods used in each sampling campaign. Briefly, elemental and organic carbon content (EC and OC) of the PM was directly measured on the quartz filters using a thermal/optical transmittance (TOT) analyzer (Sunset laboratory). Sections of filters were extracted in deionized water via sonication in a water bath and the extracts were used for water-soluble organic carbon (WSOC), metals and OP analysis. Sub-samples of the DEPs water extracts were further processed by chelation with the immobilized ligand iminodiacetate (Chelex chromatography) to remove the metal ions. The Chelex processed extracts were collected and immediately assayed for OP and subsequent elemental analysis by magnetic sector field inductively-coupled plasma mass spectrometry (SF-ICPMS).

Extracts of the filters collected from the southeast US were also passed through a C-18 column to separate the hydrophobic and hydrophilic components. Hydrophobic components are retained on the column, while hydrophilic species directly passed-through the column. The hydrophilic fraction was assayed for OP and also analyzed for metals. Thus, by difference of the original and hydrophilic fraction, OP and metals associated with the hydrophobic fraction were determined.

OP Analysis

OP of the collected particles was quantified by two different assays: 1) consumption of dithiothreitol in a cell-free system (DTT assay), and 2) in vitro exposure to rat alveolar macrophages, which are cells found in the lungs, using Dichlorofluorescin Diacetate (DCFH-DA) as the ROS probe (macrophage ROS). The former assay determines in quantitative terms the capacity of a sample to transfer electrons from dithiothreitol (DTT) to oxygen – a reaction analogous to the cellular redox reaction involving NADPH (nicotinamide adenine dinucleotide phosphate) and oxygen. The methodological procedure used for the DTT assay is described in detail by Cho, et al. (*35*). The macrophage ROS assay is a fluorogenic cell-based method to examine the production of PM-induced ROS in rat alveolar macrophages. Further details of this assay are given in Landreman, et al. (*36*).

Table 1. Summary of Sample Collection, Chemical Analysis and Statistical Analysis

Sampling campaign	*Duration*	*Filter types*	*Types of particles*	*Chemical analysis/OP analysis*	*Method employed*	*Statistical techniques used for investigating the associations of OP and chemical composition*
Dynamometer	Summer 2007	Teflon coated glass fiber HiVol filters	Total DEPs	WSOC	TOC analyzer	Simple and Multiple Linear Regressions
				OCEC	Sunset OCEC analyzer	
				metals	ICP-MS	
				OP	Macrophage ROS assay	
Los Angeles Basin	Fall 2007	Teflon coated glass fiber HiVol filters	Ambient $PM_{2.5}$	WSOC	TOC analyzer	Simple Linear Regression
				OCEC	Sunset OCEC analyzer	
				metals	ICP-MS	
				OP	Macrophage ROS assay	
					DTT assay	
	Summer 2008	Zefluor HiVol filters	Ultrafine ambient PM (<180 nm)	WSOC	TOC analyzer	Simple Linear Regression
				OCEC	Sunset OCEC analyzer	
				metals	ICP-MS	
				OP	Macrophage ROS assay	
					DTT assay	

Continued on next page.

Table 1. (Continued). Summary of Sample Collection, Chemical Analysis and Statistical Analysis

Sampling campaign	*Duration*	*Filter types*	*Types of particles*	*Chemical analysis/OP analysis*	*Method employed*	*Statistical techniques used for investigating the associations of OP and chemical composition*
Los Angeles Basin	Fall 2009	47 mm teflon filters	Ultrafine ambient PM (<180 nm)	WSOC	TOC analyzer	Simple Linear Regression
				metals	ICP-MS	
		47 mm quartz filters	Ultrafine PM (<180 nm)	OCEC	Sunset OCEC analyzer	
				OP	Macrophage ROS assay	
					DTT assay	
Southeast US	2012-2013	Quartz HiVol filters	Ambient $PM_{2.5}$	WSOC	TOC analyzer	PMF and CMB
				metals	ICP-MS	
				OCEC	Sunset OCEC analyzer	
				OP	DTT assay	

Relationship between Chemical Composition and OP

The relationship between chemical composition and OP was determined in each study using various statistical techniques, which are also summarized in Table 1. Briefly, simple linear regression was used for the ambient samples collected from Los Angeles basin; multiple linear regressions were used for the DEPs; while both multiple linear regression, PMF (EPA-PMF, version 3.0) and chemical mass balance [CMB model, version 8.2, with ensemble-based source profiles for $PM_{2.5}$ (*37, 38*)] were used for the ambient samples collected from the Southeast US. Sources in the CMB-based profile include light-duty gasoline vehicles (LDGV), heavy-duty diesel vehicles (HDDV), soil dust (SDUST), biomass burning (BURN), ammonium sulfate (AMSULF), ammonium bisulfate (AMBSULF), ammonium nitrate (AMNITR), and not otherwise apportioned organic carbon (OTHER_OC), which mostly contains secondary biogenic carbon. Contributions of each source to the measured concentration of various species were predicted on a daily basis. DTT activity was then related to these CMB-identified $PM_{2.5}$ source contributions through a multivariate linear regression with DTT activity as the dependent variable and source contributions as the independent variables.

Major Findings and Outcomes

Identifying the Bulk Fractions of the Aerosols Related with OP

The aerosol sampling and analysis efforts employed in the Los Angeles basin (i.e. ambient sampling at USC site and DEPs from the dynamometer experiments) helped to identify the bulk chemical fractions of the ambient particles (discussed below) associated with OP.

Figure 2 shows the results of the OP assays (DTT; Figure 2a and macrophage ROS; Figure 2b) conducted on samples collected during and after the fire event in Los Angeles (Fall 2008). The outcomes were reported both as per mass of PM, as well as per volume of air. While the mass-based activity is appropriate for comparing the relative PM toxicity from different sources, the volume-based activities become important in the context of the overall public exposure and the associated risks of individual contributions from various sources to ambient PM concentrations. In general, particles collected between October 24 and 28 (during fire) produced higher DTT activities, both per unit of PM mass as well as per unit of air volume, than those collected on November 01 and 14 (post-fire). In contrast to DTT, the highest macrophage ROS activity was observed for the November 1 sample, both on per mass and on per volume of air basis. Although the per mass activity was relatively high on October 26, the lower values obtained on October 24 and 27 make it difficult to associate the macrophage ROS results with the occurrence of wildfires.

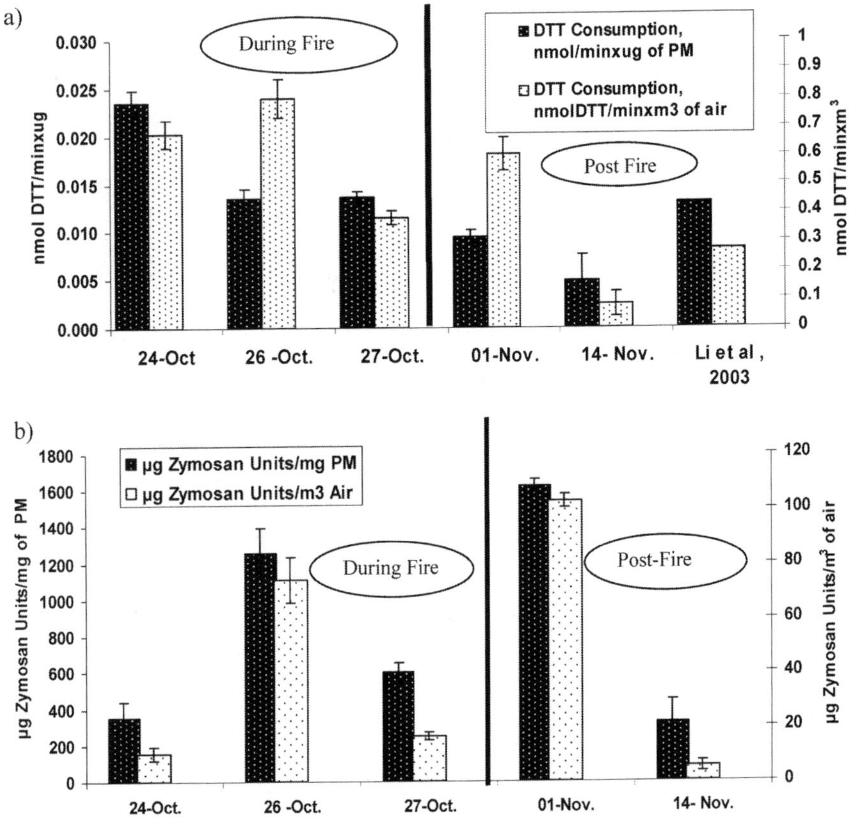

Figure 2. OP of ambient PM samples collected at the USC site during and after the fire period (a) expressed as nmol of DTT consumption/min, (b) expressed as the ROS response, or μg of Zymosan units. Reproduced with permission from reference (27). Copyright 2009, ACS

In another study, the DTT activities of primary and secondary particles collected at two different times during the day (morning and afternoon, respectively) in the summer of 2009, were compared. Figure 3 shows the results of the DTT assay for the three samples sets (S1, S2, and S3) and their average values in the morning and afternoon periods. The mass-based DTT activity of the particles is consistently higher in the afternoon period for all of the sample sets with an average afternoon-to-morning ratio of 1.8±0.7. The DTT activity per volume of air is also consistently higher for the afternoon samples, with an average afternoon-to-morning ratio of 2.8±1.1.

Figure 4 shows the results of the macrophage ROS assay in the morning and afternoon periods. There was significant variation in the macrophage ROS activity for the three sample sets and the afternoon-to-morning ratio averaged over the three grouped periods was 1.2±1.1. Although, the macrophage ROS activity of sample sets S2 and S3 was higher in the afternoon period, very high activity was observed for the morning period of sample set S1.

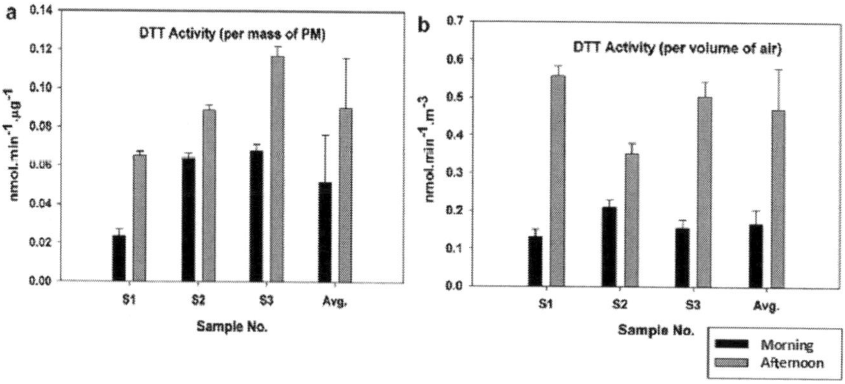

Figure 3. OP (expressed as nmol of DTT/min) of quasi-ultrafine particles collected at the USC sampling site during the morning and afternoon periods; a) per mg of particulate mass and b) per m³ of volume of air. Reproduced with permission from reference (25). Copyright 2009, Elsevier

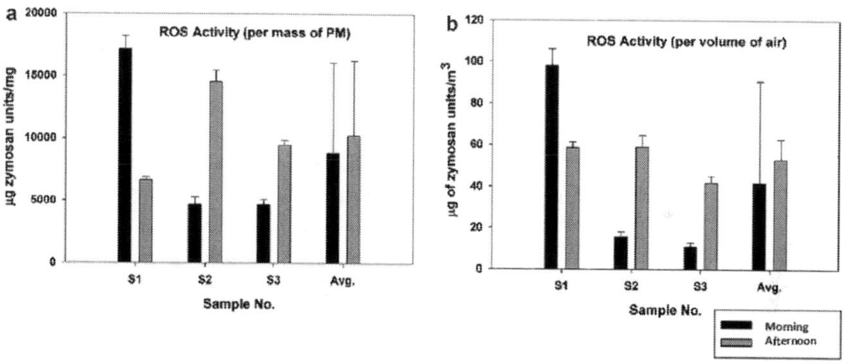

Figure 4. OP (expressed as the macrophage ROS response, or µg of Zymosan units) of quasi-ultrafine particles collected at the USC sampling site during the morning and afternoon periods; a) per mg of particulate mass and b) per m³ of volume of air. Reproduced with permission from reference (25). Copyright 2009, Elsevier

Like the biomass burning event discussed above, the dissimilar diurnal profiles of the OP obtained by DTT and macrophage ROS assays indicate the role of different chemical PM species in generating the oxidative stress. Linear regression analysis was carried out to investigate the association of the DTT and macrophage ROS activities with PM chemical composition. Table 2 shows a summary of the regression analysis for the samples collected during both sampling campaigns (i.e. Fall 2008 and Summer 2009). As shown in the table, we observed high correlation between WSOC and the DTT activity, while the macrophage ROS activity was correlated with certain water-soluble elements such as V, Cr, Fe, Cd and Ni. These metals may participate in Fenton chemistry and are capable of generating hydroxyl radicals (•OH) once coming in contact with the biological cells (*39*), and the •OH can effectively oxidize the DCFH probe used in the macrophage ROS assay.

Table 2. Summary of the regression analysis [square of correlation coefficients (r^2) and the associated levels of significance (P value)] for select species with DTT and macrophage ROS activity.[a]

Chemical Constituents	*Ambient PM*							
	Fall 2007 (Vehicular vs. Wildfires)				*Summer 2008 (Vehicular vs. Secondary)*			
	DTT		*Macrophage ROS*		*DTT*		*Macrophage ROS*	
	R^2	*p value*	*R^2*	*p value*	*R^2*	*p value*	*R^2*	*p value*
WSOC	0.65	0.1	0	0.96	0.82	0.01	0.09	0.58
PAHs	N/A	N/A	N/A	N/A	0.68	0.04	0.02	0.79
Organic Acids	N/A	N/A	N/A	N/A	0.71	0.15	0.07	0.74
V	0.25	0.39	0.79	0.04	0.01	0.84	0.91	0
Cr	0.61	0.12	0.84	0.03	0.21	0.36	0.56	0.09
Mn	0.75	0.06	0.01	0.85	0.52	0.1	0.42	0.13
Fe	0.39	0.26	0.96	0	0.69	0.04	0.28	0.28
Co	0.71	0.08	0.56	0.14	0.48	0.12	0.45	0.14
Ni	0.36	0.28	0.8	0.04	0.13	0.49	0.73	0.03
Cu	0.08	0.65	0.04	0.75	0.14	0.47	0.17	0.42
Cd	0.39	0.26	0.42	0.23	0.08	0.6	0.77	0.02
Zn	0.31	0.32	0.54	0.15	0.19	0.38	0.3	0.26

[a] N/A: Not applicable as the species was not measured in that study

Although this statistical analysis can provide some insights on the chemical species, which might be associated with the ROS generation, the inherent limitations of any regression technique, i.e. lack of an explicit causal relationship irrespective of the strength of correlation, and multicollinearity among independent variables, pose challenges in deciphering the underlying mechanisms linking particle composition to OP. Therefore, to confirm and quantify these relationships, we performed some mechanistic analysis to remove different PM constituents (i.e. organic compounds and metals) and observe the impacts on the residual OP. Two techniques were applied: heating via a TD for removing the organic compounds, and chelation for removing the transition metals (see methods section).

Figure 5 shows the comparison of OC in ambient and denuded particles collected at different TD temperature settings, i.e. 50, 100 and 200°C. OC progressively decreases with an increase in TD temperature (35±2 %, 40±3 % and 74±5 % loss at 50, 100 and 200°C, respectively). Figure 6 shows the DTT activity of ambient and thermodenuded particles at the same three TD temperatures (50, 100 and 200°C). The DTT activity decreases by 42±5 %, 47±8 % and 66±6 % as the particles were heated at 50, 100 and 200°C, respectively. The marked decline in DTT activity as the aerosols are denuded of their semi-volatile organic components highlights the significant contribution of organic compounds to the DTT-measured OP of ambient particles. Since metals are not expected to be semivolatile and lost in TD, their contributions to OP should remain unchanged by TD at any temperature. Chelation can be used to assess the role of metals.

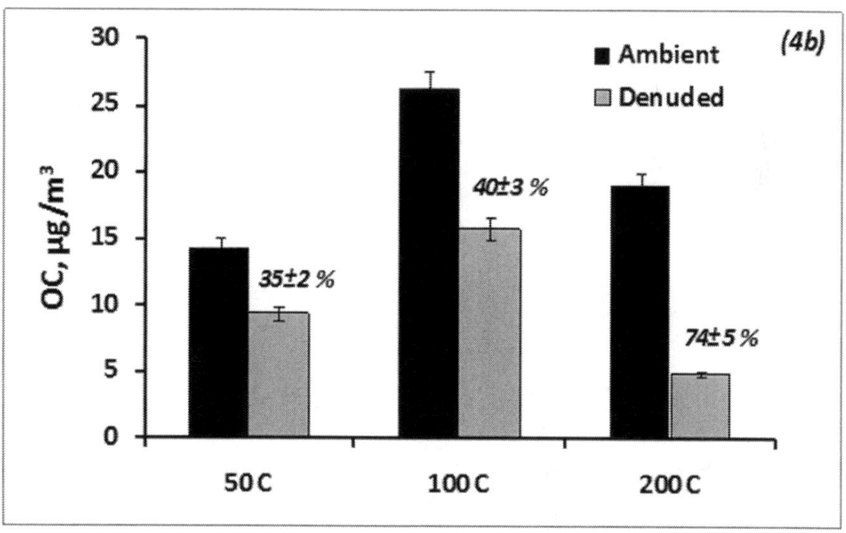

Figure 5. Concentration of organic carbon in the concentrated ambient and thermodenuded particles collected at three TD temperature configurations (50, 100 and 200°C). Percentages on top of the bars show loss in concentration on heating the aerosols. Reproduced with permission from reference (26). Copyright 2011, Elsevier

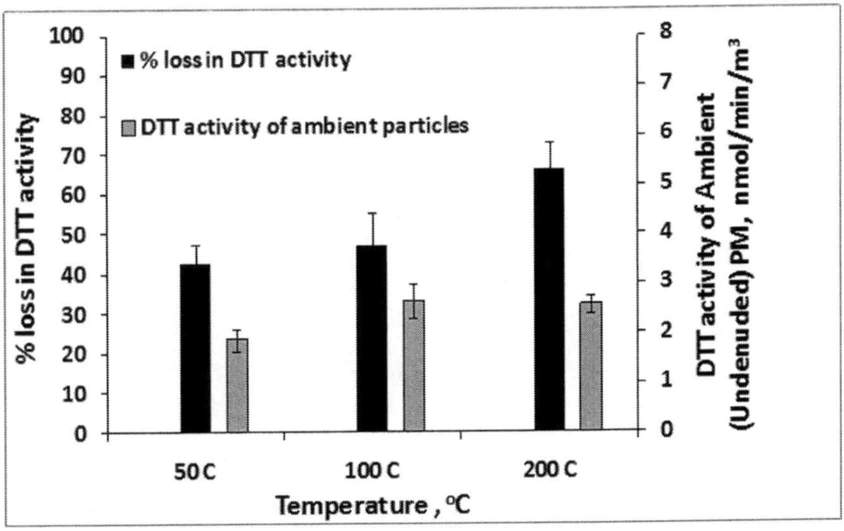

Figure 6. Percentage loss in OP, measured by DTT activity on heating the ambient aerosols at 50, 100 and 200 °C. The DTT activity of the concentrated ambient (undenuded) PM, corresponding to different temperature settings of the thermodenuder is also shown on secondary Y-axis as a reference. Reproduced with permission from reference (26). Copyright 2011, Elsevier

Figure 7 shows the percentage removal of macrophage-ROS activity after Chelex treatment of the water extracts of DEPs from various test vehicles under different driving cycles. To emphasize the correspondence between water-soluble metals and macrophage ROS activity, the percentage removal of the sum of major metals (Mg, Al, P, Ca, V, Cr, Mn, Fe, Co, Ni, Cu, Zn, As, Cd, Ba, and Pb) from respective PM samples is also shown (Figure 7). Chelex treatment of DEPs water extracts removed a very large and fairly consistent fraction of macrophage ROS activity for most of the samples. The reduction in macrophage ROS activity across all samples analyzed averages at 77 (±20) %. The substantial reduction in macrophage ROS activity after metals chelation and removal highlights the dominant contribution of water-soluble metals to the OP of PM measured by the macrophage ROS assay.

Development of the Automated Instrument for Measuring the DTT Activity

A major constraint in the routine use of OP measurements in the large-scale health studies is their labor-intensive and time-consuming protocols. To specifically address these concerns, we developed a semi-automated instrument for quantifying the OP of aerosol liquid extracts based on the DTT assay. This instrument, discussed in detail in Fang et al. (*33*), is capable of unattended analysis at 24 samples per day without any manual intervention; it has a high analytical precision (coefficient of variation = 4 % for ambient samples) and a low limit of detection (0.31 nmol min^{-1}). This instrument facilitated the analysis of a large number of PM samples (N>500) collected during subsequent campaigns (i.e. southeastern US) and generating an extensive dataset on the DTT activity of ambient particles from contrasting environments (i.e. urban, roadside, and rural), which was then used in the source apportionment studies.

Emission Sources Contributing to the DTT-Based OP

In both of the source apportionment modeling approaches (PMF and CMB), three major sources were identified contributing to the DTT activity of the particles: 1) vehicular sources, 2) biomass burning, and 3) secondary formation and processing of these primary particles and gases (see Figure 8). Note, CMB-identified vehicle sources also include road dust, which may explain, to some extent, a higher vehicular contribution to DTT activity by CMB (32 %) compared to PMF (16 %). Mechanistic tests conducted with pure ammonium sulfate for DTT assay showed that sulfate is not able to catalyze DTT oxidation (*11*). Therefore, the DTT activity apportioned to secondary sulfate in both approaches should be associated with the unidentified secondary organic compounds. Nevertheless, the contribution from biomass burning to the DTT activity is unanimously the highest and very close in both of the source apportionment approaches, which makes it the largest emission source of DTT activity in Southeast US.

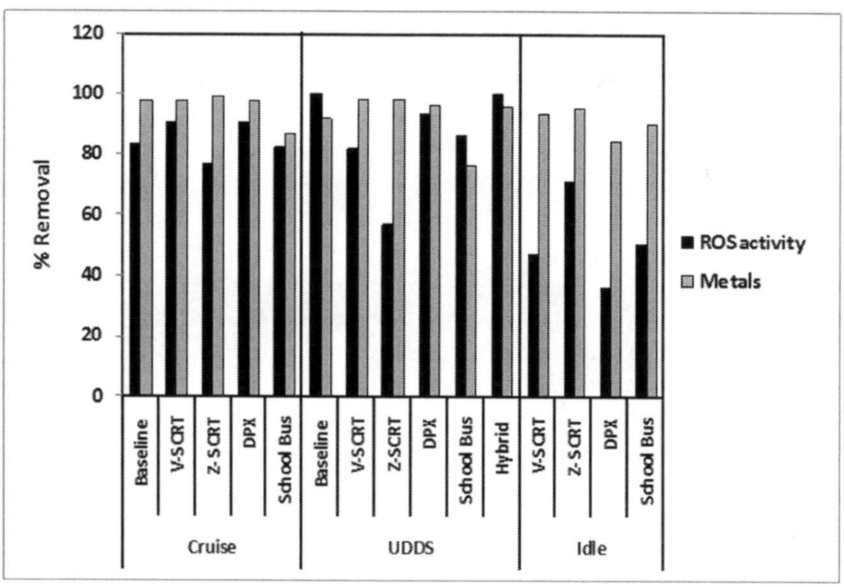

Figure 7. Percent removal of macrophage ROS activity in relation to that of aggregate water-soluble metals after Chelex treatment of the exhaust PM samples from test vehicle-configurations under different driving cycles. Reproduced with permission from reference (30). Copyright 2010, Elsevier

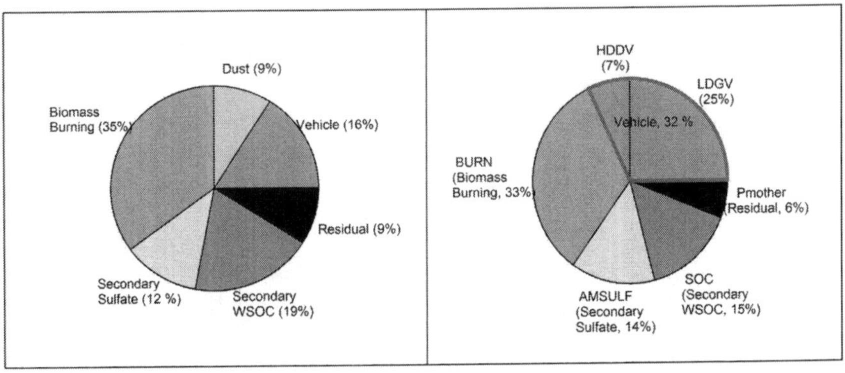

Figure 8. Annual aggregate contributions of different sources identified by PMF (left) and CMB method (right) to the water-soluble DTT activity of ambient PM2.5 in urban Atlanta. Reproduced with permission from reference (23). Copyright 2014, Copernicus

Identifying the Specific Chemical Components Associated with DTT-Based OP

Elucidating the chemical components associated with OP (DTT-assay) and their respective contributions has been a topic of research for almost a decade. Our work conducted in the Los Angeles basin has shown substantial contribution of OC (or WSOC) (*25–27, 35, 40*), in the DTT activity. Although, Cu and Mn have also been shown to be mechanistically active in the DTT assay, they generally do not show up in the correlation of measurements from collected samples; this has been mostly attributed to their non-linear relationship with the DTT oxidation (*41*). However, the relative contribution of organic compounds and metals in the DTT assay has remained a complex question. Our approach of separating the aerosol components based on their hydrophobicities (using the C-18 column, see methods section) and then applying the multiple linear regressions between the measured aerosol components and DTT activity provides some insights on this question.

First, the elemental analysis of the hydrophobic and hydrophilic fractions revealed that most metals of concern in the DTT assay, i.e. Cu and Mn, pass through the C-18 column and are collected in the hydrophilic fraction (*24*). Thus, from the perspective of the DTT assay, the separation of hydrophobic and hydrophilic components of aerosols by C-18 column can help to essentially segregate the individual contribution of organic compounds and metals. The data collected from multiple sites in Southeast US (Figure 9) shows that both metals (Cu and Mn, partitioned in the hydrophilic fraction) and organic compounds (in the hydrophobic fraction) are important in the DTT activity of ambient PM (metals~40 %; organic compounds~60 %).

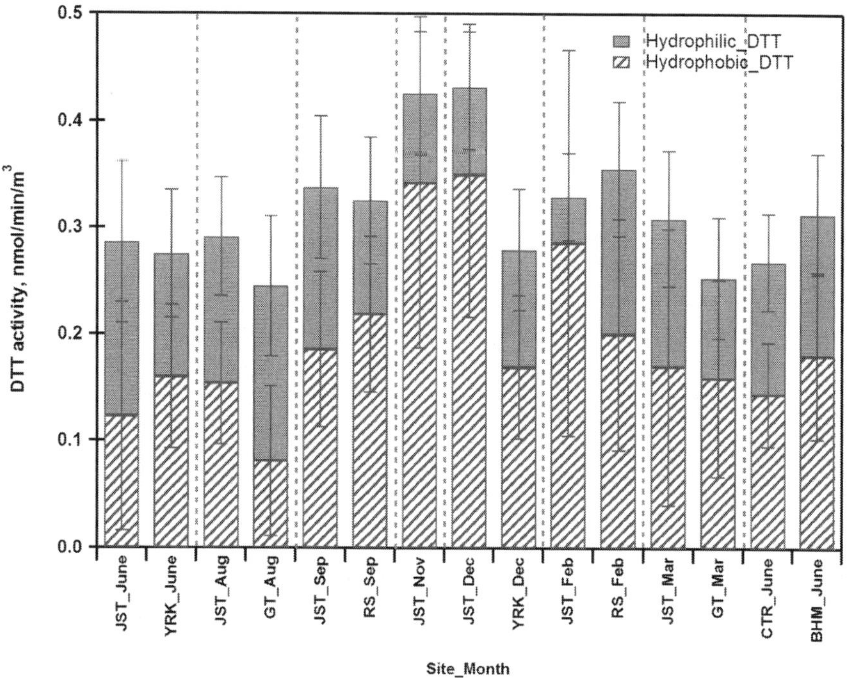

Figure 9. Hydrophobic and hydrophilic fractions of the water-soluble DTT activity (WS_DTT) measured at SCAPE sites separated via a C-18 column. Simultaneous sampling was done at paired sites. Hydrophobic fraction was calculated by subtracting the DTT activity measured on the hydrophilic fractions from the original extracts. Sample size for each bar is N > 25 (range = 25−38), except JST-November (N = 14). Error bars denote standard deviation (1σ) of the monthly averages. Reproduced with permission from reference (24). Copyright 2015, ACS

These results are helpful in evaluating the relative contribution of organic compounds and metals in DTT-based OP, but, unlike metals, which have been individually characterized for their intrinsic OP (*41*, *42*), OC represents a bulk group of species consisting of probably thousands of organic compounds, most of which are not yet speciated. Given the generic nature of OC emissions from most combustion sources, it cannot be used as a specific source tracer. Therefore, to further identify the specific components of OC associated with OP, we used a statistical approach where we conducted multiple linear regressions on the DTT activity of ambient PM and the various factors of OA resolved by the HR-ToF-AMS, i.e. isoprene-derived OA (Isop_OA), less-oxidized oxygenated OA (LO-OOA), more-oxidized oxygenated OA (MO-OOA), hydrocarbon-like OA (HOA), cooking OA (COA), and biomass-burning OA (BBOA). In the first step, a simple linear regression was conducted to screen the variables, which were significantly correlated with DTT activity. These selected variables were then included in a multiple linear regression model using a stepwise regression procedure. The coefficients of the regression equations obtained from different sites were averaged and are shown in Figure 10.

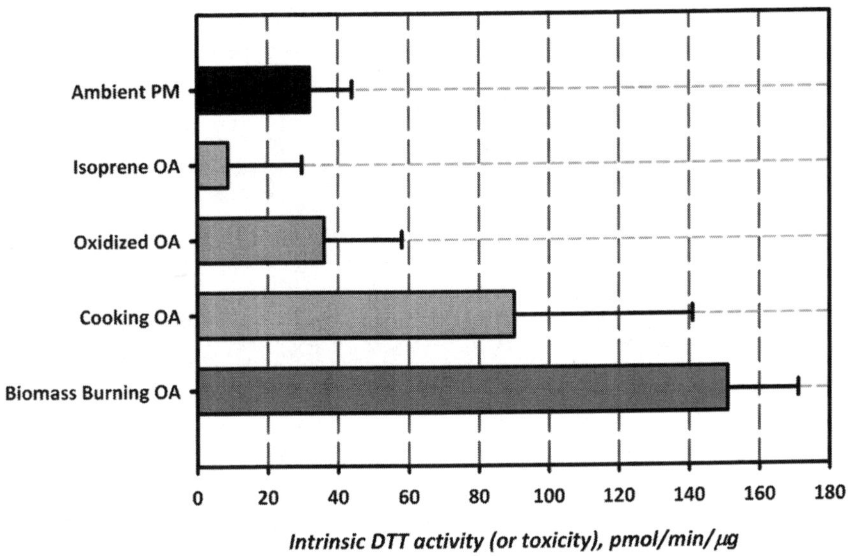

Figure 10. Intrinsic DTT activity of various organic aerosols as apportioned by an HR-ToF-AMS. Reproduced with permission from reference (24). Copyright 2015, ACS

The results show that biogenic SOA (Isop_OA or LO-OOA) does not have a significant DTT activity. Since secondary sources are shown to be responsible for a large part of the DTT activity (PMF and CMB analysis), it implies that the secondary compounds formed from the precursors of vehicular and biomass-burning emissions are the prominent DTT-active components in SOA. Among the organic aerosol factors, BBOA yielded the highest and most consistent (i.e. low standard deviation) intrinsic activity (151±20 pmol/min/µg). Cooking organic aerosol (COA) also had high intrinsic activity (90±51 pmol/min/µg), but the large standard deviation makes its estimate somewhat uncertain. The intrinsic DTT activity of MO-OOA is also significant (36±22 pmol/min/µg) and is in the typical range as observed for the ambient PM in southeast US. Note, the chemical composition of OA is highly dynamic and changes with the atmospheric processing, which could significantly alter the intrinsic DTT activities of different factors of OA. Therefore, the estimates shown in Figure 10 should be used with caution.

Estimates of the intrinsic DTT activities obtained from the statistical analysis can help to quantify the contributions of different organic aerosols in total DTT activity of the ambient PM by simply multiplying them with their respective ambient concentrations. The application of this method in the southeast region led to an important finding that all DTT-active organic compounds are hydrophobic in nature. This was revealed by an excellent agreement between the summed contributions of different organic aerosols resolved by HR-ToF-AMS and the measured activity of the hydrophobic fraction of PM separated by the C-18 column (see Figure 11). These hydrophobic organic compounds are also called humic-like substances (or HULIS), as they resemble the terrestrial/aquatic humic and fulvic acids. The chemical composition of HULIS is not well known, but limited studies (*43–45*) indicate that the large molecular weight compounds with mostly aromatic moieties comprise this fraction. Detailed chemical speciation (e.g. gas chromatography/mass spectroscopy; GC/MS) of the HULIS could provide more insights for identifying the major moieties present in ambient PM associated with the DTT-based OP.

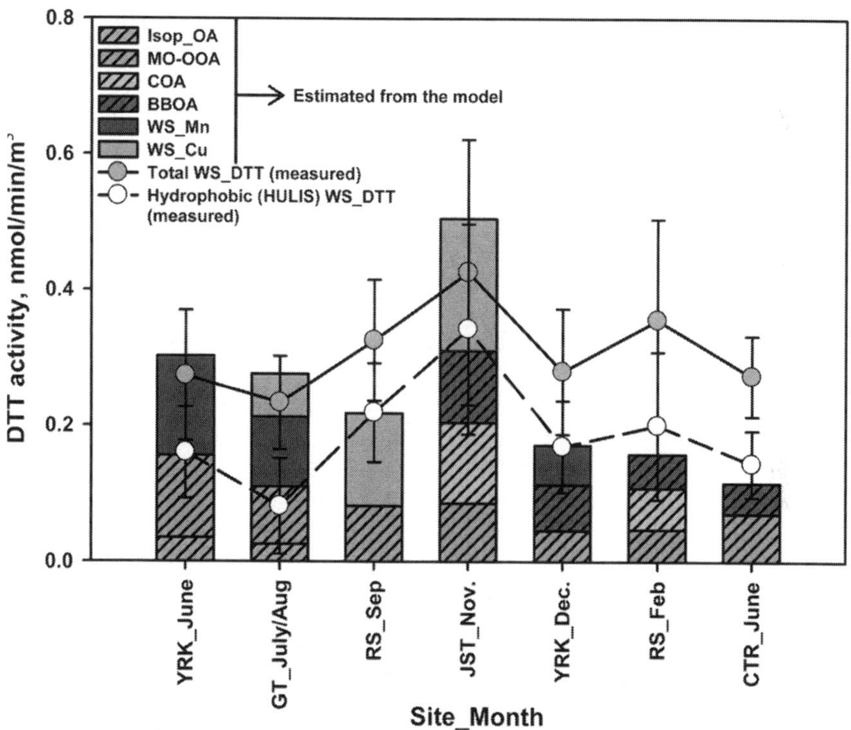

Figure 11. Contribution of various PM components (organic aerosols and water-soluble metals (i.e., Mn and Cu) to the water-soluble DTT activity, determined from the product of average intrinsic water-soluble activity (as predicted by the multiple linear regression model; Figure 10) and average ambient concentration of the PM components. The estimated water-soluble DTT activity from these components (e.g., organic aerosols (hatched region) and metals) is compared to the measured total water-soluble DTT activity (Total WS_DTT) and its hydrophobic fraction (hydrophobic WS_DTT). Reproduced with permission from reference (24). Copyright 2015, ACS

Discussion and Future Directions

Limitations of Current OP Probes and Overcoming Those Limitations

Among all the chemical assays developed so far for measuring the OP, the dithiothreitol (DTT) assay has been found to be responsive with the largest pool of chemical components, which involves many aromatic hydrocarbons, in addition to two transition metals - Cu and Mn (*41*). But there are limitations of the DTT assay as well, since it does not capture an important step of ROS cascade, which is the generation of •OH. Hydroxyl radical is a highly reactive species (*46*) with an in vivo half-life of approximately 10^{-9} seconds (*47*), which makes it the most damaging ROS. In the conventional DTT assay protocol, we measure only the rate of decay of DTT (*35*). At most, this rate corresponds to the formation of superoxide radicals ($O_2^{\bullet -}$), but the generation of •OH is not captured. This could be one of the causes for the underestimation of the role of important metals (e.g. Fe) in the DTT assay, which are otherwise known to contribute to ROS generation in biological systems (*48*). Therefore, in one of our recent studies (*49*), we overcame this limitation by measuring, for the first time, the rate of •OH generated by the ambient particles in the DTT assay. A fluorescent-based approach (disodium terephthalate for •OH and Amplex Red for H_2O_2) was used for the ROS measurement. We found that although DTT oxidation by ambient PM correlates well with H_2O_2 generation, an overall capability of the ambient aerosols to generate ROS cannot be generalized by this endpoint. H_2O_2 can further react with the organic compounds and metals to generate •OH, which is not correlated with the DTT oxidation (see Figure 12). Thus, the traditional endpoint (DTT consumption) is limited in terms of capturing the interactions among the PM components for ROS generation. The newer endpoint, i.e. •OH measurement in DTT assay, will capture such interactions. For example, the interactions among the organic compounds and Fe were additive in DTT consumption, but strongly synergistic in the •OH generation (see Figure 13). Furthermore, in one of our pilot studies (*50*), we showed that the •OH generation in DTT assay is probably a more biologically relevant endpoint, as evident from its excellent correlation ($R=-0.82;\ p<0.01$) with the cellular toxicity assessed via Chinese hamster ovary (CHO) cells (Figure 14), in comparison to the traditional DTT probe ($R=-0.58;\ p=0.02$). We attribute this improvement in the correlation to the capability of the newer endpoint for capturing the contribution from both organic compounds and metals, and their interactions in the ROS generation.

OP as a Proxy for the PM Toxicity

The current national ambient air quality standards (NAAQS) for $PM_{2.5}$ are based on PM mass concentrations. The NAAQS have immensely helped in controlling the PM emissions and reducing the concentrations of air pollutants in US cities. However, our studies clearly indicate that not all PM components are equally toxic, questioning the suitability of these standards, particularly in the current scenario of already reduced PM mass concentrations. An aerosol's OP could serve as an alternative air quality metric instead of (or in addition to) $PM_{2.5}$ mass. However, the most important constraint in assessing the suitability of OP for the regulatory standards is the lack of sufficient data. More measurements of OP need to be made and integrated in the epidemiological or toxicological studies to assess the biological relevance of OP. We recommend addressing this gap by including all possible methods of OP measurement in long-term sampling campaigns in various parts of the United States and developing the spatial and temporal profiles of OP of the ambient PM. These profiles would provide an important dataset to be integrated into future epidemiological studies, considering the associations of respiratory and cardiovascular health with the concentration of ambient particles. We anticipate that these studies would increase our mechanistic understanding of the links between the concentrations of ambient PM and the health effects, and thus evaluate the suitability of using OP as a proxy for the aerosol toxicity.

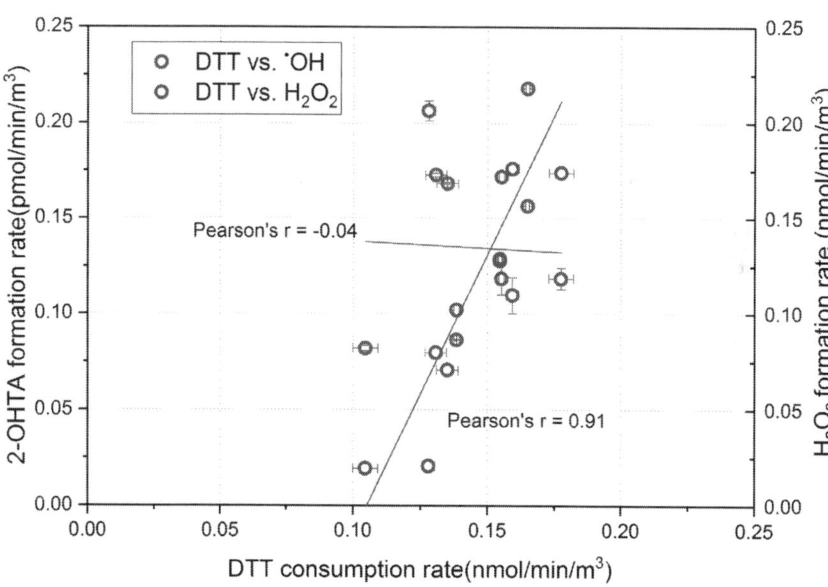

Figure 12. Rate of DTT oxidation versus ROS [H_2O_2 and •OH (expressed in terms of the concentration of 2-OHTA)] generation from ambient $PM_{2.5}$ samples. Both rates of DTT oxidation and ROS generation are normalized by the volume (m^3) of air for each sample. The samples were collected from the roof a parking deck in the Urbana campus of University of Illinois for 10 days in summer 2016. Reproduced with permission from reference 55. Copyright 2017, ACS

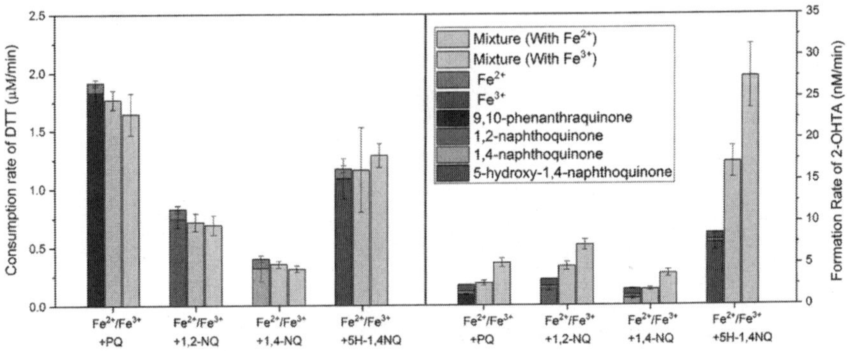

Figure 13. Rate of DTT oxidation and •OH generation (expressed in terms of the concentration of 2-OHTA) from various quinones, Fe (II), Fe (III), and their binary mixtures [i.e. PQ+Fe (II), PQ+Fe (III), 1,2-NQ+Fe (II), 1,2-NQ+Fe (III), 1,4-NQ+Fe (II), 1,4-NQ+Fe (III), 5H-1,4NQ+Fe (II), and 5H-1,4NQ+Fe (III)]. Notes: 1. The concentrations of different quinones, Fe (II), and Fe (III) in each experiment were 0.25 μM, 1 μM, and 1 μM, respectively 2. Error bars denote one standard deviation of the triplicate measurements. Reproduced with permission from reference 55. Copyright 2015, ACS

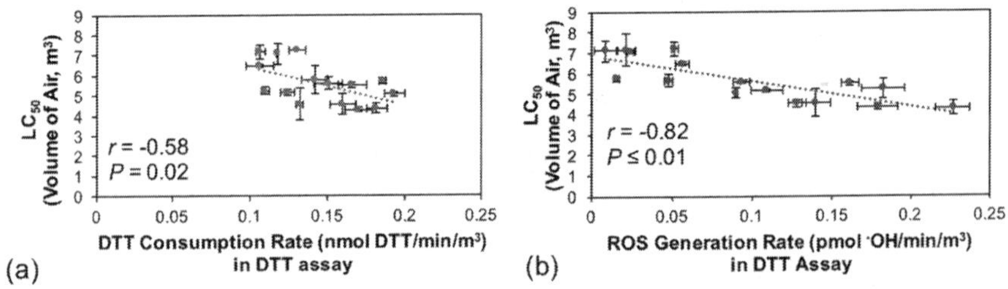

Figure 14. Correlation of cytotoxicity of ambient $PM_{2.5}$ measured by Chinese Hamster Ovary (CHO) cells with the particle OP measured by two different probes: a) DTT consumption in DTT assay, b) •OH generation in DTT assay. Ambient $PM_{2.5}$ samples (N=16) were collected from the roof a parking deck in the Urbana campus of University of Illinois between September 26, 2016 and June 12, 2017. Reproduced with permission from reference 56. Copyright 2018, Elsevier

Appendix

The appendix contains Table A1 and Figure A1 that follow.

Table A1. Details of the test vehicle-configurations

Vehicle				*Engine*			*After treatment (AT)*		*Nomen-clature*
Manufac-turer	*Year*	*Tested Wt (lb)*	*Miles*	*Model*	*Size [L]*	*Miles*	*Type*	*Miles on AT*	
Kenworth	1998	53,320	360,000	1998 Cummins M11, reflashed	11	360,000	None	NA	**Baseline**
							Vanadium based SCRT®	50,000	**V-SCRT**
							Zeolite based SCRT®	0 on SCRT	**Z-SCRT**
International	1999	20,920	40,000	1999 International DT466E	7.6	40,000	Engelhard DPX	30,000	**DPX**
Thompson-School Bus	1988	NA	325,000	2003 Cummins	5.9	50,000	Cleaire - Horizon	32,000	**School Bus**
Gillig (35ft) with Alison Hybrid	2007	NA	1,000	2006 Cummins	5.9	1,000	CCRT®	1,000	**Hybrid**

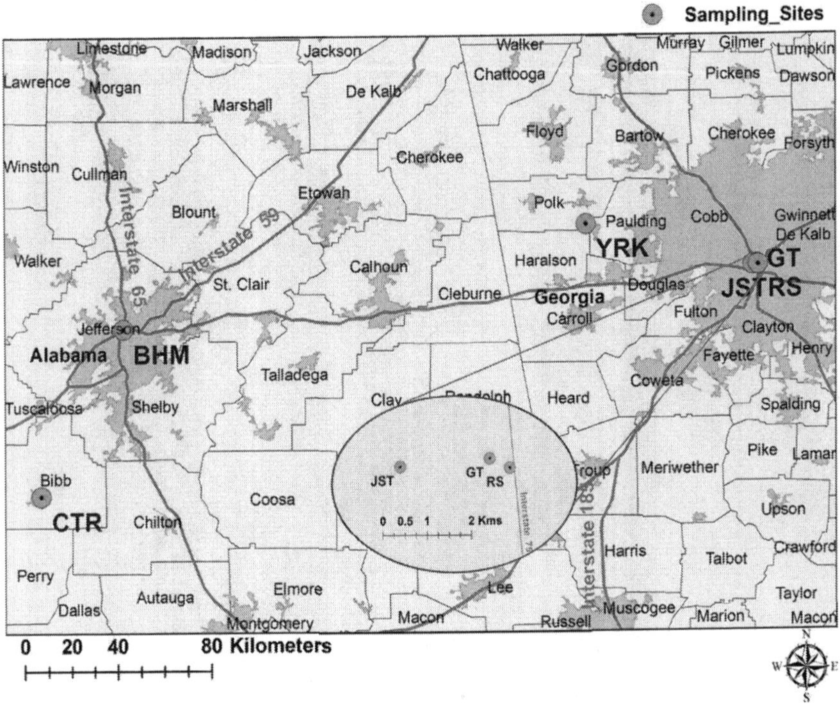

Figure A1. Map of Sampling Sites in Southeastern US

References

1. Atkinson, R. W.; Anderson, H. R.; Sunyer, J.; Ayres, J.; Baccini, M.; Vonk, J. M.; Boumghar, A.; Forastiere, F.; Forsberg, B.; Touloumi, G.; Schwartz, J.; Katsouyanni, K. Acute effects of particulate air pollution on respiratory admissions - Results from APHEA 2 project. *Am. J. Resp. Crit. Care* **2001**, *164*, 1860–1866.
2. Dockery, D. W.; Pope, C. A. Acute Respiratory Effects of Particulate Air-Pollution. *Annu. Rev. Publ. Health* **1994**, *15*, 107–132.
3. Gauderman, W. J.; Vora, H.; McConnell, R.; Berhane, K.; Gilliland, F.; Thomas, D.; Lurmann, F.; Avoli, E.; Kunzli, N.; Jerrett, M.; Peters, J. Effect of exposure to traffic on lung development from 10 to 18 years of age: a cohort study. *Lancet* **2007**, *369*, 571–577.
4. Laumbach, R. J.; Kipen, H. M. Respiratory health effects of air pollution: Update on biomass smoke and traffic pollution. *J. Allergy Clin. Immunol.* **2012**, *129*, 3–13.
5. Brook, R. D. Cardiovascular effects of air pollution. *Clin. Sci.* **2008**, *115*, 175–187.
6. Brook, R. D.; Rajagopalan, S.; Pope, C. A.; Brook, J. R.; Bhatnagar, A.; Diez-Roux, A. V.; Holguin, F.; Hong, Y. L.; Luepker, R. V.; Mittleman, M. A.; Peters, A.; Siscovick, D.; Smith, S. C.; Whitsel, L.; Kaufman, J. D.;

Epidemiol, A. H. A. C.; Dis, C. K. C.; Metab, C. N. P. A. Particulate Matter Air Pollution and Cardiovascular Disease An Update to the Scientific Statement From the American Heart Association. *Circulation* **2010**, *121*, 2331–2378.

7. Milojevic, A.; Wilkinson, P.; Armstrong, B.; Bhaskaran, K.; Smeeth, L.; Hajat, S. Short-term effects of air pollution on a range of cardiovascular events in England and Wales: case-crossover analysis of the MINAP database, hospital admissions and mortality. *Heart* **2014**, *100*, 1093–1098.
8. Riediker, M.; Cascio, W. E.; Griggs, T. R.; Herbst, M. C.; Bromberg, P. A.; Neas, L.; Williams, R. W.; Devlin, R. B. Particulate matter exposure in cars is associated with cardiovascular effects in healthy young men. *Am. J. Resp. Crit. Care* **2004**, *169*, 934–940.
9. Hoek, G.; Krishnan, R. M.; Beelen, R.; Peters, A.; Ostro, B.; Brunekreef, B.; Kaufman, J. D. Long-term air pollution exposure and cardio- respiratory mortality: a review. *Environ. Health* **2013**, *12*, 43.
10. *HEI's National Particle Component Toxicity (NPACT) Initiative*; Health Effects Institute: Boston, MA, 2013; http://www.healtheffects.org/publication/national-particle-component-toxicity-npact-initiative-report-cardiovascular-effects (accessed on June 08, 2018).
11. Bates, J. T.; Weber, R. J.; Abrams, J.; Verma, V.; Fang, T.; Klein, M.; Strickland, M. J.; Sarnat, S. E.; Chang, H. H.; Mulholland, J. A.; Tolbert, P. E.; Russell, A. G. Reactive Oxygen Species Generation Linked to Sources of Atmospheric Particulate Matter and Cardiorespiratory Effects. *Environ. Sci. Technol.* **2015**, *49*, 13605–13612.
12. Ayres, J. G.; Borm, P.; Cassee, F. R.; Castranova, V.; Donaldson, K.; Ghio, A.; Harrison, R. M.; Hider, R.; Kelly, F.; Kooter, I. M.; Marano, F.; Maynard, R. L.; Mudway, I.; Nel, A.; Sioutas, C.; Smith, S.; Baeza-Squiban, A.; Cho, A.; Duggan, S.; Froines, J. Evaluating the toxicity of airborne particulate matter and nanoparticles by measuring oxidative stress potential - A workshop report and consensus statement. *Inhal. Toxicol.* **2008**, *20*, 75–99.
13. Delfino, R. J.; Staimer, N.; Tjoa, T.; Gillen, D. L.; Schauer, J. J.; Shafer, M. M. Airway inflammation and oxidative potential of air pollutant particles in a pediatric asthma panel. *J. Exposure Sci. Environ. Epidemiol.* **2013**, *23*, 466–473.
14. Yang, A.; Wang, M.; Eeftens, M.; Beelen, R.; Dons, E.; Leseman, D. L. A. C.; Brunekreef, B.; Cassee, F. R.; Janssen, N. A. H.; Hoek, G. Spatial Variation and Land Use Regression Modeling of the Oxidative Potential of Fine Particles. *Environ. Health Perspect.* **2015**, *123*, 1187–1192.
15. Yang, A.; Janssen, N. A. H.; Brunekreef, B.; Cassee, F. R.; Hoek, G.; Gehring, U. Children's respiratory health and oxidative potential of PM2.5: the PIAMA birth cohort study. *Occup. Environ. Med.* **2016**, *73*, 154–160.
16. Pope, C. A. Respiratory-Disease Associated with Community Air-Pollution and a Steel Mill, Utah Valley. *Am. J. Public Health* **1989**, *79*, 623–628.
17. Pope, C. A.; Burnett, R. T.; Thun, M. J.; Calle, E. E.; Krewski, D.; Ito, K.; Thurston, G. D. Lung cancer, cardiopulmonary mortality, and long-term exposure to fine particulate air pollution. *JAMA-J. Am. Med. Assoc.* **2002**, *287*, 1132–1141.

18. Downs, S. H.; Schindler, C.; Liu, L. J. S.; Keidel, D.; Bayer-Oglesby, L.; Brutsche, M. H.; Gerbase, M. W.; Keller, R.; Kunzli, N.; Leuenberger, P.; Probst-Hensch, N. M.; Tschopp, J. M.; Zellweger, J. P.; Rochat, T.; Schwartz, J.; Ackermann-Liebrich, U.; Team, S. Reduced exposure to PM(10) and attenuated age-related decline in lung function. *New Engl. J. Med.* **2007**, *357*, 2338–2347.
19. Schindler, C.; Keide, D.; Gerbase, M. W.; Zemp, E.; Bettschart, R.; Brandli, O.; Brutsche, M. H.; Burdet, L.; Karrer, W.; Knopfli, B.; Pons, M.; Rapp, R.; Bayer-Oglesby, L.; Kunzli, N.; Schwartz, J.; Liu, L. J. S.; Ackermann-Liebrich, U.; Rochat, T.; Team, S. Improvements in PM10 Exposure and Reduced Rates of Respiratory Symptoms in a Cohort of Swiss Adults (SAPALDIA). *Am. J. Resp. Crit. Care* **2009**, *179*, 579–587.
20. Cakmak, S.; Dales, R.; Kauri, L. M.; Mahmud, M.; Van Ryswyk, K.; Vanos, J.; Liu, L.; Kumarathasan, P.; Thomson, E.; Vincent, R.; Weichenthal, S. Metal composition of fine particulate air pollution and acute changes in cardiorespiratory physiology. *Environ. Pollut.* **2014**, *189*, 208–214.
21. Sarnat, S. E.; Raysoni, A. U.; Li, W. W.; Holguin, F.; Johnson, B. A.; Luevano, S. F.; Garcia, J. H.; Sarnat, J. A. Air Pollution and Acute Respiratory Response in a Panel of Asthmatic Children along the U.S.-Mexico Border. *Environ. Health Perspect.* **2012**, *120*, 437–444.
22. Sarnat, S. E.; Winquist, A.; Schauer, J. J.; Turner, J. R.; Sarnat, J. A. Fine Particulate Matter Components and Emergency Department Visits for Cardiovascular and Respiratory Diseases in the St. Louis, Missouri-Illinois, Metropolitan Area. *Environ. Health Perspect.* **2015**, *123*, 437–444.
23. Verma, V.; Fang, T.; Guo, H.; King, L.; Bates, J. T.; Peltier, R. E.; Edgerton, E.; Russell, A. G.; Weber, R. J. Reactive oxygen species associated with water-soluble PM2.5 in the southeastern United States: spatiotemporal trends and source apportionment. *Atmos. Chem. Phys.* **2014**, *14*, 12915–12930.
24. Verma, V.; Fang, T.; Xu, L.; Peltier, R. E.; Russell, A. G.; Ng, N. L.; Weber, R. J. Organic Aerosols Associated with the Generation of Reactive Oxygen Species (ROS) by Water-Soluble PM2.5. *Environ. Sci. Technol.* **2015**, *49*, 4646–4656.
25. Verma, V.; Ning, Z.; Cho, A. K.; Schauer, J. J.; Shafer, M. M.; Sioutas, C. Redox activity of urban quasi-ultrafine particles from primary and secondary sources. *Atmos. Environ.* **2009**, *43*, 6360–6368.
26. Verma, V.; Pakbin, P.; Cheung, K. L.; Cho, A. K.; Schauer, J. J.; Shafer, M. M.; Kleinman, M. T.; Sioutas, C. Physicochemical and oxidative characteristics of semi-volatile components of quasi-ultrafine particles in an urban atmosphere. *Atmos. Environ.* **2011**, *45*, 1025–1033.
27. Verma, V.; Polidori, A.; Schauer, J. J.; Shafer, M. M.; Cassee, F. R.; Sioutas, C. Physicochemical and toxicological profiles of particulate matter in Los Angeles during the October 2007 Southern California wildfires. *Environ. Sci. Technol.* **2009**, *43*, 954–960.
28. Verma, V.; Rico-Martinez, R.; Kotra, N.; King, L.; Liu, J.; Snell, T. W.; Weber, R. J. Contribution of water-soluble and insoluble components and

their hydrophobic/hydrophilic sub-fractions to the reactive oxygen species-generating potential of fine ambient aerosols. *Environ. Sci. Technol.* **2012**, *46*, 11384–11392.
29. Verma, V.; Rico-Martinez, R.; Kotra, N.; Rennolds, C.; Liu, J. M.; Snell, T. W.; Weber, R. J. Estimating the toxicity of ambient fine aerosols using freshwater rotifer Brachionus calyciflorus (Rotifera: Monogononta). *Environ. Pollut.* **2013**, *182*, 379–384.
30. Verma, V.; Shafer, M. M.; Schauer, J. J.; Sioutas, C. Contribution of transition metals in the reactive oxygen species activity of PM emissions from retrofitted heavy-duty vehicles. *Atmos. Environ.* **2010**, *44*, 5165–5173.
31. Fang, T.; Guo, H.; Verma, V.; Peltier, R.; Weber, R. J. PM2.5 water-soluble elements in the southeastern United States: automated analytical method development, spatiotemporal distributions, source apportionment, and implications for heath studies. *Atmos. Chem. Phys.* **2015**, *15*, 11667–11682.
32. Fang, T.; Verma, V.; Bates, J. T.; Abrams, J.; Klein, M.; Strickland, M. J.; Sarnat, S. E.; Chang, H. H.; Mulholland, J. A.; Tolbert, P. E.; Russell, A. G.; Weber, R. J. Oxidative potential of ambient water-soluble PM2.5 measured by Dithiothreitol (DTT) and Ascorbic Acid (AA) assays in the southeastern United States: contrasts in sources and health associations. *Atmos. Chem. Phys.* **2016**, *16*, 3865–3879.
33. Fang, T.; Verma, V.; Guo, H.; King, L. E.; Edgerton, E. S.; Weber, R. J. A Semi-automated System for Quantifying the Oxidative Potential of Ambient Particles in Aqueous Extracts Using the dithiothreitol (DTT) assay: Results from the Southeastern Center for Air Pollution and Epidemiology (SCAPE). *Atmos. Meas. Tech.* **2014**, *8*, 471–482.
34. Verma, V.; Wang, Y.; El-Afifi, R.; Fang, T.; Rowland, J.; Russell, A. G.; Weber, R. J. Fractionating ambient humic-like substances (HULIS) for their reactive oxygen species activity - Assessing the importance of quinones and atmospheric aging. *Atmos. Environ.* **2015**, *120*, 351–359.
35. Cho, A. K.; Sioutas, C.; Miguel, A. H.; Kumagai, Y.; Schmitz, D. A.; Singh, M.; Eiguren-Fernandez, A.; Froines, J. R. Redox activity of airborne particulate matter at different sites in the Los Angeles Basin. *Environ. Res.* **2005**, *99*, 40–47.
36. Landreman, A. P.; Shafer, M. M.; Hemming, J. C.; Hannigan, M. P.; Schauer, J. J. A macrophage-based method for the assessment of the reactive oxygen species (ROS) activity of atmospheric particulate matter (PM) and application to routine (daily-24 h) aerosol monitoring studies. *Aerosol Sci. Tech.* **2008**, *42*, 946–957.
37. Gass, K.; Balachandran, S.; Chang, H. H.; Russell, A. G.; Strickland, M. J. Ensemble-Based Source Apportionment of Fine Particulate Matter and Emergency Department Visits for Pediatric Asthma. *Am. J. Epidemiol.* **2015**, *181*, 504–512.
38. Balachandran, S.; Pachon, J. E.; Hu, Y. T.; Lee, D.; Mulholland, J. A.; Russell, A. G. Ensemble-trained source apportionment of fine particulate matter and method uncertainty analysis. *Atmos. Environ.* **2012**, *61*, 387–394.

39. Schoonen, M. A. A.; Cohn, C. A.; Roemer, E.; Laffers, R.; Simon, S. R.; O'Riordan, T. Mineral-induced formation of reactive oxygen species. *Rev. Mineral. Geochem.* **2006**, *64*, 179–221.
40. Biswas, S.; Verma, V.; Schauer, J. J.; Cassee, F. R.; Cho, A. K.; Sioutas, C. Oxidative Potential of Semi-Volatile and Non Volatile Particulate Matter (PM) from Heavy-Duty Vehicles Retrofitted with Emission Control Technologies. *Environ. Sci. Technol.* **2009**, *43*, 3905–3912.
41. Charrier, J. G.; Anastasio, C. On dithiothreitol (DTT) as a measure of oxidative potential for ambient particles: evidence for the importance of soluble transition metals. *Atmos. Chem. Phys.* **2012**, *12*, 9321–9333.
42. DiStefano, E.; Eiguren-Fernandez, A.; Delfino, R. J.; Sioutas, C.; Froines, J. R.; Cho, A. K. Determination of metal-based hydroxyl radical generating capacity of ambient and diesel exhaust particles. *Inhal. Toxicol.* **2009**, *21*, 731–738.
43. Sannigrahi, P.; Sullivan, A. P.; Weber, R. J.; Ingall, E. D. Characterization of water-soluble organic carbon in urban atmospheric aerosols using solid-state C-13 NMR spectroscopy. *Environ. Sci. Technol.* **2006**, *40*, 666–672.
44. Sullivan, A. P.; Weber, R. J. Chemical characterization of the ambient organic aerosol soluble in water: 2. Isolation of acid, neutral, and basic fractions by modified size-exclusion chromatography. *J. Geophys. Res.-Atmos.* **2006**, *111*, D05315.
45. Sullivan, A. P.; Weber, R. J. Chemical characterization of the ambient organic aerosol soluble in water: 1. Isolation of hydrophobic and hydrophilic fractions with a XAD-8 resin. *J. Geophys. Res.-Atmos.* **2006**, *111*, D05314.
46. Halliwell, B.; Murcia, M. A.; Chirico, S.; Aruoma, O. I. Free-Radicals and Antioxidants in Food and in-Vivo - What They Do and How They Work. *Crit. Rev. Food Sci.* **1995**, *35*, 7–20.
47. Sies, H. Strategies of Antioxidant Defense. *Eur. J. Biochem.* **1993**, *215*, 213–219.
48. Dixon, S. J.; Stockwell, B. R. The role of iron and reactive oxygen species in cell death. *Nat. Chem. Biol.* **2014**, *10*, 9–17.
49. Xiong, Q. S.; Yu, H. R.; Wang, R. R.; Wei, J. L.; Verma, V. Rethinking Dithiothreitol-Based Particulate Matter Oxidative Potential: Measuring Dithiothreitol Consumption versus Reactive Oxygen Species Generation. *Environ. Sci. Technol.* **2017**, *51*, 6507–6514.
50. Wang, Y. X.; Plewa, M. J.; Mukherjee, U. K.; Verma, V. Assessing the cytotoxicity of ambient particulate matter (PM) using. Chinese hamster ovary (CHO) cells and its relationship with the PM chemical composition and oxidative potential. *Atmos. Environ.* **2018**, *179*, 132–141.

Chapter 20

Insights on Aerosol Oxidative Potential from Measurements of Particle Size Distributions

Rodney Weber,[1,*] Ting Fang,[2] and Vishal Verma[3]

[1]School of Earth and Atmospheric Sciences, Georgia Institute of Technology, 311 Ferst Drive, Atlanta, Georgia 30332, United States
[2]Department of Chemistry, University of California, Irvine, California 92697, United States
[3]Department of Civil and Environmental Engineering, University of Illinois at Urbana–Champaign, 205 North Mathews Avenue, Urbana, Illinois 61801, United States
*E-mail: rweber@eas.gatech.edu

Fine particle mass concentration is the metric used to characterize exposure of populations to ambient aerosols and to relate this exposure to the consequent health impacts. As an alternative to mass, recent research has focused on the application of potentially more relevant health-related aerosol properties, such as aerosol oxidative potential (OP). Fine particle OP determined with the dithiothreitol assay (OP^{DTT}) has been linked to emergency department visits for cardiorespiratory diseases in recent studies. While bulk measurements of OP are becoming more common, size resolved data provide unique insights on sources of aerosol OP and atmospheric processes that affect OP. Here we present contrasts in the size distributions of water-soluble OP^{DTT} and water-insoluble OP^{DTT}, along with size distributions of other key aerosol species, and show the data are consistent with two main chemical species contributing to both soluble and insoluble OP^{DTT} forms; transition metal ions and oxidized aromatic species, such as quinones. Size distributions of these two forms of OP^{DTT} differ. Insoluble OP^{DTT} follows the typical bimodal fine-coarse mode distribution, whereas the water-soluble distribution is unimodel with a peak near the fine-coarse mode minimum at 1 to 2.5 μm. In this study region, for both the soluble and insoluble forms, aerosol OP^{DTT} is largely generated by atmospheric processing. This includes oxidation of organic components and metals mobilization by acidic particles, implying that aerosol toxicity, as measured by this assay, is largely secondary and affected by varied multiphase processes.

Introduction

Exposure to ambient particulate matter (PM) is a leading cause of global disease (*1*) and can result in a range of adverse health effects (*2*, *3*). Associations found between particle mass concentrations and health effects through epidemiological studies have supported PM air quality standards (*4*, *5*), such as the National Ambient Air Quality Standards (NAAQS) (https://www.epa.gov/criteria-air-pollutants/naaqs-table) for $PM_{2.5}$ and PM_{10} (particles with aerodynamic diameters less than 2.5 and 10 μm, respectively). Factors other than bulk mass may be driving the adverse health associations. Recently, aerosol metrics beyond mass have been explored as possible better measures of the aerosols ability to cause adverse health responses. Oxidative potential (OP), defined as the capability of particles to deplete physiological antioxidants (reductants) and generate reactive oxygen species (ROS), see Figure 1, has been proposed as a more biologically relevant metric than bulk PM mass concentration. OP represents the integrated effects from multiple components in PM that can induce a state of oxidative stress; an imbalance toward an excess of ROS producing inflammation and a cascade of responses that can induce an overall adverse health response (*6*).

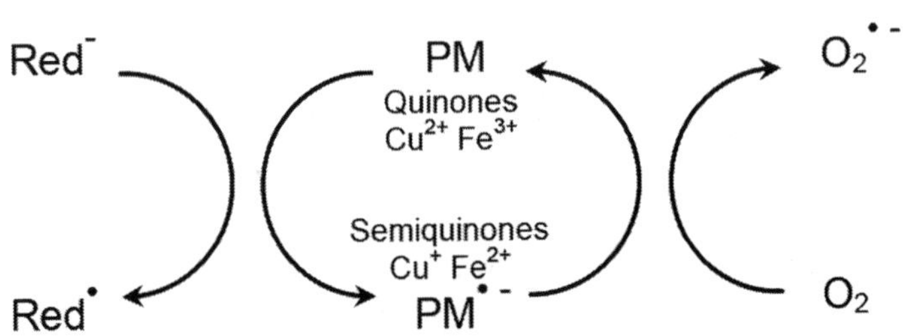

Figure 1. Schematic showing loss of antioxidants (reductants) and formation of oxidants (i.e., ROS) by aerosol components. Red- represents a generic physiological reducing agent. Specific aerosol components, such as quinones and redox-active transition metals (e.g., Fe and Cu), in vivo catalytically transfer an electron (-) from the reductant (Red-) to dissolved oxygen (O_2) leading to the formation of superoxide ($O_2^{\bullet-}$). Such loss of antioxidant and generation of oxidants can lead to oxidative stress in living systems. The catalytic ability of PM to transfer electrons from reductants to form oxidants is referred to as PM oxidative potential (OP) and in this study involves measuring the loss of DTT, a surrogate for physiological reductants (Red-).

In the early years of aerosol research, the classic work of Whitby illustrated the major sources and processes that affected ambient aerosols based on size distributions (*7*). Here we apply the same concept of using highly size-resolved data for investigating the sources and atmospheric processes affecting OP.

A number of studies have reported on the size distribution of particle OP. Several studies (*8–11*) examined OP measured with the dithiothreitol assay (OP^{DTT}) over different size fractions of PM by contrasting OP between broad aerosol modes, such as quasi-ultrafine (aerodynamic diameter, $D_p < 0.18$ μm), accumulation ($0.18 \leq D_p \leq 2.5$ μm), and coarse fractions ($D_p > 2.5$ μm). For a given size range, OP is often normalized by the aerosol mass of that size range, providing a measure of intrinsic OP^{DTT}, which is viewed as a measure of PM toxicity. For intrinsic OP^{DTT}, the smaller particles, such as those in the ultrafine mode, have very low mass and so most often dominate intrinsic OP^{DTT} (*10–18*). In contrast, a few studies show OP^{DTT} per mass is mainly in the accumulation mode size range (*19*, *20*). However, volume-normalized OP, which is more relevant to health than mass-normalized OP, as it is associated with exposure, is dominated by particles in the accumulation-mode (*21*, *22*). These past studies were limited in that segregation of OP by particle size was done only over broad size ranges. Highly size resolved OP^{DTT} data are needed to investigate the sources of OP^{DTT} and how it is transformed in the atmosphere as it ages.

A large multi-investigator air quality campaign, SCAPE (Southeastern Center for Air Pollution and Epidemiology), was conducted in Atlanta to characterize PM OP. Linkages between measures of $PM_{2.5}$ OP and cardiorespiratory health endpoints through epidemiological studies were investigated. Positive associations were found between asthma/wheeze and congestive heart failure and water-soluble OP^{DTT} (OP^{WS_DTT}) (*23*, *24*). In addition to the health studies, by sampling at multiple sites in the southeastern US over an extended period of time (June 2012 to September 2013), a variety of analyses provided insights on sources of PM OP and its atmospheric processing. This included multiple chemical analyses on approximately 500 $PM_{2.5}$ filter samples collected over the course of the study (*25*, *26*).

The major findings from the SCAPE study provided a consistent narrative. A diversity of aerosol chemical components contributed to OP^{WS_DTT}, including water-soluble metal ions and highly oxygenated organic species found in Humic Like Substances (HULIS), such as quinones (*27–30*). OP^{WS_DTT} was also most strongly related to atmospherically-processed vehicle-generated emissions and biomass burning emissions (*24*), and because it was largely secondary, OP^{WS_DTT} was more spatially uniform than known primary emissions, such as elemental carbon (EC) (*26*, *27*, *31*).

In the following we further explore the sources and processes affecting OP^{WS_DTT}, and extend the analysis to include water-insoluble OP^{DTT} (OP^{WI_DTT}), by investigating their size distributions and comparing the distributions to major aerosol chemical species. Consistent with SCAPE findings, we show that the two main components driving OP^{WS_DTT}, oxygenated aromatic species and transition metal ions, also apparently contribute to OP^{WI_DTT}, just appearing in different physical forms; either in solution or adsorbed to a solid insoluble particle surface that was co-emitted. Some fraction of the quinones and metal ions may be directly emitted, but the data indicate that they are largely formed as the aerosol ages. Differing phases are involved in these processes. The most obvious is the physical state of the OP^{DTT}; soluble versus insoluble. Atmospheric processing also involves differing phases, which includes oxidation of the aromatic species, either through gas phase or possible particle phase reactions. Particle liquid reactions are involved in the dissolution of insoluble forms of metals to form metal ions through interactions with strong acids from sulfate aerosol derived from sulfur dioxide oxidation. Hence, these data show that the PM toxicity increases with age due to multiphase processing.

Methods

Sites and MOUDI Measurements

Size resolved data of water-soluble OP^{DTT} and total OP^{DTT} used in this study were obtained from measurements at two sites, a road-side site and an urban site, both of which were located on the Georgia Tech campus in Atlanta, GA. The road-side site (RS) was within 5 meters of interstate 75/85 that passes through central Atlanta and is utilized mainly by light duty vehicles due to heavy duty vehicle restrictions (trucks with six wheels or more). In 2015 the average daily traffic was 301,000 vehicles (http://geocounts.com/gdot/). For contrast, the urban site (Urban) was located on the rooftop of the Environmental Science and Technology building roughly 420 meters from the RS site. Aerosols collected at this site were less influenced by fresh emissions and more representative of background urban aerosols; OP^{DTT} levels at this site are similar to those measured at Jefferson Street, the established Atlanta urban site in operation since 1998 (*26*).

Size resolved OP data were obtained with two Micro-Orifice Uniform Deposit Impactors (MOUDI, MSP Corp., Shoreview, MN, USA) for subsequent off line OP analysis. The MOUDIs used in this study were operated in a non-rotating mode without a back-up filter to reduce overall pressure drop. These cascade impactors divided particles into ten different size bins under ambient conditions (i.e., RH and T), since the MOUDIs were situated out-side. Aerodynamic particle diameters at 50% collection efficiency for the stages (so-called cut-off sizes) were 18, 10, 5.6, 3.2, 1.8, 1.0, 0.56, 0.32, 0.18, 0.1, and 0.056 µm. Particles with sizes between these cutoffs were collected on separate stages. Two MOUDI samples were collected at each site, with one set using impaction substrates of pre-baked quartz filters (47 mm Tissuquartz™ Filters, Pall Corp., Ann Arbor, MI, USA), the other Zefluor filters (47 mm, PTFE Membrane, 2 µm pore size, Pall Corp., Ann Arbor, MI, USA). For each MOUDI, sampling was conducted for approximately 7 days continuously to collect sufficient mass for subsequent analysis. Samples were analyzed for various chemical components, including organic carbon (OC) and elemental carbon (EC) by a thermal optical method, (NIOSH protocol) (*32*), various anions and cations by ion chromatography, and total and water-soluble Fe, Cu, and Mn by inductively coupled plasma mass spectrometry (ICPMS). Detailed descriptions of the analysis methods can be found in Fang et al. (*17*, *33*). Water-soluble and total OP^{DTT} were also measured. Additional sets of MOUDI samples were collected at each site for OP analysis only, some of which were sampled at both sites simultaneously. In all cases, collected filter samples were immediately sealed in petri dishes and stored at −18 °C until analyzed, which typically occurred within 3 days following collection, except for the metals analysis, which was done after a 2-month storage period.

Oxidative Potential

MOUDI Zefluor filters were cut in half and one of the halves used for analysis of OP. Sample preparation and the OP methods have been described in detail (*26*). Filters were extracted in DI water and then filtered with 0.45 μm syringe filters. OP measurement on Teflon MOUDI filters included water-soluble OP^{DTT} (OP^{WS_DTT}) and total OP^{DTT} (OP^{Total_DTT}). Water-insoluble OP^{DTT} was determined by difference ($OP^{WI\_DTT} = OP^{Total\_DTT} - OP^{WS\_DTT}$). The DTT assay was conducted manually or by a semi-automated instrument (*26*) on the filtered DI water extracts to determine OP^{WS_DTT}, explicitly following the method described by Cho et al. (*10*). OP^{Total_DTT} is operationally defined and was determined with the same method as that for OP^{WS_DTT}, except that the extract was not filtered and the reaction between aerosol and DTT occurred in the extraction vial with the filter present (*31*). The OP^{Total_DTT} method attempts to measure all OP^{DTT} collected on the filter, including OP^{DTT} associated with solid particle that may not be easily extracted from the filter media. For more details on methods, blanks and positive controls, see (*17*, *33*).

Lognormal distributions were fit to the various data. When investigating the OP^{WI_DTT}, measured mass distributions of elemental carbon (EC) and mineral dust species (i.e., $CaCO_3$ inferred from Ca^{2+}) are converted to surface area distributions assuming spherical particles using the following equation:

$$n_s = n_m\, 6/(\rho_p\, D_{pg}) \tag{1}$$

where, n_s and n_m are the surface area and mass size distribution functions, D_{pg} the geometric mean size for a given MOUDI stage and ρ_p is the density of the aerosol (taken to be 1.77 g cm^{-3} for EC and 2.71 g cm^{-3} for $CaCO_3$, under the assumption that the particles containing EC or $CaCO_3$ were externally mixed, since no mixing state data were available. Note that the objective is to determine the shape of the distribution and not necessarily an accurate surface area concentration).

Finally, in the analysis below, particle pH on each stage was estimated from measured aerosol inorganic species with the thermodynamic model ISORROPIA II (*34*). Since semivolatile gases in equilibrium with their aerosol counterpart were not measured in this study, an iterative solution was applied. This introduces error into the pH calculation and does not permit assessing the pH predictions by comparing measured and model-inferred gas/particle partitioning of semivolatile species. However, the pH levels predicted were in agreement with those from a more comprehensive analysis at a site in the southeastern US (*35*). Details on the pH calculation of the MOUDI data can be found in Fang et al. (*33*).

Results

Contrasting Total, Water-Soluble and Water-Insoluble OP Size Distributions at Two Sites

In the following, comparisons are made between size-resolved OP^{DTT} and specific chemical components at the roadside and urban site. The contrasts provide insights on the sources and processes that affect Atlanta OP levels. To aid in these analyses, a schematic of emissions and processes of pertinent species discussed in the following sections is shown in Figure 2.

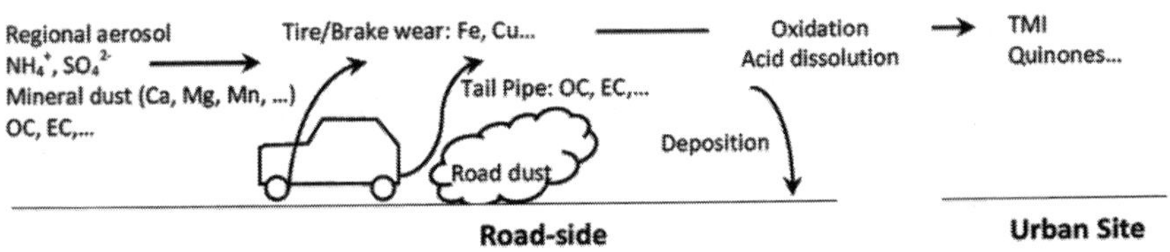

Figure 2. Schematic of various emissions from roadways that affect PM OP and how these are altered with atmospheric aging. Three main roadway emissions are: 1) coarse mode tire/brake wear, which contains metals such as Cu and Fe, 2) fine mode combustion emissions from the tail pipe that contains EC and various OC species, such as PAHs, and 3) re-entrained road dust that can contain mineral dust and tire/brake wear and tail pipe emissions associated with larger coarse mode particles that had settle out near the roadway. These emissions mix with the regional aerosol and undergo transformations as they age, which includes oxidation, (egg, PAHs converted to quinones) and metals acid-driven dissolution by mixing with SO_4^{2-}, leading to TMI (transition metal ions, including $Cu^{+}+Cu^{2+}$, $Fe^{2+}+Fe^{3+}$, etc).

Size distributions of OP^{WS_DTT} and OP^{Total_DTT} are shown in Figure 3. The difference between the two curves is insoluble OP, (OP^{WI_DTT}). As expected, OP^{Total_DTT} is larger than OP^{WS_DTT}. These plots show that levels of these two physical forms of OP vary with particle size and the proximity of the measurement to the highway. The most obvious feature is that at the road-side site, OP^{Total_DTT} and OP^{WI_DTT} are much larger in the coarse mode, consistent with a roadway source, such as mechanically generated road dust and brake wear (*7*). At the urban site, where the aerosol is more processed (aged), OP^{Total_DTT} is more uniformly higher than OP^{WS_DTT} over all size ranges, indicating there was less variability in OP^{WI_DTT} with particle size. It is also clear that neither OP^{WS_DTT} or OP^{Total_DTT} size distributions are bimodal, both lacking the separation of fine and coarse modes.

Multiple size-resolved measurements were made at these two sites, both simultaneously and at different times. For an overall comparison of the dependence of OP on particle size using all the measured distributions, average frequency distributions were determined from the lognormal fitted data (N = 5 for OP^{WS_DTT}, and N = 2 for OP^{WI_DTT} for both the road-side and urban sites). The results are plotted in Figure 4. Here, calculated OP^{WI_DTT} is plotted and shows a bi-modal distribution with a peak in the accumulation and coarse modes, as is often seen in ambient aerosol surface area or mass distributions and attributed to secondary formation and mechanic processes leading to separate fine and coarse mode particles, respectively (*7*). In contrast, OP^{WS_DTT} is unimodal with a peak or mean size near the minimum between the two OP^{WI_DTT} modes. For example, the geometric mean diameter, GMD, of OP^{WS_DTT} ranges from 0.80 to 2.51 µm at both sites. The sources and/or atmospheric processes producing OP^{WS_DTT} and OP^{WI_DTT} clearly differ.

For a more direct comparison of OP^{WS_DTT} and OP^{WI_DTT} between the sites, Figure 5 shows the results of simultaneous measurements. Contrasts in these size distributions support the above inferences. For example, OP^{WI_DTT} is bimodal with distinct coarse and fine modes, and OP^{WS_DTT} is unimodal. OP^{WI_DTT} has much higher coarse mode levels at the road-side site than at the urban site, suggestive of a major contribution from non-tailpipe traffic emissions to coarse-mode OP^{WI_DTT} (i.e., mechanically generated aerosol). Fine (accumulation) -mode OP^{WI_DTT} and OP^{WS_DTT} are more similar at both sites, indicating mainly a processed or regional widespread influence on fine-mode OP^{WI_DTT} and OP^{WS_DTT}. Overall, for all particle sizes, OP^{WS_DTT} is much more similar at the two sites and the distributions broader, indicating that a large fraction of the OP^{WS_DTT} is not as directly linked to the roadside source as OP^{WI_DTT}.

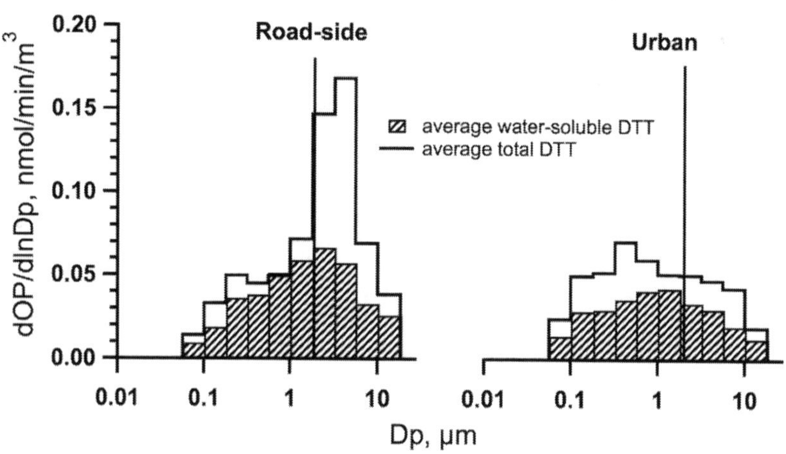

Figure 3. Average of size distributions measured at two sites (not simultaneously). The vertical line is at 2.5 µm, the upper limit of $PM_{2.5}$. Distributions are based on the average for each MOUDI stage (i.e., each stage from multiple measurements at the site are averaged). Water-soluble and total OP^{DTT} are plotted (OP^{WS_DTT} and OP^{Total_DTT}). The difference between is insoluble OP^{DTT}, (OP^{WI_DTT}).

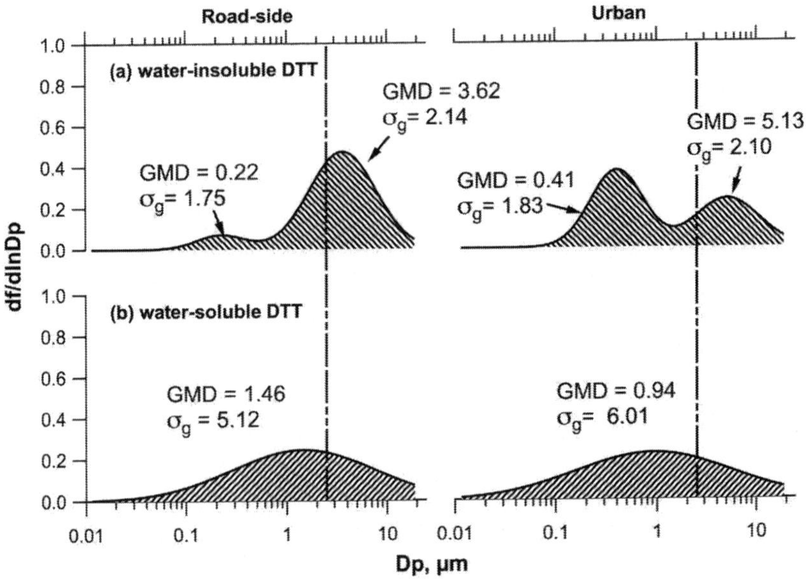

Figure 4. Average frequency distributions based on averaging the lognormal fits to the various measurements. Results for a) water-insoluble and b) water-soluble OP^{DTT} from two sites are shown. $PM_{2.5}$ upper limit is indicated by the vertical line. GMD is the geometric mean diameter and s_g the geometric standard deviation.

Sources and Processes Based on Other Measured Species

Size Distributions of Crustal Material and Metals

Ambient size distributions of chemical species provide clues to the sources and multi-phase processes that may have led to the observed OP distributions. Figure 6 shows the measured size distribution of various PM chemical components. Although the MOUDI samples at the two sites in Figure 6 were collected at different times, given the samples were averaged over 7 days, and that both were collected relatively close in time, the shape of the size distributions is taken to represent the general characteristics of the aerosol at the two sites.

First, compare the composition of the coarse and fine modes. The top panel shows that water-soluble calcium, (Ca^{2+}) is found at both sites, and is almost exclusively in the coarse mode (Dp > 2.5 μm). (Note that water-soluble Ca^{2+} is a fraction of total aerosol Ca, since most is likely in an insoluble carbonate form, such as $CaCO_3$. Ca^{2+} and inferred $CaCO_3$ are taken as surrogates of mineral dust and the shape of the mass and surface area distribution of Ca^{2+} or $CaCO_3$ is assumed to be largely similar to the ambient mineral dust distribution). Various total (or elemental) transition metals, such as copper (Cu), iron (Fe), and manganese (Mn) are also mainly in the coarse mode at both sites. Sources of coarse mode aerosols are generally attributed to mechanical processes (see Figure 2) (*7*).

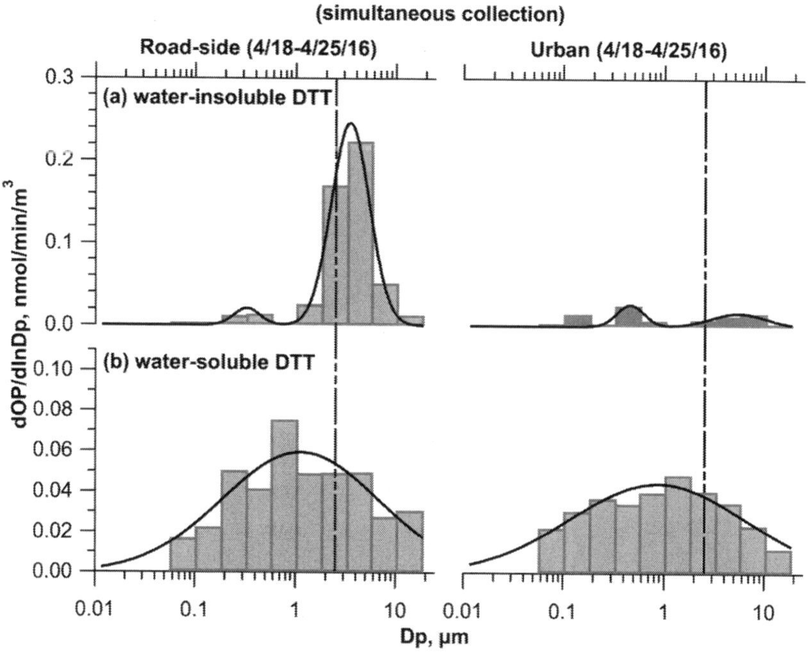

Figure 5. Results for MOUDI samples collected simultaneous at the road-side and urban sites. Results for a) water-insoluble and b) water-soluble OP^{DTT} from two sites are shown. Note the difference in vertical scales in upper and lower plots. The lognormal fit to the distributions are shown as solid lines, the data as bars. $PM_{2.5}$ upper limit is indicated by the vertical line.

Comparing concentrations between the two sites yields additional insight. Total metal concentrations also differ between sites (although, as noted above, the data were not collected simultaneously). The RS site has higher concentrations of total Cu and Fe compared to the urban site, consistent with the above fine/coarse mode comparisons, indicating a primary traffic source. Cu has been attributed to brake/tire wear (*36–38*), and Fe to brake/tire wear (*36, 39*) and mineral dust (*40, 41*). Total Mn is more uniform between the two sites, consistent with a significant source being from mineral dust (*40, 41*) and minor contributions from brake/tire wear (*42*). Ca^{2+}, also largely associated with mineral dust, is fairly uniform between the two sites.

Comparing the water-soluble metal concentrations at the two sites shows that both water-soluble Cu and Fe (although harder to see due to low concentrations) are higher at the RS site, whereas Mn levels are more similar between the two sites. Thus the spatial distribution of the water-soluble fraction of these metals is similar to that of the total metal. Based on a much larger data set, Fang et al. (*25*) also found in the Atlanta region that bulk $PM_{2.5}$ water-soluble Mn had a more regional characteristic, suggesting a more widely dispersed mineral dust source, while $PM_{2.5}$ water-soluble Cu and Fe was higher near roadways, indicating a stronger traffic-related source.

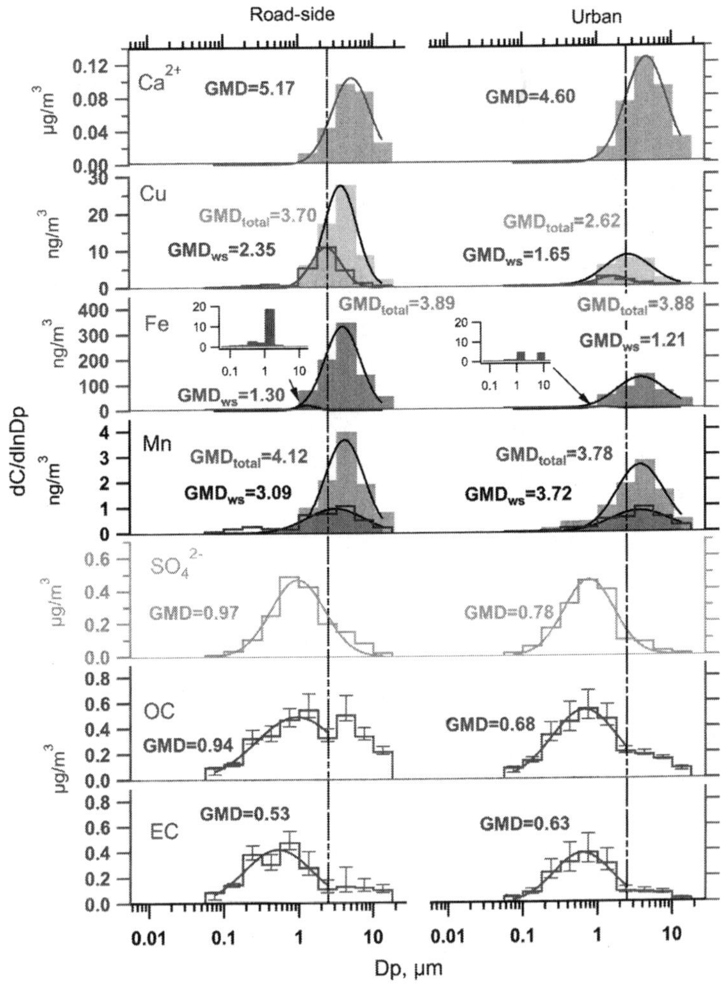

Figure 6. Size distributions of various aerosol species measured at the road-side (3/28 – 4/4/2016) and urban site (3/16 – 3/23/2016). The inserts in the Fe panel are enlarged plots of water-soluble Fe, which is a small fraction of total Fe. A particle diameter of 2.5 μm is shown as the vertical line. Lognormal fits to the data are also plotted with the corresponding geometric mean diameter (GMD). Since Ca^{2+} was measured by ion chromatography, it is the water-soluble fraction of calcium. Both total (elemental) and water-soluble (ws subscript) concentrations were measured for Cu and Fe. (see color insert)

The particle sizes of the water-soluble metals, however, differ from the total or elemental form of the corresponding metal. Contrasts between the mean size of the water-soluble transition metals relative to the total form is also shown in Figure 6. For Cu and Fe, the water-soluble metal distributions are shifted to a smaller size compared to the total metal. For example, the geometric mean diameter (GMD) for water-soluble Fe is about 2.5 μm lower than total Fe at both sites, water-soluble

Cu is about 1 to 1.5 μm lower, and Mn is shifted less than 1 μm. The trend showing shifts to smaller sizes follows the overall fractional solubility (water-soluble/total mass) of these metals. The average (± SD) water-soluble mass fraction for all size ranges from all MOUDI data collected is 13 ± 14% for Fe, 44 ± 36% for Cu, and 50 ± 30% for Mn. As expected, Fe is the least soluble among the three metals, consistent with other studies (*42*). Water-soluble Fe is also found associated with generally smaller particles relative to total Fe. Mn, the most soluble, has soluble versus total size distributions that are similar to each other. As will be discussed below, these observations are all consistent with an acid dissolution process driven by increasing acidity as one transitions from the coarse to fine modes and the acid solubility characteristics of these different metals.

Size Distributions of Carbonaceous Particles and Sulfate

Organic and elemental carbon-containing (OC and EC) particles show a typical aerosol bi-modal distribution with a clear fine-mode (≤ $PM_{2.5}$) and coarse-mode ($PM_{2.5\text{-}10}$), and a minimum between modes at about 2.5 μm diameter. Unlike the metals, generally, OC and EC concentrations are higher in the fine mode relative to the coarse mode, consistent with combustion emissions and secondary sources (*7*), see Figure 2, although significant levels of coarse mode OC is detected at the RS site. Sulfate is most abundant in the fine mode. In the region of this study, large electrical generating units, which are large point sources, are the main source of sulfur dioxide (SO_2). Oxidation of SO_2, either in the gas or liquid (cloud) phase leads to secondary sulfate, and this process results in a more spatially uniform concentration for sulfate (*43*), accounting for similar concentrations between the road-side and urban sites (Figure 4). Comparisons between size distributions of the various aerosol chemical components and OP^{WI_DTT} and OP^{WS_DTT} provide insights on the sources of PM OP.

Insoluble OP

Bimodal OP^{WI_DTT} distributions follow distributions of insoluble species in the fine and coarse modes. A main insoluble fine-mode aerosol component is EC (or soot). Figure 7 shows the measured EC mass size distributions converted to an estimated EC surface area distribution, assuming spherical externally mixed particles. The EC surface distribution matches well with OP^{WI_DTT}, consistent with OP^{WI_DTT} being associated with EC surfaces. However, OP^{WI_DTT} associated with solid OC surfaces can't be excluded, if it exists, since OC and EC have similar surface area distributions (not shown). Since much of the OC is water-soluble, it is difficult to infer insoluble OC size distributions from this data. Studies report that oxidation of PAHs (*44*) and diesel exhaust soot or black carbon (*45–48*) by ozone form quinones strongly adsorbed to soot surfaces which increased OP^{DTT}. OP^{DTT} is highly responsive to highly oxidized quinones (*30*). This process is in line with observations in this study of fine-mode OP^{WI_DTT} adsorbed to EC surfaces. Other data from Atlanta also support this conclusion. Measurements of fine-mode particle bound PAHs in Atlanta traffic were highly correlated with black carbon (r

= 0.86) (*49*). Also, fine-mode number distributions of solid particles that remain insoluble when ambient aerosol was collected in water were found to be highly correlated with EC mass concentrations (*50*) and had similar size distributions to that of EC, indicating that EC was the main insoluble species in the fine mode .

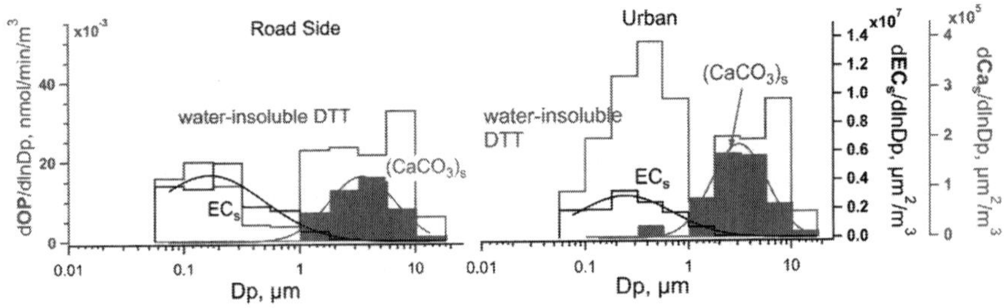

Figure 7. Measurements of water-insoluble DTT ($OP^{WI\text{-}DTT}$) size distributions and estimated particle surface area distributions, based on mass concentration measurements and assuming spherical particles, for elemental carbon (EC_s) and calcium carbonate ($CaCO_{3(s)}$, is inferred from Ca^{2+} measurements). The plots are to provide comparisons between the shapes of the distributions. (see color insert)

The current data also show the importance of atmospheric aging on fine-mode OP^{WI_DTT}. This can be seen by comparing the ratios of OP^{WI_DTT} to EC_s (EC surface area) at the two sites. Fine-mode (D_p < 1 μm) OP^{WI_DTT}/EC_s ratios are 4 to 5 times higher at the urban site; the mean OP^{WI_DTT}/EC_s is $(1.5 \pm 0.6) \times 10^{-8}$ at the urban site and $(0.32 \pm 1.6) \times 10^{-8}$ nmol min^{-1} μm^{-2} at the road-side site. PAH-coated EC being emitted by vehicles and converted to quinone coatings by O_3 oxidation is consistent with these results in that the ratios are higher for the urban site, which is less influenced by fresh roadway EC emissions. Higher O_3 levels further from the road due to less NO_x titration may also play a role.

Soluble transition metals also contribute to OP^{DTT} and are mostly found in particles larger than 1 μm (Figure 6) and so may account for much of the coarse-mode OP^{WI_DTT}. In Figure 7, the estimated $CaCO_3$ surface area distribution is seen to be similar in shape to coarse-mode OP^{WI_DTT} at both sites, which would be expected if OP^{WI_DTT} was also related to a surface property of the aerosol, such as the road dust-tire/brake wear particles. This could possibly occur by acid dissolution on surfaces of these particles, producing soluble metals. For example, for this to occur, sulfate would have to be deposited on some region of the dust surface at sufficient levels to take up water and lower the pH enough to solubilize a portion of the metals, despite the large overall pH-buffering capacity of non-volatile cations (e.g., Ca^{2+}) associated with the dust particle. A fraction of the soluble metals would then have to remain bound to these particles during the water extraction process to account for at least some of the observed coarse-mode OP^{WI_DTT}. Alternatively, quinones adsorbed to the road dust-tire/brake wear particles could be the main contributor. Figure 6 shows there is a large coarse mode OC at the roadside site. Overall, there is limited information on which chemical species (e.g., quinones or soluble metals) actually contributes to coarse-mode OP^{WI_DTT}. This is in contrast to experiments that have elucidated soot-PAH oxidation to quinones to explain fine-mode OP^{WI_DTT}.

Soluble OP

The soluble OP^{DTT} size distributions are highly unique in that they are unimodal with peak near 2.5 μm, the size that typically separates ambient aerosol mass-based coarse and fine modes (see sulfate, OC, EC, and OP^{WI_DTT} distributions, for examples). The cause for this unique shape can be explained by how fine and coarse mode interactions contribute to water-soluble metals.

Metals in forms that are largely insoluble can become soluble by acid dissociation under low pH conditions (*51–53*). (E.g., consider the dissolution of hematite by sulfuric acid-driven acidity to form water-soluble iron(III) sulfate; $Fe_2O_3(s) + 3H_2SO_4(aq) \rightarrow Fe_2(SO_4)_3(aq) + 3H_2O(l)$, or assuming ionic species are completely dissociated, $Fe_2O_3(s) + 6H^+(aq) + 3SO_4^{2-}(aq) \rightarrow 2Fe^{3+}(aq) + 3SO_4^{2-}(aq) + 3H_2O(l)$). Alternatively, soluble metals can result by forming a ligand with an organic species, such as oxalate, at higher pH (*54*, *55*). In this case, the size distribution of water-soluble metals indicates largely an acid dissolution process. Soluble Cu and Fe mass distributions are seen in Figure 6 to peak at between approximately 1 and 2.5 μm, a lower size than the total metal. This occurs at the overlap between the sulfate and the total metal size distributions. As an example, the size distributions of sulfate, and water-soluble and total Cu are plotted together in Figure 8 for both the Road-Side and Urban site. Water-soluble Cu peaks within the overlap of the lower tail of the primary total Cu distribution in the coarse mode and upper tail of the secondary sulfate distribution in the fine mode. If in this overlap area, sulfate and total Cu were internally mixed (within a single particle, consistent with the size distribution data), insoluble Cu may be mobilized over time by sulfate-driven acidity, creating a soluble form of Cu. Rates of acid mobilization of metals depend on the type of metal and the particle pH.

Calculated particle pH for each MOUDI stage is also plotted in Figure 8. The calculations show a highly acidic fine mode aerosol and near neutral aerosol in the coarse mode, which is due to the distribution of sulfate and nonvolatile mineral cations, (assuming complete internal mixing of all components for each MOUDI stage). The predominance of sulfate and lack of mineral cations in the fine mode (Dp less than approximately 1.8 μm) results in predicted pH values that are very low, ranging between 1 and 2. In contrast, low levels of sulfate and high levels of cations, such as Ca^{2+} (see Figure 6, a similar distribution is seen for Mg^{2+}), likely in the form of carbonates ($CaCO_3$ and $MgCO_3$) (*56*) in the coarse mode leads to an aerosol that is more neutral. The cations buffer the sulfate-driven acidity through the formation of calcium and magnesium sulfate, which results in the loss of a proton (i.e., increase in pH) through the formation of water and evaporation of CO_2 (e.g., $CaCO_3 + H_2SO_4 \rightarrow CaSO_4 + H_2O + CO_2$, or to illustrate the effect on pH, i.e., conversion of H^+ to water, $Ca^{2+}(aq) + CO_3^{2-}(aq) + 2H^+(aq) + SO_4^{2-}(aq) \rightarrow Ca^{2+}(aq) + SO_4^{2-}(aq) + H_2O(aq) + CO_2(g)$). The large buffering capacity of the coarse mode mineral cations leads to a predicted pH of approximately 7. Lower pH of some coarse mode particles is possible if the sulfate is not completely internally mixed with all the nonvolatile cations. However, given that some internal mixing is expected, the pH of the coarse mode will be higher than the fine mode, which has much lower concentrations of nonvolatile cations. At the tail end of the coarse mode, the cation buffering capacity diminishes as particle size decreases because the dust cation (e.g., Ca^{2+}) concentration decreases and sulfate concentration increases. The transition between these two modes of differing pH is where the soluble metals are mainly found.

For both Fe and Cu, the ratio of water-soluble to total concentration versus pH for each MOUDI stage from both sites also supports this idea. For MOUDI stages where pH is near neutral, metal solubility is low, but for stages with low predicted pH, the ratio of measured water-soluble to total Fe and Cu substantially increases (see Fang et al. (*33*)). The very low pH levels for the fine mode reported here are also consistent with more detailed calculations of pH reported for the same region (*35*, *57*), and in other locations (*58*, *59*), indicating that this mechanism of metals solubility may apply to many regions.

The dissolution of metal oxides at low pH takes time; from hours up to weeks, depending on the metal (*52*). Meskhidze et al. (*53*) found that 2-5% Fe was mobilized after 4 days at pH = 1. The dissolution is much faster at lower pH since the rate depends exponentially on pH. Cu is more easily solubilized, consistent with the observations that the fraction of soluble to total Cu was higher than that for Fe (from above, 13 ± 14% for Fe, and 44 ± 36% for Cu).

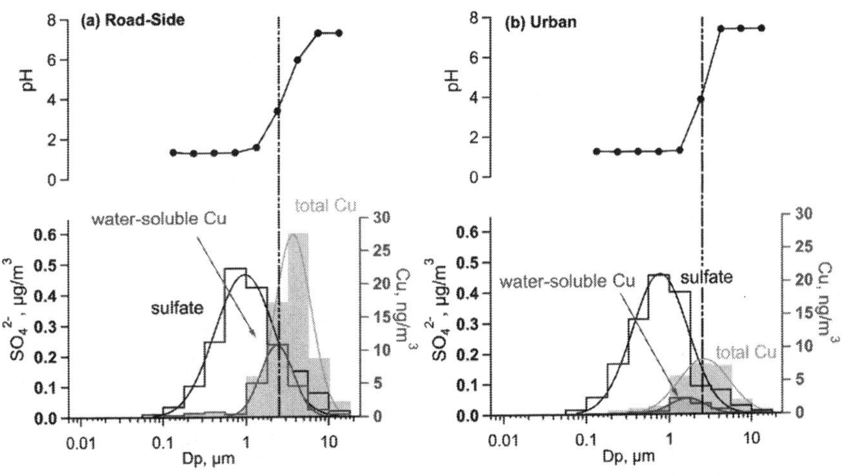

Figure 8. Evidence for the dissolution of copper (Cu) by sulfate under acidic conditions based on measured size distributions at a) the road-side and b) the urban sites. The vertical dotted line is aerodynamic diameter (Dp) at 2.5 µm, the upper limit of $PM_{2.5}$. Particle pH was estimated for each MOUDI stage from ISORROPIA-II based on measured ionic species from MOUDI samples collected on 3/28−4/4/2016 and 3/16−3/23/2016 at the road-side and urban site. Differences in geometric mean diameters (GMD) between water-soluble and total Cu can be found in Figure 4. (see color insert)

Other studies in Atlanta support the idea that metals dissolution results from sulfate-driven pH and the internal mixing assumption. Correlations have been observed between water-soluble Fe and sulfate (r^2 = 0.62 to 0.76, N=181) in summer and fall (*25*). Single particle x-ray fluorescence (XRF) measurements have shown that Fe solubility was associated with the particles sulfur content (*60*) and that approximately 50% of the sulfate within individual particles between 1 and 2.5 µm was associated with a metal cation, likely in the form of iron or copper sulfates (e.g., $Fe_2(SO_4)_3(aq)$) (*61*). These results are also consistent with source apportionment analysis for aerosols in metropolitan Atlanta. For $PM_{2.5}$, roughly 50% of the water-soluble Fe and 40% of the water-soluble Cu were associated with secondary processing (*25*). The remaining fraction of these metals was largely associated with a primary brake/tire wear source (32% of Fe and 51% of Cu) consistent with high levels at the RS site (see Figure 6). (Note that although the source apportionment attributes a significant fraction of the soluble Fe and Cu to primary brake/tire wear, it does not necessarily mean they are all directly emitted. Since these species are associated with larger particles they will deposit near the road-way due to sedimentation (Figure 2), and can later be re-aerosolized as road dust and some fraction advected from the roadway. During the period between emission and reemission as road dust, some secondary chemical processing can occur).

Because water-soluble transition metals contribute to OP, these processes play an important role in shaping the OP size distribution. For the DTT assay measurement of OP, other studies showed that Cu is a significant contributor to OP^{WS_DTT} in bulk $PM_{2.5}$ ambient samples (*62*). As noted above, humic like substances (HULIS), which includes oxidized aromatic species from combustion sources, such as quinones, are also important contributors to OP^{DTT} (*27*, *28*, *30*). If these species have a very broad size distribution, similar to OC, see Figure 6 (quinones were not specifically measured in this study), then the combined contribution from these two different aerosol components to OP^{WS_DTT} would explain its broad unimodel distribution with a peak near, but slightly below, 2.5 µm.

Conclusions

Multiphase processing of aerosols increases levels of both fine and coarse mode OP. Analysis of the OP^{DTT} size distributions are consistent with two main processes; oxidation of organic species, such as polycyclic aromatic hydrocarbons to form highly DTT-active quinones (*30*), and acid mobilization of transition metals to form transition metal ions. These processes appear to apply to both water-insoluble and water-soluble forms of OP^{DTT}, however it is not clear if these processes differ across particle sizes. For example, the insoluble size distributions of fine OP^{DTT} are consistent with laboratory experiments of oxidation of PAHs adsorbed to soot by ozone to form adsorbed quinones. As for OP^{WS_DTT}, the process may not involve surfaces. Some fraction of the PAHs may be absorbed into the particle liquid phase and undergo and oxidation, or the PAH may be oxidized in the gas phase and the resulting products partition to the particle; both result in the production of species, such as soluble quinones, that contribute to OP^{WS_DTT}. A measurement of the quinone size distribution is needed to explore their source in more detail. For the metal cations, the acid-driven dissociation must occur in the liquid phase, which will lead to OP^{WS_DTT}. The extent that soluble transition metals on insoluble particle surfaces (such as mineral dust) contribute to OP^{WI_DTT} is an open question.

Because the different physical forms of OP^{DTT} are associated with different particle sizes, they may also have distinctive health effects since they can deposit in different regions of the respiratory system (*17*). Soluble versus insoluble forms of oxidative potential species can also lead to different physiological responses once deposited.

In summary, complex interactions between species from different sources likely occur widely in the ambient atmosphere. We have shown examples of how this can affect PM OP and thus air quality. Since studies have linked OP^{DTT} to adverse health (*23*, *24*, *62*), it follows that multiphase processing needs to be considered to fully understand how emissions that produce aerosols impact human health.

References

1. Lim, S. S.; Vos, T.; Flaxman, A. D.; Danaei, G.; Shibuya, K.; Adair-Rohani, H.; Amann, M.; Anderson, H. R.; Andrews, K. G.; Aryee, M.; Atkinson, C.; Bacchus, L. J.; Bahalim, A. N.; Balakrishnan, K.; Balmes, J.; Barker-Collo, S.; Baxter, A.; Bell, M. L.; Blore, J. D.; Blyth, F.; Bonner, C.; Borges, G.; Bourne, R.; Boussinesq, M.; Brauer, M.; Brooks, P.; Bruce, N. G.; Brunekreef, B.; Bryan-Hancock, C.; Bucello, C.; Buchbinder, R.; Bull, F.; Burnett, R. T.; Byers, T. E.; Calabria, B.; Carapetis, J.; Carnahan, E.; Chafe, Z.; Charlson, F.; Chen, H.; Chen, J. S.; Cheng, A. T.-A.; Child, J. C.; Cohen, A.; Colson, K. E.; Cowie, B. C.; Darby, S.; Darling, S.; Davis, A.; Degenhardt, L.; Dentener, F.; Jarlais, D. C. D.; Devries, K.; Dherani, M.; Ding, E. L.; Dorsey, E. R.; Driscoll, T.; Edmond, K.; Ali, S. E.; Engell, R. E.; Erwin, P. J.; Fahimi, S.; Falder, G.; Farzadfar, F.; Ferrari, A.; Finucane, M. M.; Flaxman, S.; Fowkes, F. G. R.; Freedman, G.; Freeman, M. K.; Gakidou, E.; Ghosh, S.; Giovannucci, E.; Gmel, G.; Graham, K.; Grainger, R.; Grant, B.; Gunnell, D.; Gutierrez, H. R.; Hall, W.; Hoek, H. W.; Hogan, A.; III, H. D. H.; Hoy, D.; Hu, H.; Hubbell, B. J.; Hutchings, S. J.; Ibeanusi, S. E.; Jacklyn, G. L.; Jasrasaria, R.; Jonas, J. B.; Kan, H.; Kanis, J. A.; Kassebaum, N.; Kawakami, N.; Khang, Y.-H.; Khatibzadeh, S.; Khoo, J.-P.; Kok, C.; Laden, F.; Lalloo, R.; Lan, Q.; Lathlean, T.; Leasher, J. L.; Leigh, J.; Li, Y.; Lin, J. K.; Lipshultz, S. E.; London, S.; Lozano, R.; Lu, Y.; Mak, J.; Malekzadeh, R.; Mallinger, L.; Marcenes, W.; March, L.; Marks, R.; Martin, R.; McGale, P.; McGrath, J.; Mehta, S.; Mensah, G. A.; Merriman, T. R.; Micha, R.; Michaud, C.; Mishra, V.; Hanafiah, K. M.; Mokdad, A. A.; Morawska, L.; Mozaffarian, D.; Murphy, T.; Naghavi, M.; Neal, B.; Nelson, P. K.; Nolla, J. M.; Norman, R.; Olives, C.; Omer, S. B.; Orchard, J.; Osborne, R.; Ostro, B.; Page, A.; Pandey, K. D.; Parry, C.

D. H.; Passmore, E.; Patra, J.; Pearce, N.; Pelizzari, P. M.; Petzold, M.; Phillips, M. R.; Pope, D.; III, C. A. P.; Powles, J.; Rao, M.; Razavi, H.; Rehfuess, E. A.; Rehm, J. T.; Ritz, B.; Rivara, F. P.; Roberts, T.; Robinson, C.; Rodriguez-Portales, J. A.; Romieu, I.; Room, R.; Rosenfeld, L. C.; Roy, A.; Rushton, L.; Salomon, J. A.; Sampson, U.; Sanchez-Riera, L.; Sanman, E.; Sapkota, A.; Seedat, S.; Shi, P.; Shield, K.; Shivakoti, R.; Singh, G. M.; Sleet, D. A.; Smith, E.; Smith, K. R.; Stapelberg, N. J. C.; Steenland, K.; Stöckl, H.; Stovner, L. J.; Straif, K.; Straney, L.; Thurston, G. D.; Tran, J. H.; Dingenen, R. V.; Donkelaar, A. v.; Veerman, J. L.; Vijayakumar, L.; Weintraub, R.; Weissman, M. M.; White, R. A.; Whiteford, H.; Wiersma, S. T.; Wilkinson, J. D.; Williams, H. C.; Williams, W.; Wilson, N.; Woolf, A. D.; Yip, P.; Zielinski, J. M.; Lopez, A. D.; Murra, C. J. L.; Ezzati, M. A comparative risk assessment of burden of disease and injury attributable to 67 risk factors and risk factor clusters in 21 regions, 1990–2010: a systematic analysis for the Global Burden of Disease Study 2010. *The Lancet* **2012**, *380*, 2224–2260.

2. Samet, J. M.; Dominici, F.; Curriero, F. C.; Coursac, I.; Zeger, S. L. Fine particulate air pollution and mortality in 20 U.S. cities, 1987–1994. *New Engl. J. Med.* **2000**, *343*, 1742–1749.
3. Metzger, K. B.; Tolbert, P. E.; Klein, M.; Peel, J. L.; Flanders, W. D.; Todd, K.; Mulholland, J. A.; Ryan, P. B.; Frumkin, H. Ambient air pollution and cardiovascular emergency department visits. *Epidemiology* **2004**, *15*, 46–56.
4. Pope, C. A. I. Review: Epidemiological basis for particulate air pollution health standards. *Aerosol Sci. Tech.* **2000**, *32*, 4–14.
5. Pope, A. C.; Burnett, R. T.; Thurston, G. D.; Thun, M. J.; Calle, E. E.; Krewski, D.; Godleski, J. J. Cardiovscular mortality and long-term exposure to particulate pollution: Epidemiological evidence of general pathophysciological pathways for disease. *Circulation* **2004**, *109*, 71–77.
6. Nel, A. Air pollution-related illness: Efects of particles. *Science* **2005**, *308*, 804–806.
7. Whitby, K. The physical charactoristics of sulfur aerosols. *Atmos. Environ.* **1978**, *12*, 135–159.
8. Li, N.; Hao, M.; Phalen, R. F.; Hinds, W. C.; Nel, A. E. Particulate air pollutants and asthma A paradign for the role of oxidative stress in PM-induced adverse health effects. *Clin. Immunol.* **2003**, *109*, 250–265.
9. Steenhof, M.; Gosens, I.; Strak, M.; Godri, K. J.; Hoek, G.; Cassee, F. R.; Mudway, I. S.; Kelly, F. J.; Harrison, R. M.; Lebret, E.; Brunekreef, B.; Janssen, N.; Pieters, R. In vitro toxicity of particulate matter (PM) collected at different sites in the Netherlands is associated with PM composition, size fraction and oxidative potential - the RAPTES project. *Part. Fibre Toxicol.* **2011**, *8*, 1–15.
10. Cho, A. K.; Sioutas, C.; Miguel, A. H.; Kumagai, Y.; Schmitz, D. A.; Singh, M.; Eiguren-Fernandez, A.; Froines, J. R. Redox activity of airborne particulate matter at different sites in the Los Angeles Basin. *Environ. Res.* **2005**, *99*, 40–47.

11. Ntziachristos, L.; Froines, J. R.; Cho, A. K.; Sioutas, C. Relationship between redox activity and chemical speciation of size- fractionated particulate matte. *Part. Fibre Toxicol.* **2007**, *4*, 1–12.
12. Janssen, N.; Yang, A.; Strak, M.; Steenhof, M.; Hellack, B.; Gerlofs-Nijland, M. E.; Kuhlbush, T.; Kelly, F.; Harrison, R.; Brunekreef, B.; Hoek, G.; Cassee, F. Oxidative potential of particulate matter collected at sites with different source characteristics. *Science Total Environ.* **2014**, *472*, 572–581.
13. Chirizzi, D.; Cesari, D.; Guascito, M. R.; Dinoi, A.; Giotta, L.; Donateo, A.; Contini, D. Influence of Saharan dust outbreaks and carbon content on oxidative potential of water-soluble fractions of $PM_{2.5}$ and PM_{10}. *Atmos. Environ.* **2017**, *163*, 1–8.
14. Mugica, V.; Ortiz, E.; Molina, L.; Vizcaya-Ruiz, A. D.; Nebot, A.; Quintana, R.; Aguilar, J.; Alcantara, E. PM composition and source reconciliation in Mexico City. *Atmos. Environ.* **2009**, *43*, 5068–5074.
15. Jeng, H. A. Chemical composition of ambient particulate matter and redox activity. *Environ. Monit. Assess.* **2010**, *169*, 597–606.
16. Hu, S.; Polidori, A.; Arhami, M.; Shafer, M. M.; Schauer, J. J.; Cho, A.; Sioutas, C. Redox activity and chemical speciation of size fractioned PM in the communities of the Los Angeles-Long Beach harbor. *Atmos. Chem. Phys.* **2008**, *8*, 6439–6451.
17. Fang, T.; Zeng, L.; Gao, D.; Verma, V.; Stefaniak, A.; Weber, R. J. Ambient Size Distributions and Lung Deposition of Aerosol Oxidative Potential: A Contrast Between Soluble and Insoluble Particle. *Environ. Sci. Technol.* **2017**, *51*, 6802–6811.
18. Li, N.; Sioutas, C.; Cho, A.; Schmitz, D.; Misra, C.; Sempf, J.; Wang, M.; Oberley, T.; Froines, J.; Nel, A. Ultrafine particlulate pollutants induce oxidative stress and mitochondrial damage. *Environ. Health Perspect.* **2003**, *111*, 455–460.
19. Charrier, J. G.; Richards-Henderson, N. K.; Bein, K. J.; McFall, A. S.; Wexler, A. S.; Anastasio, C. Oxidant production from source-oriented particulate matter – Part 1: Oxidative potential using the dithiothreitol (DTT) assay. *Atmos. Chem. Phys.* **2015**, *15*, 2327–2340.
20. Samara, C. On the Redox Activity of Urban Aerosol Particles: Implications for Size Distribution and Relationships with Organic Aerosol Components. *Atmosphere* **2017**, *8*, 1–18.
21. Godri, K. J.; et al. Increased Oxidative Burden Associated with Traffic Component of Ambient Particulate Matter at Roadside and Urban Background Schools Sites in London. *PLoS One* **2011**, *6*, 1–11.
22. Zhang, X.; Staimer, N.; Tjoa, T.; Gillen, D. L.; Schauer, J. J.; Shafer, M. M.; Hasheminassab, S.; Pakbin, P.; Blonghurst, J.; Sioutas, S.; Delfino, R. J. Associations between microvascular function and short-term exposure to traffic-related air pollution and particulate matter oxidative potential. *Environ. Health* **2016**, *15*, 1–16.
23. Abrams, J.; Weber, R. J.; Klein, M.; Samat, S. E.; Chang, H. H.; Strickland, M. J.; Verma, V.; Fang, T.; Bates, J. T.; Mulholland, J. A.; Russell, A. G.; Tolbert, P. E. Associations between ambient fine particulate

oxidative potential and cardiorespiratory emergency department visits. *Environ. Health Perspect.* **2017**, *25*, 1–9.

24. Bates, J. T.; Weber, R. J.; Abrams, J.; Verma, V.; Fang, T.; Klein, M.; Strckland, M. J.; Sarnat, S.; Chang, H.; Mulholland, J. A.; Tolbert, P. E.; Russell, A. G. Reactive Oxygen Species in Atmospheric Particulate Matter Suggest a Link to Cardiorespiratory Effects. *Environ. Sci. Technol.* **2015**, *49*, 13605–13612.
25. Fang, T.; Guo, H.; Verma, V.; Peltier, R. E.; Weber, R. J. PM2.5 Water-Soluble Elements in the Southeastern United States: Automated Analytical Method Development, Spatiotemporal Distributions, Source Apportionment, and Implications for Heath Studies. *Atmos. Chem. Phys.* **2015**, *15*, 11667–11682.
26. Fang, T.; Verma, V.; Guo, H.; King, L. E.; Edgerton, E. S.; Weber, R. J. A Semi-automated System for Quantifying the Oxidative Potential of Ambient Particles in Aqueous Extracts Using the dithiothreitol (DTT) assay: Results from the Southeastern Center for Air Pollution and Epidemiology (SCAPE). *Atmos. Meas. Tech.* **2015**, *8*, 471–482.
27. Verma, V.; Fang, T.; Guo, H.; King, L.; Bates, J. T.; Peltier, R. E.; Edgerton, E.; Russell, A. J.; Weber, R. J. Reactive Oxygen Species Associated with Water-Soluble PM2.5 in the Southeastern United States: Spatiotemporal Trends and Source Apportionment. *Atmos. Chem. Phys.* **2014**, *14*, 12915–12930.
28. Verma, V.; Fang, T.; Lu, X.; Peltier, R. E.; Russell, A.; Lee, N.; Weber, R. J. Organic Aerosols Associated with the Generation of Reactive Oxygen Species (ROS) by Water-Soluble PM2.5. *Environ. Sci. Technol.* **2015**, *49*, 4646–4656.
29. Verma, V.; Rico-Martinez, R.; Kotra, N.; King, L.; Liu, J.; Snell, T. W.; Weber, R. J. Contribution of water-soluble and insoluble components and their hydrophobic/hydrophilic sub-fractions on the ROS-generating potential of fine ambient aerosols. *Environ. Sci. Technol.* **2012**, *46*, 11384–11392.
30. Verma, V.; Wang, Y.; El-Afifi, R.; Fang, T.; Rowland, J.; Russell, A. G.; Weber, R. J. Fractionating Ambient Humic-like Substances (HULIS) for their reactive oxygen species activity - Assessing the importance of quinones and atmospheric aging on particulate matter toxicity. *Atmos. Environ.* **2015**, *120*, 351–359.
31. Gao, D.; Fang, T.; Verma, V.; Zeng, L.; Weber, R. J. A method for measuring total aerosol oxidative potential (OP) with the dithiothreitol assay and comparisons between an urban and roadside site of water-soluble and total OP. *Atmos. Meas. Tech.* **2017**, *10*, 2821–2835.
32. Eller, P. M., et al., Eds. *M.E.C. NIOSH Manual of Analytical Methods*, 4th ed. (1st Supplement); National Institute for Occupational Safety and Health: Cincinnati, 1996.
33. Fang, T.; Guo, H.; Zeng, L.; Verma, V.; Nenes, A.; Weber, R. J. Highly acidic sulfate makes aerosols toxic by dissolving metals. *Environ. Sci. Technol.* **2017**, *51*, 2611–2620.

34. Fountoukis, C.; Nenes, A. ISORROPIA II: a computationally efficient thermodynamic equilibrium model for K^+-Ca^{2+}-Mg^{2+}-NH_4^+-Na^+-SO_4^{2-}-NO_3^--Cl^--H_2O aerosols. *Atmos. Chem. Phys.* **2007**, *7*, 4639–4659.
35. Guo, H.; Xu, L.; Bougiatioti, A.; Cerully, K. M.; Capps, S. L.; Hite, J. R.; Carlton, A. G.; Lee, S.-H.; Bergin, M. H.; Ng, N. L.; Nenes, A.; Weber, R. J. Predicting particle water and pH in the southeastern United States. *Atmos. Chem. Phys.* **2015**, *15*, 5211–5228.
36. Adachi, K.; Tainosho, Y. Characterization of heavy metal particles embedded in tire dust. *Environ Int.* **2004**, *30*, 1009–1017.
37. Sternbeck, J.; Sjödin, A.; Andreásson, K. Metal emissions from road traffic and the influence of resuspension-results from two tunnel studies. *Atmos. Environ.* **2002**, *36*, 4735–4744.
38. Garg, B. D.; Cadle, S. H.; Mulawa, P. A.; Groblicki, P. J.; Laroo, C.; Parr, G. A. Brake wear particulate matter emissions. *Environ. Sci. Technol.* **2000**, *34*, 4463–4469.
39. Hopke, P. K.; Lamb, R. E.; Natusch, D. F. S. Multielemental characterization of urban roadway dust. *Environ. Sci. Technol.* **1980**, *14*, 167–172.
40. Rodriguez, S.; Querol, X.; Alastuey, A.; Viana, M. a.-M.; Alarcoń, M.; Mantilla, E.; Ruiz, C. R. Comparative PM10–PM2.5 source contribution study at rural, urban and industrial sites during PM episodes in Eastern Spain. *Sci. Total Environ.* **2004**, *328*, 95–113.
41. Choi, J. C.; Lee, M.; Chun, Y.; Kim, J.; Oh, S. Chemical composition and source signature of spring aerosol in Seoul, Korea. *J. Geophys. Res.* **2001**, *106*, 18,067–18,074.
42. Birmili, W.; Allen, A. G.; Bary, R.; Harrison, R. M. Trace metal concentrations and water solubility in size-fractionated atmospheric particles and influence of road traffic. *Environ. Sci. Technol.* **2006**, *40*, 1144–1153.
43. Hidy, G. M.; Blanchard, C. L.; Baumann, K.; Edgerton, E.; Tanenbaum, S.; Shaw, S.; Knipping, E.; Tombach, I.; Jansen, J.; Walters, J. Chemical climatology of the southeastern United States, 1999–2013. *Atmos. Chem. Phys.* **2014**, *14*, 11893–11914.
44. Antinolo, M.; Willis, M. D.; Zhou, S.; Abbatt, J. P. D. Connecting the oxidation of soot to its redox cycling abilities. *Nat. Commun.* **2015**, *6*, 1–7.
45. Li, Q.; Wyatt, A.; Kamens, R. M. Oxidant generation and toxicity enhancement of aged-diesel exhaust. *Atmos. Environ.* **2009**, *43*, 1037–1042.
46. Li, Q.; Shang, J.; Zhu, T. Physicochemical characteristics and toxic effects of ozone-oxidized black carbon particles. *Atmos. Environ.* **2013**, *81*, 68–75.
47. McWhinney, R. D.; Badali, K.; Liggio, J.; Li, S.-M.; Abbatt, J. P. D. Filterable redox cycling activity: A comparison between diesel exhaust particles and secondary aerosol constituents. *Environ. Sci. Technol.* **2013**, *47*, 3362–3369.
48. McWhinney, R. D.; Gao, S. S.; Zhou, S.; Abbatt, J. P. D. Evaluation of the Effects of Ozone Oxidation on Redox-Cycling Activity of Two-Stroke Engine Exhaust Particles. *Environ. Sci. Technol.* **2011**, *45*, 2131–2136.
49. Greenwald, R.; Bergin, M. H.; Yip, F.; Boehmer, F.; Kewada, P.; Shafer, M. M.; Schauer, J. J.; Sarnat, J. A. On-roadway In-cabin exposure to particulate

matter: measurement results using both continuous and time-integrated sampling approaches. *Aeosol Sci. Technol.* **2014**, *48*, 664–675.

50. Greenwald, R.; Bergin, M. H.; Weber, R.; Sullivan, A. Size-resolved, real-time measurement of water-soluble aerosols in metropolitan Atlanta during the summer of 2004. *Atmos. Environ.* **2007**, *41*, 519–531.
51. Nenes, A.; Krom, M. D.; Mihalopoulos, N.; Van Cappellen, P.; Shi, Z.; Bougiatioti, A.; Zarmpas, P.; Herut, B.; Nenes, A. Atmospheric acidification of mineral aerosols: a source of bioavailable phosphorus for the oceans. *Atmos. Chem. Phys.* **2011**, *11*, 6265–6272.
52. Shi, Z.; Bonneville, S.; Krom, M. D.; Carslaw, K. S.; Jickells, T. D.; Baker, A. R.; Benning, L. G. Iron dissolution kinetics of mineral dust at low pH during simulated atmospheric processing. *Atmos. Chem. Phys.* **2011**, *11*, 995–1007.
53. Meskhidze, N.; Chameides, W. L.; Nenes, A.; Chen, G. Iron mobilization in mineral dust: Can anthropogenic SO2 emissions affect ocean productivity. *Geophys. Res. Lett.* **2003**, *30*, 1–5.
54. Schwertmann, U. Solubility and dissolution of iron oxides. *Plant Soil* **1991**, *130*, 1–25.
55. Ali, M. A.; Dzombak, A. D. Effects of simple organic acids on sorption of Cu^{2+} and Ca^{2+} on goethite. *Geochim. Cosmochim. Acta* **1996**, *60*, 291–304.
56. Ito, A.; Feng, Y. Role of dust alkalinity in acid mobilization of iron. *Atmos. Chem. Phys.* **2010**, *10*, 9237–9250.
57. Weber, R. J.; Guo, H.; Russell, A. G.; Nenes, A. High aerosol acidity despite declining atmospheric sulfate concentrations over the past 15 years. *Nat. Geosci.* **2016**, *9*, 282–285.
58. Guo, H.; Sullivan, A. P.; Campuzano-Jost, P.; Schroder, J. C.; Lopez-Hilfiker, F. D.; Dibb, J. E.; Jimenez, J. L.; Thornton, J. A.; Brown, S. S.; Nenes, A.; Weber, R. J. Fine particle pH and the partitioning of nitric acid during winter in the northeastern United States. *J. Geophys. Res. Atmos.* **2016**, *121*, 10,355–10,376.
59. Bougiatioti, A.; Nikolaou, P.; Stavroulas, I.; Kouvarakis, G.; Weber, R.; Nenes, A.; Kanakidou, M.; Mihalopoulos, N. Particle water and pH in the eastern Mediterranean: Sources variability and implications for nutrients availability. *Atmos. Chem. Phys.* **2016**, *16*, 4579–4591.
60. Oakes, M.; Weber, R. J.; Lai, B.; Russell, A. T.; Ingall, E. D. Characterization of iron speciation in single particles using XANES spectroscopy and micro X-ray fluorescence measurements: insight into factors controlling iron solubility. *Atmos. Chem. Phys.* **2012**, *12*, 1–12.
61. Longo, A. F.; Vine, D. J.; King, L. E.; Oakes, M.; Weber, R. J.; Huey, L. G.; Russell, A. G.; Ingall, E. D. Composition and oxidation state of sulfur in atmospheric particulate matter. *Atmos. Chem. Phys.* **2016**, *16*, 13389–13398.
62. Fang, T.; Verma, V.; Bates, J. T.; Abrams, J.; Klein, M.; Strickland, M. J.; Sarnat, S. E.; Chang, H. H.; Mulholland, J. A.; Tolbert, P. E.; Russell, A. G.; Weber, R. J. Oxidative potential of ambient water-soluble PM2.5 in the southeastern United States: contrasts in sources and health associations between ascorbic acid (AA) and dithiothreitol (DTT) assays. *Atmos. Chem. Phys.* **2016**, *16*, 3865–3879.

Chapter 21

Can Reactions between Ozone and Organic Constituents of Ambient Particulate Matter Influence Effects on the Cardiovascular System?

Michael T. Kleinman,[1,*] Lisa M. Wingen,[2] David A. Herman,[1] Rebecca Johnson,[1] and Andrew Keebaugh[1]

[1]Department of Medicine, University of California, Irvine, Irvine, California 92697-1830, United States
[2]Department of Chemistry, University of California, Irvine, Irvine, California 92697-2025, United States
[*]E-mail: mtkleinm@uci.edu

Exposure to fine (FP; $PM_{2.5}$) and ultrafine (UFP; $PM_{0.1}$) ambient particles is associated with adverse health effects which include changes in cardiac function, systemic inflammation, arterial dysfunction, oxidative stress and the development of atherosclerosis. Individuals with preexisting residing coronary heart disease (CHD) are among the most susceptible to these adverse effects. Our previous studies demonstrated that organic and elemental carbon constituents of ambient particles were capable of exacerbating, or possibly initiating, these health effects. We also demonstrated that on days when ambient ozone levels were more elevated, we observed more changes in electrocardiographic patterns in mice exposed to concentrated ambient particles, even though the ambient ozone had first been removed from the exposure atmosphere. This suggested that ambient ozone could modify the toxicity of particle phase components of ambient air. We tested this hypothesis by exposing mice, that were genetically modified to be prone to developing atherosclerosis, to mixtures of concentrated ambient FP (CAPs) in the presence and absence of ozone. The mice were exposed 5 hr per day, 4 days per week for 8 weeks. The mice were implanted with electrocardiographs and were monitored using telemetry before, during, and after exposures. Particle concentrations and chemical compositions were monitored during the exposures using an Aerosol Mass Spectrometer (AMS) and other particle monitors. Ozone rapidly reacted with organic components on the particles causing detectable changes in oxygenated species. Changes in the chemical composition were associated with significant cardiovascular effects. Electrocardiographic changes were more pronounced when particles contained organics with low O:C ratios and those effects were diminished when organics had higher O:C ratios.

Introduction

Residents of California have been exposed historically to high ambient concentrations of both $PM_{2.5}$ and ozone (O_3), albeit not always at the same time. Air quality in California has been greatly improved over the past two decades but National Ambient Air Quality Standards (NAAQS) and California standards for both pollutants continue to be exceeded at times. Epidemiologic studies, which are the health-related basis for the $PM_{2.5}$ NAAQS, have shown that $PM_{2.5}$-related health effects on the cardiovascular system are large and clinically significant, but there are substantial gaps and uncertainties in our understanding of how inhaled particles that are deposited in the lung can have large effects on more distal organs such as the brain and the heart. To date, mechanistic studies investigating how inhaled PM induces adverse health effects have focused on generic, non-specific modes of action (e.g., oxidative stress and inflammation) that are not unique to air pollution.

In contrast, the ozone NAAQS is primarily based on human exposure studies that have investigated the relationship between well-defined ozone exposures and changes in clinical endpoints, primarily of the respiratory system. While mechanistic pathways through which ozone exposure affects respiratory health effects have been studied, recent research suggests that ozone exposure may also have cardiovascular effects. However, little is known about potential biological mechanisms for ozone-induced cardiovascular effects.

Although humans are often exposed to both $PM_{2.5}$ and O_3 as parts of a complex mixture of ambient air pollutants, there is limited information as to the potential for interactions or synergisms among these two important ambient pollutants (*1–3*). Mixtures of PM and O_3 modified autonomic balance in mice (*4*), caused cardiovascular depression and increased inflammation and oxidative stress in rats, an effect which was potentiated by high fat (*5*) and high fructose diets (*6*). Understanding these potential interactions could provide critical data that can assist in the development of health protective air quality policies. Epidemiological and *in vivo* exposure studies demonstrate that particles (both fine and ultrafine) are important contributors to cardiovascular mortality and morbidity (*7*, *8*) and that they accelerate the development of atherosclerotic plaque, which is a major contributor to cardiovascular disease (*9*) and deaths associated with heart disease

(*10*). Heart disease is arguably a major cause of non-accidental deaths in the United States especially for people over 45 years of age; the Center for Disease Control (CDC) estimated that in 2016 approximately 30% of deaths could be attributed to heart disease (*11*). Associations of O_3 with mortality, and specifically with heart-related mortality, have been reported (*12–14*), but are less strongly established than those for $PM_{2.5}$ (*15*). This may be due in part to the co-variation of $PM_{2.5}$ and O_3 and to the seasonal variations of O_3 ambient concentrations that might obscure relationships to some disease outcomes. However, there is evidence from animal studies that inhaled O_3 can induce vascular dysfunction, mitochondrial damage, and the development of atherosclerosis (*16*). O_3 exposure can also impair pulmonary gas exchange to a degree that might be clinically important in persons with significant preexisting cardiovascular impairment (*17*). Mechanistically, both O_3 and $PM_{2.5}$ cause inflammation and can induce oxidative stress when inhaled, which suggests that in combination they might act in an additive or perhaps synergistic manner (*8*). For example, Wang and colleagues (*18*) demonstrated that concomitant O_3 (0.8 ppm) exposure potentiated the inflammatory and cardiac effects of nasally instilled $PM_{2.5}$ with some evidence of synergy (i.e. a more than additive interaction), however the results were not from a $PM_{2.5}$ inhalation exposure and the O_3 concentration was higher than typical environmental levels. Because of the use of pollutant concentrations that are often higher than representative environmental exposures, as well as the variability in results among these studies (*1*), further experiments are needed under more relevant conditions.

Reduced heart rate variability (HRV), is a marker of poor cardiac autonomic function and has been associated with exposures to both $PM_{2.5}$ and O_3 (*19–23*). Altered HRV may be one of the pathophysiological mechanisms that link PM exposure with cardiopulmonary disease (*24, 25*) and decreased HRV is recognized as a predictor of sudden cardiac death and arrhythmias (*26*). Fakhri and associates observed a small potentiation by O_3 of CAP's effects on diastolic BP and HRV in young adults (*27*).

Therefore, the overarching objectives of this study was to provide some insights into the importance of the atmospheric interactions of O_3 and ambient $PM_{2.5}$ and the effects of concomitant $PM_{2.5}$ and O_3 inhalation. To accomplish this objective, we exposed genetically modified, mice (apoE-/-) to concentrated ambient $PM_{2.5}$ (CAPs), O_3 or to a mixture of CAPs + O_3. These mice are hyperlipidemic and have high levels of circulating lipids, especially low-density lipoprotein cholesterol, which is associated with increased risk of atherosclerosis.

Methods

Exposure

This study used transgenic mice lacking the gene that codes for apolipoprotein E (apoE$^{-/-}$). The mice were obtained from The Jackson Laboratory (Bar Harbor, ME). The apoE$^{-/-}$ mice develops atherosclerotic lesions in coronary arteries and in the aorta and have been shown to be susceptible to the atherogenic effects of both concentrated $PM_{2.5}$ and UFP. These mice also have high serum levels of very low-density lipoproteins (LDL) and have been used extensively in studies of the effects of PM exposure on the heart (*28, 29*). Mice (apo$E^{-/-}$) were exposed to $PM_{2.5}$ concentrated ambient particles (CAPs) using a Versatile Aerosol Concentration Enrichment System (VACES), shown in Figure 1, to enrich the concentration of ambient particles in the size range of 0.02 to 10 μm by a factor of 10 (*30, 31*). All mice were between 6 and 8 weeks of age at the start of exposures and were conditioned to the exposure system in purified air for one week before beginning CAPs exposures. We continually monitored heart rate and electrocardiogram (ECG) waveforms in the mice using implanted telemetry devices (PhysioTel® RMC-1, Data Sciences International, St Paul, MN, USA). Exposures were started after stable baseline signals and HR levels were achieved. During exposures, the mice were placed into segmented exposure chambers that were connected to the outlet of the VACES (*32*). Exposure chambers were designed to provide a uniform distribution of particles to each chamber segment (*32*). Mice (n=16 for each group; 5 were implanted with telemetry monitors for ECG analyses; 11 additional mice were included for histology and biochemical analyses for oxidative stress and inflammation) were exposed to toxicant atmospheres of $PM_{2.5}$ CAPs, 200 ppb O_3, or to a CAPs + 200 ppb O_3 co-pollutant mixture, for 5 hours per day, 4 days per week for 8 weeks. The CAPs+O_3 mixing time was on the order of seconds. Control mice were exposed to air purified over potassium permanganate-impregnated alumina beads and activated carbon ("purified air"), and passed through a HEPA filter. under conditions identical to those of the animals exposed to CAPs. Exposure chamber temperatures were monitored every 15 minutes during the exposures and held to 75 ± 5°F. Animals were observed throughout the exposure period for signs of distress. Between exposures, mice were housed in the UCI vivarium and received water and food, ad libitum. All surgeries were performed aseptically and all animal procedures were approved by the UCI Animal Care and Use Committee (Protocol # 2001-2242). The numbers of animals used in the various bioassays were determined using a statistical analysis of power to detect a 50% change in key biological parameters at the $p \leq 0.05$ level.

Physical and Chemical Characterization of PM of Ambient and Exposure Atmospheres

Particle size distributions were measured using a TSI Scanning Mobility Particle Sizer (SMPS, Shoreview, MN, USA) for the UFP fraction of $PM_{2.5}$ and particles up to about 0.65 μm. In addition to monitoring particle mass, a TSI condensation particle counter (CPC Model 3022) was run in parallel to measure total particle number concentrations. A TSI DustTrak optical mass monitor (Model 8520) provided integrated $PM_{2.5}$ mass concentrations. Size-resolved organic and inorganic aerosol composition was measured using an Aerodyne Aerosol Mass Spectrometer (AMS, Aerodyne Research, Billerica, MA, USA) which provided size and chemical composition as well as mass concentrations in real-time for non-refractory sub-micron aerosol particles, down to about 50 nm. AMS sampling, during the exposure studies, was used to analyze exposure and ambient aerosol characteristics to determine differences in characteristics before and after addition of O_3. Ozone was monitored at the chamber inlet using a UV Absorption Monitor (Dasibi Model 1003-AH) which was checked daily against a calibrated transfer standard.

Electrocardiographic Methods

The ecgAUTO® (EMKA Technologies S.A.S., Falls Church, VA, USA) system was used to analyze heart rate, incidence of abnormal heart beats (arrhythmias), waveform abnormalities, and measures of HRV. Simply put, HRV measures the variability of the interval between normal heartbeats, or interbeat intervals. A number of statistical metrics are used to characterize HRV either in the time domain or the frequency domain. Changes in some of these parameters are associated with specific processes in the body. In the time domain, HRV is reported as 1) the standard deviation in interbeat heartbeats from which artifacts have been removed, SDNN, and 2) the root mean square of successive heartbeat intervals between all successive heartbeats, RMSSD. In the frequency domain, HR oscillations can be reported as 1) low frequency oscillations, LF HRV, 2) high frequency oscillations, HF HRV, or 3) a ratio of LF/HF power. For example, the high frequency band (HF, 0.15-0.40 Hz) of the heart period power spectrum is a measure of cardiac vagal control (*33*). Decreased cardiac vagal activity in humans is associated with an increased risk of coronary atherosclerosis (*34*). Lower frequency HRV (LF, 0.04-0.15 Hz) may represent mixed sympathetic-parasympathetic and thermoregulatory influences (*35, 36*).

In this project, mice were surgically implanted with telemetry devices and allowed to recover. Baseline readings for each mouse were acquired after recovery, for 1 week prior to the initiation of exposures. It is important to note that in all ECG analysis results are described as changes from baseline readings.

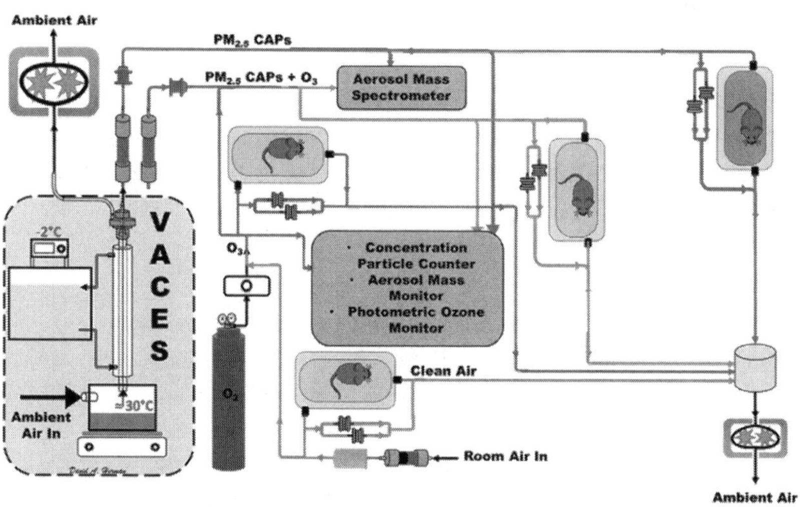

Figure 1. Schematic Representation of the VACES Exposure System

Data Analysis and Statistics

ECG and heart rate variability data were statistically analyzed via SPSS® (IBM, Armonk, NY, USA). using analyses of variance. Differences between group means were contrasted using Tukey's post-hoc test. Significance was assessed at $P \leq 0.05$.

Results

Exposures to CAPs and O_3, Alone and in Combination

Mice were exposed to CAPs, 200 ppb O_3, CAPs + 200 ppb O_3, or to purified air for 5 hours per day, 4 days per week for 8 weeks. The weekly average particle number, particle mass and ozone concentration (means ± standard error of the mean, SEM) are summarized in Table 1. Particles were concentrated approximately 10-fold over the ambient levels and ozone was titrated into the exposure atmospheres to a target concentration of 200 ppb at the chamber inlet. Particle and gas concentrations within the chambers had been tested and shown to be uniform. The particle concentrations were not significantly different between CAPs and CAPs + O_3 exposures.

Table 1. Exposure Concentrations for Particle Number, Particle Mass and Ozone (O_3)

Exposure Atmosphere	*CAPs*	*CAPs + O_3*	*O_3*	*Purified Air*	*Ambient Air*
Particle Number Concentration (cm-3)	(8.7 ± 1.2) x e^4	(8.4 ± 1.2) x e^4	≤ 3.0	5.8 ± 3.0	(6.8 ± 1.0) x e^3
Particle Mass Concentration (µg m^{-3})	133 ± 20	125 ± 15	≤ 3.0	0.3 ± 0.5	12.5 ± 0.9
O_3 (ppbv)	≤ 3	201 ± 3.5	187 ± 3.8	≤ 3	≤ 3

Heart Rate and Heart Rate Variability

Average percent changes from baseline (measurements made the week prior to exposures) for HR and HRV during the 5-hour afternoon time-period (12:00pm – 5:00pm) were averaged over the entire exposure and are summarized in Table 2. Relative to purified air exposures, HR was slightly but not significantly elevated by exposures to the O_3-containing atmospheres. This is consistent with human and other mammalian responses to inhaled O_3. HRV parameters were, however, changed by both CAPs and O_3 exposures when they were administered alone, relative to purified air, but the effects were blunted in mice exposed to the CAPs + O_3 mixture. CAPs exposure significantly reduced the root mean square of successive differences of normal beat to normal beat intervals (RMSSD) and the high frequency HRV (HF HRV), which are HRV parameters related to parasympathetic control, compared to air exposure. These measures were reduced to an even greater degree in the mice exposed to O_3 alone. However, the HRV measures related to parasympathetic influcnces in mice exposed to CAPs + O_3 were not different from those in mice exposed to purified air. The CAPs + O_3 exposure induced a significant increase in low frequency HRV (LF HRV). An increase in LF HRV could reflect a shift towards increased sympathetic nervous system (SNS) tone, however the value of LF HRV changes alone as a marker of SNS influence may be limited.

These overall averages, including the entire exposure period, might have minimized some differences because to some extent the early effects might be different from those later in the exposure. Nevertheless, the data provided clear evidence that indicators of parasympathetic tone were reduced (HF HRV and RMSSD) by exposures to CAPs and O_3 alone but that the CAPs + O_3 mixture did not produce these same effects.

Table 2. Change from Baseline in HR and HRV (Mean ± SEM; *$p \leq 0.05$ compared to Purified Air)

Parameter (% change from baseline)	*CAPs*	*CAPs + O_3*	*O_3*	*Purified Air*
HR	2.4 ± 0.3	3.7 ± 0.3	3.0 ± 0.3	2.8 ± 0.2
SDNN	3.7 ± 0.4	8.5 ± 0.7*	-0.5 ± 0.9*	4.8 ± 0.7
RMSSD	3.3 ± 0.7*	10.6 ± 1.1	0.3 ± 1.0*	9.3 ± 1.0
LF HRV	6.0 ± 1.3	18.7 ± 1.9*	-2.1 ± 1.9*	6.9 ± 2.1
HF HRV	0.4 ± 1.3*	15.5 ± 2.0	-0.6 ± 2.0*	16.2 ± 2.2
LF/HF	7.8 ± 1.5*	4.2 ± 1.3*	0.3 ± 1.5*	-7.2 ± 1.5

We also examined the week to week changes in the weekly averages of RMSSD and HF HRV, which are shown in Figure 2 and Figure 3 as averages of the percent change from baseline for each of the 5 animals within each group; bars shown represent standard errors. Weekly changes in RMSSD show that ozone and CAPs are statistically different than purified air during the last two weeks of the exposure. The individual exposures show a percent change in the opposite direction of that of the purified air control as the groups separate as the exposure progressed. The co-exposure does not statistically vary from purified air. RMSSD in CAPs and O_3-exposed mice tended to progressively decrease after the second week of exposure. Data during the first 2 weeks could have been influenced by the mice being less acclimated to exposure-related stresses at the beginning of the study. RMSSD for the O_3-exposed group exhibited a progressive downward trend after week 4.

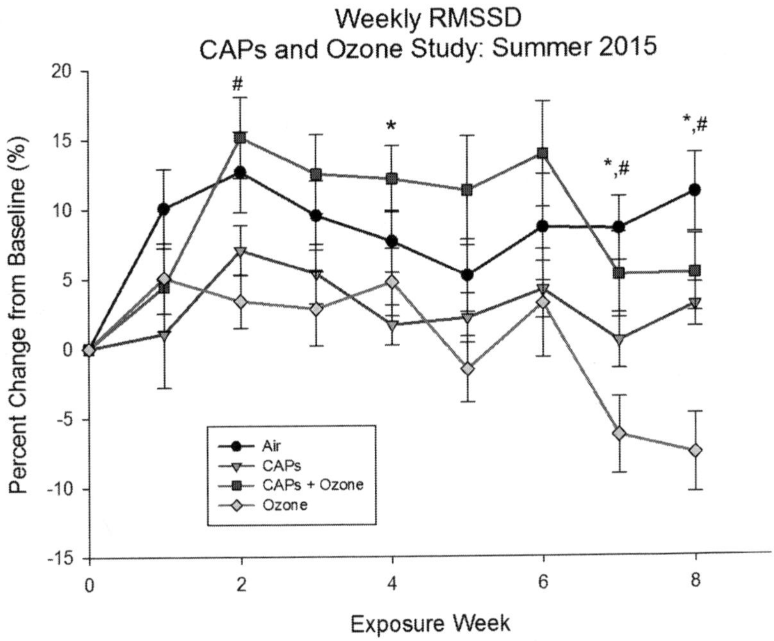

*Figure 2. CAPs and O_3 Induced Significant Decreases in RMSSD Relative to Purified Air Averaged Weekly During Exposures ($p \leq 0.05$. * = CAPs different than Air, # = O_3 different than Air)*

Weekly changes in HF HRV in Figure 3 show similar patterns as for RMSSD where CAPs and O_3 individual exposures are statistically different and in the opposite direction of purified air throughout the exposure period. The CAPs + O_3 co-exposure is not statistically different from purified air. These two measures indicate that exposures to CAPs or O_3, alone, significantly impact parasympathetic nervous system influences on the cardiovascular system, but the effects are blunted following exposure to the CAPs+O_3 mixture. For the mixture, HF HRV was reduced, relative to purified air, to a greater extent than by O_3, and exhibited a decreasing trend after week 2.

The significance of HRV changes in rodents and their relevance to human health can be debated. The ECG measurements can also be used to calculate ST segments, which represent the isoelectric period when the ventricles are in between depolarization and repolarization. ST-segment changes do have clinical relevance and in the clinical setting are used as an indicator of ischemia. Although the analysis of this parameter is not complete for the weekly changes, the average changes, relative to the purified air exposure groups have been tabulated for the three exposure groups and are summarized in Table 3.

These data support the HRV findings in that the mixture effects did not show significant changes but the CAPs and O_3 alone groups were significantly elevated, statistically. These changes were small and probably not clinically important except as an indicator of changes in a potentially adverse direction. Other analyses are currently in progress.

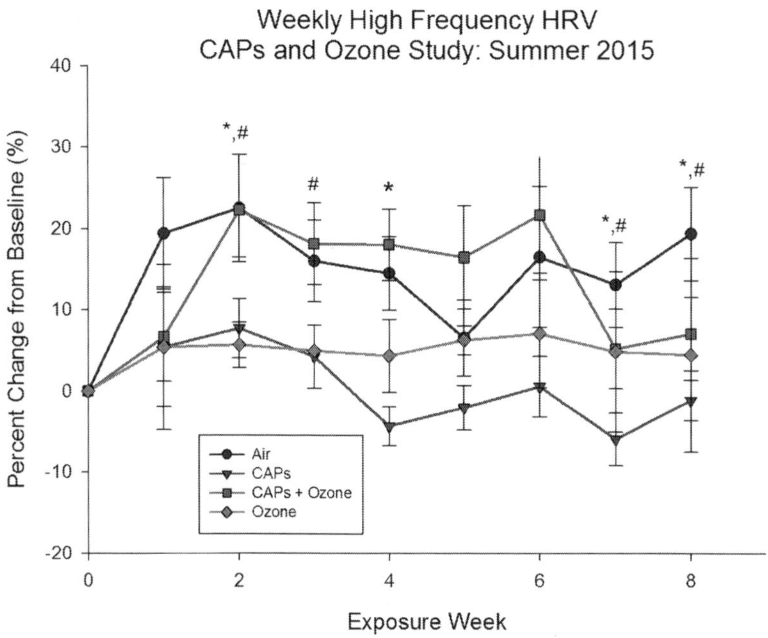

*Figure 3. CAPs and O_3 Induced Significant Decreases in High Frequency HRV Relative to Purified Air Averaged Weekly During Exposures ($p \leq 0.05$. * = CAPs different than Air, # = Ozone different than Air)*

Table 3. CAPs and O_3 Exposures Induced Significant Increases in 8-Week Average ST-segment Elevation Relative to Purified Air (mV ± SEM). *$p \leq 0.05$ compared to Purified Air

Exposure	*4 hr Post Exposure*	*12 hr Post Exposure*	*48 hr Post Exposure*
CAPs	11 ± 1 *	8 ± 1.5 *	6 ± 1 *
CAPs + O_3	-1 ± 1	0 ± 1	-1 ± 1
O_3	9 ± 1 *	9.5 ± 1 *	7 ± 1 *

Changes in Oxidation State of PM Organic Constituents Are Related to Reductions in HRV

In this study we related AMS measurements on a given day to post-exposure ECG measurements for that day. We had previously shown that organic constituents of CAPs were possibly the driving force behind HRV reductions induced by CAPs exposure (*37*). In addition, we had shown using the aerosol mass spectrometer (AMS) that the oxygen to carbon ratio (O:C) of the organic fraction of concentrated PM was inversely associated with HRV, i.e. changes in the direction of decreased HRV were greater when the O:C ratio was low. In this study, we focused our analyses on HF HRV because the data in Figure 3 show that this parameter was more influenced by CAPs than by other exposures. On a daily basis, the O:C ratios of the CAPs+O_3 mixture were slightly higher than those of CAPs.

Figure 4 shows the O:C ratio as a function of HF HRV for both the CAPs only exposure and the CAPs+O_3 exposure. Both the CAPs and CAPs+O_3 data follow the same trend, that of decreasing HF HRV with decreasing O:C. In addition to the importance of the particle O:C ratio, the presence of specific classes of organic compounds (*38*) were identified and associated with changes in HF HRV. For example, the relative strength of the signal at *m/z* 44 (or f44) represents the fraction of organic mass present from organic acids. Figure 5 shows how HF HRV varies with this signal. As the organic acid content decreases in the CAPs and CAPs+O_3, HF HRV decreases, indicating a decreased parasympathetic nervous system influence. Similarly, f43 represents the fraction of organic mass present at *m/z* 43 and has been correlated to carbonyl compounds that are non-acids; these are components of less oxidized or less aged organic aerosol (*39*). Figure 6 shows that as this carbonyl content increases in the aerosol, HF HRV changes in the direction indicative of decreased parasympathetic nervous system influence.

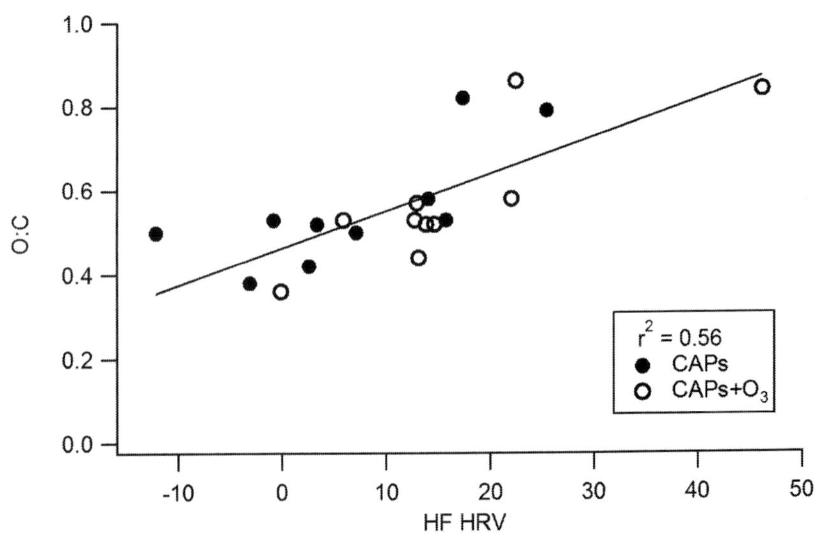

Figure 4. The Ratio of Oxygen to Carbon (O:C) in Organic Constituents of $PM_{2.5}$ is Related to Changes in HF HRV such that Lower Levels of O:C are Associated with More Intense HRV Decreases.

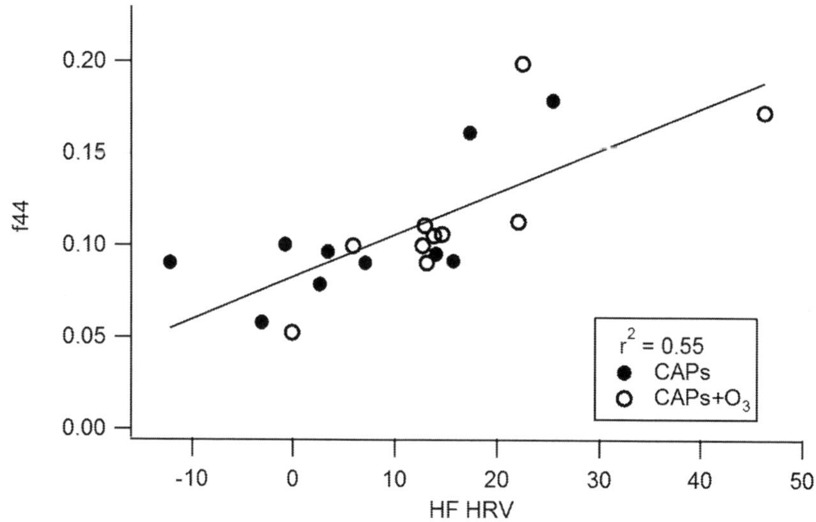

Figure 5. Days with Lower Concentrations of Organic Acids (represented as AMS f44) were Associated with Greater HRV Decreases

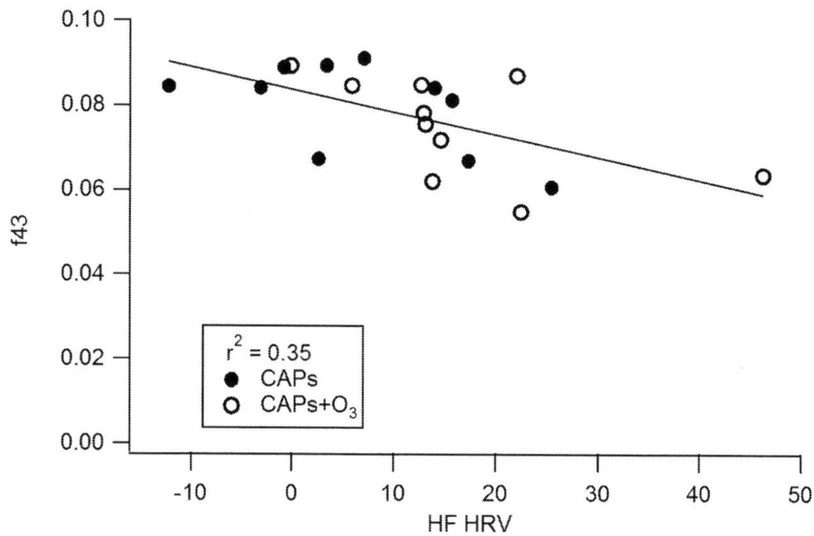

Figure 6. Days with Increased Concentrations of Carbonyl Compounds (f43) were Associated with Greater HRV Decreases

The observed compositional changes may also correspond to changes in volatility or particle diameters for the organic constituents. Figure 7 shows an example of a mass-weighted size distribution of the organic component for CAPs and CAPs + O_3 atmospheres. In general, the organic mass concentrations of CAPs + O_3 decreased relative to CAPs. Furthermore, on exposure days that exhibited a prominent UFP mode, organics present in CAPs + O_3 UFP demonstrated a larger decrease than accumulation mode organics. The UFP mode is generally much less oxidized than accumulation mode particles in urban environments and is often combustion related, i.e. emitted from vehicles (*40*). Reaction with O_3 suggests that these UFP were composed of organics such as alkenes and perhaps including some polycyclic aromatic hydrocarbons (*41–43*), since these compounds have a higher potential for reaction with O_3. Reaction products formed in the CAPs + O_3 atmosphere could have had lower molecular weights and higher vapor pressures and therefore were lost from particles, as suggested by the average decrease in mass concentration relative to CAPs.

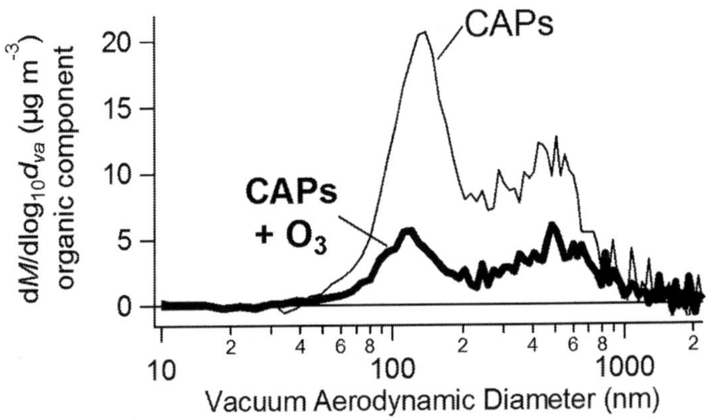

Figure 7. Mass-Weighted Size Distributions of the Organic Component of CAPs and CAPs + O_3 Aerosol (data from 9/3/2015)

Discussion

There are several potential mechanisms that are relevant to our objectives. For example, inhalation of $PM_{2.5}$ could induce (1) losses of pulmonary function (*44*), (2) pulmonary inflammation with secondary systemic effects (*45*) or, (3) direct toxic cardiovascular effects after translocation from the lung into the circulation system (*46–48*). Through the induction of cellular oxidative stress and pro-inflammatory pathways, particulate matter augments the development and progression of atherosclerosis, plaque buildup in the arteries, via detrimental effects on platelets, vascular tissue, and the myocardium. These effects seem to underpin the atherothrombotic consequences of acute and chronic exposure to air pollution (*49*). Oxidative stress and inflammation are central to both the toxicology of PM and the pathogenesis of atherosclerosis and coronary artery disease. It is possible that ultrafine particles (UFP) or soluble components of PM may translocate into the bloodstream, resulting in direct effects on atherosclerotic plaque stability, the vascular endothelium, platelet function, and thrombosis (*50*). Our findings suggest that modifications in the O:C ratio of $PM_{2.5}$ organic constituents can alter biological activity and therefore particles with lower O:C ratios may have the greatest effect on cardiac physiology.

Cardiovascular diseases and atherosclerosis in general are multifocal in nature and there are, as mentioned above, many molecular mechanisms that can play roles in their development. A schematic representation of some of the mechanistic pathways relevant to the research we performed to evaluate mechanisms of possible interaction of $PM_{2.5}$ and O_3 are summarized in Figure 8.

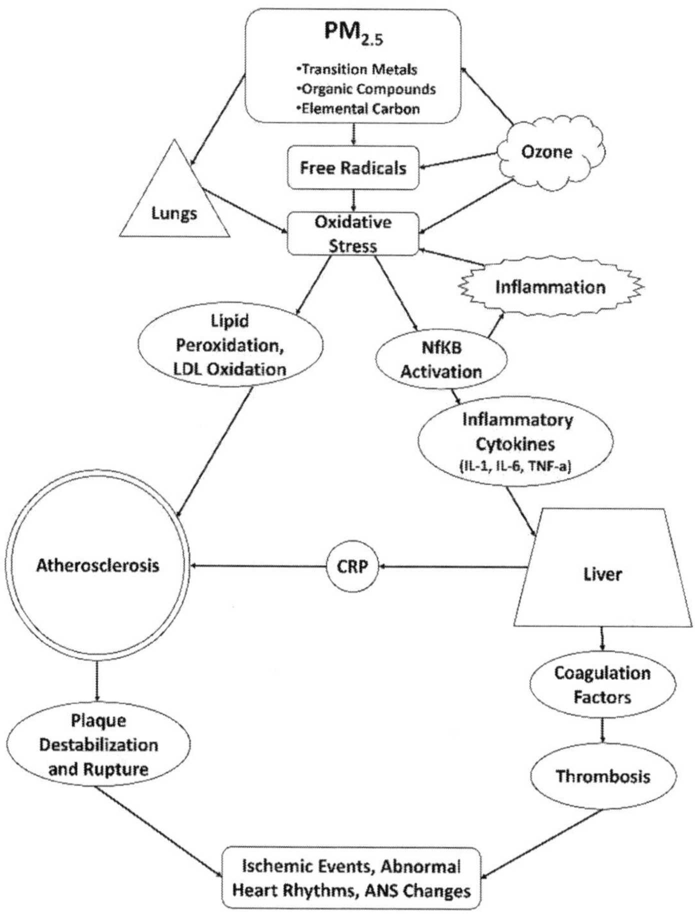

Figure 8. Mechanistic Framework for $PM_{2.5}$ and Ozone Cardiovascular Effects

The above diagram highlights several factors. Ambient $PM_{2.5}$ contains reactive components that can form, or release, free radicals when deposited in the lung. Particles that are scavenged by resident macrophages in the lung can initiate oxidative bursts that can also release free radicals and contribute to oxidative stress. Ambient ozone can interact with $PM_{2.5}$, mainly with the organic constituents, to form reactive oxygen species (ROS) and partially oxidized organics that are biologically reactive and have the potential to initiate free radical reactions. When inhaled, O_3 is a strong oxidant and can oxidize lipids and proteins in lung lining fluids and tissues and so can also contribute to oxidative stress. Oxidative stress occurs when oxidant production overwhelms the antioxidant capacity of the tissue. Free radicals and ROS can activate the NF-κB pathway, which can initiate inflammatory activities and the release of inflammatory cytokines such as IL-1, IL-6 and TNFα. Oxidant compounds released into circulation can cause lipid peroxidation and specifically oxidation of LDL lipoproteins which are taken up assiduously by circulating monocytes and macrophages to form foam cells and adherent cells that bind to the endothelial surface in the vasculature. Such changes in the vasculature can contribute to atherosclerotic plaque development and, as the disease progresses, to plaque destabilization and rupture. Inflammatory cytokines activate releases of acute phase proteins (such as C-reactive protein or CRP) and alter production of coagulation factors such as soluble P-selectin, Factor VII and plasminogen Activation inhibitor-1 (PAI-1) that can contribute to abnormal clotting (thrombosis). Material that breaks off and is released from destabilized plaques or from clots can block small arteries in the heart preventing delivery of oxygen to the tissue resulting in ischemic events including abnormal heart beats and electrocardiographic changes such as ST-segment changes.

A major goal of this study was to determine whether concurrent exposure to CAPs + 0.2 ppm O_3 mixture would elicit more adverse biological responses. However, data from these experiments indicate that concurrent exposures to CAPs + O_3 were not usually worse than the effects of exposures to the individual pollutants alone. Interestingly, indicators of HRV were decreased after exposures to CAPs alone or O_3 alone, but not after exposures to the CAPs + O_3 mixture. Indicators of another ECG abnormality, ST segment elevation changes, also decreased after exposures to CAPs or O_3 but not after exposure to the CAPs + O_3 mixture. Additional analyses of ST segment changes and measurements of arterial plaques are still in progress.

One possible explanation for the lack of observed changes in the metrics chosen is that when CAPs and O_3 are in a mixture, the two pollutants react and change the chemical composition of the particles. Data show that changes in HF HRV were strongly correlated with changes in composition of the particulate matter. Specific particle components were found to correlate, e.g. decreased organic acids, which are less biologically active, were associated with lower HF HRV. Additionally, carbonyl compounds, which are partially oxidized and thus able to initiate free radical reactions *in vivo*, were associated with higher HF HRV. We have not found other studies with exposures as long-term as ours. Most of the published PM + O_3 studies investigate effects of acute exposures. For example, Farraj et al. found that a single exposure of 800 ppb, but not 200 ppb, of ozone was sufficient to induce changes in cardiovascular endpoints (*51*) and exposure to ambient PM + O_3 mixtures caused a significant decrease in SDNN after a single 4 hr exposure (*52*). In contrast, the chronic exposures and co-exposures with ozone at 200 ppb presented here were sufficient to induce physiological modifications, underscoring the importance of more realistic long-term exposure studies. The reactions between particulate matter and ozone change the composition of the particles, and these changes can modify the responses they elicit in animals and, presumably, people.

The chemical differences in the particle-associated organic components when 200 ppb O_3 was added were consistent with a pattern of progressive oxidation, alcohols → aldehydes or ketones → organic acids. We found that reductions in HRV were associated with increased concentrations of carbonyls, while exposure atmospheres that contained more oxidized organic constituents, i.e. higher concentrations of organic acids, were less toxic.

Conclusions

HRV was decreased in apoE$^{-/-}$ mice exposed for 8 weeks to CAPs or to 200 ppb O_3 alone, but not when the two pollutants were delivered concurrently as a mixture. Preliminary data show ST segment changes in the ECGs, which are suggestive of reduced oxygen delivery to the heart. Although not discussed here, increases in blood pressure were similarly affected after exposures to CAPs or O_3, alone, but not after exposures to a mixture of CAPs + O_3. Chemical differences were documented in the particle-associated organic components when 200 ppb O_3 was added to the CAPs which were consistent with oxidation from alcohols → aldehydes or ketones → organic acids. We observed that reductions in HRV were associated with exposure to CAPs with increased concentrations of carbonyls, while exposures to CAPs containing more highly oxidized constituents, such as organic acids, were less toxic.

Acknowledgments

Andrew Keebaugh, David Herman and Rebecca Johnson played major roles in the conduct of these studies and performed much of the data analysis. Lisa Wingen collected and analyzed the Aerosol Mass Spectrometry data and provided important insights into the chemical reactions that occurred in the PM + O_3 mixture studies. Samantha Renusch, Irene Hasen and Amanda Ting provided the technical expertise and support that lent to the success of this project. We want to thank Barbara Finlayson Pitts and Sergey Nizkorodov of AirUCI for their comments, suggestions, and the loan of their SMPS which helped us to better characterize our exposure atmospheres and helped provide important validation of the use of the VACES particle concentrator. We thank Cyril McCormick and the Finlayson-Pitts research group for technical assistance with equipment. We also acknowledge the suggestions and guidance of Payam Pakbin (AQMD) and Constantinos Sioutas (USC) on the use of the VACES and the thermal denuder, especially in the early stages of this project. Laura Messina and Pam Bui provided logistical and administrative support for the project. We would also like to express thanks to our editor for many helpful suggestions in the draft of this chapter.

This project was funded under the ARB's Dr. William F. Friedman Health Research Program. During Dr. Friedman's tenure on the Board, he played a major role in guiding ARB's health research program. His commitment to the citizens of California was evident through his personal and professional interest in the Board's health research, especially in studies related to children's health. The Board is sincerely grateful for all of Dr. Friedman's personal and professional contributions to the State of California. We would like to acknowledge the National Science Foundation's Major Research Instrumentation (MRI) program (Grant no. 09923323) for funding the aerosol mass spectrometer.

References

1. Mauderly, J. L.; Samet, J. M. Is there evidence for synergy among air pollutants in causing health effects? *Environ. Health Perspect.* **2009**, *117*, 1–6.
2. Warren, D. L.; Last, J. A. Synergistic interaction of ozone and respirable aerosols on rat lungs. III. Ozone and sulfuric acid aerosol. *Toxicol. Appl. Pharmacol.* **1987**, *88*, 203–16.
3. Last, J. A.; Hyde, D. M.; Guth, D. J.; Warren, D. L. Synergistic interaction of ozone and respirable aerosols on rat lungs. I. Importance of aerosol acidity. *Toxicology* **1986**, *39*, 247–57.
4. Kurhanewicz, N.; McIntosh-Kastrinsky, R.; Tong, H.; Walsh, L.; Farraj, A. K.; Hazari, M. S. Ozone co-exposure modifies cardiac responses to fine and ultrafine ambient particulate matter in mice: concordance of electrocardiogram and mechanical responses. *Part. Fibre Toxicol.* **2014**, *11*, 54.
5. Wagner, J. G.; Allen, K.; Yang, H. Y.; Nan, B.; Morishita, M.; Mukherjee, B.; Dvonch, J. T.; Spino, C.; Fink, G. D.; Rajagopalan, S.; Sun, Q. H.; Brook, R. D.; Harkema, J. R. Cardiovascular Depression in Rats Exposed to Inhaled Particulate Matter and Ozone: Effects of Diet-Induced Metabolic Syndrome. *Environ. Health Perspect.* **2014**, *122*, 27–33.
6. Sun, L.; Liu, C.; Xu, X.; Ying, Z.; Maiseyeu, A.; Wang, A.; Allen, K.; Lewandowski, R. P.; Bramble, L. A.; Morishita, M.; Wagner, J. G.; Dvonch, J.; Sun, Z.; Yan, X.; Brook, R. D.; Rajagopalan, S.; Harkema, J. R.; Sun, Q.; Fan, Z. Ambient fine particulate matter and ozone exposures induce inflammation in epicardial and perirenal adipose tissues in rats fed a high fructose diet. *Part. Fibre Toxicol.* **2013**, *10*, 43.
7. Madrigano, J.; Kloog, I.; Goldberg, R.; Coull, B. A.; Mittleman, M. A.; Schwartz, J. Long-term exposure to PM2.5 and incidence of acute myocardial infarction. *Environ. Health Perspect.* **2013**, *121*, 192–6.
8. Beckerman, B. S.; Jerrett, M.; Finkelstein, M.; Kanaroglou, P.; Brook, J. R.; Arain, M. A.; Sears, M. R.; Stieb, D.; Balmes, J.; Chapman, K. The association between chronic exposure to traffic-related air pollution and ischemic heart disease. *J. Toxicol. Environ. Health A* **2012**, *75*, 402–11.
9. Pope, C. A., 3rd; Burnett, R. T.; Thurston, G. D.; Thun, M. J.; Calle, E. E.; Krewski, D.; Godleski, J. J. Cardiovascular mortality and long-term exposure to particulate air pollution: epidemiological evidence of general pathophysiological pathways of disease. *Circulation* **2004**, *109*, 71–7.

10. Pope, C. A., 3rd; Muhlestein, J. B.; May, H. T.; Renlund, D. G.; Anderson, J. L.; Horne, B. D. Ischemic heart disease events triggered by short-term exposure to fine particulate air pollution. *Circulation* **2006**, *114*, 2443–8.
11. CDC. *Ten leading causes of death by age group, United States - 2016*; 2016.
12. Brook, R. D.; Franklin, B.; Cascio, W.; Hong, Y.; Howard, G.; Lipsett, M.; Luepker, R.; Mittleman, M.; Samet, J.; Smith, S. C., Jr.; Tager, I. Air pollution and cardiovascular disease: a statement for healthcare professionals from the Expert Panel on Population and Prevention Science of the American Heart Association. *Circulation* **2004**, *109*, 2655–71.
13. Ito, K. Semi-long-term mortality effects of ozone. *Am. J. Respir. Crit. Care Med.* **2011**, *184*, 754–5.
14. Henrotin, J. B.; Zeller, M.; Lorgis, L.; Cottin, Y.; Giroud, M.; Bejot, Y. Evidence of the role of short-term exposure to ozone on ischaemic cerebral and cardiac events: the Dijon Vascular Project (DIVA). *Heart* **2010**, *96*, 1990–6.
15. Schwartz, J.; Morris, R. Air pollution and hospital admissions for cardiovascular disease in Detroit, Michigan. *Am. J. Epidemiol.* **1995**, *142*, 23–35.
16. Chuang, G. C.; Yang, Z.; Westbrook, D. G.; Pompilius, M.; Ballinger, C. A.; White, C. R.; Krzywanski, D. M.; Postlethwait, E. M.; Ballinger, S. W. Pulmonary ozone exposure induces vascular dysfunction, mitochondrial damage, and atherogenesis. *Am. J. Physiol. Lung Cell. Mol. Physiol.* **2009**, *297*, L209–16.
17. Gong, H., Jr.; Wong, R.; Sarma, R. J.; Linn, W. S.; Sullivan, E. D.; Shamoo, D. A.; Anderson, K. R.; Prasad, S. B. Cardiovascular effects of ozone exposure in human volunteers. *Am. J. Respir. Crit. Care Med.* **1998**, *158*, 538–46.
18. Wang, G.; Jiang, R.; Zhao, Z.; Song, W. Effects of ozone and fine particulate matter (PM(2.5)) on rat system inflammation and cardiac function. *Toxicol. Lett.* **2013**, *217*, 23–33.
19. Park, S. K.; O'Neill, M. S.; Vokonas, P. S.; Sparrow, D.; Schwartz, J. Effects of air pollution on heart rate variability: the VA normative aging study. *Environ. Health Perspect.* **2005**, *113*, 304–9.
20. Chang, C. C.; Hwang, J. S.; Chan, C. C.; Wang, P. Y.; Hu, T. H.; Cheng, T. J. Effects of concentrated ambient particles on heart rate variability in spontaneously hypertensive rats. *J. Occup. Health* **2005**, *47*, 471–80.
21. Pope, C. A.; Hansen, M. L.; Long, R. W.; Nielsen, K. R.; Eatough, N. L.; Wilson, W. E.; Eatough, D. J. Ambient particulate air pollution, heart rate variability, and blood markers of inflammation in a panel of elderly subjects. *Environ. Health Perspect.* **2004**, *112*, 339–45.
22. Holguin, F.; Tellez-Rojo, M. M.; Hernandez, M.; Cortez, M.; Chow, J. C.; Watson, J. G.; Mannino, D.; Romieu, I. Air pollution and heart rate variability among the elderly in Mexico City. *Epidemiology* **2003**, *14*, 521–7.
23. Gold, D. R.; Litonjua, A.; Schwartz, J.; Lovett, E.; Larson, A.; Nearing, B.; Allen, G.; Verrier, M.; Cherry, R.; Verrier, R. Ambient pollution and heart rate variability. *Circulation* **2000**, *101*, 1267–73.

24. Pope, C. A. What do epidemiologic findings tell us about health effects of environmental aerosols? *J. Aerosol Med.* **2000**, *13*, 335–54.
25. Whitsel, E. A.; Quibrera, P. M.; Christ, S. L.; Liao, D. P.; Prineas, R. J.; Anderson, G. L.; Heiss, G. Heart Rate Variability, Ambient Particulate Matter Air Pollution, and Glucose Homeostasis: The Environmental Epidemiology of Arrhythmogenesis in the Women's Health Initiative. *Am. J. Epidemiol.* **2009**, *169*, 693–703.
26. Franchini, M.; Mannucci, P. M. Thrombogenicity and cardiovascular effects of ambient air pollution. *Blood* **2011**, *118*, 2405–12.
27. Fakhri, A. A.; Ilic, L. M.; Wellenius, G. A.; Urch, B.; Silverman, F.; Gold, D. R.; Mittleman, M. A. Autonomic effects of controlled fine particulate exposure in young healthy adults: effect modification by ozone. *Environ. Health Perspect.* **2009**, *117*, 1287–92.
28. Araujo, J. A.; Barajas, B.; Kleinman, M.; Wang, X.; Bennett, B. J.; Gong, K. W.; Navab, M.; Harkema, J.; Sioutas, C.; Lusis, A. J.; Nel, A. E. Ambient particulate pollutants in the ultrafine range promote early atherosclerosis and systemic oxidative stress. *Circ. Res.* **2008**, *102*, 589–96.
29. Keebaugh, A. J.; Sioutas, C.; Pakbin, P.; Schauer, J. J.; Mendez, L. B.; Kleinman, M. T. Is atherosclerotic disease associated with organic components of ambient fine particles? *Sci. Total Environ.* **2015**, *533*, 69–75.
30. Kim, S.; Jaques, P. A.; Chang, M. C.; Barone, T.; Xiong, C.; Friedlander, S. K.; Sioutas, C. Versatile aerosol concentration enrichment system (VACES) for simultaneous in vivo and in vitro evaluation of toxic effects of ultrafine, fine and coarse ambient particles - Part II: Field evaluation. *J. Aerosol Sci.* **2001**, *32*, 1299–314.
31. Kim, S.; Jaques, P. A.; Chang, M. C.; Froines, J. R.; Sioutas, C. Versatile aerosol concentration enrichment system (VACES) for simultaneous in vivo and in vitro evaluation of toxic effects of ultrafine, fine and coarse ambient particles - Part I: Development and laboratory characterization. *J. Aerosol Sci.* **2001**, *32*, 1281–97.
32. Oldham, M. J.; Phalen, R. F.; Robinson, R. J.; Kleinman, M. T. Performance of a portable whole-body mouse exposure system. *Inhal. Toxicol.* **2004**, *16*, 657–62.
33. Liao, D. P.; Cai, J. W.; Barnes, R. W.; Tyroler, H. A.; Rautaharju, P.; Holme, I.; Heiss, G. Association of cardiac autonomic function and the development of hypertension - The ARC study. *Am. J. Hypertens.* **1996**, *9*, 1147–56.
34. Hayano, J.; Yamada, A.; Mukai, S.; Sakakibara, Y.; Yamada, M.; Ohte, N.; Hashimoto, T.; Fujinami, T.; Takata, K. Severity of Coronary Atherosclerosis Correlates with the Respiratory Component of Heart-Rate-Variability. *Am. Heart J.* **1991**, *121*, 1070–9.
35. Fleisher, L. A.; Frank, S. M.; Sessler, D. I.; Cheng, C. T.; Matsukawa, T.; Vannier, C. A. Thermoregulation and heart rate variability. *Clin. Sci.* **1996**, *90*, 97–103.
36. Lossius, K.; Eriksen, M.; Walloe, L. Thermoregulatory Fluctuations in Heart-Rate and Blood-Pressure in Humans - Effect of Cooling and Parasympathetic Blockade. *J. Autonom. Nerv. Syst.* **1994**, *47*, 245–54.

37. Keebaugh, A. J.; Sioutas, C.; Pakbin, P.; Schauer, J. J.; Mendez, L. B.; Kleinman, M. T. Is atherosclerotic disease associated with organic components of ambient fine particles? *Sci. Total Environ.* **2015**, *533*, 69–75.
38. Allan, J. D.; Delia, A. E.; Coe, H.; Bower, K. N.; Alfarra, M. R.; Jimenez, J. L.; Middlebrook, A. M.; Drewnick, F.; Onasch, T. B.; Canagaratna, M. R.; Jayne, J. T.; Worsnop, D. R. A generalised method for the extraction of chemically resolved mass spectra from aerodyne aerosol mass spectrometer data. *J. Aerosol Sci.* **2004**, *35*, 909–22.
39. Ng, N. L.; Canagaratna, M. R.; Zhang, Q.; Jimenez, J. L.; Tian, J.; Ulbrich, I. M.; Kroll, J. H.; Docherty, K. S.; Chhabra, P. S.; Bahreini, R.; Murphy, S. M.; Seinfeld, J. H.; Hildebrandt, L.; Donahue, N. M.; DeCarlo, P. F.; Lanz, V. A.; Prevot, A. S. H.; Dinar, E.; Rudich, Y.; Worsnop, D. R. Organic aerosol components observed in Northern Hemispheric datasets from Aerosol Mass Spectrometry. *Atmos. Chem. Phys.* **2010**, *10*, 4625–41.
40. Drewnick, F.; Schwab, J. J.; Jayne, J. T.; Canagaratna, M.; Worsnop, D. R.; Demerjian, K. L. Measurement of ambient aerosol composition during the PMTACS-NY 2001 using an aerosol mass spectrometer. Part I: Mass concentrations. *Aerosol Sci. Technol.* **2004**, *38*, 92–103.
41. Setyan, A.; Zhang, Q.; Merkel, M.; Knighton, W. B.; Sun, Y.; Song, C.; Shilling, J. E.; Onasch, T. B.; Herndon, S. C.; Worsnop, D. R.; Fast, J. D.; Zaveri, R. A.; Berg, L. K.; Wiedensohler, A.; Flowers, B. A.; Dubey, M. K.; Subramanian, R. Characterization of submicron particles influenced by mixed biogenic and anthropogenic emissions using high-resolution aerosol mass spectrometry: results from CARES. *Atmos. Chem. Phys.* **2012**, *12*, 8131–56.
42. Setyan, A.; Sauvain, J. J.; Riediker, M.; Guillemin, M.; Rossi, M. J. Characterization of surface functional groups present on laboratory-generated and ambient aerosol particles by means of heterogeneous titration reactions. *J. Aerosol Sci.* **2009**, *40*, 534–48.
43. Setyan, A.; Sauvain, J. J.; Guillemin, M.; Riediker, M.; Demirdjian, B.; Rossi, M. J. Probing Functional Groups at the Gas-Aerosol Interface Using Heterogeneous Titration Reactions: A Tool for Predicting Aerosol Health Effects? *ChemPhysChem* **2010**, *11*, 3823–35.
44. Liu, L.; Poon, R.; Chen, L.; Frescura, A. M.; Montuschi, P.; Ciabattoni, G.; Wheeler, A.; Dales, R. Acute effects of air pollution on pulmonary function, airway inflammation, and oxidative stress in asthmatic children. *Environ. Health Perspect.* **2009**, *117*, 668–74.
45. Delfino, R. J.; Staimer, N.; Tjoa, T.; Arhami, M.; Polidori, A.; Gillen, D. L.; George, S. C.; Shafer, M. M.; Schauer, J. J.; Sioutas, C. Associations of Primary and Secondary Organic Aerosols With Airway and Systemic Inflammation in an Elderly Panel Cohort. *Epidemiology* **2010**, *21*, 892–902.
46. Nakane, H. Translocation of particles deposited in the respiratory system: a systematic review and statistical analysis. *Environ. Health Prev. Med.* **2012**, *17*, 263–74.
47. Terzano, C.; Di Stefano, F.; Conti, V.; Graziani, E.; Petroianni, A. Air pollution ultrafine particles: toxicity beyond the lung. *Eur. Rev. Med. Pharmacol. Sci.* **2010**, *14*, 809–21.

48. Wallenborn, J. G.; McGee, J. K.; Schladweiler, M. C.; Ledbetter, A. D.; Kodavanti, U. P. Systemic translocation of particulate matter-associated metals following a single intratracheal instillation in rats. *Toxicol. Sci.* **2007**, *98*, 231–9.
49. Mills, N. L.; Donaldson, K.; Hadoke, P. W.; Boon, N. A.; MacNee, W.; Cassee, F. R.; Sandstrom, T.; Blomberg, A.; Newby, D. E. Adverse cardiovascular effects of air pollution. *Nat. Clin. Pract. Cardiovasc. Med.* **2009**, *6*, 36–44.
50. Mills, N. L.; Tornqvist, H.; Robinson, S. D.; Gonzalez, M. C.; Soderberg, S.; Sandstrom, T.; Blomberg, A.; Newby, D. E.; Donaldson, K. Air pollution and atherothrombosis. *Inhal. Toxicol.* **2007**, *19* (Suppl. 1), 81–9.
51. Hazari, M. S.; Callaway, J.; Winsett, D. W.; Lamb, C.; Haykal-Coates, N.; Krantz, Q. T.; King, C.; Costa, D. L.; Farraj, A. K. Dobutamine "Stress" Test and Latent Cardiac Susceptibility to Inhaled Diesel Exhaust in Normal and Hypertensive Rats. *Environ. Health Perspect.* **2012**, *120*, 1088–93.
52. Farraj, A. K.; Walsh, L.; Haykal-Coates, N.; Malik, F.; McGee, J.; Winsett, D.; Duvall, R.; Kovalcik, K.; Cascio, W. E.; Higuchi, M.; Hazari, M. S. Cardiac effects of seasonal ambient particulate matter and ozone co-exposure in rats. *Part. Fibre Toxicol.* **2015**, *12*, 12.

Editors' Biographies

Sherri W. Hunt

Sherri Hunt received a B.S. in Chemistry from the University of Georgia and a Ph.D. in Physical Chemistry from the University of Minnesota. After a postdoctoral research position with Barbara Finlayson-Pitts at University of California, Irvine, Hunt spent two years as a AAAS Science and Technology Policy Fellow. In 2006, she joined the United States Environmental Protection Agency as a physical scientist in the Office of Research and Development. Her work involves research planning and coordination for agency scientists and the extramural STAR grants program. Research expertise and topics include understanding, monitoring, and modeling air pollution and its health effects.

Alexander Laskin

Alexander Laskin received his undergraduate degree from the Polytechnical Institute, St. Petersburg, Russia, in 1991 in Physics, and Ph.D. in Physical Chemistry from the Hebrew University of Jerusalem, Israel, in 1998. Following postdoctoral research appointments at the University of Delaware, Princeton University, and Pacific Northwest National Laboratory (PNNL), he was a senior research scientist at PNNL. In 2017, he joined Purdue University as a Professor of Chemistry. His present and past research interests include physical and analytical chemistry of aerosols, environmental and atmospheric effects of aerosols, chemical imaging and molecular level studies of aerosols, microspectroscopy and high-resolution mass spectrometry of aerosols, combustion-related aerosols, combustion chemistry, and chemical kinetics.

Sergey A. Nizkorodov

Sergey Nizkorodov received his undergraduate degree in biochemistry from Novosibirsk State University and graduate degree in physical chemistry from Basel University. After doing postdoctoral research at the University of Colorado at Boulder, and in atmospheric chemistry at the California Institute of Technology, he joined the faculty of the Department of Chemistry, University of California, Irvine (UCI). His primary areas of expertise are molecular spectroscopy, high-resolution mass spectrometry, chemical reaction dynamics, and photochemistry.

Indexes

Author Index

Subject Index

B

M

N

O

P

S

T

Printed in the USA/Agawam, MA
November 27, 2019

743228.006